The Classical Hollywood Reader

The Classical Hollywood Reader brings together essential readings to provide a history of Hollywood from the 1910s to the mid-1960s.

Following on from a Prologue that discusses the aesthetic characteristics of Classical Hollywood films, Part I covers the period between the 1910s and the mid- to late 1920s. It deals with the advent of feature-length films in the US and the growing national and international dominance of the companies responsible for their production, distribution and exhibition. In doing so, it also deals with film making practices, aspects of style, the changing roles played by women in an increasingly business-oriented environment, and the different audiences in the US for which Hollywood sought to cater.

Part II covers the period between the coming of sound in the mid-1920s and the beginnings of the demise of the 'studio system' in the late 1940s. In doing so it deals with the impact of sound on films and film production in the US and Europe, the subsequent impact of the Depression and World War II on the industry and its audiences, the growth of unions, and the roles played by production managers and film stars at the height of the studio era.

Part III deals with aspects of style, censorship, technology and film production. It includes articles on the Production Code, music and sound, cinematography and the often neglected topic of animation.

Part IV covers the period between 1946 and 1966. It deals with the demise of the studio system and the advent of independent production. In an era of demographic and social change, it looks at the growth of drive-in theatres, the impact of television, the advent of new technologies, the increasing importance of international markets, the Hollywood blacklist, the rise in art house imports and in overseas production, and the eventual demise of the Production Code.

Designed especially for courses on Hollywood Cinema, the Reader includes a number of newly researched and written chapters and a series of introductions to each of its parts. It concludes with an epilogue, a list of resources for further research and an extensive bibliography.

Steve Neale is Professor and Chair in Film Studies in the School of English at Exeter University, where he teaches Introduction to Film, Hollywood and Europe, Comedy, Comedians and Romance, and Film Noir. He is an internationally renowned film studies scholar. His research focuses principally on history and theory of Hollywood cinema and he has published several publications in these areas. Since 2004, Professor Neale has been the Academic Director for the Bill Douglas Centre for the History of Cinema and Popular Culture.

The Classical Hollywood Reader

The Classical Hollywood Reader

Edited by

Steve Neale

LONDON AND NEW YORK

First published 2012
by Routledge
2 Park Square, Milton Park, Abingdon, Oxon OX14 4RN

Simultaneously published in the USA and Canada
by Routledge
711 Third Avenue, New York, NY 10017

Routledge is an imprint of the Taylor & Francis Group, an informa business

British Library Cataloguing in Publication Data
A catalogue record for this book is available from the British Library

Library of Congress Cataloging in Publication Data
The classical Hollywood reader / edited by Steve Neale.
p. cm.
Includes bibliographical references and index.
1. Motion pictures—California—Los Angeles—History—20th century. I. Neale, Stephen, 1950-
PN1993.5.U65C565 2012
384'.80979494—dc23

2011038750

ISBN: 978-0-415-57672-7 (hbk)
ISBN: 978-0-415-57674-1 (pbk)

Typeset in Perpetua and Bell Gothic
by RefineCatch Limited, Bungay, Suffolk

Printed and bound in Great Britain by the MPG Books Group

For Karen

Contents

Illustrations

Figures

Tables

Exhibits

Richard Koszarski is a Professor of English at Rutgers University. He is the editor of *Hollywood Directors, 1914–1940* (1977) and *Mystery of the Wax Museum* (1979), and author of *Fort Lee: The Film Town* (2004) and *Hollywood on the Hudson: Film and Television in New York from Griffith to Sarnoff* (2008), as well as two award-winning books, *The Man You Loved to Hate: Erich Von Stroheim and Hollywood* (1983) and *An Evening's Entertainment: The Age of the Silent Feature Picture, 1915–1928* (1994). He has been the recipient of the Prix Jean Mitry from the Giornate del Cinema Muto and is editor-in-chief of *Film History*.

Mark Langer teaches Film Studies at the School for Studies in Art and Culture at Carleton University. His articles have appeared in anthologies and in periodicals such as *Cinema Journal* and *Animation Journal* and *Screen*, and the article reprinted here was a winner of the Norman McLaren/Evelyn Lambart Award for the best scholarly article on animation. He has curated animation retrospectives for museums and festivals, among them the Museum of Modern Art, the Ottowa International Film Festival and the Cinémathèque Française, and is currently President of the Ontario Confederation of University Faculty Associations.

Howard T. Lewis was Professor of Marketing at the Harvard Business School before becoming an Assistant Dean and, in 1936, Director of the School's Bureau of Business Research. He was the author of books such as *Industrial Purchasing* (1933), as well as *Cases on the Motion Picture Industry* (1930) and *The Motion Picture Industry* (1933).

Karen Ward Mahar is an Associate Professor of History and co-director of the American Studies programme at Siena College in Loddonville, New York. Her first book, *Women Filmmakers in Early Hollywood* (2006), examined the American film industry as a workplace for women in the silent era. She is currently working on a comparative study of gender and corporate power in the US and UK in the mid-twentieth century.

Richard Maltby is Professor of Screen Studies and Executive Dean of the Faculty of Education, Humanities and Law at Flinders University, South Australia. He is the author of *Hollywood Cinema: An Introduction* (2003) and co-editor of *'Film Europe' and 'Film America': Cinema, Commerce and Cultural Exchange, 1920–1939* (with Andrew Higson) (1999), *Going to the Movies: Hollywood and the Social Experience of Cinema* (with Melvyn Stokes and Robert C. Allen) (2007), and *Cinema, Audiences and Modernity: European Perspectives on Film Cultures and Cinema-going* (with Daniel Biltereyst and Philippe Meers), *Explorations in New Cinema History: Approaches and Case Studies* (with Daniel Biltereyst and Philippe Meers) and *The New Cinema History: A Guide to Resources* (with Kate Bowles, Deb Verhoeven and Mike Walsh) (all 2011). He is a Fellow of the Australian Academy of Humanities, Series Editor of the Exeter Studies in Film History series, and author of over 50 articles and essays.

Steve Neale is Professor of Film Studies at the University of Exeter. He is the author of *Genre and Hollywood* (2000), co-author of *Popular Film and Television Comedy* (with Frank Krutnik) (1990) and *Epics, Spectacles and Blockbusters: A Hollywood History* (with Sheldon Hall) (2010), editor of *Genre and Contemporary Hollywood* (2002), and co-editor of *'Un-American' Hollywood: Politics and Film in the Blacklist Era* (with Frank Krutnik, Brian Neve and Peter Stanfield) (2007) and *Widescreen Worldwide* (with John Belton and Sheldon Hall) (2010). He co-edits the Exeter Studies in Film History series with Richard Maltby.

Brian Neve teaches politics and film in the Department of Politics, Languages and International Studies at the University of Bath. He is the author of *Film and Politics in America: A Social Tradition* (1997) and *Elia Kazan: The Cinema of an American Outsider* (2009), and co-editor of *'Un-American' Hollywood: Politics and Film in the Blacklist Era* (with Frank Krutnik, Brian Neve and Peter Stanfield) (2007). He is currently working on a book on Cy Endfield and the European blacklist diaspora.

Thomas Schatz is Professor in the Radio-Television-Film Department at the University of Texas at Austin. He has written books on American film (and edited many others), including *The Genius of the System: Hollywood Filmmaking in the Studio Era* (1989) and *Boom and Bust: American Cinema in the 1940s* (1997). His writing on film has appeared in the *New York Times*, the *Los Angeles Times, Premiere, The Nation, Film Comment, Film Quarterly* and many other publications. He is the long-time editor of the Film and Media Studies series for University of Texas Press and is currently writing a book-length study of contemporary conglomerate Hollywood.

John Sedgwick is Professor of Film Economics and Director of the Centre for International Business and Sustainability at London Metropolitan University. He researches the economic history of film and has published articles in *Business History, Economic History Review, Explorations in Economic History*, the *Historical Journal of Film, Radio and Television*, the *Journal of Cultural Economics* and the *Journal of Economic History*, as well as numerous book chapters – many co-authored with Mike Pokorny. He has also published a monograph on the film industry in Britain in the 1930s and edited an anthology of papers. He was a Leverhulme Research Fellow in 2000, a Menzies Research Fellow in 2006 and an RMIT/AFTRS Visiting Research Fellow in 2007.

Janet Staiger is the William P. Hobby Centennial Professor of Communication in the Department of Radio-Television-Film at the University of Texas at Austin. Her books include *Political Emotions* (co-edited with Ann Cvetkovich and Ann Reynolds) (2010), *Convergence Media History* (co-edited with Sabine Hake) (2009), *Media Reception Studies* (2005), *Authorship and Film* (co-edited with David Gerstner) (2003), *Perverse Spectators: The Practices of Film Reception* (2000), *Blockbuster TV: Must-see Sitcoms in the Network Era* (2000), *Bad Women: Regulating Sexuality in the Early American Cinema* (1995), *Interpreting Films: Studies in the Historical Reception of American Cinema* (1992) and *The Classical Hollywood Cinema: Film Style and Mode of Production to 1960* (with David Bordwell and Kristin Thompson) (1985).

Kristin Thompson is an Honorary Fellow in the Department of Communication Arts at the University of Wisconsin-Madison. She is the co-author of numerous editions of *Film Art: An Introduction* and *Film History: An Introduction* (with David Bordwell), and of *The Classical Hollywood: Film Style and Mode of Production to 1960* (with David Bordwell and Janet Staiger) (1985). Her most recent books are *Storytelling in the New Hollywood: Understanding Classical Narrative Technique* (1999), *The Frodo Franchise: The Lord of the Rings and Modern Hollywood* (2007) and *Minding Movies: Observations on the Art, Craft, and Business of Filmmaking* (with David Bordwell) (2011).

Ginette Vincendeau is Professor of Film Studies at King's College London. She has written widely on popular French and European Cinema and is a regular contributor to *Sight and Sound*. She is the author of *Pépé le Moko* (1998), *Stars and Stardom in French Cinema* (2000), *Jean-Pierre Melville: An American in Paris* (2003) and *La Haine* (2005), editor of *The Encyclopedia of French Cinema* (1995), and co-editor of *French Film: Texts and Contexts* (with Susan Hayward) (1990 and 2000), *Journeys of Desire: European Actors in Hollywood* (with Alistair Phillips) (2006) and *The New Wave: Critical Landmarks* (with Peter Graham) (2009).

Janet Wasko is the Knight Chair for Communication Research at the University of Oregon. Her work focuses on the political economy of the media and she is the author, co-author, editor and co-editor of 19 books, among them *Movies and Money: Financing the American Film Industry* (1982), *Understanding Disney; The Manufacture of Fantasy* (2001), *How Hollywood Works* (2003), *A Companion to Television* (2010), *The Contemporary Hollywood Film Industry* (with Paul McDonald) (2008) and *Media in the Age of Marketization* (with Graham Murdock) (2007). Her most recent collection is *The Handbook of Political Economy of Communications* (edited with Graham Murdock and Helena Sousa) (2011).

Acknowledgements

I would like first of all to thank Natalie Foster for inviting me to edit this Reader, and Ruth Moody, Eileen Srebernik, Lisa Williams and the team at Routledge for their hard and diligent work. I would also like to thank Gerben Bakker, Tino Balio, John Belton, Mark Glancy and John Sedgwick, Douglas Gomery, Patrick Keating, Kathryn Kalinak, Richard Koszarski, Mark Langer, Karen Ward Mahar, Thomas Schatz, Janet Staiger, Kristin Thompson, Ginette Vincendeau and Janet Wasko and their publishers for allowing me to reprint and edit their work, and Sheldon Hall, Helen Hanson, Scott Higgins, Lea Jacobs and Andrea Comiskey, Patrick Keating, Brian Neve and Richard Maltby for writing such superb contributions to this book. To gather together the work of such fine scholars has been a privilege as well as a pleasure. I would like in addition to thank the University of Exeter for granting me a term of study leave and the staff at the British Film Institute library, both those who remain and those who have been forced to leave or to retire, for all their help. The library is an important and precious resource. Without it I would have been unable to edit this volume or to research my own contributions. Finally, in addition to my colleagues at Exeter, and in addition to those listed above, I would like to pay tribute to a number of mentors, ex-colleagues and friends, among them Shelley Baker, Edward Buscombe, John Caughie, Catherine Constable, Jim Cook, Pam Cook, Christine Geraghty, the late Gerry Coubro, Elizabeth Cowie, Michael Grant, Jim Hillier, Frank Krutnik, Peter Krämer, Annette Kuhn, Colin McArthur, Robert Murphy, Geoffrey Nowell-Smith, Mike O'Pray, Douglas Pye, A. L. Rees, Tom Ryall, Murray Smith, Peter Stanfield, Sarah Street, Michael Walker, Paul Willemen and, last but by no means least, Ben Brewster, from whom I learned all I know about scholarship and aesthetics, nearly all I know about early and silent cinema, and much of what I know about the Classical Hollywood Cinema.

Acknowledgements

Permissions

The articles listed below have been reproduced with kind permission. Whilst every effort has been made to trace copyright holders, this has not been possible in all cases. Any omissions brought to our attention will be remedied in future editions.

1 Patrick Keating, 'Emotional Curves and Linear Narratives', *Velvet Light Trap*, No. 58 (Fall 2006), University of Texas Press, pp. 4–15. First published as the article 'Emotional Curves and Linear Narrative', by Patrick Keating, in *Velvet Light Trap*, Volume 58, pp. 4–15. Copyright © 2006 by the University of Texas Press. All rights reserved.

2 Gerben Bakker, 'The Quality Race: Feature Films and Market Dominance in the U.S. and Europe in the 1910s', from *Entertainment Industrialised: The Emergence of the International Film Industry, 1890–1940*, Cambridge: Cambridge University Press, 2008, pp. 210–28, plus bibliography. © 2008 Gerben Bakker, Published by Cambridge University Press. Reproduced with permission.

3 Richard Koszarski, 'Making Movies, 1915–28', edited chapter from *An Evening's Entertainment: The Age of the Silent Feature Picture*, New York: Scribners, 1990, pp. 99–137. From *History of American Cinema, V 3.* © 1990 Gale, a part of Cengage Learning, Inc. Reproduced with permission. www.cengage.com/permissions.

4 Kristin Thompson, 'The Limits of Experimentation in Hollywood' in Jan-Christopher Horak (ed), *Lovers of Cinema: The First American Film Avant Garde, 1919–1945*, Madison: University of Wisconsin Press, 1995, pp. 67–93. © 1996 by the Board of Regents of the University of Wisconsin System. Reprinted by permission of The University of Wisconsin Press.

5 Karen Ward Mahar, '"Doing a Man's Work" The Rise of the Studio System and Remasculinization of Filmmaking' from *Women Filmmakers in Early Hollywood*, Baltimore: The Johns Hopkins University Press, 2006, pp. 179–203, 261–68. Excerpts. © 2006 The John Hopkins University Press. Reprinted with permission of The John Hopkins University Press.

7 Douglas Gomery, 'The Coming of Sound: Technological Change in the American Film Industry' in Tino Balio (ed.), *The American Film Industry*, Madison: University of Wisconsin Press, 1985 edition, pp. 229–51. © 1979 by the Board of Regents of the University of Wisconsin System. Reprinted by permission of The University of Wisconsin Press.

8 Ginette Vincendeau, 'Hollywood Babel: The Coming of Sound and the Multiple Language Version', *Screen*, vol. 29 no. 2, Spring 1988, pp. 24–39. Copyright © 1988, Oxford University Press.

Steve Neale

INTRODUCTION

BY PUBLISHING OR REPUBLISHING A SERIES of articles, chapters and extracts from a variety of different sources, this book aims to introduce its readers to the major facets of classical Hollywood cinema and its history. In doing so, it deals with issues of style and aesthetics, design and technology, censorship and regulation, organisation and management, and economics and politics. It also deals with the policies and practices of national and international film production, distribution and exhibition pursued by a small but industrially dominant group of companies from the mid-1910s to the early 1960s.

The precise configuration of these groups and the nature and names of these and other companies changed over time. First National, for example, was a major company in the 1920s but was bought out by Warner Bros. in 1929. Warner Bros. itself was a minor company until the late 1920s, when it successfully pioneered the adoption of sound. It was joined, among others, by RKO (Radio-Keith-Orpheum), which was set up by the Radio Corporation of America in 1928. The Fox Film Corporation, which was founded in 1915, became Twentieth Century-Fox when it merged with Twentieth Century Pictures in 1935. Paramount Pictures, initially a distribution company, became part of Famous Players-Lasky in 1916. Famous Players-Lasky became Paramount-Famous Lasky in 1927 then Paramount-Publix in 1930. Cohn-Brand-Cohn Film Sales became Columbia Pictures and Metro Pictures, Goldwyn Pictures Corporation and Louis B. Mayer Pictures were merged to become MGM (Metro-Goldwyn-Mayer) in 1924. The Universal Film Manufacturing Company became the Universal Pictures Company the following year. And so on and so forth. With the exception of United Artists, which was founded in 1919 and which distributed the films produced by its members and other independents, these companies all possessed production facilities housed in studio lots in and around Los Angeles in Southern California. These lots were staffed by artists and technicians of various kinds (cinematographers, set designers, script writers, carpenters, electricians, editors, costume designers, and so on and so forth) and were organised into departments. This was one of their distinguishing features. But it was by no means the only one.

Only three of these companies, Columbia, Paramount and RKO, possessed production facilities in the suburb of Hollywood (an area of Los Angeles now approximately bounded by Western Avenue, La Brea Avenue, Franklin Avenue and Santa Monica Boulevard). However, just as 'studio' became a synonym for 'film company', so 'Hollywood' became a synonym for the mainstream film industry in the US, a term that evoked not only the companies, their production centres, their films and their stars, but the aura of glamour that surrounded them all. In doing so, however, these synonyms only partly identified the nature, structure, location and scope of the industry and its activities. For the corporate headquarters of the studios were not in Hollywood or Southern California, but in New York. It was here that finance was raised from banks and investors, that decisions about the cost, scale and nature of each season's programme of films were taken, that publicity campaigns were mounted, that national and international distribution plans were made, and that the booking and circulation of film prints were organised. It was here, too, especially in the late 1910s, the 1920s and the early 1930s, long after its dominance as a centre of production in earlier years, that some of the films produced by some of the companies and their affiliates continued to be made. And it was here that the major companies organised the acquisition, administration and programming of their cinema chains.

As industrially oriented accounts of classical Hollywood cinema have repeatedly stressed, and as will be discussed at greater length in a number of contributions to this book, it was access to cinemas and cinema chains in key city centres and markets that provided the principal film companies with a guaranteed outlet for their films and, hence, with guaranteed income. By the late 1920s, the Big Five companies (Fox, MGM, Paramount, RKO and Warner Bros.) all owned cinemas and cinema chains. With the temporary exception of United Artists and Universal, the Little Three (which also included Columbia), did not. But they did possess national and international distribution facilities; they did subscribe to most of the principles and practices subscribed to by the Big Five (including those associated with the Production Code, Hollywood's system of self-censorship); and they did cooperate with the Big Five and the Big Five with them. The Big Five were 'vertically integrated': they were involved in all three branches of an industry (manufacture, wholesale and retail); along with the Little Three, they constituted an 'oligopoly': a group of companies that colluded to control it.

Despite tensions, minor modifications and major changes in historical circumstance (notably the advent of the Great Depression and the advent of World War Two), these structures and practices remained in place for over twenty years. After that, they were modified in more fundamental ways following anti-trust rulings by the US Supreme Court, the consequent sale of the Big Five's cinema chains, the advent of suburbanisation, the advent and spread of television and other leisure pursuits, and shrinking domestic attendances. These modifications took place during the late 1940s, the 1950s and the early 1960s. By the mid- to late 1960s, with the abandonment of the Production Code, the advent of independently owned multiplex cinemas in suburban shopping malls, the sale or scaling down of studio lots and their contents, and a series of crises, mergers and takeovers involving most of Hollywood's companies, classical Hollywood cinema was dead.

Throughout the classical era, dozens of 'classic' films – films of enduring public interest or appeal – were made. However, as we have already seen, 'classical Hollywood cinema' was more than the sum of its films (classic or otherwise). Insofar as a term like 'classical' implies a supra-personal system, a long-standing set of practices and norms, it applies as much to its industrial infrastructure as it does to the nature of its films, as much to its policies of distribution and exhibition as to its practices of production. It is in this wider sense that the term has been used to guide the scope and the contents of this book. Nevertheless, it is has been the nature of Hollywood's films and its practices of production that have attracted the most attention from scholars interested in issues of classicism. This is nowhere more apparent than in David Bordwell, Janet Staiger and Kristin Thompson's groundbreaking book *The Classical Hollywood Cinema: Film Style and Mode of Production to 1960*,[1] and it is thus with debates about Bordwell's discussion of Hollywood films and their aesthetic characteristics that this Reader begins.

The Classical Hollywood Cinema was first published in 1985. Using an extensive array of published and archival sources and a sample of a hundred different films, its authors sought not just to study the managerial and organisational practices governing film production in Hollywood's studios, nor just to study the studios' deployment of technologies and craft skills (particularly those involved in scriptwriting, set design, cinematography, sound and editing), but to relate them all to the fundamental features of its films. Conceptualising these features as 'norms', Bordwell argues, first of all, that classical Hollywood films tell stories, that these stories are character-centred, and that the fundamental premises of classical Hollywood story construction are 'causality, consequence, psychological motivations, the drive toward overcoming obstacles and achieving goals. Character-centered – i.e. personal or psychological – causality is the armature of the classical story'.[2] Characters are prime causal agents. They are defined by goals and traits that are often marked by recurrent motifs and that the familiar personas of Hollywood's stars often helped support. In addition, most classical films possess 'at least two lines of action, both causally linking the same group of characters. Almost invariably, one of these lines involves heterosexual romantic love'.[3]

Various forms of motivation (compositional, realistic, generic, artistic) help to unify classical films, justify the things that happen in them, explain why their characters do what they do, and normalise the use of otherwise unusual, self-conscious or blatantly artificial artistic devices. In telling their stories, classical films can draw on any technique as long as it 'can transmit story information. Conversations, figure positions, facial expressions, and well-timed encounters between characters all function just as narrationally as do camera movements, cuts, or bursts of music'.[4] In classical films, 'the narration is omniscient, but it lets that omniscience come forward more at some points than others'. In their opening passages, 'the narration is moderately self-conscious and overtly suppressive'. As they proceed, 'the narration becomes less self-conscious and more communicative', though towards the end 'omniscience and self-consciousness are likely to be re-asserted'.[5] The order of events in classical films can be varied, especially by using character-centred flashbacks. The narration only 'shows important events and skips the intervals between them'. However, 'the classical film creates a patterned duration not only by what it leaves out but by a specific, powerful device. The story action sets a limit to how long it must last. Sometimes this means a strictly confined duration, as in the familiar convention of one-night-in-a-mysterious house films'.[6] More commonly, though, it does so by setting deadlines.

In general, in 'making narrative causality the dominant system in the film's total form, the classical Hollywood cinema chooses to subordinate space. Most obviously, the classical style makes the sheerly graphic space of the film image a vehicle for narrative', though in doing so it uses 'image composition and editing to create a powerful representation of three-dimensional space'.[7] Its compositions, whether moving or still, tend to be balanced and centred. The human face tends to be 'positioned in full, three-quarter, or profile view; the body typically in three-quarter view. ... Standing groups are arranged along horizontal or diagonal lines or half circles'.[8] Space 'is created in planes through various depth cues. To the usual cues of visual overlap (the object that overlaps must be closer) and familiar size, the classical image adds pattern, color, texture, lighting and focus to specify depth'.[9] 'Classical continuity editing ... reinforces spatial orientation. Continuity of graphic qualities can invite us to look through the "plate-glass window" of the screen. From shot to shot, tonality, movement, and the center of compositional interest shift enough to be distinguishable but not enough to be disturbing'.[10] Similar principles govern the staging, framing and editing of character positions, interactions, movements and looks.

Finally, while 'the shot is the basic unit of material' in classical Hollywood, and while the terms 'shot' and 'scene' were often used interchangeably (at least until the 1950s), scenes and sequences were important additional building blocks as well.[11] Scenes were often linked by sequences, which compress or summarise narrative space, time and action, 'but the straightforward scene – one or more persons acting in a limited locale over a continuous duration ... remains the building-block of classical dramaturgy'. Scenes usually consist of 'two distinct phases, the exposition and the development'. The former 'specifies the time, place and relevant characters' and 'must immediately reveal two things about the characters: their relative spatial positions and their states of mind'.[12] Then, when the developmental phase begins, characters 'act toward their goals, enter into conflict, make choices, set deadlines, make appointments, and plan future actions'. 'Most scenes continue or close off' an 'old line of action' before beginning another. 'Other scenes may reintroduce the old line, toy with it, suspend it again, introduce a new causal line, then close out the old and introduce yet another before the scene ends'. This 'new causal line ... motivates the shift to the next scene'.[13]

Bordwell acknowledges the extent to which norms such as these could be varied or transgressed. A number of Hollywood's genres (among them its musicals, its melodramas and its slapstick comedies) licensed deviations from some of these norms, and a number of its films experimented with causal norms and the provision of narrative knowledge. But these experiments and deviations were rarely as extreme, as marked or as systematic as they were in avant-garde or modernist art films. Overall, the principles that Hollywood claimed as its own 'rely on notions of decorum, formal harmony, respect for tradition, mimesis, self-effacing craftsmanship, and cool control of the perceiver's response – canons which critics in any medium usually call "classical."'[14]

Since the initial publication of *The Classical Hollywood Cinema*, a number of Bordwell's arguments have been subject to critique. Some of these critiques, notably those that question the dominance of linear causality and those that emphasise the importance of spectacle and emotional engagement, are discussed by Patrick Keating in 'Emotional Curves and Linear Narratives' (Chapter 1), which serves as a prologue to this reader. Keating argues that 'metaphors of "dominance" are not always helpful in understanding the relationship between narrative and other systems. Instead, narrative and other attractions can work together to produce an intensified emotional response'. In doing so, he offers a productive alternative to some of Bordwell's arguments. He also discusses the work of theorists and historians such as Rick Altman, Noël Carroll, Elizabeth Cowie, Donald Crafton, Dirk Eitzen, Richard Maltby and Linda Williams, many of whom have offered explicit or implicit criticisms of at least some of Bordwell's arguments. These criticisms have been augmented by Robert Knopf's discussion of a number of Buster Keaton's feature films and, implicitly at least, by Martin Rubin's discussion of the Busby Berkeley musical, *The Gang's All Here* (1943).[15] However, few scholars have engaged in any detail with Bordwell's work on narration, duration, the construction of sequences and scenes, shot composition and editing, or the handling of narrative space and time. To that extent, his contributions to *The Classical Hollywood Cinema*, along with those of Kristin Thompson, who traces the history of continuity editing, staging, acting and other key devices and conventions, still remain essential reading.[16]

Notes

1 David Bordwell, Janet Staiger and Kristin Thompson, *The Classical Hollywood Cinema: Film Style and Mode of Production to 1960* (London: Routledge and Kegan Paul, 1985).
2 Bordwell, Staiger and Thompson, *The Classical Hollywood Cinema*, 13.
3 Bordwell, Staiger and Thompson, *The Classical Hollywood Cinema*, 16.
4 Bordwell, Staiger and Thompson, *The Classical Hollywood Cinema*, 24.
5 Bordwell, Staiger and Thompson, *The Classical Hollywood Cinema*, 25.
6 Bordwell, Staiger and Thompson, *The Classical Hollywood Cinema*, 44.
7 Bordwell, Staiger and Thompson, *The Classical Hollywood Cinema*, 50.
8 Bordwell, Staiger and Thompson, *The Classical Hollywood Cinema*, 51.
9 Bordwell, Staiger and Thompson, *The Classical Hollywood Cinema*, 52.

10 Bordwell, Staiger and Thompson, *The Classical Hollywood Cinema*, 55.
11 Bordwell, Staiger and Thompson, *The Classical Hollywood Cinema*, 60.
12 Bordwell, Staiger and Thompson, *The Classical Hollywood Cinema*, 63.
13 Bordwell, Staiger and Thompson, *The Classical Hollywood Cinema*, 64.
14 Bordwell, Staiger and Thompson, *The Classical Hollywood Cinema*, 3–4.
15 Robert Knopf, *The Theater and Cinema of Buster Keaton* (Princeton: Princeton University Press, 1999), 76–111; Martin Rubin, *Showstoppers: Busby Berkeley and the Tradition of Spectacle* (New York: Columbia University Press, 1993), 159–70. Plotless musical revues such as *King of Jazz* (1928) and *Ziegfeld Follies* (1946), animated 'package features' such as *Fantasia* (1940) and other instances of what David Scott Diffrient has called 'episodic cinema' might be cited as exceptions to Bordwell's model of the classical Hollywood feature film as well. See David Scott Diffrient, 'Cabinets of Cinematic Curiosities: A Critical History of the Animated "Package Feature", From *Fantasia* (1940) to *Memories* (1995)', *Historical Journal of Film, Radio and Television*, vol. 26 no. 4 (2006), 505–35.
16 For Bordwell, Staiger and Thompson's own reflections on *The Classical Hollywood Cinema* and its critics, see '*The Classical Hollywood Cinema* Twenty-Five Years *Along*', http://davidbordwell.net/essays/classical.php (September 2010).

Chapter 1

Patrick Keating

PROLOGUE: EMOTIONAL CURVES AND LINEAR NARRATIVES

IN FRED NIBLO'S 1921 VERSION OF *The Three Musketeers* D'Artagnan, played by Douglas Fairbanks, first joins forces with the title characters during an extended fight scene. The scene is packed with gags and stunts as Fairbanks leaps around the set with knife and sword in hand. At one point he even throws his sword like a harpoon. While such moments of spectacle are common in Hollywood films, ranging from the gags of comedian comedy to the musical numbers of Busby Berkeley, historians have long argued about the best way to theorize Hollywood's strategies for combining narrative and other attractions.

We can usefully group the various theoretical models into three categories: a Classical model, which argues that a certain type of narrative operates as a dominant in relation to various subordinate systems; an Alternation model, which argues that the dominance of narrative alternates with the dominance of other systems; and an Affective model, which argues that linear narrative is itself subordinate to a more important goal, the production of emotion. After surveying these alternatives, I will propose my own version of the Affective model—a version that will, I hope, draw important insights from the other two models. My argument is that metaphors of "dominance" are not always helpful in understanding the relationship between narrative and other systems. Instead, narrative and other attractions can work together to produce an intensified emotional response. We can call this the Cooperation model, since the model explains how narrative and attractions can support each other.[1] Part 1 offers a brief summary of three existing models. Part 2 explains my proposal for a Cooperation model. Part 3 applies the model to a set of films that have long played a central role in debates about the status of narrative in Hollywood: the musicals of Busby Berkeley.

1 Three models

The most complete presentation of the Classical model appears in *The Classical Hollywood Cinema* by David Bordwell, Janet Staiger, and Kristin Thompson. The book places particular emphasis on the importance of linear narrative. Bordwell writes, "Here in brief is the premise of Hollywood story construction: causality, consequence, psychological motivations, the drive toward overcoming obstacles and achieving goals. Character-centered—i.e., personal or psychological—causality is the armature of the classical story" (Bordwell, Staiger, and Thompson 13). This argument is supported by a mountain of evidence, from trade journals, how-to manuals, and one hundred randomly selected films, showing that these principles were operating in film after film.

Bordwell uses the Russian Formalist notion of the dominant to explain the relationship between the narrative and other systems. He writes, "This integrity deserves to be seen as a dynamic one, with the subordinated factors constantly pulling against the sway of the dominant ... These systems do not always rest quietly under the sway of narrative logic, but in general the causal dominant creates a marked hierarchy of systems in the classical film" (Bordwell, Staiger, and Thompson 12). If given the example from *The Three Musketeers*, Bordwell might agree that the gags and stunts have an appeal all their own while arguing that their appeal is ultimately subordinated to a larger system—the system of causal logic, in which a goal-oriented character (D'Artagnan) overcomes obstacles as the story takes another step toward closure.

While acknowledging the book's accomplishment, many theorists have criticized its arguments, including the argument that linear narrative operates as the dominant in the Hollywood system. Donald Crafton argues that the Classical model does not apply to slapstick films, even when the films contain a certain amount of narrative integration. In "Pie and Chase" he writes:

> I contend that it was never the aim of comic filmmakers to "integrate" the gag elements of their movies. I also doubt that viewers subordinated gags to narrative. In fact, the separation between the vertical domain of slapstick (the arena of spectacle I will represent by the metaphor of the thrown pie) and the horizontal domain of the story (the arena of the chase) was a calculated rupture, designed to keep the two elements antagonistically apart.
>
> (Crafton 107)

In a slapstick film the forward progress of the linear narrative is constantly being interrupted by gags that do nothing to help the protagonist achieve his or her goal. Crafton describes the structure of the slapstick film as a "complex system of alternation of spectacle and diegesis" (113). For this reason we might call this approach an Alternation model. Hollywood films do have causality, deadlines, and goals, but in some genres these linear components alternate with nonlinear moments of spectacle.[2]

In *Hollywood Cinema: An Introduction* Richard Maltby proposes a similar model. He writes, "Hollywood narration must negotiate the pleasurable interruptions of performance or spectacle, before reasserting itself in order to bring them (and consumption) to an end. The two elements of storytelling and spectacle are held in an essential tension, and the movie exists as a series of minor victories of one logic over the other" (Maltby 339). Whereas Crafton's argument is restricted to one genre, Maltby argues that all Hollywood films alternate between moments of narrative dominance and moments when narrative is subordinated to other appeals. Maltby supports this view with a compelling argument that Hollywood films practice a "commercial aesthetic." The argument is powerful because it rests on a simple yet persuasive appeal to economic common sense. Suppose there are three spectators: Sam likes detective stories, Susie likes stunts, and Sally likes both. If you make a movie with a tight detective story (but no stunts), you may get money from Sam and Sally but not Susie. If you make a movie with stunts (but no story), you may get money from Sally and Susie but not Sam. If you make a movie that offers both stunts and a story, you have a good chance of getting all three spectators to give you their money. Given that Hollywood made films to make money, it would seem that Hollywood had every reason to include as many different attractions (both narrative and nonnarrative) as possible.[3] If presented with the scene from *The Three Musketeers*, Maltby might point out that the scene goes on far longer than the causal chain requires. Every time Fairbanks performs an unnecessary stunt the commercial aesthetic is allowing attractions to overturn the dominance of narrative.

Maltby argues that one of the major reasons people go see movies is the desire to experience strong emotions in a safe context (35). Here Maltby anticipates a third model, which I call the Affective model. The Classical model emphasizes the dominance of linear narrative, while the Alternation model grants that classical narration is sometimes (but only sometimes) dominant. The Affective model goes one step farther, firmly placing the features of classical narration in a subordinate position. In "Comedy and Classicism," Dirk Eitzen writes:

> [O]ne can make a movie that has all the elements that Bordwell, Staiger, and Thompson attribute to classical Hollywood fiction—goal-oriented characters, a self-effacing style, a coherent fabula, and all the rest—that is still as dry as sand. Yet, strikingly, this is not what evolved in Hollywood. What evolved are stories that are full of sex, violence, melodrama, fast action, suspense and surprise, fantasy and horror, and, as I have been pointing out here, comedy. The transparent style evolved because, in most instances, that style gives the most emotional bang for the buck. But where another kind of emotional bang could be obtained by sacrificing narrative transparency, as in the case of comedy, there was evidently little hesitation inputting transparency aside.
>
> (Eitzen 404)

Eitzen concedes the point that most Hollywood films have the features that the Classical model says they do. However, he argues that this is only a "second-order explanation" (Eitzen 403), since narrative is only one way that Hollywood achieves its true aim: the production of emotion.

In "Melodrama Revised" Linda Williams also places classical narrative (in particular, classical realist narrative) in a subordinate position. She argues that "supposedly realist cinematic *effects*—whether of setting, action, acting or narrative motivation—most often operate in the service of melodramatic *affects*" (Williams 42). The typical melodramatic narrative encourages us to feel sympathy for an innocent victim. In pathos-driven melodramas the narrative builds up to a high point of tragic suffering. In action-driven melodramas the victimized protagonist fights and often defeats the evil forces. Either way, the melodrama is characterized by its appeal to moral feeling. *The Three Musketeers* fits neatly into Williams's action-driven category. The childlike D'Artagnan has been slighted by the Cardinal's arrogant guards, and his status as an innocent victim gives his victory a powerful emotional charge.

Like Eitzen, Williams acknowledges the presence of certain classical features, such as deadlines. Similarly, Williams argues that the presence of these classical features does not establish the correctness of the Classical model, since those features are ultimately subordinated to a larger, emotion-based strategy. While Williams's complex theory of melodrama covers many issues besides emotion, her argument that "we go to the movies not to think but to be moved" (61) suggests that her model can also be classed as an Affective model.[4]

Many of these arguments rely on terms like "linearity" and "forward progress." It might be useful to clarify what these terms might mean. First, we might say that a narrative is linear when the protagonist takes a step closer to achieving a goal. Although some of the models do use the term in this sense, it should be noted that this is a fairly limited conception of "linearity." Any time a character encountered an obstacle, forward progress would come to a halt, and we would have a break in linearity. Since Bordwell, Staiger, and Thompson include obstacles in their model, they are probably not using the notion of linearity in this narrow sense. A second sense refers to the linearity of the causal chain (Bordwell, Staiger, and Thompson 64). A typical scene starts with an effect (i.e., something caused by a previous scene) and then introduces a new cause (which produces an effect in a later scene). In this sense, obstacles are indeed part of the linear structure, since they are typically caused by previous events while motivating future events. For critics of the Classical model, many of the best counterexamples are details that make no obvious contribution to the causal chain and are therefore nonlinear in this sense. The "causal chain" account of linearity is related to a third version, which stresses the spectator's forward-looking activity. For Bordwell, the classical spectator is constantly forming hypotheses about what happens next. For Bordwell's critics, musical numbers and gags serve as counterexamples to the Classical model precisely because they do not encourage spectators to wonder what happens next. Instead, they depart from linearity by encouraging spectators to focus on the present.[5] In formulating a new model we must keep these different senses of linearity in mind.

These quick summaries cannot do justice to the nuances of these arguments, but they serve to clarify a range of options for theorizing Hollywood narrative. Each model has its strengths. The Classical model is grounded in an impressive array of empirical evidence.[6] The Alternation model is supported by the highly persuasive logic of the "commercial aesthetic" argument. The Affective model is built on the highly plausible claim that people go to movies to experience emotion. My own alternative (which I call the Cooperation model) is a contribution to the Affective model. Specifically, I want to look at some of the most "classical" features of Hollywood narrative (such as goal orientation and closure) and show that they function to produce emotion just as much as they function to produce coherence. This should be a valuable contribution for two reasons. First, narrative is often studied as a system that organizes emotion-generating attractions rather than as a system that produces emotion in its own right.[7] Second, an emotion-centered analysis of Hollywood narrative will allow us to approach the issue of the dominant in a new way by recognizing that narrative and attractions can, in some cases, mutually intensify one another.

2 Anticipation, culmination, and emotion

In *How to Write Photoplays* (1920) John Emerson and Anita Loos offer the following advice to the aspiring screenwriter:

> Many amateurs are prone to cheat their audience by ending the story without some bit of action which spectators have been led to expect and are hoping to see. Let us suppose that after your hero has triumphantly rescued the heroine, you end the story with a subtitle, "And So They Lived Happily Ever After." That is too abrupt. You must add just a few scenes to satisfy the very understandable craving to see the hero reaping the rewards of heroism as the girl comes to his arms.
>
> "But this is the same old ending," you protest. True enough. But it is essential if your audience is to feel satisfied. In the same way, if the villain is finally defeated, you must gratify your audience's desire to see him dragged off to jail. Don't leave this sort of thing to the imagination.
>
> (Emerson and Loos 104)

As Bordwell, Staiger, and Thompson have pointed out, Emerson and Loos spend many pages in their book encouraging their students to strive for causal coherence. However, I suggest that "causal coherence" is not the defining value in this passage. If spectators were only interested in learning what happens at the end, they would be satisfied with the subtitle. The fact that they are not suggests that they want something more than coherence. They want to see the culmination take place. They want to experience an emotional response to the event itself.

In the happy ending we get to celebrate an event that we have been hoping to see. This seems like an innocuous claim, but a closer look reveals hidden complications. Notice that there are two kinds of emotions working here. First, we have an anticipatory emotion: hope. A Hollywood narrative typically encourages us to anticipate future events and revelations. If these anticipated outcomes are emotionally weighted (generally, by sympathy for the protagonist), we experience hope: hope that the protagonist achieves his or her goals. By throwing obstacles in the way of the protagonist the narrative can generate another anticipatory emotion: fear that the protagonist will fail. Anticipatory emotions can be called "linear" in the sense that they are directed at future events and revelations.

By contrast, when a character actually conquers an obstacle we are expected to feel *immediate* delight. When a line of action culminates in a scene of success or failure we experience feelings of celebration or commiseration. We might call these "culminating emotions." These emotions are related to the linear structure in a different way. If hope and fear are directed toward the future, then celebration and commiseration are directed toward the present. Emerson and Loos acknowledge that Hollywood offers more than just the anticipatory emotions of hope and fear. Audiences expect to experience the culminating emotions generated by experiencing the scene of the villain being dragged off to jail.

A Hollywood narrative is not a long series of anticipatory scenes followed by a single culminating scene. Typically, the protagonist has dozens of minor victories and defeats along the way. In *An American in Paris* (Vincente Minnelli, 1951) we delight when Gene Kelly and Leslie Caron first realize their love for each other. In *Murder, My Sweet* (Edward Dmytryk, 1944) we despair when Dick Powell has been captured and drugged against his will. In *North by Northwest* (Alfred Hitchcock, 1959) we delight when Gary Grant defeats the murderous crop-duster. In *My Best Girl* (Sam Taylor, 1927) we despair when the young lovers are separated. Each one of these scenes occurs well before the ending of the respective film, but these scenes provide for moments of celebration and commiseration. The metaphor of "linearity" is only partially helpful here. From the standpoint of cognitive comprehension it can indeed be useful to see the Hollywood narrative as an unbroken linear chain with a relentless forward progress. However, from the standpoint of emotional experience, it might be more useful to see it as a complex weaving together of anticipation/culmination structures in which our emotional reactions to *present* events are just as important as our anticipatory reactions to future events. Rather than use the term *linearity* to describe this complex pattern, we might do well to make a few more distinctions. The anticipation/culmination structures can run simultaneously (as when a film contains both an action plot and a romance plot); they can run successively (as when a comedian solves one problem, only to be faced by another, unrelated problem); they can run consequentially (as when the solution to a mystery produces the new problem of apprehending the criminal); or they can be nested (as when the solution to a minor problem helps the protagonist solve a major problem). These simple distinctions should help us describe the emotional structure of Hollywood narrative with more precision than the linearity metaphor will allow.

The crop-duster sequence from *North by Northwest* is only moderately important for the overall causal chain. It encourages Roger to suspect Eve, but this narrational function could be accomplished more efficiently with a shorter scene. Why does this minor link in a larger chain have so much emotional power? My suggestion is that the scene develops its own anticipation/culmination structure, nested within the larger narrative. By having the plane attack Roger, Hitchcock generates the fear that Roger will be killed. Then he produces a culminating moment when Roger manages to escape. *North by Northwest*, like most Hollywood narratives, can be described as a system for manipulating our feelings of hope and fear, delight and despair.[8] From this point of view, the crop-duster sequence is not a minor addition to the causal chain but an excellent example of Hollywood narrative doing its work: manipulating our feelings by placing a character we care about in a situation of danger. Once the larger narrative has made us care about a character, a film can give us a series of different problems, like the "nested" crop-duster problem, to play on those feelings.

This proposal is similar to Noël Carroll's proposal concerning film suspense. Carroll argues that the causal links in a narrative are not as tight as we often suppose. Instead, he suggests that Hollywood narratives produce coherence by raising emotionally charged questions and then answering them. A narrative typically involves a few macroquestions, structuring the entire film, and several microquestions, structuring scenes and sequences. We can find similar proposals in less theoretical sources. For instance, contemporary screen-writing manuals, such as *The Tools of Screenwriting*, often discuss narrative in terms of questions, answers, hopes, and fears. As the quotation from Emerson and Loos suggests, Hollywood filmmakers of the classical period may have thought of narrative in similar terms.

Does this mean that Bordwell, Staiger, and Thompson are wrong to argue that Hollywood filmmakers valued causality and coherence? No. There is far too much evidence on their side to make such a claim. Rather than dispute that evidence, I would simply propose an additional, emotion-centered explanation for the features that are highlighted in their account. As Bordwell would point out, a commitment to causality allows us to anticipate what will come next. This helps clarify the narrative, but I would add that anticipation is also a necessary condition for the experience of hope and fear. Clearly defined endings do indeed provide closure, but I would add that they provide the opportunity for celebration or commiseration. When the lovers kiss at the end of a romance, we are not happy simply because the narrative has achieved maximum coherence; we are happy because we can celebrate the culmination of a long process of hopeful anticipation. The narrative has produced a hope and then fulfilled it.

The sympathetic, goal-oriented protagonist is another common feature of Hollywood storytelling. Goal orientation clarifies problems by producing mutually exclusive alternatives (either the protagonist will succeed or the protagonist will fail). Making the protagonist sympathetic allows us to emotionally weight those alternatives. Once the film has established this basic structure it can play on our emotions with obstacles, successes, and failures. Obstacles intensify our anticipatory feelings of hope and fear. When a protagonist overcomes an obstacle we feel a culminating emotion: delight. When a protagonist gives up in the face of an obstacle we feel another culminating emotion: despair. Obstacles are not simply breaks in the forward progress of the story. They are essential tools of Hollywood narrative as a system for producing emotions.

While conventions like goal orientation are means toward the end of coherence, they are also means toward the end of emotional experience. Since they are only means to an end, it should not surprise us to learn that Hollywood filmmakers will use variations when they can achieve the same end more efficiently. For instance, Frank Capra's movies offer numerous variations on the norm of goal orientation. In *Mr. Smith Goes to Washington* (1939) neither Jimmy Stewart nor Jean Arthur ever seems to tackle the goal of winning the love of the other. Nevertheless, the spectator is encouraged to hope that they will fall in love with each other before the end of the movie—a hope that is eventually realized. *It Happened One Night* (1934) provides another variation on the norm of goal orientation. Clark Gable tackles a goal: to help Claudette Colbert meet with her fiancé. However, even though Gable plays a sympathetic protagonist, we hope that he fails completely, because the narrative has posed a more important problem, encouraging us to hope that Gable and Colbert fall in love with each other. We might call Gable's stated goal a "false problem"—he tackles it, but spectators are supposed to know that it is not his true problem. These counterexamples do not destroy the model; rather, they encourage us to move up one level of abstraction. The emotionally weighted problem with clearly defined alternatives is the center of Hollywood dramatic structure. Goal orientation is a common way to achieve clearly defined alternatives, and the sympathetic protagonist is a common way to produce emotional weighting. This explains why they are such common features, but acknowledging that they are merely means allows ample room for exceptions and variations.

At this point it might be objected that this model, with half of its focus on hope and fear, places too much emphasis on uncertainty. Don't we often know how a film is going to end? Perhaps, given the genre, it is simply inconceivable that the boy and girl will not fall in love by the end of the film. Perhaps we have an even better reason to be certain: we have seen the film before. How can we experience hope or fear for an event that we know is going to happen? To meet this objection we can note that some anticipatory emotions are based on certainty. For instance, suppose I have a daughter in school, and I have been told ahead of time that she is going to win an award at tonight's award ceremony. My emotion is one of "eager anticipation"—I await the ceremony with delight precisely because I am 100 percent certain that she will win. Alternatively, I might feel dread when I know for a fact that something terrible is going to happen. There is no reason why narrative cannot play on these kinds of anticipatory emotions. Indeed, Emerson and Loos seem to assume as much: they note that spectators have been "led to expect" a kiss and would feel cheated without one. When we feel certain that the two lovers will kiss at the end of the film we can eagerly anticipate the scene. This anticipation will intensify our feelings of joy when the culminating scene of the kiss finally occurs.[9]

I am not the first person to discuss the relevance of emotions for an analysis of Hollywood narrative. In addition to Williams, Eitzen, and Carroll we might mention writers as diverse as Marilyn Fabe and Ed S. Tan. Fabe's close analyses carefully describe Hollywood's manipulation of hope and fear, while Tan has explained our emotional response to goal-oriented narratives in psychological terms.[10] So far I have contributed to the discussion by proposing some useful distinctions, such as the distinctions between anticipatory and culminating emotions and the distinctions between emotions of certainty and emotions of uncertainty. Next, I want to make a more ambitious contribution. Shifting our attention from comprehension to emotion can give us new insights into the ways that Hollywood integrates narrative and other attractions. Specifically, I want to argue that the notion of "dominance," which plays a crucial role in all three of the models analyzed in Part 1, is only appropriate in certain contexts. The problem is that "dominance" carries unavoidable connotations of struggle, as one system seeks to control another. If narrative is seen as a rational system of organization and containment while the attraction is seen as a momentary appeal to the senses, then it seems hard to avoid the conclusion that the systems are locked in struggle. However, if narrative and attractions are both theorized as systems for producing strong emotions, then it seems much more reasonable to conclude that they can coexist peacefully. Indeed, they might even be able to mutually intensify one another.

Returning to the example of *The Three Musketeers*, it is not enough to say that the causal chain has led us to expect a display of D'Artagnan's skill. The narrative has also led us to *hope* for such a display. This scene is a culmination of those hopes. It is true that the gags and stunts have a value independent of the narrative. They would be enjoyable if we saw them performed in, say, a variety show. However, we are supposed to take an extra joy here because the stunts are being performed by *this* character—D'Artagnan, a sympathetic character who has become the vehicle for our hopes and fears. The more successfully the narrative has manipulated our hopes and fears with regard to D'Artagnan, the more delight we take in the audacious stunts.

Meanwhile, the more spectacular the stunts, the more we admire D'Artagnan and the more delight we will take in his fixture successes. We can only speak of "dominance" and "struggle" when we have competing interests. Here, the interests of narrative and the interests of spectacle are not in competition. They are not even in an uneasy truce. They are working together to produce an intensified emotional response.

As another example, consider the dance scenes in the Astaire–Rogers musicals. Often cited as examples of the "integrated" musical, these films integrate the numbers into the emotional curve as much as they integrate them into the causal chain. As *Top Hat* screenwriter Allan Scott once said, "It wasn't just mechanical. There were points where we knew we had to have it. You came to an intimate scene of some kind, whether it was flirting or what not, and that was the time for music" (quoted in Server 190). The narrative encourages us to hope that *these* two characters will dance together as part of the romance. In other words, we don't want to see just any romantic dance. We want to see a romantic dance with the two characters the narrative has encouraged us to care about. Meanwhile, the dances themselves, as marvelous moments of spectacle, can intensify our experience of the romantic narrative. The more perfectly coordinated the dance, the more graceful the movement, the more we delight in our sense of satisfaction that these two characters are right for each other.

An emotional curve can be created by manipulating goals, obstacles, and solutions to produce feelings like sympathy, antipathy, hope, fear, delight, and despair. These components can be coordinated with many different attractions. For instance, the attraction of a particular star can generate instant sympathy for a character. An act of spectacular violence can generate instant hatred for a villain. A sad song can intensify the emotions we feel when lovers are separated. A dangerous stunt can increase our fears concerning the safety of the hero. A sight gag can allow the protagonist to overcome an obstacle in an unexpected way.[11] In such cases the spectator who likes attractions can enjoy the film just as much as the spectator who likes narrative, while the spectator who likes both can enjoy the film for multiple reasons.

Here, my argument draws on Maltby's arguments concerning the commercial aesthetic. Because Hollywood made films to make money we should expect Hollywood to include as many different appeals as possible. It is not hard to find evidence for this claim. For instance, consider the Warner Bros. film *Fashions of 1934*, directed by William Dieterle, with a musical number by Busby Berkeley. In a memo to Hal Wallis producer Henry Blanke writes, "I hope you realize that in the picture, *King of Fashion* [the film's original title], we have the making of a really excellent picture ... and I know that the women will just eat it up—not only on account of its story but on account of its interesting background of fashions" (Oct. 10, 1933). When the picture was released the next year returns were disappointing. In a memo to Jack Warner, Wallis blames both the ad campaign and the new tide for placing *too much* emphasis on fashion, a strategy that he claims "excludes men entirely." Instead, he writes, "A sales campaign aimed along the lines of *42nd Street, Footlight Parade*, etc., of beautiful girls, music, comedy, and so forth, certainly should have great appeal" (Jan. 27, 1934). It is clear that Blanke and Wallis both believe that multiple attractions should increase a film's drawing power. It is also clear that they do not expect a monolithic audience. Guided by a few obvious stereotypes, they assume that different spectators want to see different things. Their aesthetic decisions are guided by commercial considerations.

I would take Maltby's argument one step farther. Suppose we make a detective film that constantly interrupts the narrative for a series of stunts. The spectators who enjoy the emotional power of the narrative may get bored during these interruptions, no matter how pleasurable the interruptions are for the spectators who enjoy stunts. Now suppose we make a detective film in which the stunts are coordinated with the narrative in such a way that the generation of narrative-based pleasures is not suspended. For instance, a dangerous stunt might intensify our fears regarding the detective's safety. The stunt fan still gets to see some stunts, while the devotee of detective stories need not suspend his or her pleasure while waiting for the stunts to conclude. Wouldn't the latter film be more appealing than the first film, burdened as it is with so many interruptions? Once we assume that narrative has some emotional rewards to offer, Maltby's argument about the commercial aesthetic turns out to be an argument in favor of fewer interruptions, not more. Ideally, the narrative and attractions produce emotions at the same time. This is why I call my model the Cooperation model. It is possible for narrative and attractions to work together.

Does this mean that all this talk about dominance has been misguided? Not necessarily. As the example from *Fashions of 1934* suggests, Hollywood did not have a magic formula for getting the balance just right. A dominance model can still help us understand certain films, since narrative and attractions do sometimes compete for dominance. My proposal is that we should expect a spectrum of possibilities. On one side of the spectrum there are mutually intensifying examples where narrative and attractions work together, as when a sad song expresses the mood of a sad moment or when a dangerous stunt increases our fears. On the other side of the spectrum there are mutually limiting examples where narrative and attractions do indeed struggle for dominance, as when a flashy musical number appears in the middle of a suspenseful detective story or when a silly sight gag appears in the middle of a tear-jerking drama. Given the commercial aesthetic, any given film may contain a range of combinations from any side of the spectrum.

In Part 3 I would like to demonstrate what such a mixture might look like using a set of films that are often invoked as evidence for Hollywood's willingness to use narrative as a mere excuse for spectacle: the Busby Berkeley musicals of the 1930s. Drawing on archival research, I would like to show that the relationship between narrative and spectacle is much more complex. In some cases they do struggle for dominance. In other cases the emotional power of the spectacle is integrated more tightly with the emotional curve of the narrative.

3 Narrative and spectacle in the films of Busby Berkeley

At first glance, a Busby Berkeley musical might seem like the perfect example of a Hollywood film that resorts to the more extreme solution of switching back and forth between narrative and nonnarrative elements (as in the Alternation model). For instance, *Footlight Parade* (Lloyd Bacon, 1933) contains about seventy minutes of fast-paced narrative action, followed by three consecutive Berkeley numbers. A glance at the screenplays to these films also suggests that the numbers were conceived independently of the narrative. For instance, consider this passage from Delmer Daves's final continuity script of *Dames* (Ray Enright, 1934):

> 183. INT. THEATRE—(FROM AUDIENCE ANGLE) The curtains open on the:
> 184. FIRST NUMBER OF SHOW . . . MUSIC SPOT 7
> 185. MED. CLOSE SHOT—EZRA'S BOX
> As number is completed—applause.
>
> (Daves 111)

Daves does not bother to give any details about the number itself. Instead, he treats the number as an independent module, the content of which is to be determined by Berkeley. In some cases, this modular construction allowed the order of numbers to be changed in editing; the final continuity of *Gold Diggers of 1933* (Mervyn LeRoy, 1933) has the "Forgotten Man" number in the middle of the script, but it appears at the end of the finished film. The nonnarrative quality of these numbers seems so obvious that it can even be parodied in the films, as in *Dames*, when Dick Powell sings to a group of middle-aged businessmen:

> But who cares if there's a plot or not
> When they've got a lot of dames?
> What do you go for? Go see a show for?
> Tell the truth. You go to see those beautiful dames.

These lyrics would seem to prove the oft-made suggestion that musicals use the plot as a mere excuse. In his insightful book on Busby Berkeley, Martin Rubin does a good job explaining the ways that Berkeley draws on a spectacle-oriented tradition.

On the other hand, as Bordwell might point out, narrative still plays an important role. On the most basic level the numbers are motivated by the theatrical setting. Of course, this motivation does not extend to the details of the numbers, since we could hardly expect the theatrical audience to see Berkeley's overhead kaleidoscopes. Nevertheless, the numbers are undeniably integrated into the causal chain, since it is the success of the numbers that provides closure to that causal chain. Furthermore, the numbers themselves often include mininarratives, such as the story about the "Honeymoon Hotel" in *Footlight Parade*. Alternatively, a number could introduce or develop themes and motifs from the "story" portion of the film. One famous example is the ironic "We're in the Money" number that introduces us to the topical Depression-era setting of *Gold Diggers of 1933*.

But we must not stop there. As I have argued throughout this essay, it is useful to think of narrative not simply as a tool for the production of coherence but also as a tool for the creation of emotions like hope, fear, delight, and despair. Rather than break the narrative down into unity-producing causes and effects, we can break it down into emotion-producing components, relating to the success or failure of the protagonists. In the Berkeley musicals the components might include "apparent failures" (which produce feelings of despair, even as the possibility of a reversal is left open), "doubt generators" (which play on our fears that the protagonist will fail), "stakes reminders" (which encourage us to remember why we are hoping the protagonist will succeed), and "nested successes" (which allow us to feel moments of delight, even as the larger problem remains to be solved).

Let us take the final half-hour of *42nd Street* (Lloyd Bacon, 1932) as an example, using these terms to demonstrate how the numbers work with the narrative to produce a heightened emotional curve. As with most backstage musicals, the main problem is "to put on a show." We are asked to hope that the protagonists will succeed and to fear that they will fail. With a nod to Linda Williams we can note that Ruby Keeler's sweet-girl persona appeals to our sense of innocence. With a nod to Richard Maltby we can note that the commercial aesthetic even affects the narrative construction: a spectator who dislikes Ruby Keeler may like Warner Baxter. About an hour into the film star Dorothy Brock (Bebe Daniels) twists her ankle, and director Julian Marsh (Warner Baxter) tells the cast that the show will not go on that night. This is more than a typical obstacle; the obstacle is so powerful that the lead protagonist seems to have given up. This "apparent failure" works to produce powerful culminating emotions: our worst fears seem to have been realized. Perhaps because it is so effective in generating these emotions, the scene where the protagonist gives up is a staple of Hollywood dramaturgy. Indeed, *Footlight Parade* has an almost identical scene. (Contemporary screenwriters have a name for such a scene: the "darkest moment.") The culminating emotions produced by such scenes are themselves valuable, but the emotional contrast also serves to intensify our feelings of delight when a reversal occurs. In *42nd Street* the reversal begins when Marsh decides to try Peggy Sawyer (Ruby Keeler) in the leading role. This produces a new chain of anticipation. Will Peggy be able to learn the role and become a star? During the grueling rehearsal the dialogue plays on our emotionally weighted curiosity about this question. At one point Marsh says, "No, it's impossible. Impossible!" Later, Peggy almost breaks down, crying, "I can't, I can't." These lines are "doubt generators"; they increase our fears. (As Noël Carroll would point out, they intensify the feeling of suspense by making it seem unlikely that she will succeed.) Eventually, Peggy manages to make it through the rehearsal. This is a nested success: solving this minor problem is part of solving the major problem of putting on a show. A nested success provides an opportunity for culminating emotions of delight, even as it intensifies our anticipatory hope that the show will be a success. Right before Peggy goes onstage *42nd Street* uses one more trick to play on our hopes and fears. We might call this a "stakes reminder"; it is exemplified by Marsh's famous speech: "Now, Sawyer, you listen to me, and you listen hard. Two hundred people, two hundred jobs, $200,000, five weeks of grind and blood and sweat depend upon you. It's the lives of all these people who have worked with you. You've got to go on, and you have to give and give and give. They've got to like you, they've got to. Do you understand? You can't fall down. You can't because, because your future's in it, my future, and everything all of us have is staked on you." What is the purpose of this speech? It works causally, to be sure, since Marsh's directorial skill helps to motivate Peggy's success. But its primary purpose, I suggest, is to prepare us emotionally for the film's climactic culmination. The speech reminds us that all of our emotionally weighted questions about the success or failure of the protagonists will soon be answered; either our hopes or our fears will soon be decisively realized. The actual culmination is the performance itself. This means that the numbers are integrated into the anticipation/culmination structure as tightly as they could possibly be—the entire structure revolves around the success or failure of the numbers themselves.

The narrative structure of anticipation has been coordinated with a nonnarrative structure of anticipation. Just as we hope that the numbers are as good as the sympathetic protagonists need them to be, we hope that the spectacular numbers are as dazzling as they have been advertised to be. Similarly, the narrative culmination is coordinated with the nonnarrative culmination: our delight at our protagonists' success merges with our astonishment at Berkeley's numbers.

In the Busby Berkeley musical narrative is not just an excuse for the numbers, it intensifies our emotional response to the numbers. Meanwhile, the numbers are much more than mere components in a causal chain. Rather than struggle for dominance, narrative and nonnarrative elements work together to produce an intensified emotional curve. This principle of emotional coordination is even hinted at in an early draft of the script for *Footlight Parade*. The screenwriters James Seymour and Manuel Seff write, "If any scene lacks incident or drags it must be cut. The picture must have the pace of a production dance number punched up with a minimum of dialogue" (1). This passage illustrates the problems with opposing the "forward progress" of narrative to the "static spectacle" of narrative interruptions. *Footlight Parade* has a famously fast-paced causal chain, but this is not to observe the dictates of classical coherence, it is to duplicate the feel of a Busby Berkeley musical number.

So far I have argued that the Warner Bros. musicals follow a mutual intensification principle. The numbers are integrated into the anticipation/culmination structure, and the emotional curve is intensified by emotional coordination. But is there no truth to the notion that the narrative is in some way sacrificed during the numbers themselves? Although it may be true that mutual intensification is an ideal, maximizing emotion may require certain sacrifices along the way. At times, narrative and attractions will indeed struggle for dominance.

Again, *42nd Street* provides an example. Perhaps because Berkeley's working methods had not yet been fully established, the writers of the first draft continuity (Whitney Bolton and James Seymour) made a strong effort to integrate the numbers on a moment-by-moment basis. For instance, Dick Powell's main number is described in detail:

> 433. CLOSE SHOT—BILLY
> singing number.

434. CLOSE SHOT—PEGGY IN DRESSING ROOM sitting relaxed—eyes closed. At distant sound of Billy's number, Peggy opens her eyes, sits up straight, listens[,] smiling happily.

435. CLOSE SHOT—BILLY
singing his heart out.

436. CLOSE SHOT—GRIM BESPECTACLED CRITIC in aisle seat, nodding with pleased smile as Billy sings, whispers to neighbor and looks at program.
CRITIC
(whispering)
Good kid—youth, personality—who is he?
Critic studies program.

437. FULL SHOT—STAGE
Billy finishes verse and girl chorus enters.

The number in the finished film does not follow the script. Most important, there are no cutaways, neither to Peggy nor to the grim bespectacled critic. Why did the screenwriters include the cutaways in the first place? The cutaway to Peggy suggests that Billy is singing the song to her. This allows the number to become a celebration of the successful resolution of the Peggy/Billy storyline. The cutaway to the critic increases our hope that the successful numbers will contribute to the success of the show. In other words, the writers were attempting to play on anticipatory and culminating emotions. Why does the finished film eliminate the cutaways? Both these cutaways intensify our emotional response to the narrative, but they do so at a cost. Every time we cut away from the number we miss out on the spectacle of the performance. Even worse, both cutaways make it more difficult to hear Billy's singing. Presumably, the filmmakers decided that the cost was too great, and the cutaways were eliminated. Instead, a compromise has been reached. Considered as a unit, each number is tightly integrated into the emotional curve, since the success of the number is integral, to the solution of the problem. Considered on a more local level, the moment-by-moment play of the anticipation/culmination structure has been sacrificed, since it might threaten the number's independent power to astonish and delight. When the leads give way to anonymous chorus girls, the anticipation/culmination structure is virtually abandoned.

Later films employ more efficient solutions. For instance, the anticipation/culmination structure is more fully integrated into *Footlight Parade*'s notorious "Shanghai Lil" number because protagonist Chester Kent (James Cagney) has unexpectedly been forced to perform the lead role without any rehearsal at all. Of course, Kent manages to pull it off, and our delight in Cagney's talent merges with our delight in the protagonist's success. Still, Berkeley does not foreground the moment-by-moment play of the narrative's emotional curve as much as he might. A director who was solely concerned with the narrative might have used more close-ups of Cagney's face during the performance to play upon our anxieties, reminding us at each moment that Cagney might not pull it off. This strategy might have aided the narrative, but it would have ruined the spectacle.

In short, there is some struggle for dominance, but the notion of struggle does not capture the entire relationship. Narrative continues to do its emotion-generating work, even during a Busby Berkeley musical number. The film balances narrative and nonnarrative elements to produce intensified emotions, although the compromise solution must ultimately work within mutually limiting factors. The balance leans toward mutual intensification when a character we care about (such as Ruby Keeler) is doing a dazzling dance. The balance leads toward a struggle for dominance when kaleidoscopic patterns are formed by anonymous dancers.

Conclusion

In this essay I have suggested that a Cooperation model can clarify some of the details of the Affective model. The Affective approach suggests that Hollywood uses both narrative and attractions to create emotion. This essay explains how that combination works in practice. Rather than look at narrative as an organizational system designed to produce comprehension, we can look at narrative as an emotional system designed to produce feelings of hope, fear, delight, and despair. Rather than emphasize the ways that narrative causes us to look ahead to unknown events, we can recognize that narrative also encourages us to enjoy culminating emotions "in the present" and even to eagerly anticipate some events that we know for certain are going to happen. Rather than see narrative structure as a linear chain of causes and effects, we can see narrative as a complex weave of anticipation/culmination structures, including simultaneous, successive, consequential, and nested structures. Finally, rather than theorizing the relationship between narrative and attractions as one of struggle, we can recognize that narrative and attractions can often cooperate to create an intensified emotional response. At the same time, we need not sacrifice the strengths of the other two models. The Classical model is right to insist on the importance of goal orientation and closure. I would add that these devices were valued in part because they perform affective functions. The Alternation model is right to insist on the importance of the commercial aesthetic. I would add that the commercial aesthetic gave Hollywood filmmakers every reason to look for ways of combining narrative and attractions.

Attractions are more than just interruptions. Narrative is more than just an organizing structure. They are both systems with the tremendous power to influence emotion. From the action films of Douglas Fairbanks to the musical films of Busby Berkeley, Hollywood has developed a range of strategies for allowing these two systems to work together.

Notes

I would like to thank Lisa Jasinski, David Bordwell, the Warner Bros. Archives, the editors of the *Velvet Light Trap*, and the anonymous reader for their help with this article. The late Frank Daniel, who taught screenwriting at USC for many years, inspired much of my thinking about narrative and emotion.

1 Thanks to the anonymous reader for suggesting that I give my model this name.

2 In his "Response to 'Pie and Chase'" Tom Gunning argues that Crafton underestimates the degree to which slapstick comedies employ narrative integration. Gunning's model of the "cinema of narrative integration" is a variant of the Classical model, but Gunning places particular emphasis on the struggle between attractions and the narrative dominant.

3 The Sam, Susie, and Sally example is my own, but I think it captures the point of Maltby's argument (see Maltby 24–35). For another argument that questions the dominance of narrative in the face of economic pressures see Cowie. Cowie also offers a subtle critique of the way *The Classical Hollywood Cinema* employs the notion of "motivation."

4 These five arguments are but a few of the many arguments that prominent scholars have made concerning the structure of the Hollywood film. I do not have the space to cover them all, but one influential article demands acknowledgment: Rick Altman's "Dickens, Griffith, and Film, Theory Today." By granting the existence of classical features while insisting that they may coexist with other, equally important systems (such as the system of melodrama), Altman strongly influenced the development of what I am calling the Affective model.

5 In "'Now You See It, Now You Don't': The Temporality of the Cinema of Attractions" Tom Gunning contrasts the temporality of the cinema of narrative integration, in which moments are absorbed into a complex pattern, with the temporality of the cinema of attractions, in which spectators experience "an intense form of the present tense" (44) While my essay focuses on Hollywood cinema of the studio period, it is important to note that studies of early cinema have inspired much of the interest in the relationship between narrative and other attractions. At the other end of the historical spectrum,

many scholars compare the "classical" Hollywood cinema with a "postclassical" Hollywood cinema. See Smith for a discussion of this scholarship.

6 Many of the critics of the Classical model attempt to problematize the model by pointing out examples of scenes that do nothing to advance the causal chain. However, with their wealth of evidence, Bordwell, Staiger, and Thompson can happily grant the existence of exceptions while pointing to hundreds of films that support their theory. Indeed, they insist that subordinate elements can be seen struggling against the dominant principle. Bordwell has also argued that many of these counterexamples do, in fact, serve the causal chain, albeit in subtle ways (see *Planet Hong Kong* 179).

7 Bordwell has excellent insights about the ways narration shapes cognitive comprehension, but he openly acknowledges that he is less interested in emotion (*Narration* 39). Maltby argues that narrative serves to generate emotion, but he is more interested in the ways that narrative organizes the pleasures of performance and spectacle (339). Similarly, Rick Altman argues that narrative serves to give Hollywood's list of appeals the appearance of internal motivation (26), while Tom Gunning argues that narrative is usefully understood as a system of containment ("Response" 121). The point is that Hollywood narrative is usually studied as a system of organization and not as a system for the production of emotion. Even Eitzen stops short of explaining how narrative itself can generate emotion. The essays by Linda Williams and Noël Carroll represent important exceptions to this trend.

8 Ed Tan's phrase "film as an emotion machine" fits nicely here (see Tan 251).

9 Would any of the authors discussed in Part 1 disagree with the analysis offered so far? Not necessarily. This model may spell out details that are implicit in their theories. For instance, Bordwell does not deny the existence of emotion; he is just more interested in comprehension.

10 It might clarify my project to say that I am not offering a psychological theory of the emotions in the manner of Tan. Rather, I am making the historical argument that Hollywood filmmakers systematically designed their films to appeal to emotions as they are understood in ordinary language. In other words, my argument does not rest on evidence about the psychology of spectators. It rests on evidence about Hollywood filmmakers and the construction of films.

11 For a careful analysis of the ways that gags are integrated with narrative in the films of Buster Keaton see Noël Carroll's two essays on *The General* in his book *Interpreting the Moving Image*.

Works cited

Altman, Rick. "Dickens, Griffith, and Film Theory Today." *Classical Hollywood Narrative: The Paradigm Wars*. Ed. Jane Gaines. Durham: Duke UP, 1992, 9–48.

Blanke, Henry. Memo to Hal Wallis. *Fashions of 1934* Box. Warner Bros. Archives, Los Angeles.

Bolton, Whitney, and James Seymour. First Draft Continuity. *42nd Street* Box. Warner Bros. Archives, Los Angeles.

Bordwell, David. *Narration in the Fiction Film*. Madison: U of Wisconsin P, 1985.

— *Planet Hong Kong*. Cambridge: Harvard UP, 2000.

Bordwell, David, Janet Staiger, and Kristin Thompson. *The Classical Hollywood Cinema: Film Style and Mode of Production to 1960*. New York: Columbia UP, 1985.

Carroll, Noël. *Interpreting the Moving Image*. New York: Cambridge UP, 1998.

— "Toward a Theory of Film Suspense." *Theorizing the Moving Image*. New York: Cambridge UP, 1996, 94–117.

Cowie, Elizabeth. "Storytelling: Classical Hollywood Cinema and Classical Narrative." *Contemporary Hollywood Cinema*. Ed. Steve Neale and Murray Smith. New York: Routledge, 1998, 178–90.

Crafton, Donald. "Pie and Chase." *Classical Hollywood Comedy*. Ed. Kristine Brunovska Karnick and Henry Jenkins. New York: Routledge, 1994, 106–19.

Daves, Delmer. Final Continuity Script. *Dames* Box. Warner Bros. Archives, Los Angeles.

Eitzen, Dirk. "Comedy and Classicism." *Film Theory and Philosophy*. New York: Oxford UP, 1999, 394–411.

Emerson, John, and Anita Loos. *How to Write Photoplays*. Philadelphia: George W. Jacobs, 1920.

Fabe, Marilyn. *Closely Watched Films: An Introduction to the Art of Narrative Film Technique*. Berkeley: U of California P, 2004.

Gunning, Tom. "'Now You See It, Now You Don't': The Temporality of the Cinema of Attractions." *The Silent Cinema Reader.* Ed. Lee Grieveson and Peter Krämer. New York: Routledge, 2003, 41–50.

— "Response to 'Pie and Chase.'" *Classical Hollywood Comedy.* Ed. Kristine Brunovska Karnick and Henry Jenkins. New York: Routledge, 1994, 120–22.

Howard, David, and Edward Mabley. *The Tools of Screenwriting: A Writer's Guide to the Craft and Elements of a Screenplay.* New York: St. Martin's P, 1995.

Maltby, Richard. *Hollywood Cinema: An Introduction.* Oxford: Blackwell, 1995.

Rubin, Martin. *Showstoppers: Busby Berkeley and the Tradition of Spectacle.* New York: Columbia UP, 1993.

Server, Lee. *Screenwriter: Words Become Pictures.* Pittstown, NJ: Main Street Press, 1987.

Seymour, James, and Manuel Seff. Script Draft. *Footlight Parade* Box. Warner Bros. Archives, Los Angeles.

Smith, Murray. "Theses on the Philosophy of Hollywood History." *Contemporary Hollywood Cinema.* Ed. Steve Neale and Murray Smith. New York: Routledge, 1998, 3–20.

Tan, Ed S. *Emotion and the Structure of Narrative Film: Film as an Emotion Machine.* Trans. Barbara Fasting. Mahwah, NJ: Lawrence Erlbaum Associates, 1996.

Wallis, Hal. Memo to Jack Warner. *Fashions of 1934* Box. Warner Bros. Archives, Los Angeles.

Williams, Linda. "Melodrama Revised." *Refiguring American Film Genres: History and Theory.* Ed. Nick Browne. Berkeley: U of California P, 1998, 42–88.

PART I

Feature films, Hollywood and the advent of the studio system, 1912–26

Introduction

In their discussions of classical Hollywood cinema, Bordwell, Staiger and Thompson and their critics focus exclusively on feature films (or, to be more precise, feature-length films): films of four or more reels that usually ran for at least an hour. (The length of 'silent' films was generally measured in reels rather than hours or minutes because projection speeds could vary; the standard length of a reel of film was 1,000 feet by the late 1900s.) Films as long as this were extremely rare prior to the 1910s. Those produced in the US or imported from abroad were usually either Biblical or boxing films (films that recorded prize fights, usually on a round-by-round basis). Their exceptional length and their equally exceptional (if very different) cultural status meant they tended to circulate outside the systems and sites of distribution and exhibition that prevailed in the late 1890s and early 1900s, and those that began to prevail with the establishment of the Motion Picture Patents Company (MPPC) in 1908.[1]

The MPPC was a cartel of patent holders, importers, and film and equipment manufacturers.[2] Its films were targeted at those who managed and frequented nickelodeons. Nickelodeons were exhibition venues. (The term 'nickelodeon' conjoined the Greek word for theatre with the US term for a five cent coin.) When introduced in or around 1905, they proved extremely successful. By 1907, it was estimated that there were over 2,500. By 1908, this number had more than doubled. It was estimated that by 1910 as many as 26 million people visited nickelodeons every week.[3] By then, some nickelodeons could accommodate over a thousand spectators. But most were much smaller. Given the relatively low admission price, profits were dependent on the rapid turnover of relatively short (half-hour or hour-long) programmes of equally short (split-reel or reel-long) films. Longer two-reel or three-reel films were produced and imported in greater numbers between 1909 and 1912, but initially most were released a reel at a time on consecutive days. Most were shown a reel at a time as well, even when, as was the norm by 1912, reels were released simultaneously and most nickelodeons were equipped with more than one projector.[4]

For all these reasons, this 'system had no place for films longer than three reels, the films that later came to be called "feature films"'. However, 'the entertainment industry had a ready-made alternative in the state-rights system, in which exclusive rights to an act were granted to a regional franchise holder, who would then book it into theatres in his or her territory, guaranteeing the theatre owner exclusive exhibition for a negotiable period, thus allowing for long runs and a run-up period for an advertising campaign'.[5] Franchise holders could also subdivide and re-sell these rights to other distributors. Either way, in facilitating long runs and extensive advertising campaigns for individual films, and in facilitating higher ticket prices, scheduled screening times, and special modes of presentation (particularly in large venues such as opera houses and legitimate theatres), state-rights franchisees not only demonstrated the commercial viability of feature films, they also shared many of the practices associated with roadshowing. Roadshowing involved the hiring of venues directly, usually on a percentage-of-the-gross basis. Roadshowing on a small scale had been used to tour film presentations in rural areas since the 1890s and it continued to be used in this way for several decades.[6] On a larger scale, on a scale akin to its use in touring major circuses and legitimate theatre productions, one or more companies, usually comprising a manager, a publicist, a group of musicians, a print of the film and a projectionist, would either tour venues throughout the country or, if there were multiple companies (as was much more common), venues in specific cities, states or regions.[7]

Large-scale roadshowing was used initially to distribute lengthy prestigious imports such as *Quo Vadis?* (1913) and *Cabiria* (1914) and lengthy and prestigious domestic productions such as *The Birth of a Nation* (1915) and *Intolerance* (1916); the state-rights system was used slightly earlier to distribute feature-length imports such *Dante's Inferno* (1909), *The Fall of Troy* (1910) and *Queen Elizabeth* (1912).[8] Roadshowing on a major scale was inherently unsuited to routine productions, though in various guises it was used to distribute exceptionally long, expensive or prestigious films for decades to come. The state-rights system offered more advantages, at least in the short term. It was used by the Famous Players Motion Picture Company (which was founded by Adolph Zukor, Daniel Frohman and Edwin S. Porter in 1912) not only to distribute *Queen Elizabeth*, but to provide the basis for distributing a regular annual programme of feature-length films. However, although it 'could be profitable for an importer or producer film by film', the state-rights system 'provided too uncertain a cash flow for the continuous production of long and expensive films'.[9] As a result, while General Film, the distribution arm of the MPPC, 'created an "exclusive service" of feature programs made up of expensive one- to three-reel films', and while the state-rights system came increasingly to be used for marginal productions of one kind or another for decades to come, the Paramount Pictures Corporation, which had been formed by W.W. Hodkinson in 1914, regularised 'the system that Famous Players had developed, combining the weekly program format of General Film with the feature marketing techniques of the state rights firms'.[10] With films supplied by Famous Players and the Jesse L. Lasky Feature Play Company, among others, 'Paramount began offering a yearly program of features, consisting of 104 films of four to six reels each, released at a rate of two per week'.[11]

By the time Hodkinson had been deposed by Adolph Zukor in 1916, Paramount (which at that point became a brand name for the newly formed Famous Players-Lasky Corporation) had been joined by other producer-distributors of feature-length films, among them the World Picture Corporation and the Fox Film Company. However, in advertising its features individually, publicising its stars, innovating 'a percentage distribution fee, whereby producer and distributor received a percentage of the gross exhibitor rentals', offering the first annual programmes of feature-length films, and subsequently building and acquiring chains of cinemas and establishing a global distribution network, Paramount became 'the model for other firms in the classical Hollywood period'.[12] By then a new cartel of companies had long since emerged to displace the MPPC and rival cartels of independent suppliers of one-reel or two-reel films, and this was formally marked by the establishment of the Motion Picture Producers and Distributors of America (MPPDA) in 1922.[13] Most of the MPPDA's members distributed short films, cartoons and newsreels as well as feature-length films. Some were produced in-house. Others were subcontracted from small-scale suppliers. But along with ownership or access to major cinemas chains, it was the production and national (and international) distribution of programmes of relatively high-cost feature-length films that marked and cemented their power.

For Gerben Bakker, an economic historian, these were key developments in the history of what he calls the 'industrialisation of entertainment'. Seeking, among other things, to explain the emerging global dominance of Hollywood, Bakker focuses on the 1910s as a key decade. He argues that during the course of the nineteenth century 'falling working hours, rising disposable income, increasing urbanisation, expanding transport networks and strong population growth resulted in a sharp increase in the demand for entertainment'. This demand was initially met 'by a surge in the quantity and variety of live entertainment ... Large cities would offer a cascade of entertainment at staggered prices and qualities, including opera, theatre, vaudeville, variety, music hall, circuses and burlesque'.[14] At this point, motion pictures 'industrialised entertainment by automating it, standardising its quality and transforming it into tradeable product', thus merging 'freshly integrated national entertainment markets into an international one'.[15] Both before and immediately after the advent of feature-length films these markets were often dominated by European companies. The US was potentially the largest national market in the world. Yet despite vociferous campaigns against foreign films and an increasingly successful export drive by the MPPC and its member companies,[16] it was European companies that initiated what Bakker calls 'the quality race', the race to establish or re-establish market dominance by escalating expenditure on the production and marketing of feature-length films and on research and development (R&D) in general.

As Bakker goes on to argue in the extract reprinted here (Chapter 2), European companies lost out, not because their films were inferior, nor because of the advent of World War One as such, but because at a time when they needed capital and access to international markets in order to sustain the quality of their feature films they found it increasingly difficult to obtain either, let alone both.[17] Further hampered by high rates of tax on admissions and tickets, the result was the collapse of a number of European companies during and after the war, and the increasing dominance not just of US companies producing US feature films, but of US companies with a production base in Southern California. Though not all these companies prospered in the long term, those that did benefited not only from clement weather, anti-union policies and a diverse range of nearby locations, but also from their proximity to one another, as Bakker points out later on in his book. Actors, directors, technicians and other creative employees generally lived close by and could easily be hired by or loaned out to other companies, thus reducing 'downtime'. This in turn meant that employees like these could be paid more, and high rates of pay meant that talented personnel could be attracted both from other fields of entertainment and from competitor industries in Europe and elsewhere abroad.[18] In addition, ready access to a range of 'inputs' of this kind meant 'fast and low-cost try-outs' in arriving at the optimal combination of creative personnel on any one production and the capacity to make changes and adjustments with relative ease.[19]

As Richard Koszarski points out in 'Making Movies' (Chapter 3), 'Southern California was clearly recognized as the major American production center by 1915', though, as he also points out, 'executive operations, newsreel production, and even much of the animation industry remained' in New York, along with studio facilities owned or opened by Goldwyn, Metro, Cosmopolitan, Vitagraph and Famous Players-Lasky.[20] He goes on to discuss the costs, the systems of management and other aspects of film production in the late 1910s and early to mid-1920s. Taking issue with the emphasis on the 'central producer system' that marks Janet Staiger's account of this period in *The Classical Hollywood Cinema*, and arguing that the stress in industry discourse on 'the tight, pyramidal control exercised by the top executives' functioned largely as a means of reassuring investors and theatre owners, Koszarski notes that the 'most memorable work of many of the key filmmakers, Griffith, Weber, von Stroheim, Cruze, Lubitsch, DeMille, Neilan, and Ingram in particular, was created via a simpler director-unit system, where projects were developed from script level through editing by individual creative directors and their personal staffs'. Either way it is clear that production management systems varied. While central producer systems were adopted by smaller studios like Cosmopolitan under William Randolph Hearst as well as by MGM under Louis B. Mayer and Irving Thalberg, Paramount tended to use a mix of smaller director-unit and producer-unit systems.[21] These and other systems are further discussed in later sections of this Reader. In meantime, Koszarski himself goes on to discuss set design and art directors, cinematography and cinematographers, and editing and acting, noting in conclusion that by 'the late silent period, exhibitors could choose alternate endings for a number of major films' and cautioning care 'when asserting that any version of a silent film is definitive'.[22]

Set design, cinematography, editing and acting are also discussed in Kristin Thompson's article, 'The Limits of Experimentation in Hollywood' (Chapter 4). Noting that innovation, stylisation and differentiation could be prized as means of improving the standards of Hollywood's films, Thompson discusses the influence of modernist design on the costumes and sets in one-reel or two-reel films such as *The Yellow Girl* (1915) and *Madame Cubist* (1916) as well as in feature-length films such *The Blue Bird* and *Prunella* (both 1918). She also draws attention to the experiments in lighting and cinematography pursued in *A Girl's Folly* (1916) and *The Last of the Mohicans* (1920), and the experiments in narrative construction and intertitle-less narration pursued in *Intolerance* and in *The Old Swimmin' Hole* (1921), respectively. The influence of German Expressionist set design and the European avant-garde on *Beggar on Horseback* (1925) and *The Last Warning* (1929) and the influence of literary naturalism on films such as *Greed* (1925) and *The Crowd* (1928) are also discussed, as is the continuing influence of modernist set and costume design on *The Affairs of Anatole* (1921), *The Thief of Bagdad* (1924) and *Our Dancing Daughters* (1928).[23] Most of these films were produced by major companies. But the most 'extreme cases' were *Camille* (1921) and *Salome* (1923), both of which starred Alla Nazimova, both of which were designed by Natacha Rambova and both of which were produced by Nazimova Productions.

As Karen Ward Mahar points out in *Women Filmmakers in Early Hollywood*, Nazimova was one of a number of women who produced or directed films in late 1910s and the early 1920s. However, as the extract reprinted in this Reader makes clear (Chapter 5), women's roles in the US film industry were severely curtailed in the mid- to late 1920s, a period marked by expansion, corporatisation, professionalisation and increasing investment by Wall Street. As Mahar points in the introduction to her book: 'Women filmmakers (broadly defined to include a range of production activities) were present in most facets of the American film industry by 1909. At the height of their activity, in the period between 1918 and 1922, women directed forty-four feature-length films, headed more than twenty production companies, wrote hundreds of produced screenplays, became the first agents and held positions as editors and heads of scenario and publicity departments'.[24] However, this boom was sustained as much by small-scale investors 'willing to "take a flyer in the films"', as it was by Wall Street.[25] Overinvestment led to overproduction, recession and the demise of a number of independent production companies. In an industry already marked by the values of the business world, it also led to an even more conservative approach to management as a means of sustaining investment from more established financial sources.[26] Thus, as Mahar explains, in the process of becoming what Tino Balio has called 'a modern business enterprise',[27] the major film companies increasingly adopted the masculine ethos that marked the worlds of big business and finance. Women's roles were not only curtailed but increasingly sex-typed. Female costume designers, editors and screenwriters were relatively common. But the only senior female executive was Mary Pickford at United Artists, and only "one female director, Dorothy Arzner, sustained a successful career in mainstream Hollywood during the so-called golden age of the 1930s and 1940s'.[28]

The collateral for Wall Street investment lay as much in the theatres and theatre chains owned or accessed by the major companies as it did in their programmes of films. Company ownership of groups of theatres and showcase cinemas in New York and other major cities dates back to the 1910s, a period that witnessed the advent of larger nickelodeons, combination vaudeville houses and picture palace theatres. Indeed, 'most of the noted major studios ... began as local chain exhibition operations during the early cinema era'.[29] In 1917, in order to counter the growing power of Paramount, a group of exhibitors set up the First National Exhibitors Circuit and began subcontracting films from the likes of Charles Chaplin, Mary Pickford, D.W. Griffith and Louis B. Mayer. By 1919, 'First National had successfully grown to include distribution and production ... Adolph Zukor made the obvious countermove in 1919: from production and distribution he would move into exhibition'.[30] The ensuing 'battle for the theatres' and its place in the development of vertical integration is detailed at length in Benjamin Hampton's book *History of the American Film Industry*.[31] While the size, scale and lavishness of picture palace theatres tended to increase,[32] the number of theatres acquired by the major companies tended to fluctuate. However, in addition to owning showcase venues in New York and other major cities, First National owned theatres in New Jersey and Pennsylvania, Paramount in Canada, New England and the South, Fox on the West Coast, and Loew's (MGM's parent company) in New York State. (At this point Warner Bros. owned no theatres, nor did other minor companies such as the Producers Distributing Corporation or the Film Booking Office.) Only in the major metropolitan centres, which housed the bulk of the population and earned the bulk of domestic box-office revenue, did these companies compete directly with one another.[33] Elsewhere they usually exhibited each other's films in each other's theatres before making them available, often in groups or 'blocks', to independent theatres and theatre chains in urban neighbourhoods, small towns and rural locations. Along with 'blind selling' (the distribution of films sight unseen), block booking was contentious. It first became a focal point of investigation by the Federal Trade Commission in 1921.[34] However, as Lea Jacobs and Andrea Comiskey point out in 'Hollywood's Conception of Its Audience in the 1920s', the distribution system as a whole was marked by differences, by variations in practice that often corresponded to variations in the nature and perceived appeal of the films themselves.

Using newspapers, magazines and industry journals as sources, Jacobs and Comiskey note that press reviews often offered 'a highly differentiated conception of the audience: as urban or rural, as belonging to the "classes" or the "masses," as male or female'. Looking in detail at a range of productions and genres (from costly roadshown specials to low-budget films made for neighbourhood theatres, from films with rural settings to sophisticated comedies made for metropolitan audiences), they go on to 'correlate trade reviewers' judgements about how films were likely to play across different markets, and about the preferences of theater audiences, with distribution patterns in a small number of Northeastern and mid-Western cities and towns'. In this way they challenge the idea that viewers of films had become a mass audience, 'duped into accepting an ersatz luxury as the real thing, and unwillingly participating in a "pseudo-democratic" effacement of class differences and allegiances'. In doing so, they also challenge the idea that vertical integration and the systems of management that now characterised the principal US film companies resulted solely in the mass production and distribution of more or less interchangeable films for more or less interchangeable tastes in more or less interchangeable venues.

As is now well documented, there were limits to Hollywood's diversity, both on screen and off. Sexual minorities and a number of ethnic minorities, among them Native Americans, Asian and African Americans, and Hispanic and Italian Americans, were largely marginalised. So too were those with avant-garde tastes, revolutionary political views or an interest in explicit sexual behaviour. Some set up small-scale production and distribution companies; others founded, toured or hired exhibition venues to cater for them. Along with the representation and employment of minorities in Hollywood, and along with the prevalence of first or second generation Jewish immigrants among the management of Hollywood's companies, the history of these cinemas has been detailed in a number of important articles and books, most of which are listed in the Bibliography.

Similar points to those made by Jacobs and Comiskey might be made about the distribution, exhibition and consumption of Hollywood films abroad. As Mike Walsh has pointed out, 'American producers attempting to move into Europe had four options available to them': 'using foreign agents or establishing their own distribution subsidiaries, and two intermediate positions, involving contracting with franchise agents or joint distribution with other American companies'.[35] By the end of World War One, US companies had gained a substantial foothold in the world market, partly by centring the export trade in New York, partly by 'dealing directly with more markets, opening more subsidiary offices outside Europe and thereby establishing a control which other producing countries would find difficult to erode', partly by enlisting the help of the US Committee on Public Information, which 'worked closely with the American commercial film industry, both domestically and abroad', and partly by factoring overseas earnings into the budgets for their films, thereby increasing their appeal abroad as well as at home.[36]

Immediately after the war, the strength of the dollar acted as a barrier to US exports. However, while the German film industry profited from the weakness of its currency, rampant inflation and the introduction of a quota on imports, countries such as Britain, France and Italy went into decline, their position exacerbated by the introduction of entertainment taxes and exhibition quotas which antagonised those who owned and operated cinemas and which, in the case of France, led to a downturn in cinema-going and a resurgence in live entertainment.[37] In response to these developments, a number of European producers began to consider the creation of a Continent market for Continental productions, a 'Film Europe' to counter the dominance of 'Film America'. Between 1926 and 1929, despite the activities of the MPPDA and the aid of the US State Department, the share of European films in European markets began to stabilise then increase, partly because, as Andrew Higson and Richard Maltby have pointed out, 'the major American companies did not necessarily act as a cartel. In each market, they were in direct competition with each other as well as with domestic producers'; partly because although 'the MPPDA was the industry's principal negotiating agent with foreign governments, it had no power to coerce its members, and could not always persuade them that their hegemony was, on occasion, best maintained through concession'.[38] In the meantime, although the principal US film companies established themselves in Canada, Australia and New Zealand, South America, India and parts of Africa,[39] their hegemony outside Europe was by no means absolute either: 'The USSR had, by dint of government regulation and support, regained its market to a large extent by this point ... Japan, depending upon several strong, vertically integrated companies, was unique in keeping a large portion of its domestic market for native productions without expensive regulation. A few small markets, like Persia, depended on films from the nearest, cheapest sources and American firms were largely content to ignore them'.[40] In an increasingly volatile era, various forms of nationalism threatened Hollywood's hegemony and the potential success of Film Europe alike. However, the most immediate threat to them both was the advent of sound.

Notes

1 Sheldon Hall and Steve Neale, *Epics, Spectacles and Blockbusters: A Hollywood History* (Detroit: Wayne State University Press, 2010), 11–12; Charles Musser, *The Emergence of*

Cinema: The American Screen to 1907 (New York, Scribners, 1990), 82–85, 94–99, 194–221; Roberta E. Pearson, 'Biblical Films', in Richard Abel (ed.), *Encyclopedia of Early Cinema* (London: Routledge, 2005), 68–71; Dan Streible, *Fight Pictures: A History of Boxing and Early Cinema* (Washington, DC: Smithsonian Institution Press, 2002), and 'Boxing Films', in Abel (ed.), *Encyclopedia of Early Cinema*, 80–81.

2 Michael Anderson, 'The Motion Picture Patents Company: A Revaluation', in Tino Balio (ed.), *The American Film Industry* (Madison: University of Wisconsin Press, 1985 edition), 133–52; Eileen Bowser, *The Transformation of Cinema, 1907–1915* (New York: Scribners, 1990), 21–36, 73–85.

3 Richard Abel, 'Nickelodeons', in Abel (ed.), *Encyclopedia of Early Cinema*, 478–79; Bowser, *The Transformation of Cinema*, 1–20; Douglas Gomery, *Shared Pleasures: A History of Movie Presentation in the United States* (London: BFI, 1992), 18–33.

4 Ben Brewster, 'Multiple-Reel/Feature Films: USA', in Abel (ed.), *Encyclopedia of Early Cinema*, 456.

5 Brewster, 'Multiple-Reel/Feature Films: USA', 456.

6 Gomery, *Shared Pleasures*, 12–13; Edward Lowry, 'Edwin J. Hadley: Traveling Film Exhibitor', in John L. Fell (ed.), *Film Before Griffith* (Berkeley: University of California Press, 1983), 131–43; Charles Musser, 'Itinerant Exhibitors', in Abel (ed.), *Encyclopedia of Early Cinema*, 340–42; Charles Musser, with Carol Nelson, *High-Class Moving Pictures: Lyman H. Howe and the Forgotten Era of Traveling Exhibition, 1880–1920* (Princeton, NJ: Princeton University Press, 1991); Calvyn Pryluck, 'The Itinerant Movie Show and the Development of the Film Industry', *Journal of the University Film and Video Association*, vol. 25 no. 4 (1983), 11–22; Mark E. Swartz, 'Motion Pictures on the Move', *Journal of American Culture*, vol. 9 no. 3 (1986), 1–7; Barbara Stones, *America Goes to the Movies: 100 Years of Motion Picture Exhibition* (North Hollywood: National Association of Theatre Owners, 1993), 11–13; Gregory A. Waller, 'Robert Southard and the History of Traveling Film Exhibition', *Film Quarterly*, vol. 57 no. 2 (2003/4), 2–14.

7 Bowser, *The Transformation of Cinema*, 192–93, 210–12; Hall and Neale, *Epics, Spectacles and Blockbusters*, 28–45.

8 Richard Abel, *Americanizing the Movies and 'Movie Mad' Audiences, 1910–1914* (Berkeley: University of California Press, 2006), 13–42, and 'The "Backbone" of the Business: Scanning Signs of US Film Distribution in the Newspapers, 1911–14', in Frank Kessler and Nanna Verhoeff (eds), *Networks or Entertainment: Early Film Distribution, 1895–1915* (Eastleigh: John Libbey Press, 2007), 85–93; Bowser, *The Transformation of Cinema*, 192–93; Hall and Neale, *Epics, Spectacles and Blockbusters*, 21–40. It should be noted here that dates given for all foreign films in this Reader are US release dates unless otherwise specified.

9 Brewster, 'Multiple-Reel/Feature Films: USA', 457.

10 Michael Quinn, 'USA: Distribution', in Abel (ed.), *Encyclopedia of Early Cinema*, 661.

11 Quinn, 'USA: Distribution', 661.

12 Quinn, 'USA: Distribution', 661. See also Douglas Gomery, *The Hollywood Studio System: A History* (London: BFI, 2005), 17–26. It should be noted that Paramount's initial attempt to introduce percentage rental fees was short lived. Aside from some of the films distributed by United Artists, and aside from roadshows, flat fees were the norm prior to the 1930s. For more on Paramount's policies and practices in the 1910s and the transition to feature-length films, see Michael Quinn, 'Paramount and Early Feature Distribution: 1914–21', *Film History*, vol. 11 no. 1 (1999), 98–113, and 'Distribution, the Transient Audience, and the Transition to the Feature Film', *Cinema Journal*, vol. 40 no. 2 (2001), 35–56.

13 Janet Staiger, 'Combination and Litigation: Structures of U.S. Film Distribution, 1896–1917', *Cinema Journal*, vol. 23 no. 2 (1984), 41–72.

14 Gerben Bakker, *Entertainment Industrialised: The Emergence of the International Film Industry, 1890–1940* (Cambridge: Cambridge University Press, 2008), 2.

15 Bakker, *Entertainment Industrialised*, 6.

16 Richard Abel, *The Red Rooster Scare: Making Cinema American, 1900–1910* (Berkeley: University of California Press, 1999), and *Americanizing the Movies and 'Movie-Mad' Audiences, 1910–1914*; Kristin Thompson, *Exporting Entertainment: America in the World Film Market, 1907–1934* (London: BFI, 1985), 1–49.

17 It should be noted that the point at which US films became dominant in Europe is subject to dispute. In *Film Style and Technology: History and Analysis* (London: Starword, 1992 edition), 114, Barry Salt argues that 'the American industry was moving into a commanding position even before the war started' and that 'when one compares a sample of American and French films made in 1913, one can see that what the European cinema-goers were already voting for in the fairly free competition for their money was: more shots per reel, more shots in each scene, more close shots; and more naturalistic acting'. However, this is to downplay the role and importance of feature-length films, the timing of their advent in the US and in Europe, and the point at which US feature-length films began to be exported in significant numbers. According to Thompson, *Exporting Entertainment*, 71: 'The big increases in American exports' occurred in 1915 and early 1916 and 'resulted primarily from the decline of the industries of other nations ... Had the war ended in mid-1916, the American film would have been in a much stronger position than before the war – yet it would not have been guaranteed any long-term hold on world markets'. Thompson does not here distinguish between feature-length and shorter films. Indeed, aside from large-scale international roadshows such as *The Birth of a Nation* and *Intolerance*, the history of the distribution of feature-length US films in Europe and in other foreign markets in the early to mid-1910s has yet to be written.

18 In addition to stars such as Emile Jannings, Greta Garbo and Pola Negri, Hollywood recruited directors such as Michael Curtiz, Paul Fejos, Paul Leni, Ernst Lubitsch, F.W. Murnau, Victor Sjostrom and Mauritz Stiller and producer Eric Pommer from Europe during the course of the 1920s, thus expanding its stylistic repertoire, the range of its films, and the national and international scope of its films' appeal. As well as Kristin Thompson's article, 'The Limits of Experimentation in Hollywood' (Chapter 4), see Ursula Hardy, *From Caligari to California: Eric Pommer's Life in the International Film Wars* (Providence, RI: Berghahn Books, 1996), 94–114; Graham Petrie, *Hollywood Destinies: European Directors in America, 1922–1931* (London: Routledge and Kegan Paul, 1985); Kristin Thompson, *Herr Lubitsch Goes to Hollywood: German and American Film after World War I* (Amsterdam: Amsterdam University Press, 2005); Kristin Thompson and David Bordwell, *Film History: An Introduction* (New York: McGraw-Hill, 2003 edition), 158–62.

19 Bakker, *Entertainment Industrialised*, 258–59.

20 For more on the importance of New York in the studio era, see Diana Altman, *Hollywood East: Louis B. Mayer and the Origins of the Studio System* (New York: Birch Lane Press, 1992).

21 For details on Cosmopolitan, see Louis Pizzitola, *Hearst Over Hollywood: Power, Passion and Propaganda in the Movies* (New York: Columbia University Press, 2002), 195–229. For details on MGM, see Thomas Schatz, *The Genius of the System: Hollywood Filmmaking in the Studio Era* (New York: Pantheon Books, 1988), 29–46, and Mark A. Viera, *Irving Thalberg: Boy Wonder to Producer Prince* (Berkeley: University of California Press, 2010), 35–88. Production supervision at Paramount in the 1920s is discussed by Matthew Bernstein in 'Hollywood's Semi-Independent Production', *Cinema Journal*, vol. 32 no. 3 (Spring 1993), 43–44. According to Jackson Schmidt, 'On the Road to MGM: A History of Metro Pictures Corporation, 1915–20', *The Velvet Light Trap*, no. 19 (1982), 48–49, Metro established a 'double director system' in 1917: 'Each production unit was assigned two directors, one of whom would take charge of a unit for a month and go shoot a film. At the end of 30 days the first director would relinquish that unit to the other director and then head to the editing room to assemble the footage just shot and prepare the script for the following month's shooting. In this manner, the two directors rotated and, most interestingly, maintained full control over their films from beginning to end'. How long this system lasted is unclear.

22 For more on set design in the silent era, see Beverly Heisner, *Hollywood Art: Art Directors in the Days of the Great Studios* (London: St James Press, 1990), 7–23; Salt, *Film Style and Technology*, 133. For more on cinematography, see Patrick Keating, *Hollywood Lighting from the Silent Era to Film Noir* (New York: Columbia University Press, 2010), 15–104; Salt, *Film Style and Technology*, 115–32, 148–63, 169–70. For more on editing, see Bordwell, Staiger and Thompson, *The Classical Hollywood Cinema*, 194–240; Salt, *Film Style and Technology*, 136–47, 170–78. Bordwell, Staiger and Thompson stress the importance of continuity and classicism, suggesting that their fundamental features were more or less in

place by 1917. However, like Salt (and Koszarski), they also suggest that these features were more prevalent and more securely codified by the mid-1920s. Their initial stress on 1917 may be due to a wish to synchronise industrial developments such as the advent of studio-based systems of supervision and the routine production of feature-length films with stylistic ones.

23 For more on Hollywood naturalism in the 1920s, see Lea Jacobs, *The Decline of Sentiment: American Film in the 1920s* (Berkeley: University of California Press, 2008), 25–78.

24 Karen Ward Mahar, *Women Filmmakers in Early Hollywood* (Baltimore: Johns Hopkins University Press, 2006), 2.

25 Benjamin Hampton, *History of the American Film Industry* (New York: Dover Publication, 1970, originally published in 1931), 212.

26 One mark of this was the establishment of a course on film in the Graduate School of Business Administration at Harvard University. Joseph P. Kennedy, president and chairman of the Film Booking Office and a Harvard alumnus, invited a number of Hollywood luminaries to contribute to the course and some their contributions were published in Joseph P. Kennedy (ed.), *The Story of the Film* (Chicago: A.W. Shaw, 1927). As Peter Decherney points out in *Hollywood and the Culture Elite: How the Movies Became American* (New York: Columbia University Press, 2005), 66: 'The Harvard Business School's courses were one indication that the film industry was evolving from an entrepreneurial field to a mature business. In Kennedy's opening remarks, for example, he insisted that the development of "depreciation tables" and "estimates of residual values" were the key to transforming movies into stable investments'.

27 Tino Balio, *Grand Design: Hollywood as a Modern Business Enterprise, 1930–1939* (New York: Scribners, 1993).

28 Mahar, *Women Filmmakers in Early Hollywood*, 2. For more on women filmmakers in the 1910s and the 1920s, see Kaye Armatage, 'Sex and Snow: Landscape and Identity in the God's Country Films of Nell Shipman', in Gregg Bachman and Thomas J. Slater (eds), *American Silent Film: Discovering Marginalized Voices* (Carbondale: Southern Illinois University Press, 2002), 125–47; Jennifer Bean and Diane Negra (eds), *A Feminist Reader in Early Cinema* (Durham, NC: Duke University Press, 2002); Cari Beauchamp, *Without Lying Down: Francis Marion and the Powerful Women of Early Hollywood* (New York: Scribners, 1997); Mark Garrett Cooper, *Universal Women: Filmmaking and Institutional Change in Early Hollywood* (Urbana: University of Illinois Press, 2010); Alison McMahon, *Alice Guy Blaché: Lost Visionary of the Cinema* (New York: Continuum, 2002); Giuliana Muscio, 'Clara, Ouida, Beulah, et al.: Women Screenwriters in American Silent Cinema', and Shelley Stamp, 'Lois Weber, Star Maker', in Vicki Callahan (ed.), *Reclaiming the Archive: Feminism and Film History* (Detroit: Wayne State University Press, 2010), 289–308, and 131–53, respectively; Thomas J. Slater, 'June Mathis: The Woman Who Spoke through Silents', in Bachman and Slater (eds), *Discovering Marginalized Voices*, 201–16. For more on Dorothy Arzner, see Judith Mayne, *Directed by Dorothy Arzner* (Bloomington: Indiana University Press, 1994). For more on Mary Pickford, see Tino Balio, *United Artists: The Company Built by the Stars* (Madison: University of Wisconsin Press, 1976), and Eileen Whitfield, *Pickford: The Woman Who Made Hollywood* (Toronto: Macfarlane Walter & Ross, 1997).

29 Douglas Gomery, 'Cinema Circuits or Chains', in Abel (ed.), *Encyclopedia of Early Cinema*, 121.

30 Richard Koszarski, *An Evening's Entertainment: The Age of the Silent Feature Picture, 1915–1928* (New York: Scribners, 1994), 74. For more on First National and its rivalry with Paramount, see Hampton, *History of the American Film Industry*, 170–96.

31 Hampton, *History of the American Film Industry*, 252–80.

32 Ben M. Hall, *The Best Remaining Seats: The Golden Age of the Movie Palace* (New York: Da Capo, 1988, originally published in 1975); Charlotte Kopac Herzog, *Motion Picture Theater and Film Exhibition, 1896–1932* (PhD dissertation: Northwestern University, 1980), 109–30; Stones, *America Goes to the Movies*, 35–61.

33 Tino Balio, *United Artists*, 63–64. For an account of national theatre chains in the late 1910s and the 1920s, see Gomery, *Shared Pleasures*, 34–56.

34 Michael Conant, *Antitrust in the Motion Picture Industry* (New York: Arno Press, 1978), 27–28; Howard T. Lewis, *The Motion Picture Industry* (New York: D. Van Nostrand, 1933), 142–80.

35 Mike Walsh, 'Options for American Foreign Distribution: United Artists in Europe, 1919–30', in Andrew Higson and Richard Maltby (eds), *'Film Europe' and 'Film America': Cinema, Commerce and Cultural Exchange, 1920–1939* (Exeter: University of Exeter Press, 1999), 133.

36 Thompson, *Exporting Entertainment*, 71, 93 and 103.

37 For more on Germany and Hollywood, see Thomas J. Saunders, *Hollywood in Berlin: American Cinema and Weimar Germany* (Berkeley: University of California Press, 1994).

38 Andrew Higson and Richard Maltby, '"Film Europe" and "Film America": An Introduction', in Higson and Maltby (eds), *'Film Europe' and 'Film America'*, 7.

39 Ian Jarvie, *Hollywood's Overseas Campaign: The North American Movie Trade, 1920–1950* (Cambridge: Cambridge University Press, 1992), 26–76; Thompson, *Exporting Entertainment*, 137–40, 144–47.

40 Thompson, *Exporting Entertainment*, 147.

Chapter 2

Gerben Bakker

THE QUALITY RACE: FEATURE FILMS AND MARKET DOMINANCE IN THE US AND EUROPE IN THE 1910s

IN THE EARLY FILM INDUSTRY, FILM technology automated and standardised live entertainment and made it tradeable, thus leading to a process of market integration, which was reaching its heights during the 1910s, when features made cinema an ever better substitute.[1] Many of the lower-value-added local entertainments were now replaced by entertainment produced in one location in the nation and distributed nationwide. Second, a basic network of fixed cinemas, an essential distribution delivery system, was more or less complete by the early 1910s, guaranteeing a large potential market. Third, consumer demand for entertainment increased substantially between the 1900s and the 1920s. Finally, before 1913, the MPPC [Motion Picture Patents Company] companies had formed a cartel and effectively limited each others' expenditure on film production, leading to an artificial constraint on the R&D/sales ratio. When the federal government started to prosecute the MPPC in 1912, its power quickly diminished, and many rival companies were formed. Firms do R&D to improve their products' quality and steal market share away from competitors. Under the cartel, this incentive to do R&D was absent, and as a consequence, when the cartel collapsed a further reason for escalating R&D emerged.

Only a handful of entrepreneurs massively escalated their production expenditure. From the late 1900s, several firms experimented with different types of film and discovered that some consistently yielded more revenue than others. In the mid-1900s, Adolph Zukor (later Paramount's president), who owned fourteen Nickelodeons in large cities, had problems obtaining films he wanted, and made every effort to search for 'better' pictures. When he exhibited the three-reel, hand-coloured *Passion Play* of Pathé (most films were under half a reel at the time), it was an enormous success, bringing in customers for months. Zukor wanted to move further into these bigger pictures.

> We stayed on with that picture for months and did a land office business . . . Then it occurred to me that if we could take a novel or a play and put it on the screen, the people would be interested . . . I did approach all the producers then in the business and tried to sell the idea of making big pictures. . . . They were so busy turning out [one and two reel pictures] that they would not undertake anything else. In fact, they did not believe that people would sit through pictures that ran three, four, five reels.[2]

In 1909, Zukor sold all his theatres to Loew's (later part of MGM), and studied the motion picture industry for three years, travelling, watching many different types of films, and especially adopting the habit of sitting in the first row, not watching the screen but the faces of the audience as they were watching. 'In 1911', Zukor recalled, 'I made up my mind definitively to take big plays and celebrities of the stage and put them on the screen'. In November of that year he advanced $35,000 for a film starring the French actress Sarah Bernhardt. Competitors were surprised Zukor was willing to pay such a large sum. In March 1912, Zukor released the picture through the chain of vaudeville theatres Klaw and Erlanger, because the MPPC did not allow it a licence. However, after its initial success, the MPPC gave it a licence and the film grossed about $60,000 – not enough to recoup costs but leading Zukor to remark that his first experiment was not too costly. That is exactly how he saw it: as an experiment to test the market. 'We did gain the knowledge that made us absolutely certain that pictures of the right type had a great future.'[3]

In the early 1910s, other entrepreneurs noted the success of longer Italian films, starting with *The Fall of Troy* and *Dante's Inferno.*[4] These films contained expensive historical sets and mass scenes, and the films lasted three to four times as long as the standard film, about forty-five to sixty minutes.[5] Production costs were high but ticket prices were also higher, and the films became widely popular.[6] In showing these first films, often 'road shows' were used; travelling companies of projectionists, publicity personnel and administrative staff would rent theatres or equivalent buildings in cities and show the film until revenues fell, and then moved on to the next city. In New York and Boston *Dante's Inferno* played for two weeks, while the average American MPPC-film lasted only two days. Moreover, it played in rented 1,000-seat theatres, at a price of $1, while American films normally played in 200-seat Nickelodeons at 5–10 cents. The difference in potential revenue per show was an order of magnitude: $1,000 versus between $10 and $20.[7] This discovery that higher costs, length and ticket prices could disproportionately improve profitability showed some smart American producers the way to the feature film.

By the end of 1913, many cinemas showed short films for six days and a feature film for one night, often a Sunday night. Features' disproportionate popularity was clear from the rental prices: a six days' supply of shorts programmes through one of the three main shorts distributors (General Film Company, Mutual and Universal, at that time) cost $45, while top-rated features rented for $50 for a single night. By the summer of 1914, the smaller cinemas were still showing eight to nine reels of short films for 5 cents admission, while the newer and larger houses offered features for 10–20 cents.[8]

The strong effect of rising sunk costs on consumers' willingness-to-pay was not only visible when comparing shorts with features, but also when comparing low-quality with high-quality features. In 1915, Vitagraph-Lubin-Selig-Essanay calculated that a cinema showing average features changing daily had daily sales of $300 and a 42 per cent margin ($125). Film rental was 8.3 per cent of sales, advertising 16.7 per cent. If the cinema shifted to a weekly, highly publicised quality film, and doubled expenditure on both film rental and advertising, daily sales would grow to $550, the margin to 54 per cent ($300), while film rental and advertising would be 9.1 and 18.2 per cent of sales.[9]

Industry veteran Benjamin B. Hampton recalled how in the mid-1910s many cinema-owners discovered the profitability of features.

> Owners of theatres, who had been cautiously advancing their admittance rates, learned that their patrons would pay twenty-five cents, or in a few cities as high as thirty-five cents, to see the best pictures; and at these prices the profits were larger than ever before, even though the exhibitor had to pay somewhat higher rentals.[10]

In Europe, the feature film appears to have had a similar revenue generating capacity. Particularly good British evidence, from a somewhat later date than the American data above, actually shows the escalation phase at work at the micro-level and confirms the superior profitability of the feature film: its revenue per metre was nearly three times that of other films. In 1918–19, Pathé Exchange Ltd., a British film distributor, reported revenues of feature films that were 32 per cent higher than revenues of all film formats taken together, when measured in revenue per metre of film.[11] Since feature films were at least several times longer than other formats, in absolute terms the difference must have been even higher. In the next season, 1919–20, the gap in revenue between feature films and other films had more than doubled: feature films now yielded 67 per cent more revenue per metre than all films taken together.[12] If for the latter season, revenues of feature films are not compared to revenues for all films, but to revenues for all *other* films, revenue per metre for feature films was 2.9 times the revenue for all other films.[13]

An advantage of feature films over a popular format such as the newsreel was that revenue did not decline as much each day after release. The rental price of newsreels, for example, halved three days after release, and nine days after release was only a quarter of the initial price, while the number of copies rented had halved on the sixth day after release.[14] These figures lend support to the notion that feature films were disproportionately and increasingly profitable, and that therefore an escalation of outlays on sunk costs on feature film production could be a profitable strategy. It also shows that entrepreneurs certainly found out the profitability of features, and that those in a position most able to do so were the ones active in distribution or exhibition.

Changes in distribution practices formed a vital element in the strategies of the companies that started the increase in sunk costs. The film industry was an industry with large fixed costs, and the marginal revenue brought in by the marginal cinema-goers equalled marginal profits to the cinema owner. Initially, when films were sold, and later rented for a flat fee, the producer saw little of these marginal revenues, but during the 1910s, the changes in distribution practices translated ever more of the marginal distributor and cinema revenues into profits for producers, first by percentage-based producer-distribution contracts, later by similar distributor-cinema contracts.[15] The result was that a producer would actually get part of the additional cinema and distributor revenues that an increase in perceived quality of a film generated. Without these changes in distribution practices, an escalation strategy could hardly have been profitable, as it would be the producer who incurred the costs but the cinema-owners who got the additional marginal revenue, equalling profit. The change also increased distributors' incentive to increase advertising outlays, until the last dollar spent equalled the distributor's marginal profits. So during the escalation phase there was a double effect on producers' revenues: marginal cinema revenues increased sharply because of price increases and capacity utilisation increases, while producers also received more of these marginal cinema revenues.[16]

Hampton underlines the way in which cinemas profited from the unskimmed marginal revenues before the change in distribution practices:

> The exhibitors were buying service at prices that made intelligent theater operation extremely profitable; net return of twenty-five to fifty percent per annum on the capital invested in movie houses was assumed to be the prevailing rate, but some theaters ran this up to a hundred percent.[17]

Firms' strategies

The previous section showed how entrepreneurs increasingly became aware of the disproportionate effect an increase in production outlays could have on revenue. This section is more preoccupied with how they acted on the information that came out of this discovery process, to what extent they followed deliberate strategies to increase their sunk costs, how they did it, and how they got hold of the vast amounts of capital needed to embark on such an escalation strategy. The section will also examine whether any differences existed between winners and losers.

Most 'escalators' began as investors in real estate and pioneered the Nickelodeons, which they discovered could maximise the return on property.[18] This is in contrast to MPPC-members, who started as equipment manufacturers and by necessity branched out into film production. These first movers were not technology-driven, nor creatively driven, but financially/real-estate-driven, and this was possibly important in their role as escalators. They knew how important it was for films to increase the return on city-centre real estate, and had experience in obtaining capital. In the mid-1910s there even was a small investment boom in film stocks. Nearly every company with the word 'motion picture' in its name was able to launch an initial public offering.

Adolph Zukor was one of the first and most successful entrepreneurs who escalated sunk costs. From 1912, his Famous Players Film Company was producing feature films, and in 1914 he was the driving force in the merger of five regional distribution organisations into one national one, called Paramount, which negotiated a twenty-five-year supply contract with Famous Players, the Jesse L. Lasky Company, and later Bosworth Inc. and Pallas. This Paramount group became the US market leader throughout the rest of the 1910s. For a full year, in 1914–15, it was the only company that could provide cinemas with a full year's supply of features (104). During 1916, all these companies merged in several stages into the Famous Players-Lasky Corporation (hereafter called Paramount).[19]

Besides increasing outlays on film production, Zukor's strategy consisted of changing distribution practices to get more of the marginal cinema revenues. The contracts between Paramount and its producing companies gave the producers 65 per cent of gross rentals instead of the then customary flat fee, thus ensuring they obtained marginal distributors' revenue. Gradually, Paramount would introduce the same type of contracts with the larger cinemas in the key cities it supplied.[20] Paramount also started national advertising campaigns for its stars and its features, which now showed disproportionate returns, because Paramount was a national organisation and because it received more of the marginal cinema revenues brought in by the increased advertising. Paramount's gross revenues increased from $10.3 million in 1917 to $17.3 million in 1918, to $12.0 million for the first six months of 1919.[21]

Paramount obtained capital through both debt and equity. Wall Street bank Kuhn, Loeb & Co. arranged a $10 million preferred stock issue for Paramount in 1919, to finance its expansion into city-centre cinemas.[22] According to Hampton, Zukor set the example which was followed by scores of competitors. 'Practically all principal manufacturers concluded to adopt Zukor's compromise – pay large prices, if need be, to directors, novelists, dramatists and continuity writers, hoping thereby to find something novel and startling to attract the crowds to their own photoplays.'[23]

Several other entrepreneurs increased production outlays at about the same time, most notably William Fox and Carl Laemmle. Fox started in film as a Nickelodeon-owner, and by the late 1900s ran a film distribution company in New York, one of the few that successfully defied the General Film Company [GFC], and whose lawsuits instigated the prosecution of the MPPC and GFC for violation of the Sherman Act. In 1914 and 1915, Fox embarked on a massive film production programme, increasing outlays from $92,000 in 1914 to $4 million in 1917. Fox was financially backed by a group of New York investors, led by William F. Dryden, president of the Prudential Life Insurance Company, which itself also invested in Fox Film Corporation.[24] Just like Paramount, Fox set up a nationwide distribution organisation synchronously with increasing production costs, enabling him to advertise nationwide, and to receive more of the marginal distributors' revenues. Fox set up a British distribution subsidiary in 1916, followed by many more foreign subsidiaries during the 1910s, just as Paramount did.

Carl Laemmle started in film in 1905 when he bought and set up a string of store-front theatres. Some years later he founded a distribution company. When in the late 1900s the MPPC/GFC tried to monopolise film production and distribution, Laemmle set up the Independent Motion Picture Company (IMP), a major competitor of the trust. In 1912 with several other entrepreneurs he formed the Universal Film Company, a nationwide distributor, which distributed the output of a string of independent companies. In 1913, Laemmle acquired full control of Universal, and took care that an increasing part of Universal's supply was made by its own production company. In 1915, he set up Universal City Studios, near Culver City, California, where the land alone cost half a million dollars.[25] Universal got its capital mainly through the stock market. In early 1916, it was reported that Universal's common stock had paid an annual dividend of 20 per cent, on average, between its foundation in 1912 and 1916.[26] In the early 1910s, IMP and Universal were leaders in increasing production outlays but, later in the escalation phase, Laemmle became more cautious and did not follow the path of Paramount and Fox. Instead, he focused on supplying secondary cinemas outside the main population centres.

Paramount and Fox would become two of the five major Hollywood studios that dominated international film production and distribution from the 1920s. The other three did not originate in the escalation phase but were based on companies that existed at that time. Metro-Goldwyn-Mayer was based upon Loew's, a theatre circuit stemming from the 1900s and Goldwyn Pictures, a not-so-successful escalator. The Warner brothers, Nickelodeon entrepreneurs from the 1900s, had entered feature film distribution in the 1910s, and bought Greater Vitagraph (the remnant of the MPPC companies) and First National (a producer-distributor set up by cinema chains in the late 1910s) in the mid-1920s, backed by Goldman, Sachs & Co. Finally, Radio-Keith-Orpheum (RKO), financially backed by Merrill Lynch and for some time managed by Jack Lynch, was mainly based on the acquisition of Pathé Exchange in 1921, and of the Keith-Orpheum theatre circuit. During the late 1910s and the 1920s, these emerging Hollywood majors started to buy large cinemas in key cities of the US, thus further increasing their share of marginal cinema revenues.[27]

Where the above entrepreneurs succeeded, many others failed. Several film companies grew rapidly and joined the jump in outlays, but did not prove successful. Triangle Film Corporation was founded in the summer of 1915. Its strategy was to set up a national distribution network (with twenty-two exchanges), to buy prestigious city-centre theatres, and to spend huge sums on feature films made by the star producers D. W. Griffith, Thomas Ince and Mack Sennett.[28] Unlike its competitors, it aimed at films with famous stage stars based on classical plays, paying Sir Herbert Tree, for example, $100,000 for three months of his services, mainly used on Shakespeare's plays. Investors initially had great confidence in the strategy and Triangle stock jumped by 40 per cent almost immediately after its flotation in July 1915, reaching a high of 78 per cent over its $5.00 offer price by October 1915. After that, everything went downhill. The American public did not like stage actors and filmed stage plays,[29] and Triangle was underspending on feature production. It spent on average about $30,000 per picture, while Paramount was spending in the range of $65,000 to $75,000. By October 1916, its shares were trading at only 40 per cent of their issue price, and it was forced to sell its distribution network, which cost $1.5 million annually to operate. It fetched $600,000, with the buyer remaining contractually obliged to distribute Triangle's output. In 1917, the company was reorganised by a new production manager, who cut costs by $2.5 million annually. This was not enough to avert receivership in early 1919.

Another loser was the Balboa Amusement Company of Long Beach, California.[30] Founded in 1913 by local businessmen, the company operated a large studio in which it invested $400,000, thus becoming the largest employer and largest tourist attraction in Long Beach. It contained twenty buildings on eight acres and an eleven-acre outdoor shooting area. Part of the studio was rented to other producers, and part was used by Balboa itself. In the fall of 1914, it made two long-term distribution agreements with Fox Film Corp. and Pathé Exchange, both national distributors. Balboa's strategy was to turn out as large a number of reasonable quality films as possible. Subsequently, Balboa expanded production, turning out 6,000 feet of negative stock and 150,000 feet of positive copies weekly (its printing department alone represented an investment of $50,000). But the massive increase in outlays on its film portfolio overextended the company. In early 1917, Balboa's distributors could not place its films, and with seventeen unreleased negatives in its vaults, it filed for voluntary bankruptcy. Without its own distribution network and without good contracts, Balboa saw little of the marginal distributor and cinema revenues. Moreover, its major focus was to increase its quantity of output rather than its films' perceived quality, which ultimately did not prove the optimal strategy.

The Mutual Film Company, founded as a distribution organisation in spring 1912, also tried to join the race. It released the films of several smaller enterprises, and acquired the Majestic and Reliance companies. It only half-heartedly joined the escalation phase. In 1916, with its stock below the 1912 issue price,[31] it tried to catch up in one jump, offering Charlie Chaplin $670,000 a year to come to Mutual. Chaplin's films were extremely successful and extended the life of Mutual, but after a year he departed, and Mutual finally collapsed in spring 1919. One of its problems was that it had complicated financial relationships with the many different production companies for which it distributed.

World Film Corporation was a company that spent massively on producing and distributing features during 1914–17 and then went spectacularly bankrupt. In the fiscal year ending June 1915, it had made a profit of $329,000, roughly 20 per cent of its outstanding share capital but paid no dividends.[32]

The eight original cartel members of the MPPC all went bankrupt or were marginalised, as their voluntary restraint of R&D expenditures had left them with few capabilities to make and market films. The MPPC companies' success was based on short films and they seemed unable to adapt. In 1914, as the feature film was starting its run to dominance of the film industry, one of its members, William Selig, remarked 'That the single reel photo-drama is the keystone of the motion picture industry becomes more apparent daily'.[33]

The only member that survived and made significant profits until the 1920s was Pathé Frères, which had left the cartel and set up its own national distribution network.[34]

Besides the winners and the losers, there were also some 'wallflowers', companies that did not join the escalation phase. Four of the eight MPPC members formed the Vitagraph-Lubin-Selig-Essanay (VLSE) distribution company, which later merged with the Vitagraph production company, forming Greater Vitagraph, financed by investments from American Tobacco.[35] Although Vitagraph survived and remained profitable, and eventually started to make feature films, it remained a minor player. When Warner Brothers bought it in the mid-1920s, it was solely to acquire its US and foreign distribution networks, through which Warner planned to pipe its sound films. The other MPPC producers disappeared from the market. Edison quit production, Lubin went bankrupt and the other companies also dissolved or went bankrupt, or continued as small, insignificant outfits.

Benjamin Hampton, vice-president of American Tobacco, who negotiated an (eventually botched) merger between Paramount and VLSE, summed it all up:

> In every other business than motion pictures, caution contributes to success, but in the movies the conservative has almost invariably disappeared from the industry. History had repeated itself each time radicals and conservatives came into conflict. In 1917, while prudent manufacturers, unable to adjust their minds quickly to the high-pressure methods necessitated by the new conditions, were thoughtfully analysing wage demands, venturesome competitors rushed in and took their celebrities away from them, agreeing to salaries that generally were declared 'impossible'.

> A few of these daring showmen got possession of stars that the public loved, and, although the prices paid seemed very high, they were able to pass the additional costs – through the theatres – to audiences, who paid the bill cheerfully.
>
> These producers made money, but others, more reckless than wise in the bidding contest, found themselves saddled with players who failed as box-office attractions. Such producers sustained severe losses in 1917–18, and some of them soon succumbed and joined the ultra-conservative producers in fading from the screen.[36]

The many mergers, dissolutions and bankruptcies during the escalation process fit well into the sunk-costs framework: when the quality race started, sunk costs increased and concentration rose, either through consolidation by mergers, by exit, or both.

The decline of the European film industry

Above, it has been argued that theory on sunk costs and market structure can explain the sharp jump in production outlays during the 1910s, the emergence of the feature film, and the increase in concentration in the industry. The question remains, however, how this can explain the decline of the *European* film industry. Nothing put forward in the explanation above would have prevented European companies from participating in the escalation phase.

The explanation for European companies not fully participating points back to the First World War. Although the war was not a direct cause of the decline of the European film industry, as has been argued above, and many companies remained profitable, the war can explain Europe's relative decline. The war made it very difficult for companies to participate in the escalation phase. On the supply side, it was difficult, if not impossible, to obtain the vast amounts of venture capital needed for a jump in production outlays.

Also, other products, especially newsreels, were very popular and profitable during the war. Pathé Exchange, for example, Pathé's US subsidiary, made large profits on war documentaries shot in Europe. *Battle of the Somme* had receipts of at least $122,000, *The Tanks* of $67,000 and *Retreat of the Germans* of $116,000.[37] Revenues per booking were $16, $25 and $21.[38] The Hearst-Selig *Official War Review*, which ran in thirty-one weekly instalments in 1918 and 1919, had total rentals of $335,000, with an average rental of $48 per booking.[39] These revenues, admittedly exceptionally high for non-fiction films, were not insubstantial compared to those of feature films. The average rental of a Fox feature film in 1917, for example, was $102,000, while the rentals of the three features made by Paramount's star producer Cecil B. de Mille ranged between $242,000 and $446,000.[40] Given the far lower cost of war films, including the inability of creative inputs such as actors and directors to extract rents, initially newsreels and non-fiction films may well have been more profitable to European firms than feature films. Over time, however, it appears that features became more and more profitable, and they probably also faced far less declining returns as firms turned out more of them. Sales figures of Pathé Exchange in Britain showed a sharp increase in the revenue per metre for feature films compared to other formats during the 1919–20 season, to about three times the revenue of other formats, although the costs of feature films were probably also substantially higher.[41]

An implication of the above theory is that for products with high sunk costs, market size was all-important, since total costs (nearly equal to sunk costs) would be determined by potential revenue. The European film industry could thus be so strong initially simply because sunk costs were low but, as sunk costs in the film industry grew, market size became ever more important and the European companies found themselves increasingly at a disadvantage in their small home markets. Charles Pathé underlined the growing importance of market size for film production just as the European market was disintegrating:

> The big countries, above all when they are very wealthy, are endowed more favourably than France, because their capacity to amortise is infinitely more significant and faster than ours.... Having the advantage of their huge interior market, which, concerning box office revenue, represents forty to fifty times the French one, thus three quarters of the world market, the Americans can engage considerable sums in the production of their negatives, amortise them completely in their home market, and subsequently conquer the export markets all over the world, especially those of countries that cannot afford the luxury of having their own national film production. The prime cinematographic importance of France in the world rested solely on its initial advantage and would have to disappear the day the building of the American film industry was finished. This day has come.[42]

As long as products are easily exportable, market size is closer to the size of the world market rather than to a particular domestic market. Unfortunately for the European companies, as sunk costs increased, so did both cultural and legal trade barriers. Before 1914, European countries were relatively open to each others' products but during and after the war consumers became more hostile to products of enemy countries, especially with cultural products like films. This resentment was reflected in legislation, as German films were not allowed to be shown in France and Britain until the early 1920s and Germany responded with a reciprocal policy. Before 1914 legislation concerning film trade was minimal as films were relatively new products. After 1914, however, taxes and duties increased, and in the mid-1920s most European governments introduced special legislation controlling the number of foreign films that could be shown.

Access to the US market remained unchanged and therefore cannot have been a reason for Europe's decline in the US.[43] Also, as European film companies became more dependent on their own national markets, the growth of these relative to the American market was hampered because of the imposition of large amounts of entertainment tax on cinemas, varying from 20–50 per cent. Thus, synchronously with the disintegration of the European home market, the individual countries' film markets also diminished, leading to a dead weight loss, and, moreover, decreasing the relative price of live entertainment, thus further reducing the film market.[44] Faced with suboptimal growth of home and export markets, an escalation strategy was a difficult option for European film companies.

Nevertheless, several European companies started a strategy that came close to escalation. Two major European 'escalators' were the French Pathé and the Danish Nordisk. Pathé had obtained its capital from the Paris stock exchange, from a few industrial families from the Lyon region and from a few French banks, such as the Banque Bauer et Maréchal. Still, Pathé's American operations must have needed large amounts of capital during the war, and it is unlikely that French sources would have handed out more capital, even if it had been allowed to leave the country.

When Pathé arrived in New York in September 1914, his American subsidiary was on the verge of bankruptcy and it was rumoured that William Fox had offered to take it over. Pathé managed to reorganise the whole American company in a matter of a few weeks, to fend off creditors and to obtain new capital. He profited greatly from having a national distribution organisation and from several expensive and highly advertised serials, films in weekly instalments. Pathé foresaw the serial becoming the industry standard and was cautious about spending too much on producing feature films, which he thought would be a passing fad. He also made profits with his newsreel, which had been the first regular newsreel in the US when it was introduced in 1909. The result was that, although Pathé Exchange remained profitable, and Pathé eventually switched to making feature films, it became somewhat marginalised and grew more slowly than the market. Nevertheless, as the profit figures in table 2.1 show, it remained profitable during the 1910s, except for 1914, and was a minor major player. In the early 1920s it was sold to Merrill Lynch, and eventually became part of RKO.

Table 2.1 *Profits of Pathé Frères, in dollars, 1911–19*

	US	*Total*	*US/Total*
1911	388,374	852,980	45.5
1912	714,595	1,210,681	59.0
1913	899,491	1,401,020	64.2
1914	–357,082	–258,799	–138.0
1915	247,037	597,590	41.3
1916	460,484	1,253,724	36.7
1917	282,234	1,107,582	25.5
1918	269,280	1,070,611	25.2
1919	667,046	1,224,433	54.5

Note: total profits exclude the profits of Pathé's phonograph subsidiary. Original figures in francs, converted using exchange rate.

Charles Pathé wrote that in the late 1910s he had faced the choice between moving his company to Hollywood or divesting his foreign business and focusing on distribution in France. He chose the latter.

> They have asked me a lot: 'Why did your American subsidiary Pathé Exchange not become a fully American company?' I have dreamed a lot of that possibility. Maybe I would have done it, had I been much younger, but this transformation would have meant the permanent relocation of our business from France to America, and my permanent residence in that country. I was too old to occupy myself with that project.[45]

The other large European company that adopted a strategy that came close to escalation, the Danish Nordisk, was financially strong and was one of the first to release feature-length films. In 1909, it even held talks with Pathé on the possible formation of a European film cartel, similar to the MPPC, which eventually came to nothing. During the 1910s, Nordisk rapidly expanded feature film production. Nordisk's home market was Germany, which must have been very profitable until 1917, and it exported its films throughout the world. Nordisk bought distributors and cinema chains in Germany, Switzerland and Austria. However, it held only a small market share in the United States where it had problems with its sales manager and complained continuously that the American market was unprofitable.[46] In 1917, Nordisk's expansion came to a halt when the government forced it to merge its German assets into the UFA company.[47] While the sudden loss of the German home market did not threaten Nordisk's continuity, it made continuation of an escalation strategy difficult. From being a potential European market leader, Nordisk changed into a small Danish film company and went bankrupt in the late 1920s.[48]

Of the smaller Italian production companies, two made attempts to increase outlays on film production costs and film portfolios. Cines appeared to have been making preparations for an escalation strategy. It had received capital from a group of US investors and had acquired German distributors and cinemas to increase its share of the marginal cinema revenues that its films generated. However, when the war started most of these investments were lost and Cines' future as a European market leader was shattered.[49]

The second attempt was led by George Kleine, a Chicago film importer and member of the MPPC who made huge profits importing foreign films into the US, using his MPPC licence to sell the films. In the early 1910s he took several of the big Italian spectacle films, such as the *Last Days of Pompeii*, on road shows. In 1913, he agreed with Pasquali, an Italian production company, to jointly set up a firm that was to make large and expensive Italian historical spectacle films of the kind that were so popular in the US, but now combined with US star actors and actresses. Preparations were made and construction of a large studio complex outside Turin was started. However, the advent of the war caused the cancellation of the whole project. All that remained were the abandoned remnants of the never-finished studio compound. They still stand there today as a silent reminder of an era long-forgotten, a future never realised.[50]

Several other French companies did not attempt an escalation strategy. Gaumont and Éclair, Pathé's main competitors in France and also quoted on the Paris stock market, both set up US and other foreign subsidiaries, but these firms were cautious about increasing outlays on film production costs. In 1914, when the US escalators were spending at least $10,000 to $20,000 a film, Léon Gaumont, for example, instructed his US manager not to spend more than $10,000 on the entire 1914 production portfolio. Both companies eventually went bankrupt, Éclair in 1919, Gaumont in the 1930s.[51] Smaller French companies such as Méliès and Lux were possibly in a less advantageous position to embark upon an escalation strategy, and would have had difficulties in obtaining capital.

One might be surprised that just four years made such a difference for Europe's film industry, a difference that was never challenged, when US dominance in many other industries has been. However, the essential difference was that films were (copy)rights, and therefore, contrary to manufacturing industries, protection could be easily evaded, no foreign production plants were needed, and 'dumping' was easy because reproduction was costless. In the car industry, for example, production costs, transportation costs, tariffs, and the need for foreign production plants all were obstacles to absolute US dominance. In the car industry no similar jump in endogenous sunk costs took place, since most costs were exogenous and dictated by technology. The creative inputs in the car industry (i.e. experts working in R&D) were less scarce than in the film industry. Even so, since cars were goods and not rights, no vertical integration was needed to maximise profit, since 'perfect' selling contracts could be written, and foreign distribution networks were therefore less of an advantage. In the film industry high endogenous sunk costs, costless reproduction, easy tariff evasion, and the absence of foreign production plants led to high scale economies and increasing returns in international trade.

Conclusion

This chapter has shown how initially the rapidly growing film market resulted in more firms entering the film business and, as a consequence, declining concentration. From the mid-1910s, however, as the market kept expanding rapidly, concentration stabilised and then increased as the market continued to grow. The cause was the rising pay-off for improving film quality because of increasing market integration, made possible by the industrialisation (automation, standardisation, tradeability) of entertainment by the completion of the distribution delivery system (a network of fixed cinemas), by the rapidly increasing demand for low-priced entertainment, and by the decline of the MPPC trust.

The industry, and several firms in particular, became part of a process of discovery of the higher pay-offs for improved quality. At the firm level, changes in the vertical structure, both through ownership and through contracts, translated ever more cinema marginal revenues into producers' marginal profits, further increasing the pay-offs of improved quality.

During the quality race, four important and partially distinct film industries or industrial districts existed. The two mature ones were the industries in Europe and on the US East coast in New York and New Jersey. Two new industries were emerging in Florida and in Southern California. The three Atlantic industries lost out to Hollywood in two stages. This chapter dealt with the first stage, in which European firms' participation in the quality race was hampered by a declining effective market size because of a disintegrating European film market. This was brought about by war, tariffs and protection, and a high amusement tax that disproportionately weighed on cinema, lowering the relative price of live entertainment. The First World War also made it difficult for European firms to obtain the large sums of venture capital needed for an escalation of quality. The war probably also made other products, such as the newsreel, initially more profitable than feature films.

By the 1920s, most large European companies had given up film production altogether. Pathé and Gaumont sold their US and international business, left film making and focused on distribution in France. Éclair, their major competitor, went bankrupt. Nordisk continued as an insignificant Danish film company, and eventually collapsed into receivership. The eleven largest Italian film producers formed a trust, which failed terribly and one by one they fell into financial disaster. The famous British producer, Cecil Hepworth, went bankrupt. By late 1924, hardly any films were being made in Britain. American films were shown everywhere.

Notes

1 Bakker (2004).
2 Zukor (1930).
3 Ibid.
4 Cherchi Usai (1989).
5 This was 3,000–4,000 feet.
6 Ibid.
7 Gomery (1986: 46).
8 Bowser (1990: 213–14). The special qualities of the feature rather than programme length enabled higher admission prices, even for a two- or three-reel film.
9 Bakker (2001a: 472); Koszarski (1990: 34).
10 Hampton (1931: 164).
11 Revenues per metre for feature films were 33 pence, vs. 24 for serials, 29.5 for comics, 15 for pictorials, 9 for newsreels (*Pathé Gazette*), and 25 pence for all films, on average. 'Report by Hedly M. Smith on the business of Pathé Ltd. for the period 1st December 1919 to 31st May 1920', Beaverbrook Papers, hereafter BP, file H274.
12 Ibid. The same six-month periods for 1918–19 and 1919–20 are compared. Revenue per metre for feature films was now about 47.5 pence, vs. 25 for serials, 36 for comics, 19 for pictorials, 8.5 for newsreels, and 28 pence for all films, on average. Newsreel revenue decline was probably caused by the peace. Total revenue for the six months was £166,994. Despite the high revenue per metre, feature films accounted for only 23% of total revenue, vs. 41% for serials, 22% for newsreels, 9% for comics, and 5% for pictorials. The large serial share was probably due to Charles Pathé's bet that it would become the industry standard, the large newsreel share to Pathé's specialisation in that genre and its world-wide correspondent network. It had been the inventor of the weekly newsreel.
13 Unfortunately, the disaggregated profits per metre of film are unavailable. Profit per metre for all films taken together was 3.2 pence in 1918–19 and 2.3 pence in 1919–20.
14 Letter Frank Smith to Lord Beaverbrook, 2 February 1920, 3, BP file H274; letter H.M. Smith to Lord Beaverbrook, 8 October 1924, BP file H279. The latter letter discusses a change in the pricing structure over time for the *Pathé Gazette*, and thus also discusses the earlier pricing structure.
15 During the silent period, revenue-sharing contracts between distributors and cinemas were mainly used in a few large city-centre theatres (which accounted, though, for a disproportionate share of revenue), and for a studio's most expensive films, which were often 'road-shown'. Vertical integration into exhibition by the five Hollywood majors combined with their collusion (of which they were found guilty by the 1948 Supreme Court Paramount Decision) was another way to gain access to marginal cinema revenues during the silent period. For a detailed analysis of the contractual evolution of film bookings see Hanssen (2000, 2002).
16 So the changes in the distribution practices did not merely involve producers attracting away profits from distributors and cinemas, but also a new incentive structure that induced producers to sink far more costs.
17 Hampton (1931: 164).
18 Lyons (1974). See also the case of real estate speculation in Pittsburgh, where forty-two Nickelodeons were opened during 1906, as discussed in Musser (1990: 420). In London from 1906 onwards, real

estate owners were able to charge a premium of up to 50 per cent for the rent to Nickelodeon operators (Burrows 2004).

19 Zukor (1930) later remarked about the merger: 'We found interest between producer and distributor was not one.' Paramount's dominant position is illustrated by the Department of Justice's prosecution during the 1920s, and later, in the 1940s, when Paramount was the first-named defendant in the Paramount case.

20 Paramount executives remarked that initially this could only be done for large cinemas, preferably with exclusive supply contracts or if Paramount owned part of them, as accounting and monitoring costs were too high for smaller cinemas.

21 Profits as percentage of gross revenue fluctuated, though, from 22.3 to 7.1 to 15.7 per cent, respectively. 'Gilmore, Field and Company. Investment bankers', in Lewis (1930: 76).

22 Wasko (1982).

23 Hampton (1931: 218).

24 Wasko (1982).

25 Bowser (1990).

26 Common monthly dividend fluctuated between 0.5 and 3 per cent (Davis 1916).

27 Contracts often were not optimal, given monitoring costs. The forced divestment of their cinema chains by the US Supreme Court in 1948 did not diminish the dominant position of the Hollywood studios in film production and distribution.

28 All Triangle information is based on Lahue (1971).

29 Triangle's literary, theatrical cinema was possibly more popular in Europe. The company held a substantially larger British and French market share in the time series analysed.

30 All information about Balboa is based on Jura and Bardin (1999).

31 Davis (1916).

32 Ibid.

33 *Moving Picture World*, 11 July 1914, quoted in Bowser (1990: 214).

34 See below.

35 Hampton (1931).

36 Ibid.: 169–70.

37 The exact figures are $122,020, $71,485 and $116,167. Pathé Exchange, Inc. 'Statements of collections ...' (1918); Charles Urban Papers, Science Museum, London, URB4/1–153. See also McKernan's (2002) detailed study, which shows that these films were rather exceptional and their costs probably high because of idiosyncratic circumstances and government involvement. The rental figures still show the revenue potential of news films, however, and Pathé Exchange did make large profits on the films.

38 Calculated using the number of bookings reported in McKernan (2002).

39 Ibid.: 384. See also Fielding (1972).

40 On De Mille, see Pierce (1991: 308–17).

41 See the Pathé Exchange UK figures discussed above.

42 Pathé (1940: 98, 92–93); see also Wilkins (1994).

43 Tariff rates for 1,000 feet of positive film declined from $207 in 1899 to $160 in 1909, to $97 in 1913, to $57 in 1922. Rates for 1,000 feet of negative film changed from $206 in 1899 to $290 in 1913, to $172 in 1922 (all rates in 1982 dollars) (Thompson 1985: 20–22). Tariffs per se only had a limited impact on foreign revenues, because the number of copies and viewings that a film generated within a foreign country could not be taxed by customs. Unlike Europe, the US did not introduce legislation limiting the showing of foreign films.

44 The US also introduced an Admission Tax, but this was only a flat 10 per cent, levied on live and filmed entertainment alike.

45 Pathé (1940: 97).

46 Mottram (1988).

47 Kreimeier (1992).

48 It was revived about a year later and still exists today.

49 Kallmann (1932: 3).

50 Cherchi Usai (1989).

51 Both companies were revived and still exist today. Gaumont was bailed out and managed by the French state for some time during the 1930s.

Chapter 3

Richard Koszarski

MAKING MOVIES, 1915–28

New York and Hollywood

Southern California was clearly recognized as the major American production center by 1915, although the generic use of the term "Hollywood" to describe nearly all such activity had not yet developed. With studios scattered from Santa Monica to Edendale to Pasadena (and north to San Francisco and east to Phoenix), most commentators spoke of the area "in and about Los Angeles" as the "Mecca of the Motion Picture." By 1924, however, a Hollywood mythos was clearly emerging. Perley Poore Sheehan, a popular screenwriter, issued a bizarre tract called *Hollywood as a World Center*, which combined elements of small-town boosterism, industry braggadocio, and occult transcendentalism (known locally as "new thought"). For Sheehan, "'The rise of Hollywood' and its parent city, Los Angeles, has world-wide significance. It is a new and striking development in the history of civilization . . . This flooding of population to the Southwest has its origins in the dim past. It is the culmination of ages of preparatory struggle, physical, mental and spiritual. In brief, we are witnessing the last great migration of the Aryan race." Going beyond traditional American disdain for the eastern cities, Sheehan saw the birth of Hollywood as the dawn of the Aquarian age and described a New Jerusalem that would reveal to all mankind the "Universal Subconscious."[1]

In the absence of any traditional moral, intellectual, or religious framework, such philosophizing was quite popular in Hollywood in the boom years of the immediate postwar era, especially among the circle around Nazimova, Valentino, and June Mathis. Before it faded from fashion, one could see its traces in such curious films as the Mathis–Valentino YOUNG RAJAH (1922) or Sheehan's original story THE WHISPERING CHORUS (1918) for Cecil B. DeMille. But its representation in specific films is far less important than the gauge it provides to help measure the emotional and intellectual distance between Hollywood and the traditional centers of American culture. The physical isolation of the place—five days by rail from the corporate home offices—very quickly inculcated a special "Hollywood" way of looking at life that generations of audiences would instantly recognize. European films tended to be produced in and around traditional centers of national culture—Paris, London, Vienna, Rome, Berlin—but this community of desert exiles made something very different of the American cinema. Of course, such films won the contempt of American critics for many years, but audiences worldwide (and some foreign critics, such as Louis Delluc) felt otherwise. Movies from Hollywood would become the first American cultural export to conquer the world.

Equating American film production in this period with Hollywood production is, however, somewhat misleading. That colony was a factory town, producing the motion picture industry's major product, while executive operations, newsreel production, and even much of the animation industry remained in the East. The fact that Los Angeles was renowned as "the nation's leading open-shop, nonunion city" also did not hurt, keeping wages to half what they might have been in the East.[2]

The concentration of feature production in Hollywood during the teens was widely noted at the time, especially because few could have predicted the rapidity of this migration. Although visiting companies had worked in Los Angeles since 1908, much of the outlying area was still largely a wilderness in 1915, lacking equipment houses, prop and costume shops, a steady supply of professional actors, or even basic sanitation and safety. Thomas H. Ince and a group from his studio were robbed "stage-coach style" by four masked gunmen on "the lonely road near Inceville" in March 1915. Three years later, Erich von Stroheim was still carrying a revolver under the seat whenever he and his fiancée drove the dangerous Cahuenga Pass to Universal City. In this light the stories told by Cecil B. DeMille, Allan Dwan, and others, regarding the need for sidearms to protect themselves from Patents Company thugs, might conceivably have a more mundane explanation.[3]

It was generally accepted that the West Coast studios were producing 125 reels of film per week in 1915, including everything from split-reel comedies to THE BIRTH OF A NATION. Knowledgeable estimates put this figure at between 62.5 and 75 percent of total American production.[4] Despite the fact that the bulk of domestic production had already moved to the West Coast, it was only in 1914–15 that the makeshift facilities used by most of the producers were replaced by permanent installations comparable to the eastern studios, most notably the American Studio in Santa Barbara and Carl Laemmle's Universal City.

The comparatively large segment of production remaining in the East maintained its position throughout the early feature years, when the proximity of Broadway plays and players caused firms such as Goldwyn and Famous Players to increase the output of their New York and Fort Lee studios. This situation survived until the winter of 1918–19, when problems with coal rationing forced nearly all the companies then operating in the East to consolidate operations in their West Coast facilities. Paramount-Artcraft closed in Fort Lee but kept operating at their Fifty-sixth Street studio in Manhattan; Universal closed all their eastern studios except their Coytesville, New Jersey, operation; Goldwyn, Metro, and Fox moved everyone to the West Coast; Vitagraph kept only a small operation in the East. Only World Film remained, because "their coal is not only all purchased, but delivered."[5]

This forcible relocation was strongly resisted by many filmmakers who found California unsuitable as a production center, but not until after the Armistice could anything be done about it. As soon as wartime restrictions were lifted, a boom in studio construction swept the New York area (for various reasons this renaissance bypassed Fort Lee, which soon disappeared as a production center). D. W. Griffith, who had filmed Hollywood's most stupendous films, THE BIRTH OF A NATION and INTOLERANCE, created a new studio for himself on the Flagler estate in Mamaroneck, just north of New York City. The East was where "the money and the brains" were, he said, as he proceeded to film WAY DOWN EAST (1920), ORPHANS OF THE STORM (1922), AMERICA (1924), and other costly features in Mamaroneck.[6]

William Randolph Hearst transformed Sulzer's Harlem River Park and Casino into the vast Cosmopolitan studio, a beer-hall-to-movie-lot conversion that also occurred at the smaller Mirror studio in Glendale, Queens. At Cosmopolitan, Marion Davies starred in such lavish epics as WHEN KNIGHTHOOD WAS IN FLOWER (1922), LITTLE OLD NEW YORK (1923), and JANICE MEREDITH (1924), an underrated Revolutionary War spectacle.[7]

Vitagraph increased the pace of its New York operations, while Goldwyn and Metro reopened their eastern studios. Fox came back as well, opening a large new studio on West Fifty-fifth Street in Manhattan, where Pearl White and Allan Dwan worked from 1920. This facility continues in operation today as the Cameramart stages. The Talmadge sisters had their studio on East Forty-eighth Street, and various rental studios abounded, hosting such New York-based stars as Richard Barthelmess. By September 1920 the *Exhibitors Trade Review* headlined "Producers Say California Has Been Filmed Out—Are Looking for New Producing Centers." The paper reported that all the major companies were now making films in the New York area and that California's weather remained its sole compelling asset.[8]

That month the most important of the new eastern studios opened, the Famous Players-Lasky studio in Astoria. Over the next seven years, this studio would produce 127 silent features, serving as home base for those Paramount stars who preferred to work in the East, such as Gloria Swanson (nine films) or Bebe Daniels (fourteen films). Such recruits from the stage as W. C. Fields and Louise Brooks began their "Hollywood" careers in Astoria, while Rudolph Valentino and D. W. Griffith saw the studio as a means of escaping the factory conditions on the West Coast. Astoria was at its most active in 1926, when the 26 features produced there constituted 40 percent of the entire Paramount program.[9]

However, by 1922 Hollywood's share of American production stood at 84 percent, with 12 percent remaining in New York and 4 percent filming elsewhere. The East Coast had several large studios and ample rental space for the smallest companies. What it lacked was a significant group of middle-level producers, an area almost completely monopolized by Hollywood. While the advantages of California sunlight were no longer crucial, problems with New York's rain and snow remained an issue, especially for a system that made heavy use of large standing sets. "Weather destroyed sets on the back lot time and again," wrote Jesse Lasky of the Astoria operation, which he temporarily closed in 1927. Other New York studios had already reduced operations, and by the close of the silent era, Hollywood was unchallenged as the center of American production.[10]

But what of that 4 percent produced outside either Hollywood or New York? With 748 features released in 1922, this suggests that some 30 feature pictures (and a proportionate number of shorts) were filmed elsewhere. Even allowing for a handful of imports, the number is still significant. The bulk of these films were made by small local producers without access to national distribution. Many remained unseen. Some were sold on a states rights basis and never played in a key theater or won the attention of an urban reviewer. For historians, these regional productions are the *terra incognita* of the American film industry.[11]

Writing pictures

No matter what the location of the studio, the tremendous quantity of film generated each week required vast amounts of original (or at least semi-original) story material. The ad hoc practices of the earliest days of filmmaking had long been abandoned by most producers, although even as late as 1915 some directors could not resist a lucky opportunity. Henry Otto, directing for Flying A in Santa Barbara, noticed several hundred blackbirds sitting on telephone wires. He filmed them, then concocted a script in which the birds caused "wire trouble."[12]

In general, though, it was well understood that regular release schedules demanded a dependable flow of production and that scenario departments were needed to process scripts and synopses. A few writers, such as Ince's C. Gardner Sullivan or Thanhouser's Philip

Lonergan, had steady positions generating large amounts of story material to order, but in 1915 the free-lance scenario market was still quite significant.

The small amount paid for original scenarios was hardly conducive to high-quality submissions. The *Photoplay Author* complained that writers accepting three dollars for a two-reel script were depressing the market: "At the present scale $35 is fair, $50 is better and $100 per reel good money. Most of the purchases are made at $50 or less per reel."[13]

As late as 1923, Douglas Brown reported to the Society of Motion Picture Engineers that "the completed script of a feature picture costs the producer less than two thousand dollars." This sum did not, however, include the cost of any story rights involved, and beginning in 1919–20 such costs began to soar for any property considered a sure success (essentially Broadway hits that already seemed to work in scenario form). "Apparently a season's run in New York automatically makes a play worth about $100,000 to the film producers," said the *New York Times* in 1920, with only slight exaggeration. Even before talkies, the existence of a usable Broadway playscript made a property far more interesting to film producers. For example, F. Scott Fitzgerald's *The Great Gatsby* reached the screen in 1926 via a 1925 stage version by Owen Davis, not directly from the original novel.[14]

Prominent screenwriter Frances Marion reported in 1924 that the average price for a successful play was $20,000. She offered the following list of high-priced properties:[15]

Turn to the Right	$225,000
Way Down East	175,000
A Tailor-Made Man	105,000
The First Year	100,000
Tiger Rose	100,000
Daddies	100,000
The Gold Diggers	100,000
Merton of the Movies	100,000
The Virginian	90,000
Dorothy Vernon of Haddon Hall	85,000

The virtual elimination of the free-lance market was among the most significant production changes of the immediate postwar era. Writing credits went to such contract employees as Jane Murfin, Lenore Coffee, Charles Kenyon, Jeanie MacPherson, Waldemar Young, and Jules Furthman, skilled wordsmiths with backgrounds as reporters or short-fiction writers. The flood of Broadway playwrights that would engulf Hollywood in the talkie years was hardly in evidence before 1927, when dialogue skills were not a requirement.

The producer system

Janet Staiger describes the central producer system as "the order of the day" by 1914 and invokes Thomas H. Ince's operation for the New York Motion Picture Company as the traditional example. Script material was recast into continuity form, which allowed careful preplanning of all production activities. Actors needed to appear only when required; props and costumes could be scheduled on a dependable basis, and the logistics of complicated location trips (or studio shoots, for that matter) might be clearly predetermined. By closely monitoring the scripting and editing process, a central producer like Ince—or Mack Sennett—could guarantee a uniform standard of quality without having to attend personally to the filming of each scene.[16]

While this pioneering demonstration of organizational efficiency does mark Sennett and Ince as important innovators, their systems were primitive in comparison to those employed later in the silent period by more mature studios such as Paramount or MGM. In fact, the collapse of Ince's entire operation on the death of its central producer suggests that his studio was more an extended workshop than a true factory. Systems that could outlast their innovators reflected a higher level of organization and took several more years to develop.

In 1925 the new Metro-Goldwyn-Mayer studio produced a thirty-minute promotional film to demonstrate the power and scale of their factory operations. In true industrial-film fashion, they lay out the shape of their physical plant, boast of the capacity of their electrical powerhouse, and awe us with a staggering array of statistics. What is most interesting, however, is the way the film organizes and presents the studio workers. Dozens of cinematographers line up on a studio lawn, cranking away on Mitchells and Bell & Howells. They are matched by an equally formidable array of writers, directors, scenic artists, carpenters, electrical workers, cutters, even shippers packing MGM prints off to distant exchanges. Seen in control of this army of artists and technicians are three men, each busy behind a desk—Louis B. Mayer, Irving Thalberg, and Harry Rapf. Finally, we see a telegraph key that allows them to stay in constant touch with New York, where an unseen Marcus Loew and Nicholas Schenck call the ultimate shots.[17]

This little film is especially revealing because it consciously deemphasizes the glamour of the studio's employees and underscores the tight, pyramidal control exercised by the top executives. MGM's stars are reduced to a few charming close-ups. MGM's directors appear in a vast and nearly anonymous group. A title card announces, "Browning, Seastrom, Vidor, Niblo, von Stroheim, von Sternberg, . . ." but the names and faces do not really connect. All that matters is *studio and system*: clearly that is the message being communicated to the stockholders or theater owners who made up the film's original audience.

Production of fodder for the nation's movie screens—many hundreds of pictures annually—was clearly generated by just such a system. Yet even Staiger suggests that the leading works of the age, the product of the most powerful stars and filmmakers, remained under individual control to a great degree.[18] The most memorable work of many of the key filmmakers, Griffith, Weber, von Stroheim, Cruze, Lubitsch, DeMille, Neilan, and Ingram in particular, was created via a simpler director-unit system, where projects were developed from script level through editing by individual creative directors and their personal staffs. In the final analysis, these were the films that created the models for new styles or genres that were then mass-produced (often more lucratively) by the factory studios. While the central-producer system certainly generated the bulk of American production in this period, those films that really mattered, to audiences of the time as well as to posterity, were often the dogged creations of an antiquated workshop system that somehow managed to survive well into the 1920s.

How much to make a picture?

There was great interest, both inside and outside the trade, in establishing some sort of "average cost" for a standard program feature. In July 1917 the *Motion Picture Classic* published a "conservative average" budget based on information from three different production companies. The *New York Dramatic Mirror* [*NYDM*], a trade paper, offered its estimate for a similar production that same month (table 3.1).

In retrospect, the budget proposed by the *Dramatic Mirror* seems to represent less an average price than a rock-bottom one. In any case, generalizations about-average picture costs in this period are of little practical use because budgets rose so rapidly during the first dozen

Table 3.1 "Average cost" estimates, 1917

	Motion Picture Classic		*New York Dramatic Mirror*	
Corporate salaries	5,000		$	–
General manager	3,000			–
Director	2,500		3,000	
Assistant directors	1,300	(two)	300	(one only)
Cameramen	1,200	(three)	300	(one only)
Assistant cameramen	600	(three)"	100	(one only)
Stage carpenters	250	(three)	–	
Other employees	1,800	(twenty)	–	
Overhead expenses	900		–	
Star	12,000	(two)	6,000	(one only)
Near stars	5,000	(three)	1,200	(one leading man)
Prominent characters	2,500	(three)	1,300	("other principals")
Other salaried players	3,000	(40)	–	
Supers, etc.	1,100		300	
Transportation	1,200		500	(transportation)
& location exp.			200	(location rental)
Scenario handling	–		1,000	
Studio settings	–		1,000	
Negative stock	–		1,000	
Costumes	–		1,000	
Incidentals	–		500	
Total cost	$41,350		$17,700	

Source: Motion Picture Classic, July 1917, p. 19; *New York Dramatic Mirror*, 7 July 1917, p. 9.

years of features, while even within a single studio or genre, costs could vary widely on different productions. As Samuel Goldwyn, who moved from Famous Players–Lasky to First National and then to United Artists during these years, told the *New York Times* in 1926, "In the old days the average negative cost of Famous Players was about $15,000. The distributors gave us an advance of $25,000. Today the average negative cost of Paramount productions is $300,000. The average negative cost of United Artists productions is from $750,000 to $800,000. The other two big companies are confronted with an average cost of between $240,000 and $260,000."[19]

Actors' salaries	25%
Director, cameraman, and assistants	10
Scenario and stories	10

The average cost of Famous Players–Lasky releases had increased twenty-fold over thirteen years, but knowing a studied "average" production cost may not in fact, be very revealing. In 1921, for example, Universal spent $34,211.79 on THE WAY BACK, a five-reel program feature. The cost of FOOLISH WIVES, an unusual "special jewel feature," made that same year, was thirty times this amount (see tables 3.2 and 3.3).

Table 3.2 Universal Film Manufacturing Co., Pacific Coast Studios. Statement of production costs for week ending 2–1–22

DIRECTOR: Paton
TITLE: "THE WAY BACK"
FEATURING: Frank Mayo

PICTURE NO: 3723 REELS: 5
DATE STARTED: 12–23–21
DATE FINISHED: 1–16–22

Charges	*Previously reported*	*This week*	*Total to date*
Stock Talent Salaries	3205.00		3205.00
Pict. Talent Salaries	1979.20		1979.20
Extra Talent Salaries	895.25		895.25
Directors Producing "	1333.30		1333.30
" Writ'g & Edit'g "	1666.70	400.00	2066.70
" Staff	1298.75	120.45	1419.20
Continuity Writers' "	525.00		525.00
Editors' & Cutters' "	267.45	102.85	370.30
Negative Raw Stock	1450.89	118.65	1569.54
" Laboratory Chgs.	322.13	31.73	353.86
Positive Raw Stock	992.25	38.43	1030.68
" Laboratory Chgs.	321.40	14.88	336.28
Wardrobe Pur. & Mfg.	98.20		98.20
" Hire & Expense	58.54		58.54
Rent Studio Ward. Equip.	252.41		252.41
Prop Pur. & Mfg.	44.92	3.75	48.67
" Hire & Exp.—Misc.	653.47	13.64	667.11
" & " Horses—			
" & " Special	40.00		40.00
Rent Studio & Prop. Equip.	606.90	17.09	623.99
Scenery	4744.24	256.12	5000.36
Auto Transportation	585.22		585.22
Traveling & Maintenance	330.70	2.00	532.70
Location Fees & Expense	25.00		25.00
Scenarios	2500.00	1000.00	3500.00
Light, Labor & Current	2153.31	226.90	2380.21
Rent Studio Elec. Equip.	1056.60	195.00	1251.60
Ranch Salaries & Exp.	1.00		1.00
Directors' Bonus			
Arsenal Salaries & Exp.	45.40		45.40
Misc. Unclassified Exp.	179.02		179.02

(Continued)

Table 3.2 Universal Film Manufacturing Co., Pacific Coast Studios. Statement of Production Costs for Week Ending 2–1–22 *(Continued)*

Charges	*Previously reported*	*This week*	*Total to date*
New York Title Charges			
Overhead Charges	4000.00		4000.00
Still Prints & Negatives	38.05		38.05
TOTAL	31670.30	2541.90	34211.79

Source: Author's collection.

Table 3.3 Universal Film Manufacturing Co., Pacific Coast Studios, daily memorandum picture costs

DIRECTOR Von Stroheim PICTURE NO. 3322 EPISODE NO. __ DATE July 7, 1921

	Estimate	*Amount today*	*Total amount to date*	*Amount UNDER estimate*	*Amount OVER estimate*
Overhead	$	$	$57,761.74	$	$
Stock Talent					
Picture Talent					
Extra Talent, Story and Continuity			257,283.20		
Director and Staff		167.00	101,009.18		
Film Editing			2,299.60		
Negative			19,287.85		
Sample Print			10,066.66		
Wardrobe			20,258.79		
Preps (Arsenal, Drops, Props)			46,226.62		
Scenery, Sets, Etc.		4.00	351,913.53		
Transportation (Automobile)			33,178. 88		
Traveling			10,469.00		
Maintenance (Lunches and Hotels)			41,209.54		
Location Fees and Expenses			1,848.11		
Lighting and Labor			77,010. 23		
Ranch and Zoo			9,347.97		
Special Automobiles			814.90		
Miscellaneous Supplies		6.00	13,305.08		
TOTAL	$	$ 177.00	$1,053,290.80	$	$
		Net Amount Under or Over Estimate			$

Source: Moving Picture Weekly, 30 July 1921, p. 28.

Despite the tremendous differences in scale between the production of THE WAY BACK and FOOLISH WIVES, it is clear that staff salaries were a significant part of the budget of both pictures. In 1924 *Barron's* offered the following statistics to explain where the production dollar went, figures that were generally accepted throughout the industry:[20]

Sets (manufacturing)	19
Studio overhead (including cutting, titling, etc.)	20
Costumes, gowns, etc.	3
Locations (and transportation)	8
Raw film	5

This negative cost factored into the total profit picture as follows:

Negative cost	40%
Distribution (U.S. and foreign)	30
Cost of positive prints	10
Administration and taxes	5
Profit	5

With salaries so large a part of production costs, much attention was devoted to limiting, or at least controlling, their growth.[21] A major reason for the introduction of the continuity script was to better manage personnel resources, but competition for top talent continued to force these figures higher and higher.

Photoplay reported in 1916 that salaries of $1,000 per week had recently become common, while the highest figure for a single picture had reached $40,000 (Billie Burke's fee for PEGGY). Leaving aside the well-known Chaplin and Pickford figures, they offered the following sampling of weekly star salaries:[22]

William Collier	$2,500	Keystone
Raymond Hitchcock	2,500	Keystone
Sam Bernard	2,500	Keystone
Eddie Foy	2,000	Keystone
Weber and Fields	3,000	Keystone
DeWolf Hopper	125,000/year ($60,000 guarantee)	Keystone
Henry B. Walthall	500	Essanay
Blanche Sweet	750	Lasky
Marguerite Clarke	1,250	Lasky
Fannie Ward	1,200	Lasky (one picture only)
Valeska Suratt	5,000	Fox
Victor Moore	500	Lasky
Frank Keenan	1,000	Ince (one picture)
William S. Hart	300	Ince
Francis X. Bushman	750 (plus a percentage of the profits)	–
Beverly Bayne	350	–

These figures reflect a relatively brief period when Broadway headliners were able to command salaries five times those of reliable film favorites such as William S. Hart. More typical of the era were weekly salary statistics offered by theater owner William Brandt following the announcement of the Famous Players–Lasky shutdown in 1923. Brandt's concern was that the exhibitors would bear the brunt of carrying these stars at half-salary while they were between pictures.[23]

Norma Talmadge	$10,000	Conway Tearle	$2,750
Dorothy Dalton	7,500	Lewis Stone	2,500
Gloria Swanson	6,500	Milton Sills	2,500
Larry Semon	5,000	James Kirkwood	2,500
Constance Talmadge	5,000	Wallace Beery	2,500
Pauline Frederick	5,000	House Peters	2,500
Lillian Gish	5,000	Elaine Hammerstein	2,500
Tom Mix	4,000	Richard Barthelmess	2,500
Betty Compson	3,500	Betty Blythe	2,500
Barbara La Marr	3,500	Florence Vidor	2,000
May McAvoy	3,000	Elliott Dexter	2,000
Mabel Normand	3,000	Viola Dana	2,000
Priscilia Dean	3,000	Lon Chaney	1,750

Brandt's figures exclude United Artists' stars, and those, like William S. Hart and Harold Lloyd, whose incomes were tied to significant participation deals. By 1926 total earning figures to date for those stars had reached truly fabulous heights:

Harold Lloyd	$1.5 million per year
Charles Chaplin	1.25 million
Douglas Fairbanks	1 million
Mary Pickford	1 million
Gloria Swanson	Refused Famous Players–Lasky's offer of $1 million to sign with United Artists

The highest-paid star then on straight salary was Tom Mix, earning $15,000 per week at Fox.[24]

Scenic art

After salaries, the highest fixed costs were those related to set construction. Wilfred Buckland, Ben Carré, and Anton Grot were already established as art directors by 1915, but most settings were still designed and constructed by carpenters. Cameraman Arthur Miller remembered Grot as "the first art director I ever worked with who hadn't come up from the ranks of the construction department." Grot created charcoal illustrations of the sets that displayed the scale and perspective of various motion-picture lenses. This technique enabled him to build only those segments of the set that would actually be used and resulted in substantial savings in construction costs. (He taught this skill to William Cameron Menzies.) Grot had received his training at the Akademie Sztuki in Cracow, Poland; Ben Carré came from the Paris Opera; and Wilfred Buckland was long associated with David Belasco.[25]

In his very thorough 1918 study *How Motion Pictures Are Made*, Homer Croy was still giving all of the credit for set design and construction to gangs of carpenters:

> With the scene locations determined upon, a list of the interior sets is handed the chief carpenter, who promptly starts building the necessary wood work. The list tells him in what order they will be wanted, with the date on which the

> first one will be needed impressed on his mind. He is held responsible for the finishing of the scene by the time specified . . . From a bare wooden platform the carpenters, under the direction of their chief, may start to work.
>
> (Homer Croy, *How Motion Pictures Are Made* [New York: Harper and Brothers, 1918], p. 110)

It should be remembered that, in this fashion, the master carpenter Huck Wortman constructed Griffith's Babylon in INTOLERANCE, albeit under the supervision of theatrical designer Walter Hall. These two traditions—that of the graphic artist or scene painter on the one hand, and the practical carpenter on the other—remained dominant in American studios during the early feature period.

This playground of graphic artists and carpenters lasted until around 1919, when producers began to turn to professional architects to create their settings. Robert M. Haas, who began designing for Famous Players–Lasky in New York in 1919, came to films after eight years as a practicing architect. His new position won two approving articles in the *American Architect*, which announced the fact that film design was now seen as "structural," not merely decorative, as in stage work. Haas was praised for the solidity of a town he had constructed in Elmhurst, New York, for THE COPPERHEAD (1920), where the details were aged to show the passage of time from 1846 to 1904. A rare full-card credit for the art direction on this film reads, "Robert M. Haas, Architecture; Charles O. Seesel, Decorations." His use of a ceiling "built to show" for ON WITH THE DANCE (1920) was also cited as an example of a new kind of structural realism unlike anything previously seen on the screen. Of course, this new approach to design only underscored the prevailing stylistic mode. "The men who design the 'sets' are constantly striving for the better effect of actuality," observed the *American Architect*.[26]

What Haas added to the Famous Players–Lasky art department was not just a new sense of creativity but a way of organizing the work flow that recalled the offices of top architectural firms rather than the workshops of fashionable graphic designers. In the years that followed, teams of draftsmen, sketch artists, and model-makers would set a standard for studio efficiency at Paramount, MGM, and the other great Hollywood studios.

Behind the camera

Unlike designers, cinematographers were not easily organized into hierarchical departments and in this period often contracted their services much as writers did. Some of the best were under contract to various studios on either a per-week or per-picture basis (for example, Joseph Ruttenberg with Fox from about 1915 to 1926). Some had long-term relationships, not always contractual, with the units of individual directors or stars (most notably G. W. Bitzer with D. W. Griffith from 1908 to 1928). But most drifted in and out of various jobs, hoping, especially, before unionization began to offer a modicum of job security, for a decent run of steady employment.[27]

The American Society of Cinematographers (ASC), a professional organization designed to exchange useful information and promote the qualifications of its members, was chartered in Hollywood in 1919 (although earlier component organizations date from 1913). Most top cameramen quickly joined and as a group resisted the unionization they saw developing in the eastern studios. Dan Clark, ASC president in 1926, explained this anti-union position in terms of their perceived status as artists, not artisans, as well as a resistance to the fixed-wage scales they saw as detrimental to their own members. Or, as Hal Mohr remembered, "I made a pretty good reputation for myself by 1928 and I was pretty much in demand and considered a pretty fine cameraman, getting a high salary. I was getting around $350 a week then. So I figured, what the hell do I want with a union organization?" But Mohr did join, serving during a long career as president not only of the ASC but of the union local as well.[28]

Largely because of the existence of the ASC and its house organ, the *American Cinematographer*, artistic and technical problems of cameramen were given relatively sophisticated discussion in a public forum, something that was not often the case for other film workers. During this period, these problems included the use of the close-up, "soft" versus "sharp" photography, the influence of German films and filmmakers, and the use of color.

While not necessarily a photographic issue, the close-up was a matter of some controversy throughout this period and of special concern to the cameramen because one of their prime responsibilities was the lighting of glamorous star portraits. Some historians argue that nickelodeon audiences resisted the introduction of the close-up, and patrons of early features in 1915 had their doubts as well. In reviewing DAVID HARUM (1915), the *New York Dramatic Mirror* commented on one sequence of Harum at dinner, which showed only his hands and the food he was eating. This "caused a spectator behind us to say at once that it was 'a poor picture because you cannot see his face.'"[29]

Audiences soon grew sufficiently sophisticated to accept the existence of offscreen space, but various stylistic complaints continued. Extreme close-ups were so rare as to be beyond general notice. The *American Cinematographer* defined the two forms of the close-up in 1923 as the waist-to-head two-shot and the chest-to-head single figure. Full-face was clearly too rare to consider, despite its dramatic use in INTOLERANCE and other Griffith films. Welford Beaton, an otherwise perspicacious critic, waged a lengthy war against the close-up. In a review entitled "Submerging the Production Under Senseless Close-Ups," he attacked Alexander Korda's THE YELLOW LILY (1928) as a "close-up orgy" and contended, "In this picture we have elaborate sets which flit across the screen to give place to an endless parade of utterly senseless close-ups. Ordinary business sense would dictate that the sets should be shown for a longer time to justify their cost." As for von Stroheim's THE WEDDING MARCH, Beaton told readers, "I would estimate that there are between seven and eight hundred close-ups in the entire picture, proving that von Stroheim treated Griffith's discovery as wildly as he did Pat's bankroll." (The "Pat" in Beaton's review is Pat Powers, producer of THE WEDDING MARCH.) Similarly, Frank Turtle, director of a self-described "artistic" film production unit called the Film Guild, wrote in 1922, "The-close-up mania is like the drug habit. It grows upon the afflicted company at a constantly accelerating pace until the whole studio is mortally ill of it."[30]

A wide array of diffusion effects suddenly became popular after 1919, when Henrik Sartov and G. W. Bitzer filmed close-ups of Lillian Gish that recalled Photo-Secessionist portraiture. The style soon spread even to war films. "In THE FOUR HORSEMEN I made all the exteriors I could on dull days in order to use an open lens and get a softer image," explained director Rex Ingram in 1921. "I wanted to get away from the hard, crisp effect of the photograph and get something of the mellow mezzotint of the painting; to get the fidelity of photography, but the softness of the old master; to picture not only the dramatic action, but to give it some of the merit of art."[31]

Some years later, cameraman Henry Sharp, then in New York filming THE CROWD (1928), found it useful to study Rembrandt's work in the Metropolitan Museum of Art. "Rembrandt's great strength was his use of one positive light scale. One central 'light perspective' was always used, rather than a multiplicity of attempted effects, and the results were contrasts in light and shade," he noted. The invocation of Rembrandt was used to cover a variety of pictorial effects, from Alvin Wyckoff's "smash of light from one side or the other" in DeMille's THE WARRENS OF VIRGINIA (1915) to Lee Garmes's celebrated "north light" effect. "Ever since I began, Rembrandt has been my favorite artist," Garmes told interviewer Charles Higham. "I've always used his technique of north light—of having my main source of light on a set always come from the north. . . . And of course I've always followed Rembrandt in my fondness for low key."[32]

By the close of the silent period, many of the cameramen most involved in the use of incandescent Mazda lighting, notably Lee Garmes and George Barnes, had eliminated much of the diffusion from their work, but Ernest Palmer, Karl Struss, and Oliver Marsh still continued to make heavy use of the more pictorial style.

The most dramatic outside influence in silent Hollywood was certainly the importation, after 1925, of an entire generation of German filmmakers. Karl Struss, one of the American cameramen on F. W. Murnau's SUNRISE (1927), a film with a large number of Germans on the design team, saw the picture as "the fore-runner of a new type of picture-play in which thought is expressed pictorially instead of by titles."[33] Such cameramen as Gilbert Warrenton (THE CAT AND THE CANARY, 1927), Ernest Palmer (THE FOUR DEVILS, 1928), and Hal Mohr (THE LAST WARNING, 1929) were especially involved in this style which featured complicated moving camera effects, frequent use of trick shots and superimposition, and stylized, low-key lighting schemes. But the death of the most important German technicians (notably Murnau and Leni), photographic problems attendant on the introduction of sound, and a general misuse of the techniques involved muted the impact of this movement after 1929. Looking back at the period, Hal Mohr, who photographed some exceptionally "Germanic" late silent films for Michael Curtiz, Paul Fejos, and Paul Leni, remembered it simply as "the era of the goofy ideas in film":

> I'll never forget one thing we did on BROADWAY [1929; an early talkie that proved to be one of the last great examples of the style]. We had this camera swinging around during one of the musical numbers, just rotating, swinging around the camera, photographing everyone on those sets all at one time like a big merry-go-round type of thing. If you make them dizzy enough they'll think it's a great scene, you know.
>
> (quoted in Richard Koszarski, "Moving Pictures," *Film Comment*, September–October 1974, p. 48)

Performance

One special problem of silent-screen acting was the need to accommodate the over-speeding of projection, which was the inevitable fate of every projected "performance." Milton Sills acknowledged this in the entry on "Motion Picture Acting" he prepared for the fourteenth edition of the *Encyclopaedia Britannica*. Admitting that performances recorded at 60 feet per minute were typically projected at 90 feet per minute, Sills felt it "necessary for the actor to adopt a more deliberate *tempo* than that of the stage or real life. He must learn to time his actions in accordance with the requirements of the camera, making it neither too fast nor too slow."[34]

Controlling this tempo was ultimately the job of the director, who would elicit a performance in one of two very different ways. As Gloria Swanson put it:

> There's the director who allows the actor to give his own interpretation of a part, and becomes a conductor of an orchestra, moving it down or bringing it

up. Then there's the other kind where the director is a thwarted actor, a ham as we call him, who wants to show the actor how to do it.

(quoted in Rui Nogueira, "I Am *Not* Going to Write My Memoirs," *Sight and Sound*, Spring 1969, p. 59)

Here, Swanson was praising her longtime collaborator Cecil B. DeMille, who would respond to actors' questions with an abrupt "I'm not running an acting school!" This approach was very much the opposite of another of Swanson's directors, Erich von Stroheim, who was notorious for indicating exactly how he wanted every gesture delivered. Von Stroheim would keep filming retakes until he saw the performance he had conceived in his mind's eye, a technique Swanson accuses Chaplin of employing as well (and with some justification, given the evidence provided in the Kevin Brownlow–David Gill Thames Television series *The Unknown Chaplin*). As for D. W. Griffith, Swanson claimed:

> You could always tell when an actor had been working with Griffith, because they all had the same gestures. All of them. They'd cower like mice when they were frightened, they'd shut not their fingers but their fists, they'd turn down their mouths.... Lillian Gish, Dorothy Gish, all of them. Even the men had a stamp on them (Nogueira, p. 59).

The onscreen evidence indicates that von Stroheim, Chaplin, and Griffith were able to use this Svengali approach to good effect. Unfortunately, far less talented directors often employed equally intrusive techniques:

> Well do I remember watching J. Searle Dawley direct Pearl White in an intensely dramatic scene, in which he played all the parts before the rehearsal was over.... Mr. Dawley did everything. He reclined on the floor, as Miss White was to do, and leaned back in the villain's arms. He played the villain, and snatched her to him despite her struggles. And then, as the big red blooded hero, he burst into the room and hurled the villain back against the wall so forcibly that it shook.
>
> (Inez and Helen Klumph, *Screen Acting: Its Requirements and Rewards* [New York: Falk, 1922], pp. 185–86)

By the late 1920s this technique had generally fallen out of favor, although some, including Ernst Lubitsch, continued to employ it successfully. Raoul Walsh, another of Swanson's directors, summed up the case for the DeMille approach:

> I should like to know how the silent drama is to develop great talent if the practice of curbing the players is adhered to. Granted the director must play an important part, he must supervise the players and see that they are getting the right stuff into the scene. There should, however, be a happy medium, as overdirection causes players to be self-conscious, mechanical and as colorless as dolls or marionettes on a string.
>
> ("Spontaneity in Acting," *New York Times*, 27 April 1924)

Editing

As soon as filming had started, the director and editor would screen the footage processed each day ("the dailies"), make initial decisions on the best takes, and begin assembling the work print. This would be relatively easy for scenes where action was limited and only a few characters appeared, but complicated sequences presented far greater problems. Because editing was acknowledged as crucial by everyone from Griffith to Sennett, it was considered important for the director to establish an intimate working relationship with his cutter. Griffith worked for many years with Jimmy Smith, and Rex Ingram had his most successful pictures cut by Grant Whytock.

Ingram would film spectacular scenes for THE FOUR HORSEMEN OF THE APOCALYPSE with twelve or fourteen cameras grinding simultaneously. It was impossible for him to analyze so much footage each night, so he depended on Whytock to make an initial selection—something he did by ignoring all but two or three of the master cameras! The editor also assembled another original negative for foreign markets. Shots were recorded twice, either by adjacent cameras or through an additional take. Kodak duplicating stock was not available until 1926, and previous attempts at producing foreign negatives by duping the original resulted in such poor print quality that overseas audiences complained. For THE FOUR HORSEMEN OF THE APOCALYPSE, a third "original" negative was assembled by Whytock after heavy print demand wore out the primary domestic negative. Produced from third-best takes, this version lacked a few key shots and sequences, and Whytock considered himself lucky that "it all made sense."[35]

When a film was completed, it was previewed at a local theater to gauge public reaction. Harold Lloyd, one of the most aggressive users of the preview system, remembered:

> Even back in the one-reel days, I would take a picture out to a theater when I knew the picture wasn't right. And the manager used to always have to come out and explain what was going on. When we were doing two-reelers, he came out in white tie and tails to do it, and it was quite an event for him and the audience would listen attentively.
>
> ("The Serious Business of Being Funny," *Film Comment*, Fall 1969, p. 47)

By the time he moved to features, Lloyd would take "scientific" readings of audience laughter and plot them on large graphs, using the information to refine the way his comedies played.

After the first preview of THE PHANTOM OF THE OPERA, a decision was made to put some comic relief into the film, and a considerable amount of footage with Chester Conklin was inserted. Later preview screenings led Universal to cut Conklin out of the picture entirely.[36] For modern audiences, Conklin is back in again, courtesy of a 1930 reissue and some questionable "film restoration."

Usually, of course, the preview process was far less traumatic, just another step in the production and polishing of any film. Henry King's rough cut of STELLA DALLAS (1925) was 26,000 feet, which he trimmed to 18,000 before previewing it in San Bernardino. Although nobody walked out on this 3½-hour cut, he trimmed another 3,000 feet before running it in Pasadena. At the screening of this fifteen-reel version, some 800 postcards were distributed asking for suggestions of which 420 were returned. It was discovered that Stella's dialogue titles were "resented by the public," so these were all done over. After this version was previewed in another small town, two weak sequences were excised, and the film went into general release at 10,157 feet.[37]

The director most taken with the preview idea was probably D. W. Griffith, who considered it an extension of the theatrical out-of-town tryout. Even the opening-night version was just another cut to Griffith, who continued refining his work well after the initial public showings. "As was his habit," Eileen Bowser tell us, "D. W. Griffith accompanied *Intolerance* on its first runs in the major cities, cutting the prints at the theater, striving to improve it. The result was that the print shown in Boston was not necessarily the same as that shown in New York, and it may be that neither was matched exactly by the original negative.[38]

Cameraman Hal Sintzenich's diaries show that Griffith not only recut his films after opening night but often continued shooting new footage as well. Notes Erik Barnouw:

> On February 10, 1924, Snitch mentions a [preview] showing of *America* in South Norwalk, Connecticut. On the same day he is shooting scenes for its Valley Forge sequence in nearby Westchester, impelled by the arrival of ideal blizzard weather for "the men in bare feet in the snow trying to pull the big wagon" (2/10/24). The Battle of Princeton is shot the following day, in time to be included four days later in a showing in Danbury, Connecticut, when Snitch is shooting Washington's inauguration. Two days later they are remaking "the stockade scenes." On the day of the New York premiere, February 21, Snitch is still doing close-ups of Caro Dempster, and during the following week does retakes of "Miss Dempster & Hamilton" and new "stockade scenes with 30 extras."
>
> ("The Sintzenich Diaries," p. 326)

Years later, at the Museum of Modern Art, Griffith was still eager to recut his pictures, running the now-antique prints in the Film Library's screening room, never satisfied that he had achieved a definitive version.[39]

The implications of such behavior for film scholarship are considerable. What is the authentic version of such a work? Griffith's incessant adding and subtracting of footage implies that he saw these films as essentially open texts, capable of showing one face to Boston and another to New York. Yet he copyrighted the montage of THE BIRTH OF A NATION and INTOLERANCE by submitting a frame from each shot to the United States Copyright Office. Did he see those versions as definitive? And if so, why the inevitable "improvements" that followed?

The problem goes beyond Griffith. By the late silent period, exhibitors could choose alternate endings for a number of major films. Some audiences, viewing Garbo as Anna Karenina in Clarence Brown's LOVE (1927), saw Anna throw herself under a train. Other theaters showed Anna happily reunited with Count Vronsky. King Vidor shot seven endings for THE CROWD and apparently issued it with two. Griffith's DRUMS OF LOVE (1928) still exists with a pair of different endings, and when it is screened by archives, audiences sometimes see them one after the other. Why producers suddenly lost confidence in their ability to make such basic decisions is unclear, but if they were moved to think twice about these films, later generations should be at least as careful when asserting that any version of a silent film is definitive.[40]

Notes

1 George Blaisdell, "Mecca of the Motion Picture," *MPW* [*Moving Picture World*], 10 July 1915, p. 215; Perley Poore Sheehan, *Hollywood as a World Center* (Hollywood: Hollywood Citizens' Press, 1924), pp. 1, 102–15.

2 Kevin Brownlow and John Kobal, *Hollywood: The Pioneers* (New York: Knopf, 1979), pp. 90–107; Sklar, *Movie-Made America*, p. 68. See also Murray Ross, *Stars and Strikes* (New York: Columbia University Press, 1941).

3 *NYDM*, 24 March 1915, p. 26 (Ince; the road was probably Sunset Boulevard); Valerie von Stroheim, in an interview with the author, Los Angeles, June 1978. Actors' Equity claimed that there were only 1,200 actors in the entire state of California at the start of 1915; this number was spread among films, vaudeville, and legitimate theater (*NYDM*, 3 February 1915, p. 9).

4 The lower figure is from *MPW*, 10 July 1915, p. 216; the higher figure, *MPN* [*Motion Picture News*], 3 April 1915, p. 39.
5 "The Grand March," *Photoplay*, November 1918, pp. 86–87.
6 In addition to outside pressures, the city fathers also seem to have done their best to rid Fort Lee of filmmakers; see Rita Altomara, *Hollywood on the Palisades* (New York: Garland, 1983). "Sunless Temple of New York's Movies," *NYT* [*New York Times*], 7 November 1920 (Griffith).
7 The Mirror studio was a small, independent operation typical of many scattered around the New York area. A number of Johnny Hines comedies were shot there.
8 *ETR* [*Exhibitor's Trade Review*], 25 September 1920, p. 1822.
9 Richard Koszarski, *The Astoria Studio and Its Fabulous Films* (New York: Dover, 1983), p. 8.
10 Edward Van Zile, *That Marvel—the Movie* (New York: Putnam, 1923), p. 216; "Movies as Investments," [*Barron's*, 14 April 1924], p. 10; Lasky, *I Blow My Own Horn*, p. 195.
11 For a rare discussion, see Kathleen Karr, "Hooray for Wilkes-Barre, Saranac Lake—and Hollywood," in *The American Film Heritage* (Washington, D.C.: Acropolis, 1972), pp. 104–9. The most attention to these filmmakers can be found in locally produced documentary films such as ALL BUT FORGOTTEN: HOLMAN F. DAY, FILM MAKER, directed by Everett Foster, and WHEN YOU WORE A TULIP, about regional production in Wisconsin, directed by Steven Schaller. Florida is one of the only regional centers adequately discussed. See Richard Alan Nelson, "Florida and the American Motion Picture Industry, 1898–1930" (Ph.D. diss., Florida State University, 1980), as well as various articles by Nelson.
12 "Gossip of the Studios," *NYDM*, 3 March 1915, p. 26.
13 Gorenflot, "Thinks and Things," *Photoplay Author*, December 1914, p. 183.
14 Douglas Brown, "The Cost Elements of a Motion Picture," *SMPE* [*Transactions* [later *Journal*] *of the Society of Motion Picture Engineers*] 17 (October 1923), p. 141; "Film Rights, and What They Are Worth," *NYT*, 11 August 1920.
15 "High Price of Stories," *NYT*, 30 March 1924.
16 Bordwell, Staiger, and Thompson, *The Classical Hollywood Cinema*, p. 136.
17 "MGM Studio Tour" is available in 16 mm from the Em Gee Film Library, Reseda, Calif., which also has other such promotional reels.
18 Cf. Bordwell, Staiger, and Thompson, *The Classical Hollywood Cinema*, p. 137: "those with greater marketability (e.g., Hart) could demand certain working conditions." United Artists, of course, was entirely based on this premise.
19 "Present Production Costs Contrasted with Low Figures of the Past," *NYT*, 19 December 1926.
20 "Motion Pictures and Finance," *Barron's*, 19 May 1924, p. 5. The figures are repeated by James Spearing in "A New Phase Opens in the Film Industry," *NYT*, 3 July 1927, and by *MPN* editor William A. Johnson in *SMPE* 32 (September 1927), p. 667.
21 William S. Holman, "Cost-Accounting for the Motion-Picture Industry," *Journal of Accountancy*, December 1920, p. 420.
22 Alfred A. Cohn, "What They Really Get—NOW!" *Photoplay*, March 1916, p. 27.
23 "Star Salaries," *FDY* [*Film Daily Yearbook of Motion Pictures*] (1924), p. 299; also "The High Cost of Film Stars," *NYT*, 4 November 1923.
24 "Harold Lloyd Heads List of Huge Earnings of Stars and Directors," *NYT*, 16 May 1926.
25 Fred Balshofer and Arthur Miller, *One Reel a Week* (Los Angeles: University of California Press, 1967), pp. 130, 134. For information on this early period, see John Hambley and Patrick Downing, *The Art of Hollywood* (London: Thames Television, 1979).
26 Leon Barsacq and Elliott Stein, *Caligari's Cabinet and Other Grand Illusions* (Boston: New York Graphic Society, 1976), p. 212; "The Architecture of Motion Picture Settings," *American Architect*, 7 July 1920, p. 1; James Hood MacFarland, "Architectural Problems in Motion Picture Production," *American Architect*, 21 July 1920, p. 65, THE COPPERHEAD is preserved in the archives of the George Eastman House/International Museum of Photography, Rochester, N.Y.
27 Other teams included John Seitz and director Rex Ingram (1920–26), Oliver Marsh photographing Mae Murray (1922–25), and Charles Rosher photographing Mary Pickford (1917–27).
28 "A Half Century of Loyalty, Progress, Artistry," *AC* [*American Cinematographer*], January 1969, p. 46; "An Open Letter," *AC*, August 1926, p. 8; Hal Mohr interview.
29 *NYDM*, 3 March 1915, p. 28.
30 Stephen S. Norton, "Close-Ups," *AC*, July 1923, p. 8; Welford Beaton, "Submerging the Production Under Senseless Close-Ups," *The Film Spectator*, 12 May 1928, p. 7; idem, "Is Eric von Stroheim a Really Good Director?" *Film Spectator*, 17 March 1928, p. 8; "Now the Close-Up," *NYT*, 1 October 1922.
31 "Ideal Directors," *NYT*, 13 February 1921.

32 "Camera Expert Studies Old Masters for Effect," *NYT*, 12 June 1927; Cecil B. DeMille, "Motion Picture Directing," *SMPE* 34 (April 1928), p. 295; Higham, *Hollywood Cameramen*, p. 35.

33 John Baxter, *The Hollywood Exiles* (New York: Taplinger, 1976), pp. 19–53; Karl Struss, "Dramatic Cinematography," *SMPE* 34 (April 1928), p. 317.

34 *The Theatre and Motion Pictures*, Britannica Booklet no. 7 (New York: Encyclopaedia Britannica, 1933), p. 29.

35 Whytock—interview with author; "Why Some American Films Prove Failures in England," *NYT*, 27 April 1924.

36 Robert E. Sherwood, "The Phantom Jinx," *Photoplay*, January 1926, p. 113.

37 "Mr. Goldwyn Describes Try-Outs of Pictures," *NYT*, 25 October 1925.

38 Eileen Bowser, *"Intolerance": The Film by David Wark Griffith. Shot by Shot Analysis* (New York: Museum of Modern Art, 1966), p. iii.

39 Robert M. Henderson, *D. W. Griffith: His Life and Work* (New York: Oxford University Press, 1972), p.288. At the Cinémathèque Française a decade later, Erich von Stroheim not only recut THE WEDDING MARCH but inserted stock footage from MERRY-GO-ROUND which he preferred to that used in THE WEDDING MARCH. This synchronized print is now shown as the standard version (Koszarski, *The Man You Loved to Hate*, pp 194–95).

40 *The America Film Institute Catalogue: Feature Films* 1921–30 (New York: Bowker, 1971), p. 454 (LOVE). King Vidor, *A Tree Is a Tree* (New York: Harcourt, Brace, 1953), p. 152. newpage

Chapter 4

Kristin Thompson

THE LIMITS OF EXPERIMENTATION IN HOLLYWOOD

TRADITIONALLY, HISTORIANS HAVE CREATED an opposition between the commercial Hollywood cinema and the American avant-garde. One goal of the International Congress of Independent Cinema held at La Sarraz, Switzerland, in 1929 was to define the "independent" cinema, a term that was preferred to "avant-garde" or "experimental." The implication was that independent filmmakers worked outside commercial cinema industries. By the 1940s, Lewis Jacobs found the independently made shorts of Paul Strand, Robert Flaherty, Robert Florey, Charles Klein, and James Sibley Watson and Melville Webber to exemplify a cinema solely concerned with "motion pictures as a medium of artistic expression."[1]

The distinction between commercial cinema and the avant-garde is, in general, valid. Certainly many separate institutions have grown up to support avant-garde filmmaking, and virtually any spectator would perceive a vast difference between, say, Watson and Webber's *The Fall of the House of Usher* and King Vidor's *The Big Parade* (1925). Other films, however, escape easy categorization. Paul Fejos' experimental, feature-length *The Last Moment*, for example, was produced independently of the Hollywood studio system. Critical praise brought the film to the attention of Charles Chaplin, who arranged its premiere; its success on that occasion led to its being picked up for distribution by Zakoro Films.[2] Similarly, Florey's shorts had commercial distribution and played in regular theaters as well as art houses.[3]

Beyond such tangential commercial links between a few avant-garde films and the mainstream, a small number of silent films made by the Hollywood studios themselves were perceived as unusual by critics, audiences, or exhibitors. They might be described as highbrow, arty, obscure, box-office poison, novel, or technically sophisticated, depending on the source.

Most of these films are not "experimental" by the standards of the avant-garde films discussed elsewhere [. . .], but occupy a fuzzy middle ground between commercial and experimental work. By examining their unusual traits, we can get a better sense of the limits of experimentation within the classical studio system. Those limits were relatively narrow, and violation of them often caused important films to fail and major directors to leave Hollywood.

Early films of the pre-1909 "primitive" era often contain anomalous, even bizarre, devices that fascinate many modern viewers, including avant-garde filmmakers and film historians. Yet such devices usually resulted from the exploration of a new medium rather than from an effort to set up an alternative to the commercial cinema. It seems to me that it is only after the formulation of classical Hollywood norms was well advanced that we can meaningfully speak of an avant-garde alternative. The earliest set of films I have been able to identify as overturning established norms in some way came in 1916, a point at which the classical cinema was approaching its complete formulation.

What counted as avant-garde for Hollywood between 1916 and 1930? In order to investigate this question, I have chosen to survey a number of films that were initially perceived as unusual.[4] These choices are based primarily upon a survey of trade journals of the period and on viewings of surviving films that these journals identified as exceptional. Some of these films were produced by major studios like Famous Players–Lasky. Others were made on a shoestring with private money, and some of these were distributed to regular theaters.

The films also present a wide range of stylistic traits. Some, like *The Last Moment*, which borrowed heavily from contemporary European avant-garde movements, were highly experimental by Hollywood's standards. Others, like *The Old Swimmin' Hole* (1921), a Charles Ray film that used no intertitles, would probably be perceived as close to a normal classical film by modern viewers. This range of "experiment" across the films should give us a sense of the possibilities that existed in mainstream American cinema. I cannot discuss each film in detail, so I shall analyze a few and refer to others briefly.

How can we compare the types of experimentation in films so different from each other? David Bordwell offers one solution in *The Classical Hollywood Cinema*, where he distinguishes among three levels of Hollywood's style: *devices*, *systems*, and *relations among systems.* Devices are isolated techniques, such as editing or camera movement. Clearly films may experiment with these, as when *The Old Swimmin' Hole* eliminates intertitles. The next level up, systems, consists of the functions that individual devices perform and the relations among the devices and their functions. Fiction films, the type we are dealing with here, can be considered to have three basic systems: narrative logic, cinematic time, and cinematic space. Devices are motivated by their relationships to one or more of these systems. One device may replace another, as long as the function remains virtually the same; for example, a track-back beginning on a close framing may establish a room's space as well as a static long shot would. The third level, as noted, involves the relations among systems—here, the relationships between narrative logic, cinematic time, and cinematic space.

In *The Classical Hollywood Cinema*, Bordwell argues that in the classical style, the spatial and temporal systems remain subordinate to the narrative logic; that is, individual devices that function to create time and space should also aid in making the ongoing action clear. At the same time, the narrative logic is likely to make unusual individual devices serve familiar functions, as when disjunctive cutting conveys a character's perception of an unsettling event.[5] The systematic quality of the classical cinema allows Hollywood to experiment in a limited way with new techniques and functions, and to assimilate those which prove useful into its overall filmmaking style.

The classical style was standardized during the 1910s. Its systems and relations among systems have changed relatively little since, and even many of its devices, like shot/reverse shot, have changed only slightly. Why would the Hollywood studios want to experiment? As Janet Staiger argues, the Hollywood system built into itself a need to balance standardization and differentiation. Unless films could be distinguished from each other, it would be hard to publicize them: "Examples of appeals to *novelty* manifest themselves continually in the catalogues and later newspaper advertising. The words 'novelty' and 'innovation' become so common as to be clichés: 'one of the genuine motion picture novelties,' 'an innovation in picture making.'" Staiger points out that the need for differentiation had two effects on style and production practices: "an encouragement of the innovative work and the cyclical innovation of styles and genres." Various "innovative workers," such as John Ford, Cecil B. De Mille, and Gregg Toland, throve within the Hollywood system: "The promotion of the innovative worker who might push the boundaries of the standard was part of Hollywood's practices. Normally, furthermore, innovations were justified as 'improvements' on the standards of verisimilitude, spectacle, narrative coherence, and continuity."[6] Of course, not all innovations pleased the public, and some directors of the experimental films under discussion here had short careers in Hollywood, or, as with Josef von Sternberg, adopted a more conventional style.

Hollywood could thus, in a limited way, risk assimilating avant-garde art. During the 1920s, for example, the German cinema influenced the work of the American studios, but, as Bordwell points out, the borrowings were selective. The moving camera, bizarre camera angles, and special effects to convey subjectivity were picked up freely. "Other formal traits of Expressionist cinema—the more episodic and open-ended narrative, the entirely subjective film, or the slower tempo of story events!—were not imitated by Hollywood; the classical style took only what could extend and elaborate its principles without challenging them."[7]

Experimentation in the early classical era

Most early experimentation in Hollywood seems to have resulted from influences entering the cinema from the other arts. Modern trends in painting and theatrical design, derived in particular from Art Nouveau and Art Deco and typified by the stage designs of Edward Gordon Craig, led to mildly avant-garde mise-en-scène in a few films during the second half of the teens and the beginning of the 1920s. In some cases, such experimentation affected the films only on the level of devices; primarily those of set design and costume. In rare instances, the stylization also extended to changes in the narrative system.

An early example of modernist design in films comes with *Madame Cubist* (1916), a two-reel film starring Mary Fuller. No print is known to survive, but a contemporary synopsis makes the plot seem quite conventional. The hero bets that he can win the love of a society woman nicknamed "Madame Cubist" for her striking clothes; he does so but falls in love with her. After a tiff resulting from her discovery of the bet, the couple reconciles. According to the *Motion Picture News*, "It was eccentric in many ways, being an eccentric drama, with eccentric settings and Miss Fuller wore very eccentric and sensational gowns."[8] The *Moving Picture Weekly* ran a fashion-oriented two-page spread; photographs of the star in "cubist" gowns accompanied a brief rundown on the Armory Show and current art trends, including Futurism, Cubism, Vorticism, and Synchronism: "If you can't visit a New Art exhibition, then go to see Miss Mary Fuller in 'Madame Cubist.'" Clearly, however, the excesses of modernism had been tamed for American consumption. The article's conclusion begins:

> Perhaps it is as well for our sartorial peace that the Cubist-Futurist, and other "ist" movements, had spent its *[sic]* first force before it reached our shores. Maybe there is a disadvantage to living where the new things start, after all. We escape some of the wildest vagaries in that way.[9]

The attention accorded this short film made by a minor production company suggests that novelty had a publicity value.

Madame Cubist apparently motivates its modernist mise-en-scène by presenting a heroine interested in the latest fashion or art style. A few American directors, however, used modern style to create the entire diegetic world of their films. A remarkable 1915 Vitagraph one-reeler, *The Yellow Girl*, seems at first to motivate its stylized costumes and decors through the characters' participation in the New York art scene. The hero is a painter; his model is a dancer in an arty stage show; his jealous girlfriend works in an ultra-modern boutique where the model buys her clothes. Later, however, the characters walk through a "real" park represented by the same kind of modernist decor (fig. 4.1: the incongruous blob of white at the lower right is a live rabbit). Here the realistic "chic" motivation outruns the plot: the artist's studio and the dancer's stage performance are no more stylized than are the landscapes. The *Moving Picture World* considered the costumes and settings "a pleasing novelty": "They are inspired by the futuristic school of art and are faithful copies of the odd figures and scenes that are produced by the followers of this method."[10] Again, however, the experiment appeared in a relatively minor one-reel film. More extensive exploration in this direction soon followed from one of the era's major filmmakers.

Maurice Tourneur gained fame for experimentation, and some of his films departed from the norms of mise-en-scène that were being established in Hollywood. In 1918 he made two ambitious films that used stylized decor extensively. The sets in both *The Blue Bird* and *Prunella* displayed strong influence from simplified, fantasy-oriented stage design, and, indeed, both were adaptations of successful plays. *The Blue Bird*, Maurice Maeterlinck's allegorical fantasy, had been a phenomenal stage success in the United States. (Reviews of the film often refer to the universal familiarity of the story and characters.) Tourneur's version used spectacular special effects, and the sets often incorporated spare, almost cartoonlike backdrops typical of contemporary stage productions (fig. 4.2). It also moved away from the typical narrative pattern of the classical cinema by stressing allegory; several of the characters visible in the illustration are personified objects (sugar, milk, and so on), and the goal of the central fantasy is to find the bluebird of happiness.

Figure 4.1 Frame enlargement from *The Yellow Girl* (1915), directed by Edgar M. Keller, with Corinne Griffith and Florence Vidor

Figure 4.2 Frame enlargement from *The Blue Bird* (1918), directed by Maurice Tourneur

Reviewers loved the film for its innovativeness and pictorial beauty. The *New York Dramatic Mirror* stressed how the obscure qualities of the narrative were smoothed over by the technical innovations: "All thoughts of symbolism, however, are forgotten in recalling the colossal splendor and magnitude of the production. Famous Players–Lasky Corporation have achieved a triumph in artistic and technical effects. In fact, since seeing the screen 'Blue Bird,' we believe in magic—for only magic could have conjured such spectacles of fact and fancy."[11] The *Moving Picture World* advised exhibitors on phrases to use in publicizing the film: "Great Stage Success Exceeded in Beautiful Photoplay," "Pictorially Beautiful Reproduction of Famous Classic," and "Mammoth Photo-Dramatic Spectacle of Lavish Splendor."[12] It is interesting to note, however, that Artcraft's trade advertisements for *The Blue Bird* featured photographs of the two lead child actors, with little suggestion of the stylized sets or the oddly costumed allegorical characters.[13] It was promoted partly as a children's or family film (just as the initial Broadway run included a special 10:00 A.M. matinee for children).[14] Even so, it was not a commercial success.

Only fragments of Tourneur's next film, *Prunella*, survive. They suggest that the odd settings were not limited to a fantasy scene, as in the earlier film. The story was not allegorical; it was based upon a recently successful symbolist play by Granville Barker, and its star, Marguerite Clark, originated the title role on Broadway. The press found it an experiment in high art, "best appreciated by persons of culture and taste."[15] Indeed, this film engendered a sort of commentary that I have not encountered in earlier American film reviewing. Several writers praised the film while regretfully suggesting that its unusual qualities might harm its box-office potential. *Variety* remarked, "'Prunella' may not break any b.o. records nor call forth any great outburst of hard applause, but it is refreshingly sweet when compared with the deluge of sickly, sentimental and maudlin romantic subjects that have swept the photoplay of late."[16] While the *New York Times* opined that most people would be "delightfully entertained," its reviewer added: "Those who do not like fanciful poetry, costumes, and people of the imagination will probably be helplessly bored. But that won't be 'Prunella's' fault."[17] Such fears proved realistic, arid the film did poorly. Many exhibitors thought it too highbrow.[18]

Tourneur's main contributions to the Hollywood cinema arguably came in his less avant-garde films, such as *The Wishing Ring* (1914), *Alias Jimmy Valentine* (1915), *A Girl's Folly* (1916), and *The Last of the Mohicans* (1920). Even here his artistry was recognized as distinctive, but in ways that did not draw so much attention away from the narrative. The cinematography in several of his films is strikingly beautiful, and he has become famous for using depth shots with framing elements in silhouette in the foreground.[19] At the level of production, however, Tourneur resisted the rise of the classical studio system in Hollywood, preferring to work autonomously.[20] In 1926 a dispute with MGM concerning *The Mysterious Island* caused Tourneur to return to Europe, the first of a string of major directors who abandoned Hollywood filmmaking.[21]

During these early years of experimentation within the classical system, one film played with all three levels of film form: devices, functions, and relations among systems. D. W. Griffith's *Intolerance* (1916) uses continuity editing within scenes, but several other devices break up the clear, linear flow of events. Most notably, parallel editing mixes four stories taking place in vastly different times and places; the function of such editing is to make conceptual, rather than causal, connections among events. Another device, the periodic use of nondiegetic inserts for the "Out of the cradle endlessly rocking" motif, also disrupts the causal flow, creating a symbolic pattern that holds the whole film together conceptually. Localized symbolic touches carry through this de-emphasis of narrative primacy: the allegorical ending, the huge, barren office of the financier, the dove cart. Although it is possible to follow the individual stories, their clarity is diminished in favor of a larger conceptual goal. Hence the normal use of the spatial and temporal systems primarily to enhance the narrative system is shifted in a radical way. (Indeed, it is possible to argue that Griffith has to a considerable degree failed to pull the four plots together into a unified whole.) A few years after the film's financial failure, Griffith tried to recoup his losses by releasing the Babylonian and Modern stories as separate, more conventionally structured films. Like other directors, Griffith was able to experiment until the rise of the studio system forced him to work in a more standard fashion, and eventually he was squeezed out of filmmaking prematurely.

The early 1920s

In the early 1920s, there was still little overt experimentation in Hollywood cinema. The slight trend toward innovative design that had begun around 1916 continued, and during the second half of the decade, a mild modernism was standardized as a major option for art directors and costume designers.

Two extreme cases were *Camille* (1921) and *Salome* (1923), both produced by and starring Alla Nazimova and designed by Natacha Rambova. Both women were thoroughly acquainted with modernist styles in theater and painting. *Camille*, an adaptation of Alexandre Dumas *fils*' novel and play, employs interiors depicting sophisticated Paris. These were based primarily on Art Deco (fig. 4.3), though the opera staircase set seems to echo Hans Poelzig's expressionist Grosses Schauspielhaus in Berlin. Once Marguerite moves to the country to live a simple life with Armand, however, the design reverts to straightforward Hollywood sets. Stylization returns only late in the film, for Marguerite's feigned renunciation of Armand and her return to Paris.

Figure 4.3 Frame enlargement from *Camille* (1921), directed by Ray C. Smallwood

Camille would seem to be a case where the novelty value of the unusual design counterbalanced the perception of the film as arty, highbrow, and hence potentially unpopular. The *Moving Picture World*'s review was reserved:

> There is no denying that the Nazimova production of *Camille* is interesting. It is filled with modern symbolism, which is expressed in the settings and the acting of the star and her supporting company. There are scenes that suggest the lost souls in Dante's Inferno, and the human interest that gave the stage play its long life doesn't get much of a chance to make itself felt. In other words the story, in its present form, stands a good chance of shooting over the heads of the common mob.[22]

Camille, like *Salome*, seems to have succeeded mainly as a result of the star's popularity.[23]

Salome shifts the usual classical emphasis on the primacy of the narrative system. The stylized mise-en-scène lasts throughout the film, and both the spatial and temporal construction function as much to display the spectacle of the sets and costumes as to serve the narrative. These innovations are motivated largely by the simple and familiar story. Some of the editing functions mainly to show off portions of the single large set or the oddly costumed minor characters who stand about in exaggerated postures; Rambova based her designs on Aubrey Beardsley's drawings for the original edition of Oscar Wilde's play. The rhythm of the action is also slower than in the usual classical film. In combination with the sets and costumes, the stylized acting seeks to create an overall tone of decadence appropriate to the play, and this tone becomes at times more prominent than any narrative progression.

Rambova designed these two films in a situation that allowed her considerable control: they were independently produced by her friend Nazimova. By the same token, however, she had little impact on modernist style in Hollywood. It was at the major studios that outside influences entered the Hollywood cinema on a standardized basis. Most histories of set design credit Joseph Urban (a Ziegfeld Follies designer hired by William Randolph Hearst as art director of Cosmopolitan) with using the first realistically motivated modernist set, a Viennese Secessionist room in *Enchantment* (1921). *Couturier* Paul Iribe did Art Nouveau sets for *The Affairs of Anatole* (1921), and William Cameron Menzies was influenced by the Ballet Russe when he designed *The Thief of Bagdad* (1924). A few films of the mid-1920s contained some Art Deco sets (e.g., *Fig Leaves*, 1926; *So This Is Paris*, 1927). By decade's end, Art Deco became the dominant modern style in the Hollywood cinema, with such films as *Our Dancing Daughters* (1928) and *The Kiss* (1929).[24] Except for *The Thief of Bagdad*, which was a fantasy, these films used modernism with a realistic motivation, to suggest a fashionable style appropriate to the characters' social situations. Similarly, Expressionism was conventionalized as appropriate to the horror genre, as in *The Cat and the Canary* (1927) and *The Old Dark House* (1932). In such cases modern design functioned within the classical system, supporting the narrative.

A very different approach to experimentation came in *The Old Swmmin' Hole* (1921), a star vehicle for Charles Ray, directed by Joseph De Grasse and produced by Ray's own company. Its only innovative element was its lack of intertitles. (At the time, it was taken to be the first titleless Hollywood feature.) Here we can see a clear case of play purely on the level of devices.

One might at first think it odd that this particular technique should be suppressed. After all, few silent-film devices promote narrative clarity more strongly than does the intertitle. In fact, however, the makers of *The Old Swimmin' Hole* balanced the novel elimination of intertitles by two simple tactics. On the level of devices, they introduced diegetic written texts into the mise-en-scène whenever vital narrative information could not be conveyed without language. On the systemic level, they used an extremely simple narrative with situations that would be recognizable to most audiences. Thus all the customary devices *besides* intertitles would suffice to convey the basic narrative information.

That the elimination of the intertitles created some difficulties for the filmmakers is evidenced by the clumsiness of their attempts to introduce written texts into the action. In the first scene we see the schoolboy hero fishing from a bridge. Some other boys join him, and one writes in chalk on the bridge's wooden surface: "Last in swimmin is a sissy." The motivation for this action is insufficient: any boy throwing out such a challenge would shout it and run immediately, not linger to write it out while his companions got the jump on him. Other written texts are somewhat better motivated, as when the hero, Ezra, writes in his diary that he will be going to a picnic on Tuesday (fig. 4.4) In general, the device of eliminating intertitles remains strangely artificial, since the scenes are shot in continuity style. Characters converse in shot/reverse shot, yet we never learn what they are saying. (By way of contrast, the German titleless films of this era adjust the plot to emphasize physical action, and the characters seldom speak.) Moreover, in order to make the action clear, Ray has to play a caricature of his usual small-town-boy character.

The plot of *The Old Swimmin' Hole* is undoubtedly comprehensible, and the film as a whole remains classical. When it was released, several critics pointed out that the plot *had* to be simple and universal to make comprehension possible. The *Moving Picture World*'s reviewer pointed out, however, that "it might be termed a picture without a plot."[25] In Hollywood parlance, a "plotless" film is simply a film with a scant, rambling, or episodic plot. The film was also sometimes seen as mildly progressive. For *Variety*'s reviewer, the simplification of the plot vitiated any experimental value the lack of intertitles might have created (and note here that the author uses the term "experiment"):

> For years the film psychologists—principally those abroad—have claimed the

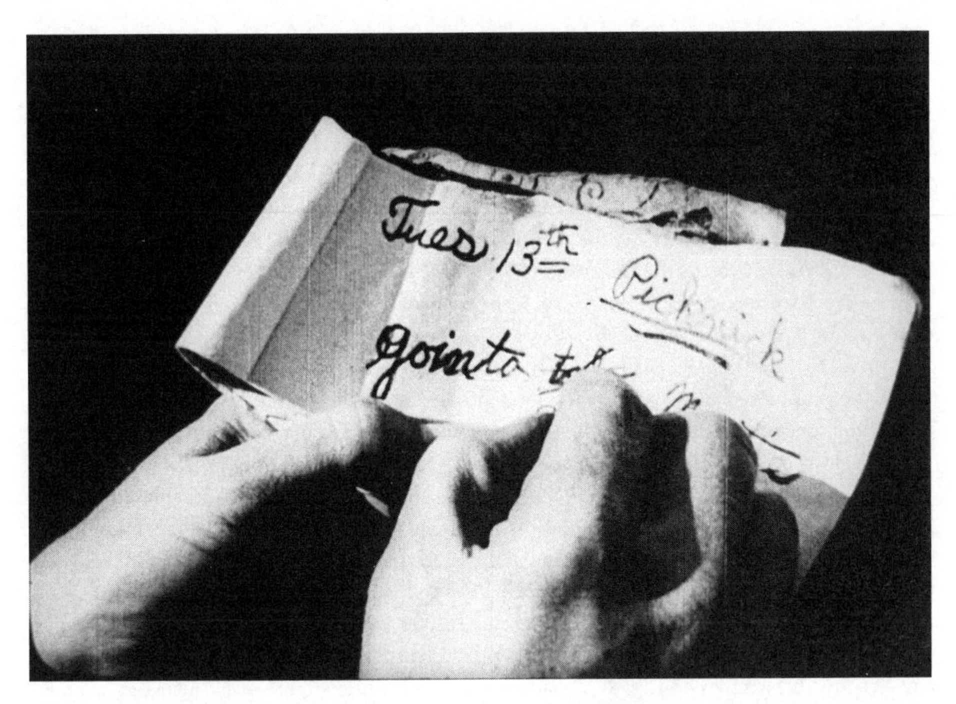

Figure 4.4 Frame enlargement from *The Old Swimmin' Hole* (1921), directed by Joseph De Grasse

> day would come when it was possible to produce a full-length feature upon the screen without the aid of subtitles. The forthcoming Charles Ray release by First National, a film adaptation of James Whitcomb Riley's poem, "The Old Swimmin' Hole," is making the experiment, and the outcome should be watched with some interest. In some respects it is hardly a fair test, inasmuch as there is little or no story to the picture, merely a series of incidents in the life of a bucolic youth. . . .
>
> The proposition, therefore, is producing pictures and showing them without the assistance of subtitles, remains unsolved *[sic]*, and if we never arrive at such a point in motion photography, we are at least moving closer and closer to it.[26]

Here Hollywood's claim to continual progress through innovation is invoked. The reviewer for the *New York Times*, however, was more enthusiastic about the film's progressive move, calling it "a unique work in kinetic photography. It is a signpost, and points a way sooner or later to be generally followed."[27] Here a description appropriate to the most avant-garde of European films is used of a classical work.

Did the lack of titles make any difference to exhibitors or audiences? It is difficult to judge, but some scraps of information suggest that it did. At about this time, the *Moving Picture World* began printing short reports from exhibitors on how various recent films had fared at their theaters. The manager of the Majestic Theatre in Greenfield, Tennessee, commended *The Old Swimmin' Hole*: "Pleases ninety-five per cent. The no-titles angle is novel. I worked the feature up with that alone," resulting in good attendance.[28] But a manager in Old Lyme, Connecticut, advised exhibitors: "Lay off this; did not please our audience. Nothing to it, no sub-titles."[29] Although *The Old Swimmin' Hole* seems to have had virtually no influence on the Hollywood cinema, its mild experimentation could generate some publicity as a novelty.

Naturalism in Hollywood

About 1924 a strain of naturalism surfaced in Hollywood cinema, spawning several films that were often regarded as stretching convention. Early cases included Griffith's *Isn't Life Wonderful?* and the trend can be seen as continuing with Erich von Stroheim's *Greed*, von Sternberg's *The Salvation Hunters*, Karl Brown's *Stark Love* (1928), William K. Howard's *White Gold* (1929), Friedrich Wilhelm Murnau's *Sunrise* (1927), Paul Fejos' *Lonesome* (1929), Vidor's *The Crowd* (1928), and Victor Seastrom's *The Wind* (1928). Tame though the subject matter of these films may seem today, several of them were seen as sordid, gruesome, or, at best, obscure by contemporary reviewers and, perhaps, by audiences.

Greed, for example, began as a project for the independent Goldwyn Company in 1922. What the completed film would have looked like if released by Goldwyn must remain a matter for speculation. Shortly after Stroheim completed a twenty-two-reel version in early 1924, Goldwyn became part of the merger that created MGM, and the film was reduced to the truncated version we see today.[30] Despite the elimination of a considerable amount of symbolism and several emblematic subplots, *Greed* did poorly upon its initial release.[31]

Greed was perhaps the most gruesome of the naturalist films, involving as it did McTeague's biting Trina's fingers (which she subsequently loses to infection), her murder at his hands, and the climactic deadly confrontation between McTeague and Marcus in Death Valley. Violence and sordidness were commonly perceived as prominent features of naturalist films of this era, and despite the happy endings that graced most such films after *Greed*, reviewers seemed squeamish when faced with subject matter so far outside the Hollywood norm. In *White Gold*, the owner of a sheep farm tries to drive his son and new daughter-in-law apart; as a result, she is nearly raped (or perhaps really raped—the situation is left ambiguous) by a farmhand whom she kills. When the father claims *he* killed the man when he found the wife committing adultery, the son believes him. Disgusted at his lack of faith, the heroine walks away to start a new life.

The Wind presents a similar plot, as the heroine travels to the West, marries a naive young farmer, and then must kill her would-be rapist. In her nightmare, the ever-present wind uncovers the body she has buried in the sand, but the film ends with the prospect of a happy life with her husband (an ending imposed on Seastrom by MGM).

Other naturalist films of the era were less violent or sordid, but they often shared a propensity toward heavy symbolism. Even in its reduced form *Greed* retained some of the characters' visions of gold, and the finished film retained the symbolism associated with the sewer and the contrast of wedding banquet with funeral procession (fig. 4.5). The very titles of several "avant-garde" films of the 1920s—*Greed*, *The Salvation Hunters*, *Sunrise*, *The Crowd*, *The Wind*—are based on elemental, emblematic concepts. Weather often provided potent symbolism. In *White Gold*, the action occurs on a sheep farm during a drought, and the failure of the young couple's marriage is indicated by the heat and lack of water. An early intertitle puts it succinctly: "Heat ... Drought ... Nerves ... Heat ..." In *The Wind,* the elements represent the heroine's neurosis, and the wind is pictured as a white horse superimposed in the sky.

Perhaps the most controversial film among this group was *The Salvation Hunters*. Certainly it was the one made with the least assistance from the Hollywood establishment. According to von Sternberg, the film cost less than five thousand dollars, and he financed it with assistance from George K. Arthur. Arthur was apparently responsible for getting Charles Chaplin, Mary Pickford, and Douglas Fairbanks to view *The Salvation Hunters*, and their enthusiasm won it a release by United Artists.

The film has gained a reputation for being slow, even static. In fact, there is plenty of action. The protagonists are a shiftless young man and a woman who take under their wing a homeless boy; the heroine is lured by a pimp into prostitution to support them. Her first

Figure 4.5 Frame enlargement from *Greed* (1925), directed by Erich von Stroheim

client takes pity on them and supplies money, and finally the pimp's abuse of the child galvanizes the hero into fighting him. The trio ends with a hopeful future.

Today the film may not seem revolutionary, yet at the time it stirred up extensive controversy. Under the subheadline "New Director Offers Symbolical Picture That Is Radically Different," the *Moving Picture World* published a revealing discussion of the film's originality:

> How is "The Salvation Hunters" different? In story, technique, acting and direction it gets away from commonly accepted standards. So radical is this departure that it is bound to provoke a variety of opinions, both pro and con, as to its artistic values, its effect on future production and its box office possibility.
>
> Just what has this new director, Josef von Sternberg, sought to do? According to the subtitles he has endeavored to present a picturization of a thought—that through unswerving faith we can eventually find the light and rise above the forces which tend to cast us down. . . .
>
> How does his treatment of this theme differ from accustomed handling? By the employment of symbolism, exceptional simplicity and straightforwardness in direction and unusual restraint in the acting of his players; by reducing his sets to the minimum, discarding the obvious as far as possible and avoiding the theatric *[sic]*.

The review then debates whether the film succeeds. On the one hand, it is seen as too sordid and as involving uninteresting characters:

> As to Mr. von Sternberg's employment of symbolism. Much has been said about this. Some of this is obvious and hammered home with sledgehammer blows by

> titles and repetition, such is the case with the dredge bringing up the mud from the darkness of the harbor bottom to the sunlight, doubly symbolical of the surroundings and the striving of the hero's struggle for the light. Also the fight which regenerates him culminates behind a real estate sign reading, "Here's where dreams come true." There are other symbolical touches, some so subtle that a great majority will not get them.[32]

The symbolism in *The Salvation Hunters* is indeed quite obvious. To be sure, many standard Hollywood films employ heavy-handed symbolism, as when the villainous German officer played by von Stroheim in *The Heart of Humanity* (1918, director Allen Hollubar) is equated with a spider by means of a cut-in to a (diegetic) arachnid. This remains an isolated moment within a largely nonsymbolic narrative, however, whereas *The Salvation Hunters* frequently employs overt symbolism, as when some game-trophy horns behind the pimp are lined up so as to render him diabolic.

The *Moving Picture World* reviewer concluded:

> While Mr. von Sternberg does not appear to have achieved his ideal goal, he has succeeded in traveling a long way toward it. Certainly he has a picture that will interest the student of the motion picture and the intelligent patron, for it will make them think and cause a diversity of opinion. So radical a departure is it that its box-office effect is problematical. Its appeal to the average theatregoer is doubtful but its very "difference," if sufficiently exploited may serve to bring them in.[33]

Despite considerable advance publicity and a record-breaking premiere at the California Theatre in San Francisco,[34] the few big-city runs recorded by *Variety* were disastrous. The reasons given for its failure in New York suggest that its naturalism disturbed some spectators: "Although this picture was lauded heavily by the daily papers it failed to pull at the box office. In some quarters it was stated this production was disgusting. The business proved that the public wants to be entertained rather than shown life in its sordidness."[35] It was not until von Sternberg mixed "sordidness" with the gangster genre in *Underworld* (1927) that he found success.

Few naturalistic films did well at the box office. Some directors made deliberately pared-down tales of typical or everyday couples: *Sunrise*, *Stark Love*, *Lonesome*, *The Crowd*. Most involved romances and happy endings. *Lonesome* was financially successful, and *The Crowd* made a modest profit. While *Sunrise* did not realize Fox's high hopes, it was not a disaster.

The "poetic" or symbolic aspirations of Hollywood naturalism in this era can be seen in several attempts to universalize a film by giving characters no names or very common ones. *The Salvation Hunters* involves The Boy, The Girl, The Child, The Man, and so on. In *Sunrise*, the triangle involves the Wife, the Man, and the "woman from the city." The *American Film Institute Catalogue* gives names for the characters in *White Gold*, but these come from the original play; they are not used in the film itself. The silent version of *City Girl* (1929) has no named characters. The central couple in *The Crowd* are simply John and Mary. This practice found its archetypal usage in the "Modern Story" of *Intolerance*, where the main characters are the Dear One, the Boy, the Musketeer of the Slums, and the Friendless One. Similarly, the filmmakers often sought to de-glamorize their characters by choosing relatively minor actors (e.g., Gibson Gowland in *Greed*, Barbara Kent and Glen Tryon in *Lonesome*), unknown aspiring actors (James Morrison in *The Crowd*), or even nonactors (the cast of *Stark Love*).

In general, naturalism or a variant of it that might be called poetic realism remained a minor trend within Hollywood, fostering films that were often considered unconventional. It is interesting to note that with the coming of sound, some directors who had worked in this vein turned to more glamorous projects (as did Sternberg); others had already been forced out of Hollywood (von Stroheim, Seastrom). Despite exceptions like *Our Daily Bread* (1930), naturalism diminished during the 1930s.

European avant-garde influences

Although a few European avant-garde films like Abel Gance's *J'Accuse* and Robert Wiene's *Das Cabinet des Dr. Caligari* reached the United States by 1921, it took some years before any direct influence became apparent. *Beggar on Horseback* (1925) was widely seen as the first major Hollywood film influenced by German Expressionism. Produced by Famous Players–Lasky, it was an adaptation of George S. Kaufman and Marc Connelly's successful 1924 play about a composer who gives music lessons while trying to write a symphony. Since he is too poor to marry, his selfless girlfriend suggests that he marry one of his pupils, Miss Cady, a *nouveau-riche* businessman's daughter. The composer falls asleep and dreams of such a marriage. In the dream, the grotesquely stylized mise-en-scène represents the hideous traits of his wife and in-laws: his sedentary mother-in-law has a rocking chair tied to her body; the women wear clothes emblazoned with dollar signs; the millionaire smokes a huge cigar and speaks into oversized phones. After the hero kills the Cadys in disgust, he is tried before a jury garbed as undertakers; when he asks that his case be taken to a higher court, the judge cranks his bench up several feet. (This gag presumably imitates the town clerk's tall stool in *Caligari*.)

Sadly, only a portion *of Beggar on Horseback*, comprising the opening and about five minutes of the crucial dream, survives. This segment, along with contemporary descriptions and photographs, allows us to draw only provisional conclusions.

The lengthy opening portion, set primarily in the hero's shabby apartment where he teaches piano, is shot in the lustrous big-studio style of the mid-1920s, with perfect continuity. As the overworked hero falls asleep, a change in the lighting signals the beginning of the dream. A stop-action effect transports him to a grotesque, though nonexpressionistic, wedding scene. During the ceremony, everyone dances to a jazz band. The attendants carry bouquets of dollar signs, and a pattern of the same signs appears on the altar. In the next scene, the honeymoon departure at a train station, the set is stylized (fig. 4.6).

These few scenes, in addition to publicity stills from the rest of the dream, suggest that this portion of *Beggar on Horseback* was grotesque and caricatural, but hardly obscure or challenging—or expressionistic. To many modern viewers, the opening sequences would seem to provide thorough preparation for the dream, and its comic qualities motivate the dream's stylized mise-en-scène. As a result one might also assume that the film would be popular (especially since the young Edward Everett Horton was widely praised for his performance as the composer).

Nevertheless, some reviewers viewed the film as European-influenced or highbrow. The *Moving Picture World* referred to its "futurist courtroom" set.[36] *Variety* praised the film's quality ("a fine production made with much intelligence and insight") but predicted dire box-office consequences ("miles over the head of picture audiences").[37] The prophecy was accurate. Since the film was distributed by Paramount, it played widely, and no "avant-garde" film was so widely commented upon in the exhibitors' column of the *Moving Picture World*. The manager of a small Iowa theater was sympathetic but complained, "It was so deep that some could not follow it. It was a money loser here. Good only for those who grasp the director's idea."[38] An exhibitor in a small town in Texas was more adamant: "I didn't get to see this one, but my people didn't fail to tell me they thought it 'a piece of cheese.'"[39] From a modern perspective, *Beggar on Horseback* would seem to be trying to balance Hollywood-style narrative motivation and stylization of individual devices. The fact that such an apparently playful film failed

Figure 4.6 Frame enlargement from *Beggar on Horseback* (1925), directed by James Cruze

suggests that by 1925 the Hollywood distribution system and publicity institutions (including reviewers) could not market such an anomalous film.

The increasing influence of European cinema over the next few years improved Hollywood's tactics for exploiting stylization. The expressionist nightclub set and vast craning camera movements in *Broadway* (1929) and the opening montage sequence in Paul Leni's *The Last Warning* (1929) exemplify devices adopted from the European avant-garde into mainstream films.

Fejos' *The Last Moment* may have been the most radically experimental American-made feature film released within the commercial American distribution network during this era. Unfortunately, no prints are known to exist, but descriptions make it sound distinctive both stylistically and narratively. Lewis Jacobs has provided us with an extensive outline of its action:

> The story was "a study in subjectivity" based on the theory that at a moment of crisis before a person loses consciousness, he may see a panorama of pictures summarizing the memories of a lifetime. The film opens with a shot of troubled water. A struggling figure and a hand reach up, "as if in entreaty." This is followed by a rapid sequence of shots: the head of a Pierrot, faces of women, flashing headlights, spinning wheels, a star shower, an explosion which climaxes in a shot of a children's book.
>
> From the book, the camera flashes back to summarize the drowning man's life. Impressions of schooldays, a fond mother, an unsympathetic father, a birthday party, reading Shakespeare, a first visit to the theatre, the boy scrawling love notes, an adolescent affair with a carnival dancer, quarrelling at home, leaving for the city, stowing away on a ship, manhandled by a drunken captain, stumbling

> into a tavern, acting to amuse a circle of revellers, reeling in drunken stupor and run over by a car, attended by a sympathetic nurse, winning a reputation as an actor, marrying, quarrelling, divorcing, gambling, acting, attending his mother's funeral, enlisting in the army, the battle-front. No attempt was made to probe into these actions but to give them as a series of narrative impressions.
>
> The concluding portions of the film were likewise told in the same impressionistic manner. The soldier returns to civilian life and resumes his acting career, falls in love with his leading lady, marries her, is informed of her accidental death, becomes distraught, finally is impelled to suicide. Wearing his Pierrot costume, the actor wades out into the lake at night.
>
> Now the camera repeats the opening summary: the troubled waters, the faces, the lights, the wheels, the star shower, the explosion. The outstretched hand gradually sinks from view. A few bubbles rise to the surface. The film ends.[40]

Contemporary reviews make it clear that the "rapid sequence of shots" that introduces the central flashback and returns at the end must have consisted of very fast montage: "The flashing summary of a lifetime first shown is mere meaningless phantasma, but when it is repeated, it does take on some significance, and the whole business, although merely a theatrical device, has a kick, even if the subject matter and mode of telling is *[sic]* morbid."[41] It would seem that Fejos was influenced by French Impressionism, with its use of rapid bursts of shots for subjective purposes. Gance had used the "life flashing before one's eyes" device in *La Roue* (1918), in the scene of Elie hanging above an abyss and recalling his life with Norma just before he falls; there the shots accelerate to a series of single-frame images. Working in Paris from 1921 to 1923, Fejos very probably saw *La Roue* and other Impressionist films.

The Last Moment was experimental in other ways. Its only intertitle appeared at the beginning, setting forth the premise of a life flashing before one's eyes in moments of crisis.[42] As with *The Old Swimmin' Hole*, this lack of intertitles meant that the narrative had to be simple, stressing obvious, recognizable points in the hero's life. In addition, *The Last Moment* apparently used all the tricks of contemporary European camerawork, including superimpositions and a rapidly moving camera.

The film, however, went beyond experimenting at the level of individual devices. Descriptions suggest that it was structured as a series of chronologically presented impressionistic scenes rather than as a linear, causally linked progression. Jacobs hints that this difference was, in my terms here, a downplaying of narrative logic in favor of a play with time and space:

> For America it was a radical departure in structure, deliberately ignoring narrative conventions of storytelling and striving for a cinematic form of narrative. Instead of subduing the camera and using the instrument solely as a recording device, the director boldly emphasized the camera's role and utilized all of its narrative devices. The significant use of dissolves, multiple exposures, irises, mobility, split screen, created a style which though indebted to the Germans, was better integrated in terms of visual movement and rhythm and overshadowed the shallowness of the film's content.[43]

Any final judgment must be reserved until a print of the film is found. It does seem, however, that *The Last Moment* may have involved experimentation even on the level of the relations among the film's formal systems.

Not surprisingly, the film was made independently. Fejos has described how, when he was trying to get work in Hollywood, he was offered a ride by a rich man's son while hitchhiking. After hearing a description of Fejos' project, the young man provided five thousand dollars, and Fejos wangled the rest on credit. Georgia Hale, a star since *The Gold Rush* (1925), agreed to defer her salary; the Du Pont company was persuaded to furnish free raw stock in return for acknowledgment at the film's beginning and end; and the Fine Arts Studio let Fejos rent bits and pieces of filming time in the standing sets from other productions. Fejos claimed that he wrote the script day by day, depending on the sets available to him. Some of the film's experimental techniques served to cover time lapses. A marriage's drift toward divorce was shown through shots of a bird cage with doves dissolving to two crows fighting, a sink with dirty dishes, and the like: "The whole film was made this way. The aim was to make it in kaleidoscope-like speed, as it might seem in actuality to the hero when he was dying."[44] Fejos' description raises the intriguing possibility that *The Last Moment* was essentially a series of montage sequences.

The finished film seems to have gone through a process similar to that which won *The Salvation Hunters* a distributor. Two prominent Hollywood critics, Welford Beaton and Tamar Lane, agreed to see it. Both wrote rave reviews and arranged for Chaplin to view it privately. After a preview at the Beverly Theatre, with Chaplin in attendance, the Zakaro Film Corporation took over the film's distribution.[45] This was an unusual way of handling a film; *Variety* remarked on the lack of a trade showing when the film had its world premiere in Boston in January 1928. Handled as an art film, it played at Symphony Hall for two dollars[46] and, apparently, did relatively well.[47]

The reviews were mostly positive. The *New York Times* commented:

> It is a film that is said to have cost $13,000, but this is in its favor, for the result is infinitely more absorbing than many pictures on which twenty times that amount has been expended. The impressionistic theme is one that brings to mind Guy de Maupassant's works, and, although the beginning and end of the narrative are far from cheerful, the events that are brought out in the meantime are so stirring and so artistically pictured that it is not depressing.[48]

The Last Moment was among the runners-up in the 1928 *Film Daily* survey of critics on the "Ten Best Pictures of 1928."[49]

Conclusions

This chapter serves mainly to confirm the longstanding belief that Hollywood has only a very limited tolerance for unconventional filmmaking. Avant-garde filmmakers may gain recognition through experiment, but if they hope to keep working within a commercial system, they must soon learn to accommodate their work, whether on the level of devices, systemic functions of devices, or relationships among systems, to the prevailing norms. In our own era we can see this same process at work, as when John Waters moves from *Pink Flamingos* to *Hairspray*, or David Lynch from *Eraserhead* to *Wild at Heart*.

Hollywood experimentation largely died out with the introduction of sound in the late 1920s. Odd camera angles, superimpositions, and nondiegetic symbolism could serve the narrative flow within the context of montage sequences. Theater programs included short films, arid while these were usually newsreels and studio cartoons, they occasionally included such items as Oskar Fischinger's abstract *An Optical Poem*, released as a regular MGM animated one-reeler in 1937. Government documentaries like Pare Lorentz's *The River* (1937) also appeared on regular theatrical bills. The presence of such films within the mainstream system should remind us that the avant-garde and the commercial cinemas are not as widely separated as we might believe. The reasons for this propinquity are not that the avant-garde can subvert the mainstream cinema. Most probably the reverse is usually true: the commercial cinema can turn *almost* anything to its own uses. In that "almost" lies the survival of the avant-garde.

Notes

I would like to thank David Bordwell, Jan-Christopher Horak, Bruce Jenkins, and Lea Jacobs for their comments on drafts of this article. Thanks also to the staffs of the Cinémathèque Royale de Belgique, the Library of Congress, the National Film Archive, and George Eastman House for help in the research.

1 Lewis Jacobs, "Avant-Garde Production in America," in *Experiment in the Film*, ed. Roger Manvell (London: Grey Walls Press, 1949), p. 113.
2 John W. Dodds, *The Several Lives of Paul Fejos: A Hungarian-American Odyssey* (New York: Wenner-Gren Foundation, 1973), pp. 33–36.
3 See Brian Taves, "Robert Florey and the Hollywood Avant-Garde," in Jan-Christopher Horak (ed.), *Lovers of Cinema: The First American Film Avant-Garde, 1919–1945* (Madison: University of Wisconsin Press, 1995), 94–117.
4 The films analyzed or mentioned include only films that were produced or distributed by a regular commercial company and that could thus be considered acceptable to at least some elements of the mainstream American film institutions.
5 David Bordwell, Janet Staiger, and Kristin Thompson, *The Classical Hollywood Cinema: Film Style and Mode of Production to 1960* (New York: Columbia University Press, 1985), p. 3.
6 Ibid., pp. 99, 109–10.
7 Ibid., pp. 70, 72–73.
8 "'Madame Cubist' Is Eccentric Play for Mary Fuller," *Motion Picture News* 13, no. 3 (22 January 1916): 377.
9 Mlle. Chic, "Mary Fuller's Eccentric Gowns," *Moving Picture Weekly*, 12 February 1916, pp. 18–19. My thanks to Richard Koszarski for giving me this source.
10 "Comments on the Films," *Moving Picture World* 29, no. 11 (9 September 1916): 1688.
11 H. O. R., "'The Blue Bird,'" *New York Dramatic Mirror*, no. 2051 (13 April 1918): 522.
12 "Advertising Aids for Busy Managers," Moving *Picture World* 36, no. 1 (6 April 1918): 134.
13 See, for example, advertisements in *New York Dramatic Mirror*, no. 2050 (6 April 1918): 2; *Moving Picture World* 35, no. 12 (23 March 1918): 1591–92.
14 "'The Blue Bird' a Hit on Screen," in *New York Times Film Reviews*, vol. 1 (New York: New York Times/ Arno, 1970), p. 34.
15 Edward Weitzel, "Prunella," *Moving Picture World*, 36, no. 10 (8 June 1918): 1472.
16 Mork, *"Prunella"* (7 June 1918), in *Variety Film Reviews*, vol. 1 (New York: Garland, 1983), n.p.
17 "Crowd Gets Thrills in 'Stolen Orders,'" in *New York Times Film Reviews*, 1: 37.
18 Jan-Christopher Horak, "Maurice Tourneur and the Rise of the Studio System," in *Sulla via di Hollywood: 1911–1920*, ed. Paolo Cherchi Usai and Lorenzo Codelli (Pordenone: Edizioni Biblioteca dell'Immagine, 1988), p. 278.
19 In 1921 Burns Mantle wrote: "Tourneur differs from most of the directors in his class in that he can achieve great beauty of background without sacrifice of story value, and while he does permit a certain repetition of his favorite shots, the views from a darkened cave through to the blazing firelight or sunlight or moonlight beyond, for example, with silhouetted figures against the light, they seldom interfere with the spectator's interest in the tale." See Burns Mande, "The Shadow Stage," *Photoplay* 19, no. 5 (April 1921), reprinted in George C. Pratt, *Spellbound in Darkness: A History of the Silent Film* (Greenwich, Conn.: New York Graphic Society, 1966), p. 281. Here we find the critic acknowledging that the director's stylistic experimentation does not deflect from the primacy of story logic.
20 Horak, "Tourneur and the Rise of the Studio System," pp. 288–90.
21 Rex Ingram moved his Metro production unit to France after the 1924 merger that formed MGM, but he continued to produce through that company for several years. MGM also caused problems for eccentric directors like von Stroheim and Seastrom.
22 Edward Weitzel, "Camille," *Moving Picture World* 25, no. 4 (24 September 1921): 446.
23 For example, *Variety*'s review of *Camille* dwelt primarily on Nazimova's performance and failed to mention the design at all. Samuel, *"Camille"* (16 September 1921), in *Variety Film Reviews*, vol. 1, n.p.
24 See Léon Barsacq, *Caligari's Cabinet and Other Grand Illusions: A History of Film Design* (New York: New American Library, 1978), p. 55; Donald Albrecht, *Designing Dreams: Modem Architecture in the Movies*

(New York: Harper & Row/Museum of Modern Art, 1986), pp. 39, 43; Mary Corliss and Carlos Clarens, "Designed for Film: The Hollywood Art Director," *Film Comment* 14, no. 3 (May–June 1978): 29; Howard Mandelbaum and Eric Myers, *Screen Deco* (New York: St. Martin's Press, 1985), pp. 7–8, 10, 48–49, 85.

25 Robert C. McElravy, "The Old Swimmin' Hole," *Moving Picture World* 49, no. 1 (5 March 1921): 44.

26 Jolo., *"Old Swimmin' Hole"* (25 February 1921), in *Variety Film Reviews*, vol. 1, n.p.

27 The same reviewer also described why the film remains thoroughly classical: "And it is entirely pictorial, cinematographic. It has no subtitles, or captions. It uses words only in a few inserts, which are integral parts of its story. Its moving pictures follow one another in logical sequence and the story comes directly from them. This is something new and important. In addition to giving pleasure to those seeking only to be entertained, this photoplay sheds a great deal of light on what may be done in pure cinematography. To this some may reply that the story in question is exceedingly simple. It may be said that it sticks so close to common human experience that its action might be autobiographical for almost any one. It may be pointed out that a story more complex in plot and characterization would not lend itself to wordless treatment as "The Old Swimmin' Hole" does! And this is all true." "The Screen," in *New York Times Film Reviews*, 1: 90.

28 "Straight from the Shoulder Reports," *Moving Picture World* 60, no. 3 (20 January 1923): 244.

29 "Straight from the Shoulder Reports," *Moving Picture World* 55, no. 9 (29 April 1921): 965.

30 Richard Koszarski, *The Man You Loved to Hate: Erich von Stroheim and Hollywood* (New York: Oxford University Press, 1983), pp. 115, 142.

31 See "Chi Takes Dive; McVicker's $19,000; Chicago, $40,300," *Variety* 78, no. 2 (25 February 1925): 26; "$20,000 at Fox's, Philly, with Big and Imposing Program," *Variety* 78, no. 2 (25 February 1925): 27; "$50 Separates L.A. Leaders; 'Devil's Cargo' at $26,550 Tops," *Variety* 78, no. 1 (18 February 1925): 26; "Straight from the Shoulder Reports," *Moving Picture World* 76, no. 4 (26 September 1925): 317.

32 C. S. Sewell, "The Salvation Hunters," *Moving Picture World* 72, no. 7 (14 February 1925): 701.

33 Ibid. *Variety* took a far less charitable approach to the film's artistic pretensions: "Word came a month or so ago from the west that 'The Salvation Hunters' was the work of a new cinema genius, that it was a revolutionary picture, that its theme was original and the treatment masterful. On top of that Charlie Chaplin, Doug. Fairbanks and Mary Pickford endorsed it as being the ultra-ultra in picture production. Applesauce. 'The Salvation Hunters' is nothing more nor less than another short cast picture to express an apparently Teutonic theory of fatalism." Sisk, *"The Salvation Hunters"* (4 February 1925), in *Variety Film Reviews*, vol. 2 (New York: Garland, 1983), n.p. The term "short cast" here apparently means that the film contained no stars.

34 "'Salvation Hunters' Has Big Premiere at the California," *Moving Picture World* 72, no. 7 (14 February 1925): 708.

35 "Super-Specials Begin Spring Battle; 'Salvation Hunters' Flop at Strand," *Variety* 77, no. 13 (11 February 1925): 27.

36 C. S. Sewell, *"Beggar on Horseback," Moving Picture World* 74, no. 8 (20 June 1925): 859.

37 *"Beggar on Horseback"* (10 June 1925), in *Variety Film Reviews*, vol. 2, n.p.

38 "Straight from the Shoulder Reports," *Moving Picture World* 76, no. 4 (26 September 1925): 318.

39 "Straight from the Shoulder Reports," *Moving Picture World* 76, no. 6 (10 October 1925): 504.

40 Jacobs, "Avant-Garde Production in America," pp. 119–20.

41 Rush, *"Last Moment"* (14 March 1928), in *Variety Film Reviews*, vol. 3 (New York: Garland, 1983), n.p.

42 "Exceptional Photoplays: *The Last Moment,*" *National Board of Review Magazine* 3, no. 2 (February 1928), reprinted in *Spellbound in Darkness*, p. 502.

43 *Spellbound in Darkness*, p. 502.

44 Information from a 1962 interview with Fejos, quoted in Dodds, *Several Lives of Paul Fejos*, pp. 28–33.

45 Ibid., pp. 33–38.

46 "Foreign *[sic]* $2 Film Starts in Boston," *Variety* 90, no. 1 (18 January 1928): 4.

47 Jacobs states, "Exhibited in many theatres throughout the country, *The Last Moment* aroused more widespread critical attention than any American picture of the year." Jacobs, "Avant-Garde Production in America," pp. 120–21.

48 Mordaunt Hall, "The Screen: An Impressionistic Study," in *New York Times Film Reviews*, 1: 430.

49 "Ten Best Pictures of 1928," in *Film Daily 1929 Year Book* (New York: John W. Alicoate, 1929), p. 9.

Chapter 5

Karen Ward Mahar

"DOING A 'MAN'S WORK'": THE RISE OF THE STUDIO SYSTEM AND THE REMASCULINIZATION OF FILMMAKING

IN JULY OF 1923 *PHOTOPLAY* PROFILED Grace Haskins, "Girl Producer." At the age of twenty-two Haskins earned her moniker by writing, directing, and producing her first film, *Just Like a Woman.* In the space of five years Haskins moved from working in a Hollywood hotel, to answering fan mail, to "talking herself into a job in the cutting room," and then becoming a continuity writer. All the while her ultimate aim was to direct, but "she knew enough of the game to know that no producer was ever going to give her her chance. Not for a long, long time, anyway." Not one to give up, Haskins turned to "several moneyed men," "dusted off" a scenario she was working on, and secured a deal with independent distributor W. W. Hodkinson. With check in hand Haskins collected a company of actors, found a ready-made set, and began making her film. Suddenly "people who had assured her that she couldn't, possibly, hope for success, began to take an unwelcome interest in the proceedings." Damaging rumors surfaced, but Haskins persevered. *Just Like a Woman* reached the screen on March 18, 1923.[1]

Haskins intended "to keep right on producing," but like most small independent filmmakers in 1923, she vanished. Unlike many male directors, however, Haskins did not resurface in any of the major, or even the minor, studios. But this scenario in itself could not account for the disappearance of female directors in the 1920s, for the new "majors" hired plenty of experienced filmmakers to grind out the features and programmers that fed their theater chains.[2] Rather, something changed in the very definition of a filmmaker. Although industry writers praised women directors for their "deft touches" and "finesse" as late as 1921, by 1928, the year that the sound film triumphed over the silent film, there was only one working female director, Dorothy Arzner, a filmmaker noted for her masculine persona and approach. By 1928 filmmaking in Hollywood was unquestionably "man's work."[3] Although Haskins could not have guessed it, she was one of the last female filmmakers of her era.[4]

In many ways Hollywood epitomized modern heterosociability. Mixed-sex groups could be found after a day's shooting at late-night "watering holes" like the Green Room, where in 1913 one might find Jack Holt, Warren Kerrigan, Bob Leonard, Ella Hall, Francis Ford, Grace Cunard, Otis Turner, and Cleo Madison. And writers, split evenly between men and women, formed the heterosocial Screen Writer's Guild and its social arm, "The Writer's Club," in

1920. It boasted a membership roster that was 25 percent female and a clubhouse with a pool and spaces for dances, billiards, and cards.[5] But by and large professionalization segregated the geography of Hollywood by sex. The industry's first trade associations modeled themselves on fraternal societies. In 1915, when the Los Angeles equivalent of the Screen Club moved into "one of the most popular clubhouses on the West Coast," men who belonged to the club could choose from an English bar, a billiard or a pool room, a lounging room, and a dining hall known as "the Stein Room," which held a "weekly good-fellowship dinner, with toasts, vaudeville acts, and music." "Ladies' night" occurred one evening a month, when the club held a tango dinner.[6] By 1917, as filmmakers migrated to Hollywood, the Los Angeles Athletic Club surpassed the Screen Club in popularity. According to *Photoplay* the Athletic Club was "the capital of the screen rialto—the Lambs, Players, and Friars rolled into one."[7]

Such organizations were not new. During the last third of the nineteenth century, while middle-class work was still almost exclusively male, clubs and lodges rose to supplement taverns and restaurants as places for masculine recreation and business dealings. But by the early twentieth century, as women entered the world of middle-class work, these all-male organizations took on added significance. Since work itself was "no longer a male club," the spittoon migrated to fraternal organizations. Forced to be genteel at home and at work, middle-class men retired to their clubs, where they might drink, smoke, swear, and revitalize themselves by indulging in the rituals of manhood. But the sexual exclusivity of the Screen Club and similar organizations was not just to insulate men from the encroachment of women in the workplace. The project of professionalization required masculinization. Men needed common ground to put aside their differences and band together for the good of their field. "The exclusion of women," argues Anthony Rotundo, "linked the bitterest of rivals in the solidarity of male professions."[8] Not surprisingly, then, when the Motion Picture Directors' Association (MPDA) was founded in February of 1915, it was described as "a fraternal order." As such, it claimed that "its rituals render impossible the idea of coercion and eliminates [*sic*] any element of partiality or unfairness."[9] It is true that in 1916 the MPDA admitted Lois Weber as an honorary member, but the "rules of the organization" had to be "set aside for this purpose," and just to be certain, the directors added that "no other of the gentler sex will be admitted to membership."[10] Although the MPDA defied its own order in 1923, adding director Ida May Park as an honorary member, it also added actress Lottie Pickford (Mary's sister), theatrical impresario Daniel Frohman, and playwright Augustus Thomas, indicating that the definition of honorary membership meant recognition for good work; it did not confer professional rank as a film director. Thus near the height of their numerical strength, female directors were excluded from the MPDA, which policed its professional neutrality with the rituals of a male lodge.[11]

The rise of these clubs and organizations denied women filmmakers valuable contacts. The list of Athletic Club members in 1917, for example, was a male who's who of Hollywood: Charlie Chaplin, producers Oliver Morosco and William Selig, director Al Christie, and noted actors Ford Sterling, Fred Mace, Tyrone Power, and William and Dustin Farnam. Ironically, a member of the Athletic Club looking for investors to back a new film project might find himself at a gathering of the "The Uplifters," which no longer referred to reformers but to an internal organization of film industry magnates, "democratic millionaires, jurists, doctors, and real estate impresarios." "Every so often they get together in one of the period dining rooms of the club," claimed *Photoplay*, "and lift their voices in song and their arms in—well, we might call it homage, to the spirit of good fellowship."[12] Good fellowship that was by definition off-limits to women.

Despite the inherently masculine project of professionalization, women still found opportunities behind the camera as a result of simple pragmatism. Thanks to "doubling in brass," actresses were often well-versed in various duties behind the camera, and it was practical and cost-effective to use all able-bodied employees to their fullest capacity. Thus, at first the reorganization of film production along more efficient lines did not exclude women from behind the camera.

The nascent efficiency movement was not innately masculine. We know that scientific manager Wilbert Melville began his career at Alice Guy Blaché's Solax studio in 1911. In 1913 he visited the Western Lubin plant (which was laid out like "a well-planned kitchen") and introduced to the industry the principle of cost accounting and a new kind of scenario department, where "the scripts are prepared for the directors in such shape that they can be produced as written." "Sooner or later," claimed *Moving Picture World*, "other motion picture manufacturers throughout the world must follow" these innovations "if they wish to meet competition and survive."[13]

According to Janet Staiger, the new sort of script, known as a continuity script, laid the foundation for the central-producer system. It was this system, which first arose in the mid-1910s, that led to the dissolution of the impromptu collaborative style of filmmaking that allowed so many women to gain experience behind the camera.

At first the rise of the continuity script boosted the status of women on the lot. Continuities emerged from the scenario department, and writing was the least sex-typed of all studio crafts. According to Lizzie Francke half the scenarios produced in the silent era were written by women, and female scenario department heads were common.[14] Indeed, many women writers in the 1910s literally defined the craft. Catherine Carr's *The Art of Photoplay Writing* (1914), Marguerite Bertsch's *How to Write for Moving Pictures* (1917), and Anita Loos and John Emerson's *How to Write Photoplays* (1920) were all handbooks on the special requirements of writing for the movies.[15]

Thus, as the continuity script was developed, women were well-positioned to become these new technical experts.[16] Continuity writers became the "brains" of the production process. Creating continuities based on budget, available properties, and available personnel, the continuity writer assumed some of the previous duties of the film producer.[17] The director now worked hand in hand with the continuity writer on the development of the story before a single scene was shot. Indeed, in some studios the director was not allowed to make any changes once the final continuity was approved. He or she had to shoot the film as written.[18]

The role of the continuity script gave writers, many of them female, a new creative authority behind the camera. Many writers took advantage of their new authority by becoming directors themselves.[19] As continuity writer Marguerite Bertsch remarked in 1916 when she was asked how it felt to direct for the first time: "You know I never wrote a picture that I did not mentally direct. Every situation was as clear in my mind as though the film was already photographed."[20] From the point of view of continuity writers the shift from "mentally directing" to physically directing was a natural one.

Even women who were not continuity writers appeared to benefit from the early implementation of efficiency methods. Regular scenario writers, those who originated the stories but not the continuities, began to direct more frequently in the mid-1910s as the script in general increased in importance. Writer-actress Nell Shipman's initiation as a Universal director is illustrative. According to Shipman's autobiography, when the director and leading lady ran off together while she was working on location at Lake Tahoe in 1914, "the Universal star [Jack Kerrigan] said I must take up the megaphone. 'You wrote this mish-mash,' he said, 'so you can direct it.'"[21] At about the same time, Universal actress-writer Jeanie MacPherson was said to have won her chance to direct by "pestering" Laemmle when a film she wrote, *The Tarantula*, had to be reshot after the print was accidentally destroyed and the original director was no longer available.[22]

In addition to Nell Shipman and Jeanie MacPherson, Universal writers Ruth Ann Baldwin, Ida May Park, and E. Magnus Ingleton became directors between 1916 and 1917.[23] At least four women directed at Vitagraph after 1916: Marguerite Bertsch, Lucille McVey Drew, Lillian Chester, and Paula Blackton.[24] At other studios, too—large and small—women with writing experience became directors after the introduction of the continuity system, whether they wrote continuities or scenarios. Scenarist Julia Crawford Ivers directed *The Majesty of the Law* (Bosworth, 1915), *The Call of the Cumberland* (Pallas, 1916), *The Son of Erin* (Pallas, 1916), and *The White Flower* (FPL/Paramount, 1923).[25] Writer Frances Marion directed *Just Around the Corner* (Cosmopolitan, 1921) and a Mary Pickford vehicle, *The Love Light* (United Artists, 1920).[26] In the mid-1920s scenarist Lillian Ducey directed *Enemies of Children* (Fisher Productions, 1924); writer Dot Farley directed comedy shorts for Mack Sennett; scenario editor Miriam Meredith worked as an assistant director at the Thomas H. Ince studios; and Elizabeth Pickett, a Wellesley graduate, wrote and directed shorts for Fox, becoming West Coast supervisor for the Fox Variety series.[27]

The implementation of efficiency measures did not initially prevent actresses from becoming directors either. Economy encouraged Carl Laemmle of Universal to allow actresses to begin directing in the mid-1910s. Ultimately, Universal employed more female directors than all other studios combined. The first actress to become a director while at Universal was Cleo Madison, who, according to *Photoplay*, cajoled studio manager Isadore Bernstein into allowing her to direct in 1915. Madison had worked in production for the legitimate stage prior to joining the film industry. She directed a few two-reelers for Universal, and in 1916 she directed two five-reel features.[28] In 1917 three more actresses became Universal directors: Ruth Stonehouse, Lule Warrenton, and Elsie Jane Wilson.[29]

One reason Universal hired so many women to direct was because its pictures were relatively low-cost, low-risk ventures.[30] "By the late teens," observed Richard Koszarski, "Universal was known as a giant factory where work was easily attainable but working conditions (especially salary) remained substandard."[31] But despite the growing tendency to entrust women with only low-budget films, in the context of an open market Universal's female directors gained enough experience to join the growing legion of independent filmmakers. Most had been employed by Universal as writers or actresses for years but left within months after becoming directors to try their luck as independents. Ruth Stonehouse, for example, directed for less than a year at Universal before she signed a contract with the Overland Film Co. to produce six features a year for states' rights release.[32] Like most new independents these companies typically failed. Cleo Madison, who picked up the megaphone to much fanfare in 1915, could not get her own company off the ground in 1917.[33] Lule Warrenton had a single critical success under her own company in 1917 but then returned to the Universal fold and gave up directing entirely.[34] Ida May Park was more successful. After directing for three years at Universal, she created Ida May Park Productions in 1920. She released two features, including the well-reviewed *Butterfly Man* (1920), but her company quickly faded.[35]

Other women directors joined the independent movement as well. Margery Wilson, a Griffith alumnus, directed short comedies and one feature, *That Something* (1921), for Margery Wilson Productions but then lost her company to her lender.[36] Vera McCord directed one film, the poorly received *Good-Bad Wife* (1921), for states' rights release before she too lost her company in a dispute with her lender.[37] In 1921 writer Marion Fairfax wrote and directed *The Lying Truth*, "a story of the newspaper world." Although trade papers announced that she was preparing another film, *The Lying Truth* was her only production.[38] A company formed by Lillian and George Randolph Chester, writer and director, also came and went in 1921.[39] The Cathrine Curtis Corporation, first announced in 1919, appeared to operate along similar lines. Curtis, a "society girl" who starred in one film, described herself as a "screen interpreter." Her company, which ran an impressive advertisement listing officers, a board of directors, and counsel, appeared to act as a literary agency as well. Curtis's first and only production, *The Sky Pilot* (1921), was an expensive "super-special" starring Colleen Moore and directed by King Vidor for First National. After *The Sky Pilot*, which was a success, the company vanished.[40] All of these companies were probably victims of the 1921 recession that forced even the largest studios to suspend production.

Some companies created by women writers after 1916 were not production companies per se but freelance writing services. The Eve Unsell Photoplay Staff, Inc., offered "everything from script to screen": continuities, synopses, "opinions and revisions," subtitling and editing, and representation of authors and publishers. Although literary agents abounded by 1921 (many of them women), Unsell hoped to bring filmmakers and authors to a new level of mutual understanding. When officially unveiled, Unsell's company already held contracts with Famous Players–Lasky and star Katharine McDonald's First National company. Also an apparent victim of the 1921 recession, Unsell's company disappeared by June of that year.[41]

The initial implementation of efficiency measures, then, most evident in the use of the continuity script, opened up new opportunities for women to direct and produce movies in the 1910s. Indeed, the feminization of filmmaking may well have seemed imminent to some observers, for the number of women directors more than doubled between 1915 and 1919. But even as women were becoming directors and producers in greater numbers than ever before, efficiency measures took on new meaning as film costs skyrocketed after 1916. Although business was booming and well-heeled patrons were attending the movies, studios were experiencing some difficulty generating enough internal financing to cover the budgets of their most expensive stars and feature productions. What the industry needed to continue to grow were reliable sources of outside capital. Thus far nearly all banking interests scorned film producers as fly-by-night operators. The one exception was the Bank of Italy, a California bank founded by two immigrants, A. P. and Attilio (the "Doc") Giannini, that looked favorably on small, struggling businessmen. The Gianninis made several successful loans to exhibitors during the nickelodeon era and by the 1910s had created a system by which they loaned money to found production of a film but held the negative in a vault until the loan was repaid.[42] Other bankers were frightened off, however, when the two ventures that attracted outside capital in the mid-1910s, Triangle and the World Film Corporation, both collapsed. This "initial lack of recognition by the financial elite," argues Janet Wasko, "drove the movie leaders even harder to build a legitimate industry, in order to be accepted in the financial world."[43]

As the film industry began remaking itself in earnest to attract Wall Street investors, the presence of women in its ranks came under greater scrutiny. Women in the American film industry had thus far enjoyed more latitude and leverage than women in any other industry, including the stage. But women in powerful and visible positions were not the norm for most industries, particularly the financial industry.[44] As the film industry began to look at itself through the eyes of the financial community, the theatrical legacy that encouraged the participation of women behind the camera seemed as archaic, and perhaps embarrassing, as the haphazard production methods of the nickelodeon era.

As the theatrical origins of the film industry gave way to cinema-specific efficiencies, the transition exposed a fundamental conflict surrounding the female filmmaker, a conflict that had existed since the nickelodeon era: the day-to-day job of a director or producer was to instruct and correct, indeed, to "boss" women and men face-to-face. In any other industry this would easily define the position as masculine. Even a few women who were offered directing opportunities cited the work as too masculine. Ida May Park initially refused to direct features in 1918 "because directing seemed so utterly unsuitable to a woman."[45] When D. W. Griffith asked Lillian Gish in 1919 to write and direct her sister Dorothy's next comedy, she was "dumbfounded." The *Ladies' Home Journal* agreed, noting that "the Lillian Gish temperament is hardly the kind one would expect to see in executive command."[46]

Even in the allied world of theater, where women had worked as producers and managers for decades, women sometimes hesitated to assume what they believed to be men's work. Musical comedy star Emma Carus professed reluctance when thrust into the job of stage producer in 1911. "I had many hard tussles with refractory working crews," she informed *Green Book* readers, "who did not relish the idea of a woman 'bossing' them," and she found it expedient to adopt "a regime of strict, though not severe, discipline" for her own company after encountering disrespect from some chorus girls.[47] Women in the film industry who pursued filmmaking careers adopted a similar approach. Louis Reeves Harrison asserted in his 1912 profile of Alice Guy Blaché that "she handles the interweaving of movements like a military leader might the maneuvers of an army [yet] she accomplishes gently what a man would attempt by stinging sarcasm."[48] Even during the uplift movement some commentators described women filmmakers as transgressing proper gender boundaries or as women who somehow adopted aspects of masculinity, suggesting that success was not possible without masculine traits. In praise of Weber one source noted that she "has the masculine force combined with feminine sympathies and intuition which seem the peculiarly combined gifts of women of genius."[49] L. H. Johnson of *Photoplay*, visiting Lois Weber on the set of *Hypocrites!*, described Weber as a "demon-ess" who "works like a man," turning out films with "super-masculine virility and 'punch'." Yet she was attractively dressed in "a silk shirt-waist and a smart skirt and chic tan boots'." as she issued commands to her "chief subject and vassal, a perspiring camera man, crank[ing] as though Old Nick, instead of [a] pretty woman, were a yard behind him."[50]

Lois Weber provided the archetype of the feminine director by assuming a maternal persona. When the somewhat cynical Frances Marion interviewed for a job in 1914, she was surprised by Weber's approach: Weber offered to protect and guide Marion under her "broad wing."[51] When Weber finally got her own studio in 1918, she deliberately made it look as much like a home as possible, distancing her studio and herself from "that business air which pervades studios generally." When arriving at Weber's studio for an interview, one writer thought he had the wrong address. Only "a very modest little sign" indicated that the large house with fruit trees and flowers was in fact Lois Weber Productions. Even her production methods were unorthodox; Weber shot her scenes in sequence (to allow for better character development) long after other directors took all the shots requiring a particular background at once to save money. Although Weber said in 1917 that she was not an "idealist," adding that all filmmakers were in business to make money, her methods sacrificed efficiency for art. "Efficiency?" Weber exclaimed in 1918. "Oh, how I hate that word!"[52]

As a highly successful director Weber was excused from the economics of film production. Other women were not. But Weber's fame and her outspokenness on the issue of filmmaking and gender reinforced the view that all women filmmakers were intrinsically different from men in their approach. Many women filmmakers were already being relegated to genres that would later be called "women's films." When Frances Denton of *Photoplay* visited the Universal lot in 1918, she found Ida May Park working on "a melodrama," and Elsie Jane Wilson directing some "sob stuff."[53] Universal directors Ruth Stonehouse, Lule Warrenton, and Elsie Jane Wilson all made films centering on children.[54] Lillian Gish's first and only picture as a director, *Remodeling Her Husband* (1920), was described as "a woman's picture. A woman wrote it, a woman stars in it, a woman was its director. And women will enjoy it most."[55] While many women, like Weber, Gish, and Warrenton, appeared to genuinely prefer "feminine" genres, the overall effect was to increase the tendency to define all women as suitable filmmakers *only* when the subject was germane to women.

The marginalization of the woman filmmaker as either a feminist novelty or a feminized artiste appeared to have had only a limited impact on the activities of women filmmakers in the 1910s. But the ideological consequences of this gendered view became clear in the context of the rapid-fire changes that occurred in the film industry during and after World War I. World War I changed the film industry in two major ways: first, the American film industry finally drew positive attention from government and big business, and second, the decimation of European filmmakers created room for the industry to extend its dominance to the far corners of the world. With these shifts the industry ended its quest for cultural legitimacy and became a bona fide big business. By 1923, when Grace Haskins completed her picture, Hollywood no longer particularly needed nor desired female directors or producers.

High-ranking politicians, like high-ranking bankers, kept their distance from the moving picture industry until a few years before the United States entered the war. In 1915 the film industry gained a major boost with Woodrow Wilson's alleged exclamation after a White House screening of *Birth of a Nation*: "It is like writing history with lightning." By the time that the United States finally declared war in 1917, Wilson marshaled the forces of the film industry as he did other major industries. The major contribution to be made, of course, was in propaganda, both at home and abroad, to be accomplished through the Creel Committee on Public Information. Adolph Zukor and Marcus Loew served the war effort alongside Wall Street tycoon Bernard Baruch, now head of the War Industries Board, and food administrator Herbert Hoover. Meanwhile, Charlie Chaplin, Mary Pickford, and Douglas Fairbanks popularized liberty bonds through live appearances and even short films. The icing on the cake came in 1918, when the government declared filmmaking "an essential industry," so vital to winning the war that it had to keep operating even in the face of shortages. Movies were vital not only to keeping up morale at home but to export American values and goods abroad. "Trade Follows Film" was the new Hollywood mantra.[56]

On top of this sea change in the industry's reputation came staggering new levels of cost and scale. A failure could no longer be so easily forgiven. To meet the pressures of filmmaking at this new level studio managers began to impose the central-producer system even more stringently. The first to do so was Irving Thalberg, a young man hired to assist Carl Laemmle at Universal in 1918. Thalberg imposed strict compliance, "wherein shooting scripts, production schedules, and detailed budgets were seen as requisites." Directors who demonstrated particular talent as writers were allowed to continue to both write and direct, but for most Universal directors, responsibility for scenarios and continuities rested in the scenario department, freeing them to concentrate on shooting the film.[57] Although the use of the continuity script had earlier opened up opportunities for women in the early central-producer system, the new "businesslike" approach to filmmaking taken by Thalberg imposed a stricter sexual division of labor and enhanced the power of the central producer. After Thalberg became manager at Universal, only one woman director was hired, and she was a special case.[58] Florence Turner, financially destitute after her British company was destroyed by war, was hired to direct a series of short comedies in 1919. Given the minor genre this was a gesture of kindness.[59]

Thalberg's methods became standard as the studio system coalesced after World War I. As we already know, the rise of the studio system, with its hold on first-run theaters, made it nearly impossible for independent companies, many of them headed by women, to survive in the 1920s. The studio system also curtailed the leverage enjoyed by stars, a critical avenue to power for women, through oligarchical control of the industry and the seven-year contract. There was only one route left for women to continue to participate as filmmakers: as employees of the majors. But the rise in the scale and scope of the film industry just after World War I finalized the masculinization of filmmaking by implementing a strict division of labor, closing this avenue as well. As the studio system emerged, the shift from a theatrically informed model of production and progress to one informed by American industrial norms was completed. Although movies were still a creative product, under the studio system of the 1920s the movie industry became, above all else, a Big Business.

Three events occurring between 1921 and 1923 ushered in the studio system: (1) a brief but devastating recession, (2) Wall Street investment and participation, and (3) the subsequent rise of huge new producer–distributor–exhibitor combinations—the new "majors"—that owned or controlled their own theaters. The recession of 1921 shook the complacency of studios that did well during the war years and during the immediate postwar boom. It was during this brief halcyon period before the recession that independent companies proliferated and the budgets of the most extravagant feature films reached an astonishing $150,000 to $350,000. When the recession hit, the country as a whole suffered from deflation, but the moving picture industry was already facing the consequences of bloated expenses and overproduction. Rental prices fell drastically while budget-conscious patrons stayed home. Larger studios could afford to suspend production, but most of the new independent companies were closed permanently.[60]

The financial health of the larger studios, though wobbling in 1921, rapidly improved as Wall Street investors cast a new eye on the movie business. Otto Kahn, of Kuhn, Loeb, and Company, already showed his favorable take on the movies with a cameo in Reliance's upscale serial *Our Mutual Girl* (1913). A German immigrant and Wall Street financier with a taste for night life, Kahn commissioned a study of the entire industry before granting $10 million in preferred stock to Famous Players–Lasky in 1919 to allow that company to buy its own theaters. From Wall Street's perspective real estate is a good investment—it becomes collateral for future loans.[61] The imprimatur of Kuhn, Loeb, and Company attracted competing Wall Street investment firms such as Goldman, Sachs, who now also desired a toehold in the burgeoning film industry. But Wall Street cast a wary eye on the still-haphazard—and financially dangerous—methods of film production.[62] Demanding stars, profligate directors and producers, and "unbusiness-like methods" fell away under the scrutiny of Wall Street advisers hired not only by Famous Players–Lasky, who won a berth on the Stock Exchange, but by companies that hoped to do so. "There is no room for such items in a report to stockholders," said one banker.[63]

Although the industry itself had long attempted to rationalize production by hiring outside efficiency experts and imposing a modicum of centralized control, the unique culture of moviemaking always mitigated against complete success.[64] Individual directors still retained a great deal of creative control, and quasi-independent director units survived despite the implementation of a central producer. But investment bankers were particularly inspired to see what they could do to rationalize production, and to this end they sent representatives to Hollywood to ensure efficient production and the safety of their investments. Along with its $10 million stock issue, Kuhn, Loeb, and Company imposed rigid scientific management economies on Famous Players. One cost-saving measure had Famous Players personnel dropping friendly salutations like "Regards" from telegrams, transforming requests into rude edicts.[65] All employees were given numbers and color-coded badges (the latter rendered comically ineffective by color-draining klieg lights).[66] According to Lewis Jacobs, the Wall Street "producer supervisor" assumed "more and more power" in the 1920s, "making the director, stars, and other movie workers mere pawns in production, of which he assumed full charge."[67] It seems likely that these Wall Street producer supervisors brought with them their own masculine work culture and traditional ideas regarding women and business. Uplift, for example, fell by the wayside. "This is certainly not a campaign to make the world safe for high-brow pictures," said Kahn; "any such effects one way or the other will be entirely accidental."[68] In 1927, on the occasion of its twentieth anniversary, *Moving Picture World* condemned "the absolute domination of the financier" and the "peculiar, steel-mill efficiency in a business that thrives through art."[69]

It was insider Irving Thalberg who became the architect of a regimented production mode when he joined MGM in 1923. One-half to three-quarters of MGM's revenues were generated by its first-run theaters. The job of the production supervisor was to make sure that the studio kept the company's first-run screens humming with enough product. To that end Thalberg centralized production on a new scale, implementing a mode of production characterized by "meticulous scheduling and script development, close collaboration with the various department heads to ensure efficiency and to maintain production values, and careful supervision of each picture."[70] In the mid-1920s Thalberg had five male supervisors under him, each assigned projects well before the director or other creative personnel. They closely developed the project, and once in production, "monitored shooting, keeping an eye on budget and schedule as well as the day-to-day activities on the set." Directors were assigned just before production began but were able to contribute to the final script and to the first cut.[71] Similar changes were occurring in other studios.

The boundaries between film crafts solidified under the studio system, and as they did, each craft became sex-typed. Some positions experienced little change. Art directors were male.[72] Costume designers were mostly female. And screenwriting remained opened to women throughout the 1920s. But directing, producing, and editing became masculinized. The average director was now a "glorified foreman," chiefly valued for his administrative ability rather than his artistic leanings. He was no longer the sole creator but the "representative of a creative team; he is the man in authority, the field commander who accepts responsibility."[73] According to Cecil B. DeMille the director in the 1920s was an administrator who "never sleeps": "Because if he superintends a staff of brilliant and infallible scenario writers, temperamental stars and untemperamental actors, helpless extra people, nut cameramen, artistic artists, impractical technical directors, excitable designers, varied electricians and carpenters, strange title writers, the financial department and the check signers; if he endeavors ultimately to please the exhibitors, the critics, the censors, the exchangemen, and the public, it's a perfect cinch he won't have time to sleep."[74] Directors need to be "dominating," a quality DeMille believed to be "rare in men and almost absent in women."[75] The masculine image of the film director that characterized associational life in the Screen Club and the MPDA emerged full-blown. Cecil B. DeMille and Erich von Stroheim, in particular, honed this image in the 1920s. Both were among the chosen few who were partly excused from the economies of the studio system on the basis of their past success, and both enjoyed a measure of creative control far out of reach for most directors. But despite their exceptionalism, DeMille and von Stroheim became the popular archetypes of the Hollywood director, placing an insurmountable ideological distance between the "domesticated directress" of the 1910s and the masculinized ideal of the 1920s.

By the 1920s the most important qualities that the director brought to the set were leadership and discipline. Female directors fell outside these parameters, stereotyped as soft, emotional, and intuitive. The arguments put forward in the 1910s—that the movies needed a "woman's touch"—now served to exclude women from becoming directors in the major studios. Even as independent filmmakers, women ran into interference on the grounds of gender. When Margery Wilson rented studio space from Robert Brunton in 1920 to make *That Something*, he warned her that the set would be "bedlam" because as a woman she could not control the cast and crew.[76] As production values swelled after the war, requiring huge casts and dozens of specialized workers, the quiet, refined, and even domestic style of filmmaking associated with women seemed to have no purchase in the new Hollywood.[77] Even Lois Weber joined the chorus. In a 1927 article about directing entitled "The Gate Women Don't Crash," the author introduced Weber as the woman who "retired from motion pictures with about $2,000,000 in cash and property, a nervous breakdown, and the record of being the only woman who had been able consistently to stand the gaff of directing." When asked if she would recommend filmmaking as a potential career for girls, Weber issued a stern warning: "If you feel a heaven-sent call, take careful stock of your qualifications," she advised. "If you haven't got a superabundant vitality, a hard mind that can be merciless in shutting off disturbances, and the ability to keep going from sunrise to midnight, day after day, don't try it. You'll never get away with it."[78]

Although directors garnered the lion's share of publicity, producers became the most powerful individuals on the lot under the studio system. Not surprisingly, production duties were masculinized as well. As budgets reached into the hundreds of thousands of dollars, if not millions, writer-producer Jane Murfin claimed that studios assumed women were simply not smart enough or experienced enough to handle this level of fiscal responsibility: "Men don't expect women to understand the intricacies of business, the cost of production and distribution, the percentage of overhead, locked up capital and liquid assets, and especially the complications of banking transactions. I admit I've sometimes wondered just how clearly the men themselves understood them, and one or two unwisely frank gentlemen have even admitted that they were congenitally hazy about 'earned and unearned profits' and the 'circuit velocity of money,' doubtless due to the parental influence of their mothers."[79] Cecil B. DeMille represented both production and direction when the film-friendly Gianninis of the Bank of Italy made him vice president of one of their branches: the Commercial National Trust and Savings Bank in Los Angeles. The Gianninis "packed the board" of each of their branches with industry insiders—producers, directors, actors, and actresses—but industry insiders were likely drawn from the new major studios, so they likely favored each other over independent filmmakers when extending credit. DeMille illustrated this when he made a $200,000 unsecured loan to Samuel Goldwyn as one of his first acts as a banking official.[80]

A few women did find work as producers in the 1920s. Paramount hired Elinor Glyn, author of the scandalous novel *Three Weeks* and the woman who coined the phrase "IT," or sex appeal, as an all-round production adviser in the 1920s.[81] Jane Murfin wrote and coproduced five films for First National (most of them featuring a dog, Strongheart, predecessor to Rin Tin Tin) between 1921 and 1924, and codirected one film for First National, *Flapper Wives* (1924), listed as "her own production." Murfin did not direct again, but she become RKO's first female production supervisor in 1934.[82]

Much more significant was June Mathis, one of the most influential figures in 1920s Hollywood and the woman who might have set a precedent for female producers under the studio system. Mathis became chief of Metro's scenario department in 1919 after working as a screenwriter for only one year. At the age of twenty-seven Mathis made script selections, adaptations, and continuities for the studio. In 1921 she adapted Ibanez's war novel *The Four Horsemen of the Apocalypse*, insisting that the studio hire Rex Ingram to direct it and cast bit player Rudolph Valentino as the male lead. *The Four Horsemen* became known as one of the best films of the year. Already well known, Mathis became a celebrity.[83] After her triumph with *The Four Horsemen* she was hired by Samuel Goldwyn as editorial director of Goldwyn Pictures. She set studio policy and handled continuities and scenarios. It was during her term that some of the most prominent directors in the movies worked for Goldwyn: King Vidor, Victor Seastrom, Marshall Neilan, and Erich von Stroheim.[84]

As direction and production became rapidly masculinized after World War I, the increasingly important craft of film editing was still open to women, at least for a while. With the rise of the continuity script, the position of film cutter, like that of other workers, became a specialized element of the postproduction process. Using the continuity script and the slate numbers as a guide, the cutter could assemble a rough cut, and even a final cut, often without the director's personal instruction. By 1922, editing had reached the status of a creative craft, one that was just below that of director in terms of its creative impact on the final product.[85] It was true that inexperienced "boys" were often hired to piece together films according to continuities, but in the 1910s and early 1920s many studios recruited women from the joining room to become "cutter girls." Margaret Booth recalled her move from joiner to cutter in the mid-1910s in a matter-of-fact manner: "Irene Morra was the negative cutter and she took me to help her and showed me how to cut." Viola Lawrence, a former film polisher, learned how to cut film at Vitagraph in 1915. By 1918 the "master cutter" gained recognition as an important creative force behind the finished film. Rose Smith cut Griffith's *Intolerance* with her husband James Smith, and at least a dozen other women were counted among the first editors in Hollywood, among them Anne Bauchens, Blanche Sewell, Anne McKnight, Barbara McLean, Alma MacCrory, Nan Heron, and Anna Spiegel.[86] Editor Adrienne Fazan recalled that in the early 1920s "every studio had a few women editors . . . [A] woman could get started then."[87]

Women who learned to edit in the 1910s and early 1920s enjoyed long careers. Viola Lawrence became head editor at Columbia in 1925, Margaret Booth became MGM's supervising film editor in 1936, and Cecil B. DeMille employed editor Anne Bauchens for more than forty years.[88] According to Douglas Gomery, "all filmmakers from the late 1930s through the late 1960s who worked for MGM had, in the end, to go through Margaret Booth to have the final editing of sound and image approved."[89] But editing was masculinized as well. Viola Lawrence recalled having "all boy assistants" (but for one) in the 1920s and 1930s, as did Adrienne Fazan.[90] Even female editors who began their careers in the 1910s and early 1920s ran into hostility from male editors. Viola Lawrence's husband, Frank, who taught her to cut film in 1915, was "mean" to the female assistant editors he supervised at Paramount in the 1920s. "He just hated them," she claimed. "If any of the girls were cutting—if they did get the chance to cut—he'd put them right back as assistants," but he "broke in a lot of boys." In the early 1930s, editor Adrienne Fazan recalled, "MGM didn't want me to become a feature cutter." Production head Eddie Mannix told her that film editing was "just too tough work for women," who "should go home and cook for their husbands and have babies." (It was Dorothy Arzner, the only woman to survive the purge of female directors in the 1920s, who took Fazan out of the short film department by asking specifically for a female editor.)[91] As editing became recognized as a critical step in the production of what were now often million-dollar films, it, like direction and production, became a masculinized craft.

By the mid-1920s, the power of stardom diminished, the independent movement ended, and the gendered studio emerged. Directors and producers were almost exclusively male, and film editing was rapidly becoming masculinized. Of the creative crafts behind the camera, only screenwriting, a job that often paid poorly and was chronically disrespected, remained open to women. By 1928, when the "talkies" triumphed, the reorganization of studio work that followed merely codified the sexual division of labor that had emerged after World War I. The era when actresses and writers easily slipped into the director's chair, when a woman was one of America's most critically acclaimed and successful directors, and when "America's Sweetheart" Mary Pickford was one of the most powerful producers in Hollywood was over. Yet the very presence of this generation of women filmmakers demonstrates that the male domination of Hollywood moviemaking was not a foregone conclusion but rather the outcome of a historical struggle that might have had a different ending.

Notes

1 Sydney Valentine, "The Girl Producer," *Photoplay*, July 1923, 55, 110; Joseph Danning, ed., *Film Year Book* (Hollywood: Film Daily, 1924), 63.
2 Lewis Jacobs, *The Rise of the American Film: A Critical History* (New York: Columbia University Press, 1968), 296–97.
3 Untitled review of an unidentified Lois Weber film, *MPN* [*Motion Picture News*], Aug. 27, 1921, n.p., Claire Windsor Collection, SC-USQ review of Ida May Park's *The Butterfly Man*, *Variety*, May 21, 1920, n.p. For a discussion of director Dorothy Arzner and her "butch style" see Judith Mayne, *Directed by Dorothy Arzner* (Bloomington: Indiana University Press, 1994); for an early description of filmmaking as "man's work" see Frances Denton, "Lights! Camera! Quiet! Ready! Shoot!" *Photoplay*, Feb. 1918, 48–50.
4 Feature films directed by women from 1922 through 1928 (with distributor): 1922: Nell Shipman, *The Girl from God's Country* (F. B. Warren); May Tully, *Our Mutual Friend*, *The Old Oaken Bucket* (Wid Gunning, Inc.); Marion Fairfax, *The Lying Truth* (American Releasing Corp.); Lois Weber, *What Do Men Want?* (Film Booking Offices of America/Robertson-Cole); Mr. and Mrs. George Randolph Chester, *The Son of Wallingford* (Vitagraph); Ruth Bryan Owen, *Once upon a Time* (states' rights [?]). 1923: Julia Crawford Ivers, *The White Flower* (Paramount); Lois Weber, A *Chapter in Her Life* (Universal-Jewel); Grace Haskins, *Just like a Woman* (W. W. Hodkinson). 1924: Jane Murfin, *Flapper Wives* (Selznick); Lillian Ducey, *Enemies Of Children* (Mammoth Pictures). 1925: No feature films directed by women. 1926: May Tully, *This Old Gang of Mine* (states' rights [?]); Lois Weber, *The Marriage Clause* (Universal-Jewel). 1927: Dorothy Arzner, *Fashions for Women* (Paramount), *Ten Modern Commandments* (Paramount), *Get Your Man* (Paramount); Lois Weber, *Sensation Seekers* (Universal-Jewel), *Angel of Broadway* (DeMille pictures/Pathé). 1928: Dorothy Arzner, *Manhattan Cocktail* (Paramount). After 1928, with only two exceptions, Dorothy Arzner was the only woman to work as a Hollywood director until the 1970s; information from *Film Year Book* (Hollywood: Film Daily, 1922–28).
5 Wendy Holliday, "Hollywood's Modern Women: Screenwriting, Work Culture, and Feminism, 1910–40" (PhD diss., New York University, 1995), 174–76.
6 "Photoplayers Now in Its Third Year of Success," *MPN*, April 3, 1915, 171.
7 K. Owen, "The Club, James!" *Photoplay*, Feb. 1917, 67.
8 Anthony Rotundo, *American Manhood: Transformations in Masculinity from the Revolution to the Modern Era* (New York: Basic Books, 1993), 199–203, 250.
9 Lillian R. Gale, "Motion Picture Director's Association," in *Film Year Book, 1920–21*, ed. Joseph Dannenberg (New York: Wid's Films and Film Folks, [1921?]), 211.
10 "Directors' Association Honors Lois Weber," unidentified source, Nov. 21, 1916, Lois Weber clippings file, MOMA.
11 Motion Picture Directors Association, *Souvenir: Annual Ball, Feb. 17, 1923, Alexandria Hotel, Los Angeles*, AMPAS.
12 Ibid., 69–70.
13 "Studio Efficiency: Scientific Management as Applied to the Lubin Western Branch by Wilbert Melville," *MPW* [*Moving Picture World*], Aug. 9, 1913, 624.
14 Lizzie Francke, *Script Girls: Women Screenwriters in Hollywood* (London: British Film Industry, 1994), 6. See also Cari Beauchamp and Mary Anita Loos, *Anita Loos Rediscovered: Film Treatments and Fiction* (Berkeley: University of California Press, 2003).
15 Francke, *Script Girls*, 18.
16 Clara Beranger, "Feminine Sphere in the Field of Movies," *MPW*, Aug. 2, 1919, 662.
17 Staiger, "Blueprints for Feature Films," 190.
18 Louis Reeves Harrison, "Directorial Censorship," *MPW*, April 12, 1913, n.p.; Louis Reeves Harrison, "A Wondrous Training School," *MPW*, Aug. 16, 1913, 720; "Studio Efficiency: Scientific Management as Applied to the Lubin Western Branch by Wilbert Melville," *MPW*, Aug. 9, 1913, 624; Captain Leslie T. Peacocke, "Studio Conditions as I Know Them," *Photoplay*, June 1917, 127–30.
19 Captain Leslie T. Peacocke, "The Scenario Writer and the Director," *Photoplay*, May 1917, 112; Jacobs, *Rise of the American Film*, 219.
20 From *Moving Picture Stories*, n.d., n.p., quoted in Anthony Slide, *Early Women Directors* (New York: A. S. Barnes, 1977), 104.
21 Nell Shipman, *The Silent Screen and My Talking Heart: An Autobiography*, 2nd ed. (Boise, ID: Boise State University, 1988), 43–44; Peter Morris, "The Taming of the Few: Nell Shipman in the Context of Her Times," in Shipman, *Silent Screen*, 215.
22 Alice Martin, "From 'Wop' Parts to Bossing the Job," *Photoplay*, Oct. 1916; 96–97.

23 Slide, *Early Women Directors*, 54–55.
24 Ibid., 105–8; "Creator of 'J. Rufus' Will Write Reliance Serial," *MPW*, Dec. 5, 1914, 25–26; "George Randolph Chester Forms Company; Plans to Make Two Big Films Yearly," *MPW*, Nov. 19, 1921, 290.
25 Obituary, *Los Angeles Times*, May 9, 1930, Ivers biography file, AMPAS; *Majesty of the Law*, reel 4 of 5, viewed at MPBRSD-LC; "The Shadow Stage," review of *Call of the Cumberlands*, *Photoplay*, April 1916, 104; Patricia King Hansen, ed., *The American Film Industry Catalog of Motion Pictures Produced in the United States: Feature Films, 1911–20* (Berkeley: University of California Press, 1988), 117–18, 567, 863; review of *The White Flower*, *Variety*, March 8, 1923, n.p.; Slide, *Early Women Directors*, 110.
26 Marion codirected one more film, *The Song of Love* (1923), after its original director became ill, but that was the end of her directing career. Dewitt Bodeen, "Frances Marion," *Films in Review* 20, no. 2 (Feb. 1969): 75–87; *MPW*, April 5, 1919, 75; The Screen, *NYT* [*New York Times*], Jan. 19, 1920, 16; review of *The Love Light*, *Variety*, Jan. 14, 1921, n.p.; "Frances Marion," in *American Screenwriters*, ed. Robert E. Morsberger, Stephen O. Lesser, and Randall Clark (Detroit: Gale Research, 1984), 232.
27 E. Leslie Gilliams, "Will Woman's Leadership Change the Movies?" *Illustrated World*, Feb. 1923, 38, 860, 956; Slide, *Early Women Directors*, 114–15.
28 See *Photoplay*, Dec. 1915, 159; William F. T. Henry, "Cleo the Craftswoman," *Photoplay*, Jan. 1916, 110. The two five-reel features were *A Soul Enslaved* and *Her Bitter Cup*; see "Cleo Madison in *Her Bitter Cup*," *Moving Picture Weekly*, April 16, 1916, 24–25, 32.
29 "Such a Little Director," *Moving Picture Weekly*, March 24, 1917, 19; "One of the Universal Faithful—Lule Warrenton, Character Leads," *Universal Weekly*, Feb. 7, 1914, 5; "Mrs. Warrenton Starts Children's Photoplays," *MPW*, Feb. 17, 1917, 1030; Cal York, Plays and Players, *Photoplay*, Nov. 1917, 81; Slide, *Early Women Directors*, 56–58.
30 "Where Work Is Play and Play Is Work, Universal City, California, the Only Incorporated Moving Picture Town in the World," *Universal Weekly*, Dec. 27, 1913, 4–5.
31 Richard Koszarski, *An Evening's Entertainment: The Age of the Silent Feature Picture*, 1915–28 (Berkeley: University of California Press, 1994), 87.
32 "Such a Little Director," *Moving Picture Weekly*, March 24, 1917, 19; Cal York, Plays and Players, *Photoplay*, Feb. 1918, 118.
33 Cal York, Plays and Players, *Photoplay*, Jan. 1917, 83; Cal York, Plays and Players, *Photoplay*, Oct. 1917, 114.
34 "Mrs. Warrenton Starts Children's Photoplays," *MPW*, Feb. 17, 1917, 1030; Cal York, Plays and Players, *Photoplay*, Nov. 1917, 81; Slide, *Early Women Directors*, 56–56.
35 Slide, *Early Women Directors*, 60; Film Comment, Nov. 1972, n.p., in Ida May Park biography file, AMPAS; *Wid's Year Books* for 1918, 1919–20, 1922, n.p.; review of *The Butterfly Man*, *Variety*, May 21, 1920, n.p.
36 Wilson directed one more film, *Insinuation* (1922), but it was not released widely, if at all; see *Wid's Year Book* (1921–22), 323; Slide, *Early Women Directors*, 62–66; *Insinuation* does not appear in *Wid's Year Book*, the industry's official chronicle of the year's productions.
37 "Vera McCord's 'The Good-Bad Wife' Arousing Big Interest among Buyers," *MPW*, Aug. 28, 1920, 1168; "Vera McCord Productions Asks Receiver for Walgreene in Suit over Receipts," *MPW*, Sept. 24, 1921, 402; review of *The Good-Bad Wife*, *Variety*, Jan. 21, 1921, n.p.
38 "Marion Fairfax, Dramatist, Is with Lasky," *MPW*, June 12, 1915, 64; "Initial Marion Fairfax Picture Is Completed On the West Coast," *MPW*, June 11, 1921, 634; "Preview of Marion Fairfax Film," *New York Morning Telegraph*, July 10, 1921, n.p., in Marion Fairfax biography file, AMPAS; ad for "Marion Fairfax Productions," *Wid's Year Book* (1921–22), 40.
39 "*The Son of Wallingford* a Winner," *Exhibitor's Herald*, Oct. 8, 1921, 45; "Creator of 'J. Rufus' Will Write Reliance Serial," *MPW*, Dec. 5, 1914, 25–26; "George Randolph Chester Forms Company; Plans to Make Two Big Films Yearly," *MPW*, Nov. 19, 1921, 290.
40 "A Society Girl in Filmdom's Whirl," *Photo-Play Journal*, Oct. 1919, n.p.; advertisement, *Wid's Year Book* (1919–20), n.p.; *Wid's Year Book* (1920–21), 85, 91, 314, 394; Hansen, *AFI Catalog ... 1911–20*, 730; *MPW*, Jan. 1, 1921, 92; "Brought into Focus," *NYT*, Feb. 27, 1921, 2; "'The Sky Pilot' Prologue Conceived by First National and Strand Scores Hit," *MPW*, May 7, 1921, 81; review of *The Sky Pilot*, *Variety*, April 22, 1921, n.p.
41 By July Unsell had accepted employment as head of the Robertson-Cole scenario department. "Eve Unsell Creates Scenario Service Bureau: Will Also Dispose of Film Rights to Books," *MPW*, Jan. 15, 1921, 288; advertisement, "Eve Unsell Photoplay Staff, Inc.," *MPW*, Jan. 15, 1921, 254. For the recession see *Wid's Year Book* (1921–22), 99, 105, 107; "Eve Unsell Will Sail for England," *MPW*, July 26, 1921, 515; "Eve Unsell Heads Scenario Department of Robertson-Cole on the West Coast," *MPW*, July 23, 1921, 412.

42 David Puttnam, with Neil Watson, *Movies and Money* (New York: Knopf, 1998), 94.
43 Wasko, *Movies and Money*, 11–12.
44 According to Angel Kwolek-Folland the banking industry's approach to women was firmly grounded in nineteenth-century gender ideals; see Angel Kwolek-Folland, *Engendering Business: Men and Women in the Corporate Office,* 1870–1930 (Baltimore: Johns Hopkins University Press, 1994), 171–72.
45 Denton, "Lights! Camera! Quiet! Ready! Shoot!" 48–50.
46 Henry MacMahon, "Women Directors of Plays and Pictures," *LHJ* [*Ladies Home Journal*], Dec. 1920, 140; *Remodeling Her Husband* (1920) was not well reviewed, but it made money; see Lillian Gish, with Ann Pinchot, *The Movies, Mr. Griffith, and Me* (Englewood Cliffs, NJ: Prentice-Hall, 1969), 222–37; Anthony Slide, "Remodeling Her Husband," in *Magill's Survey of Cinema*, ed. Frank N. Magill (Englewood Cliffs, NJ: Salem Press, 1982), 1: 918–20.
47 Emma Carus, "The Feminine Stage Producer," *Green Book*, Nov. 1911, Robinson Locke Collection, Scrapbook, ser. 2, NYPL-PA.
48 Louis Reeves Harrison, "Studio Saunterings," *MPW*, June 15, 1912, 1007–10.
49 Unidentified clipping, with photos, 152, Kevin Brownlow private collection.
50 L H. Johnson, "'A Lady General in the Motion Picture Army,' Lois Weber-Smalley, Virile Director," *Photoplay*, June 1915, 42.
51 Bertha H. Smith, "A Perpetual Leading Lady," *Sunset, the Pacific Monthly*, March 1914, 635; Frances Marion, *Off with Their Heads!* (New York: Macmillan, 1972), 12.
52 Fritzi Remont, "The Lady Behind the Lens," *Motion Picture Magazine*, May 1918, 59–61, 126; "we're all in business" quote from Arthur Denison, "A Dream in Realization," *MPW*, July 21, 1927, 417–18.
53 Denton, "Lights! Camera! Quiet! Ready! Shoot!" 48–50.
54 "Such a Little Director," *Moving Picture Weekly*, March 24, 1917, 19; "One of the Universal Faithful," 5; "Mrs. Warrenton Starts Children's Photoplays," 1030; Cal York, Plays and Players, *Photoplay*, Nov. 1917, 81; Slide, *Early Women Directors*, 56–58.
55 *Photoplay*, Sept. 1920, quoted in Anthony Slide, "Remodeling Her Husband," in *Magill's Survey of Cinema*, ed. Frank N. Magill (Englewood Cliffs, NJ: Salem Press, 1982), 1: 920.
56 Puttnam and Watson, *Movies and Money*, 76–79.
57 Thomas Schatz, *The Genius of the System: Hollywood Filmmaking in* the *Studio* Era (New York: Pantheon, 1988), 23–25.
58 Ibid., 16–25; *MPW*, Jan. 25, 1920, 6.
59 Slide, *Early Women Directors*, 103.
60 Benjamin B. Hampton, *A History of the American Film Industry from Its Beginnings to* 1931 (New York: Dover, 1970), 248–51; *Wid's Year Book* (1921–22), 99, 105, 107.
61 Puttnam and Watson, *Movies and Money*, 97, 101.
62 Wasko, *Movies and Money*, 14; Tino Balio, *United Artists: The Company Built by the Stars* (Madison: University of Wisconsin Press, 1976), 36.
63 Helen Bullitt Lowry, "Wall Street's Heel on the Prodigal Movies," *NYT*, July 24, 1921, 6, 36.
64 Janet Staiger, "The Hollywood Mode of Production to 1930," in *The Classical Hollywood Cinema: Film Style and Mode of Production to* 1960, by David Bordwell, Janet Staiger, and Kristin Thompson (New York: Columbia University Press, 1985), 135–36.
65 Puttnam and Watson, *Movies and Money*, 97.
66 "Efficiency," *Wid's Daily*, March 7, 1921, 1.
67 Jacobs, *Rise of the American Film Industry*, 295; Hampton, *History of the American Film Industry*, 244; Wasko, *Movies and Money*, 295; Balio, *United Artists*, 36.
68 Lowry, "Wall Street's Heel on the Prodigal Movies," 6.
69 "1907–27, Too Much Efficiency," *MPW*, March 26, 1927, 277.
70 Schatz, *Genius of the System*, 31–36, 39, 44–45.
71 Ibid., 46.
72 The Motion Picture Art Directors Association first appeared in *Wid's Year Book* (1920–21), 215; there were forty-two names, none of them female; in *Wid's Year Book* (1921–22), 282, there were forty-four names, again, no women; for development of art direction, see Staiger, "Hollywood Mode of Production to 1930," 147–48.
73 Brownlow, *The Parade's Gone By*, 79.
74 Ibid.
75 Charles S. Dunning, "The Gate Women Don't Crash," *Liberty*, May 14, 1927, 31, 33.
76 Slide, *Early Women Directors*, 62–66.
77 Brownlow, *The Parade's Gone By*, 71.
78 Dunning, "The Gate Women Don't Crash," 31, 33.

79 Hughes, *Truth about the Movies*, n.p.
80 Puttnam and Watson, Movies and Money, 95.
81 Zvan St. Johns, "Glyn and Glynne," *Photoplay*, May 1924, 53, 129.
82 Cal York, Plays and Players, *Photoplay*, May 1923, 87; "Our Foremost Woman Director," *Photoplay*, July 1924, 74.
83 Nora B. Geibler, "Rubbernecking in Filmland," *MPW*, June 18, 1921, 719.
84 Koszarski, *An Evening's Entertainment*, 240–41.
85 Thompson, "Formulation of the Classical Style," 152; Kristin Thompson, "Film Style and Technology to 1930," in *The Classical Hollywood Cinema; Film Style and Mode of Production to 1960*, by David Bordwell, Janet Staiger, and Kristin Thompson (New York: Columbia University Press, 1985), 278.
86 Margaret Booth, interview by Rudy Behlmer, in *An Oral History with Margaret Booth* (Beverly Hills, CA: AMPAS, 1991); Francis Humphrey, "The Creative Woman in Motion Picture Production" (master's thesis, University of Southern California, 1970), 123–25; Barrie Pattison, "Thirty Leading Film Editors," in *International Film Guide*, ed. Peter Cowie (New York: A. S. Barnes, 1973), 67; Dick Cantwell, "Anna Spiegel's Story in Series on Old-Timers," *Metro-Goldwyn-Mayer Studio Club News*, Dec. 24, 1938, 12, editors general file, AMPAS; for references to male cutters see Captain Leslie T. Peacocke, "Studio Conditions as I Know Them," *Photoplay*, June 1917, 127–30; and Starr, "Putting It Together," 52.
87 Humphrey, "Creative Woman," 111.
88 "Film Editor [Lawrence] Retires; Put in 50 Years," *Los Angeles Herald-Examiner*, Aug. 9, 1962, A-17; see also Anne Bauchens biography file, AMPAS.
89 Douglas Gomery, "Margaret Booth," in *International Dictionary of Films and Filmmakers*, ed. Christopher Lyon and Susan Doll (New York: Putnam, 1985), 4: 53.
90 Humphrey, "Creative Woman," 114–15, 125–26, 130.
91 Ibid., 110–12, 114–17, 123–27.

Chapter 6

Lea Jacobs and Andrea Comiskey

HOLLYWOOD'S CONCEPTION OF ITS AUDIENCE IN THE 1920s

> The term 'prominent' . . . does not express real artistic accomplishments, but is part of the grandiloquent phraseology affected by all advertising in the entertainment industry, with its insincere slogan that nothing is too good for the public. This kind of prominence is determined by the fabulous salaries paid to those whom the publicity agencies elect to build up—the prominence of Radio City, the Pathé Theatre in Paris, or the Ufapalast am Zoo in Berlin. It belongs to the realm that Siegfried Kracauer called *Angestelltenkultur*, culture of the white-collar workers, of supposedly high-class entertainment, accessible to recipients of small pay checks, yet presented in such a way that nothing seems too good or too expensive for them. It is a pseudo-democratic luxury, which is neither luxurious nor democratic, for the people who walk on heavily carpeted stairways into the marble palaces and glamorous castles of moviedom are incessantly frustrated without being aware of it.
>
> (Adorno and Eisler, "Composing for the Films")

IN THEIR SCORNFUL EVOCATION OF THE prominence accorded the conductor in the picture palaces of the 1920s, Adorno and Eisler evoke a conception of the audiences which frequented those theaters: it is a mass audience, duped into accepting an ersatz luxury as the real thing, and unwittingly participating in a "pseudo-democratic" effacement of class differences and allegiances. While we believe this idea of the mass audience for 1920s cinema still has some currency in our field, a variety of film historians have sought to modify or nuance it. Some have argued that early film may be distinguished from the 1920s (and after), and that the nickelodeons, by virtue of their placement in proximity to working-class neighborhoods and their function as public spaces, allowed for some differentiation among audiences.[1] In addition, scholars concerned with the classical period have turned their attention to the history of exhibition and/or reception in specific viewing communities such as Lexington, Kentucky or Sacramento, California, and have thereby emphasized regional differences in movie-going.[2]

In addition to these exhibition-based accounts, two recent distribution-based accounts have argued that the film industry's own conception of its audience in the 1930s was far from monolithic. Based on an analysis of the trade press (primarily *Variety* and the *Motion Picture Herald*), Richard Maltby argues that:

> From 1929 to 1933, Hollywood's audience was classified by the discourse of the exhibition sector into a series of overlapping binary distinctions between 'class' and 'mass', 'sophisticated' and 'unsophisticated', 'Broadway' and 'Main Street'.

> These terms were not exact synonyms, but their differences were matters of emphasis. The terminology distinguished between the clientele in different types of theatre, classifying viewers as part of the process of classifying theatres, and it served consistently to indicate that different audiences required different products. Overlaying this typology of theatre audiences was a terminology designed to distinguish between groups of viewers within any given audience by gender and age, as a guide for exhibitors to the most effective exploitation of a particular movie.[3]

In a study based on the monthly tables of picture grosses in key cities published in *Variety* between October 1934 and October 1936 (and reprinted elsewhere in this Reader; Chapter 11), Mark Glancy and John Sedgwick examine how grosses varied by city. They also conclude that the industry understood its audience as geographically and culturally differentiated: ". . . there was apparently little interest among *Variety*'s readers in national box-office grosses; the box-office earnings that were reported so carefully each week were not added together to form a national gross. The national level simply was not important. Rather, it was important to report what was happening in individual cities and cinemas, and how audiences were responding to specific programmes, promotions and pricing strategies. The cinemagoing experience itself varied widely, even at the local level, and the task of *Variety* was to report on the success or failure of any number of variations."[4]

Although Maltby suggests that the distinctions made in the trade press among audiences and types of theater were the product of the coming of sound, in *The Decline of Sentiment*, which deals with changes in taste across the 1920s, Lea Jacobs demonstrates that very similar judgments about audience preferences predominated in the trade press discourse in the late silent period.[5] Many of the critical judgments found in the trade press in this period were framed in terms of a film's potential profitability and appeal within the market, and the nature of this market gives us some insight into the way in which the trade press constructed its idea of the audience. While those involved in film distribution and exhibition did not conduct audience research in the 1920s, they did know a great deal about theaters.[6] By the early 1920s, three major film companies, Paramount, First National and Loew's, were vertically integrated, encompassing film production units, distribution exchanges, and theater chains. Ticket sales for affiliated theaters would have been carefully monitored by distribution personnel. All theaters were classified according to their location and the population that they served, and on this basis assigned a "run" and a minimum ticket price, and thereby a place in the distribution hierarchy. Reviewers for the trade press, particularly *Variety*, tried to estimate where a film would fit within this hierarchy: in the major downtown picture palaces, in the subsequent-run theaters in urban neighborhoods (the "nabes" in *Variety* parlance), or in small towns and rural areas. The reviewers would also estimate about how long a film would play—whether it would last only a week or be "held over" on Broadway; whether a film was appropriate for a split week (three or four days) in a neighborhood theater, or, even worse, only a single day in what was called a "grind" house.[7] Some films were deemed only fit for the second half of a double bill. There was frequently an estimation of the budget spent on the film and its worth relative to this: thus, *Variety* would sometimes praise a cheaply made independent film, on the grounds that it was a "good independent," and able to hold an audience despite its low cost. Similarly, reviewers would make an estimation about whether a big-budget film was worth roadshowing at special prices—what was called a \$2 special.[8] Sometimes studios were chastised for trying to sell an ordinary big-budget film as a special, or praised for refraining from trying to elevate a simple "programmer" to this status.

Variety's judgments as to where a film would play, how long it would play, and at what price, were linked to judgments about its potential audience. For example, some films were deemed unlikely to play well "west of the Hudson" while others were dubbed "applesauce outside of New York." *Variety* often referred to the audiences in the nabes in terms which implied that they were less sophisticated and/or working-class. There were, however, occasions in which the neighborhood theaters were associated with female working-class viewers specifically. Trade press reviews thus offer us a subtle, professional estimate of the market for a given film and a highly differentiated conception of the audience: as urban or rural, as belonging to the "classes" or the "masses," as male or female. Nonetheless, it seemed to us that a comparison of these estimates with actual distribution practices might prove instructive. In this study we seek to correlate trade reviewers' judgments about how films were likely to play across different markets, and about the preferences of theater audiences, with distribution patterns in a small number of Northeastern and mid-Western cities and towns. The sample of twenty-two films is drawn from research on the critical reception of films and film genres conducted for *The Decline of Sentiment*. The sources employed include *Variety, Film Daily* (called *Wid's Daily* until 1922), *Moving Picture World*, the *Exhibitor's Trade Herald*, the fan magazine *Photoplay*, the more refined *Exceptional Photoplays*, the New York humor magazine *Life*, and the *New York Times*.

It should be noted that *Variety* did not publish picture grosses for key cities in the 1920s. In any case, our interest is not confined to how films performed in the metropolitan centers: we seek to understand the relative timing of the movement of films from run to run, from metropolitan to neighborhood theaters and from larger to smaller cities and towns. Online searches of Proquest Historical Newspapers and the Newspaper Archive site permitted us to trace where our sample films played week by week (largely on the basis of theater advertisements and published lists of local screenings).[9] We examined six markets: Manhattan (unfortunately the *New York Times* did not cover film exhibition in the boroughs and we have not located a computerized source for them); Chicago (for which we have information both on the major theaters in the Loop and in the nabes); Syracuse, New York, a mid-sized city of a dozen theaters; Davenport, Iowa, a small town with three movie theaters and a vaudeville house that also showed movies; Fayetteville, Arkansas, a small town with four theaters; Bradford, Pennsylvania, a small town with three theaters, one of which was a legitimate theater that also showed movies. Syracuse had an advanced "sub-run" (subsequent run) structure, maintaining second, third and occasionally fourth runs. While the smaller towns sometimes showed films in their subsequent run, they did not have an orchestrated sub-run structure as was the case in Syracuse. Theater ownership in Chicago was dominated by the Balaban & Katz [B&K] chain which, after 1925, was affiliated with Paramount Publix theaters and was managed by Sam Katz.[10] Although there were some changes in specific theater affiliations over the course of the decade, three of the four theaters in Davenport were also controlled by Paramount Publix. Theater ownership in the other small towns was diversified and all of the major studios had theaters on Broadway. Thus, although restricted in number, the regional markets chosen for analysis are differentiated in structure as well as size.

Roadshows

The two most successful films in our sample both derive from the play *What Price Glory* by Maxwell Anderson and Laurence Stallings, which was produced by Arthur Hopkins at the Plymouth Theatre in New York in September 1924. The play was an enormous success, running on Broadway for 299 performances and launching a touring company.[11] Hollywood attempted to capitalize on its popularity in a number of films. Fox's *What Price Glory*, directed by Raoul Walsh, began its roadshow run in December 1926 and went into general release in August 1927. It was preceded in 1925 by MGM's *The Big Parade*, based on a story that director

King Vidor worked out with Stallings after the success of his play. *Variety* predicted that both of these films would be box office knockouts. Its review (2 December 1925: 40) of the New York opening of *The Big Parade* reported: "The first Sunday night [November 22] saw them standing six deep in the Astor, and they've been on their feet to view it ever since . . . it's a cinch Jeff McCarthy knew what he was talking about when he took one flash at this picture long before it was finished and said 'road show'."[12] *What Price Glory* was similarly assessed by *Variety* (1 December 1926: 12): "The chances are that 'What Price Glory' will be just as big at the box office as 'The Big Parade' was, providing it is as deftly handled as a road show. One thing the Fox people do not want to do and that is to rush in all over the country with road shows right off the bat. The thing to do with this one is to lay back, pick the spots and play about six of the big cities this season. Philadelphia, Boston, Chicago added to New York and Los Angeles already opened, and possibly San Francisco should be all that are hit this season, and then late next August strike out with about 12 companies in the week stands and get the money."

The release patterns for both these films confirm *Variety*'s predictions. As noted above, *The Big Parade* opened on Broadway at the Astor, a 1500-seat legitimate theater that had recently been acquired by Loew's and converted to film screenings on a permanent basis.[13] It played at the Astor for nearly two years, from mid-November 1925 through mid-September 1927. It then moved into general release at the Capitol, Loew's 5230-seat theater, where it stayed for two weeks.

The Big Parade began its roadshow run in Chicago at the Garrick, a 1000-seat legitimate theater, in late December of 1925 (according to a review in the *Chicago Daily Tribune*, 29 December 1925: 13, the opening had been "billboarded all up and down the highways of our fair city" for a month). It stayed through the end of April 1926 at the Garrick and went into general release in summer 1927, playing at Balaban & Katz's McVickers, a 2200-seat downtown theater, from 27 July to 4 September. By mid-September, it started playing the biggest neighborhood houses in Chicago. It went first to Orchestra Hall, a large Lubliner & Trinz house (L&T was a separate chain affiliated with B&K).[14] It then moved, in quick succession, to other large B&K and L&T houses (Uptown, Harding, Tivoli, Norshore). At the end of October, it moved back downtown to the 300-seat Castle and into some large neighborhood houses owned by smaller chains. It appeared in a large number of neighborhood houses in the second week of November and spread throughout the neighborhood houses in four discernible waves throughout the month of November.

There were isolated screenings of *The Big Parade* in small cities and towns while it was still in its roadshow release, probably to help build word of mouth. Bradford got it for three days in April 1926 (local advertisements boasted that "Bradford is one of only four cities in the State of PA where the famous production will be seen this season"). It arrived in Syracuse in March 1926 where it played at the Wieting, a legitimate theater, and returned there for a second roadshow in March 1927. It went into general first-run release at the Strand in mid-September (this was actually its third Syracuse engagement). It had sub-runs in October and December 1927, as well as March, May, and June 1928. The film arrived at the Davenport-area Fort Armstrong theater in early March 1927, where it played at roadshow prices, and received a general release in downtown Davenport in late September (and a sub-run at the Garden in May 1928; "first time in tri-cities at $.25"). Fayetteville finally got the film in early November 1927. Thus, it took the film at least two years to go through the distribution hierarchy, and to work its way from the biggest to the smallest markets.

A similar pattern marked the release of *What Price Glory*, though it was not continuously roadshowed for as long. It played in New York City at the Sam Harris, then a legitimate 1200-seat theater, for seven months from November of 1926 to May of 1927. It roadshowed at the Garrick in Chicago from late December 1926 to mid-March 1927. The film moved into general release at about the same time in both cities, in August 1927 in New York (at the Roxy) and in September in Chicago (at the Monroe). Concurrent with its general release in Chicago, it opened in Syracuse and Davenport. A month later it opened in Fayetteville. It did not get to Bradford until January 1928, at which point its runs in the other cities in our sample were coming to an end. It thus took about 14 months to move through the system.

Sometimes films were not roadshowed in multiple cities but were pre-released at special prices on Broadway in order to build subsequent demand. Frank Borzage's *7th Heaven* (Fox, 1927) is a good example. The film was extremely well reviewed in the trade and popular press. *Variety* (11 May 1927: 14) wrote: "It is a great big romantic, gripping and red-blooded story told in a straight to the shoulder way and when the last foot of some 11,000 or so feet is unwound, if there is a dry eyelash on either man, woman or child, they just have no red blood. This Frank Borzage production is an out-and-out hit and one on the $2 order." The film opened at the Sam Harris (not at $2.00, but at $1.50 to $.50, all seats reserved) from May to September 1927. After a week's clearance, it moved into general release at the Roxy for two weeks. It opened in Chicago, Syracuse and Davenport in November. While the runs in Syracuse and Davenport were brief (as was normal for these markets), the run in Chicago was substantial. The film played for a month in the Loop at the Monroe, a 950-seat theater that seems to have served as the standard theater for Fox's product. Like other major films, it then received a downtown sub-run at the 300-seat Castle. During January 1928, after a clearance of a little over two weeks, it moved into a few large neighborhood houses, including several National Playhouses theaters and one L&T house. It saturated the Chicago nabes in the last week of January and the first week of February 1928. Concurrently, it also had a three-day sub-run at the Syracuse Theater and arrived at the Ozark in Fayetteville for a split week. It played for three days at Shea's in Bradford in February 1928. The release pattern of *7th Heaven* was thus much briefer than for more extensively roadshowed films: it was held on Broadway for five months, then moved through the rest of its markets in four months. While extremely successful, it was not of the same order as the first two in our sample.

Herbert Brenon's *Sorrell and Son* (United Artists, 1927) offers an example of a relatively prestigious film that opened without a roadshow of any sort. Adapted from a novel by George Warwick Deeping, the film starred the accomplished Broadway actor H. B. Warner. *Variety* (16 November 1927: 21) praised the film highly and it won Herbert Brenon the first Academy Award for direction. While it had a gala opening at the 2000-seat Rivoli theater on Broadway, it opened at popular prices. After playing the Rivoli for about a month, it went to the Brooklyn Strand at the end of December. It had downtown New York sub-runs at the Loew's State in January 1928 and at the Plaza in February. For reasons that are not entirely clear, the film opened in smaller cities and towns before it premiered in Chicago. It played in Syracuse, Davenport and even Bradford in February and March before arriving in the Loop in April. Its Chicago run was respectable but not stellar. It played at the 1700-seat United Artists Theatre from 13 April to 10 May. After a month's clearance, it moved to two large L&T neighborhood houses in June, followed by a large B&K neighborhood house at the end of the month. It began saturating the nabes in the beginning of July and wound down at the end of the month.

The nabes

In contrast to the high-budget prestigious A films, which typically received relatively long Broadway runs, were the films largely made for the nabes and/or for rural markets. For example, *Variety* (12 November 1927: 24) moderately praised *The Thirteenth Juror* (Edward Laemmle for Universal, 1927):

> For the not inconsiderable portion of the fan public who take their drama straight and unsmiling, "The 13th Juror" ought to prove first-rate entertainment. To the sophisticated the play takes itself rather seriously and absurdities creep in, but the story is direct and absorbing in its naive way, working up to a melodramatic climax that for theatric effect is not without its kick. . . . Of course, it's old stuff, but here presented with intensity and obvious aiming for effect that carries it through. A Sunday afternoon crowd at the Roxy might giggle at it, but the sentimental customers of a Washington Heights neighborhood will love its emotional splurge.

The Thirteenth Juror opened for one week at the Colony on Broadway in November 1927, followed by a week at Loew's New York and a week at the Plaza. These were far from distinguished Broadway houses. Loew's New York is described by the Cinema Treasures web site as a subsequent-run theater with the lowest admission prices on Broadway.[15] The Plaza is also described as a subsequent-run theater, and this claim is borne out by the fact that a large number of films in our sample showed at the Plaza after their initial Broadway run, usually preceded by a clearance of two to three weeks.[16] In December, while it was playing at the Plaza, *The Thirteenth Juror* opened in the Chicago neighborhood houses—it never played in the Loop. It opened sporadically in other markets over an extended span of time. A notice in the Syracuse paper suggests that it played at a Keith circuit theater in late April 1928, followed by two brief sub-runs at the Regent and the Harvard (double-billed with *The Big Killing* and *Nevada* respectively) in November. It played for two days at the Bradford Grand in mid-July 1928 and for two days at the Capitol in Davenport later that summer. The scattered nature of the opening dates suggests that there was not much of an attempt to capitalize on the advertising or word of mouth generated by the New York and Chicago screenings, presumably because there was not much to capitalize upon.

Hook and Ladder No. 9 (F. Harmon Weight for FBO, 1927), a drama about firemen, is a similar case. *Variety* (21 December 1927: 25) judged that: "This picture will be most appreciated by unsophisticated customers. Best for the neighborhoods and small towns. In the best places it would encounter tough sledding." We could find no record of this film playing on Broadway. It did not open in the Loop in Chicago nor in any of the major B&K or L&T neighborhood theaters. It first appeared for two days in the Jeffery, a National Playhouses theater in Chicago, and thereafter played brief runs (most just one day) in about twenty neighborhood theaters in February 1927. It played along with six acts of vaudeville at the Temple in Syracuse for a brief run in mid-November 1927, followed by sub-run showings in January (with *Dog of the Regiment*) and March 1928. It played two days at the Grand in Bradford in late May 1928 and does not seem to have shown at all in Fayetteville or Davenport.

A better-made and relatively high-budget programmer such as *Twelve Miles Out* (Jack Conway for MGM, 1927) received a more prominent if brief opening in the metropolitan theaters followed by much more time in the nabes. The film was highly praised by *Variety* at the time of its release for delivering more of both humor and suspense than expected, while *Film Daily*'s review (31 July 1927: 9) described it as: "A real romance without any fancy boudoir trimmings. A story with guts. . . . They'll have to can the drawingroom stuff for [John] Gilbert after this one. . . ." Largely on the strength of its star, it opened on Broadway at the Capitol in July 1927 where it was held over a second week. It made its Chicago debut at the Chicago, B&K's flagship 3500-seat Loop theater, where it also played for about two weeks. After a one week clearance, it moved to one of B&K's large neighborhood houses, the Tivoli, for a week followed by a week at the Uptown, another large B&K nabe house. After moving to two large L&T houses in succession, then to two Orpheum houses, it began saturating the nabes around 16 October, winding down in the first few days of November.

In late September, at the same time it was moving around the large nabe houses in Chicago, the film arrived in Syracuse playing a week at the Empire followed by two-day sub-runs at the Rivoli, the Regent, and the Harvard in rapid succession. It opened early in Davenport, showing for three days at the Columbia in July. It got to Fayetteville in mid-December 1927, sharing the bill with the University Band, and to Bradford the first week of January 1928. Thus despite good reviews and approximately two-week openings on Broadway and in the Loop, *Twelve Miles Out* remained largely a film for the neighborhood theaters and the smaller cities and towns, though it spent longer in those venues than did *Hook and Ladder No. 9* or *The Thirteenth Juror*.

We would argue that the distribution pattern followed for films like *The Thirteenth Juror* and *Twelve Miles Out* was not simply a matter of budget (though these are all clearly rather modest films in comparison with a roadshow, or even a semi-roadshow like *7th Heaven*). As the review of *The Thirteenth Juror* makes clear, the industry's press distinguished between the audiences for these films and for those that played for longer in the major metropolitan markets. The distinction between sophisticated and naïve taste, between "Broadway and Main Street," is a fundamental one for the industry trade press, as noted both in *The Decline of Sentiment* and Maltby's study.[17] The divide between metropolitan and other audiences is alluded to frequently in relation to the films in our sample. For example, *Variety* (16 November 1927: 21) worried that *Sorrell and Son*, despite the distinguished performance elicited from the star, might lack appeal in the "10 and 20 cent," that is, the smaller exhibition venues. While both *The Big Parade* and *What Price Glory* contain "rough" characters and what would have been considered "low" humor (e.g., a running gag involving a raspberry in *What Price Glory*), these films were nevertheless considered to be at the vanguard of literary taste. Thus, in a column entitled "On War Books" published in the *Chicago Tribune* (30 May 1926: D1), H. L. Mencken praises the realism of the play and of Vidor's film (presumably he had not yet seen the film version of *What Price Glory*):

> The sentimental war tale seldom, if ever, survives the war it celebrates: I can think of no example. Its sentimentality quickly kills it, and it is forgotten. The moment the bands cease to play a market begins to open for such harsher and sounder things as the novel "No More Parades," the drama "What Price Glory" and the movie "The Big Parade." For a while, perhaps, their harshness is too shocking to be borne. But soon or late they are embraced.[18]

The differentiation between the metropolitan theaters and the nabes is not apparent in every case, and it did not always pose problems for the major producer-distributors. For example, it does not seem to have been problematic for either *The Big Parade* or *What Price Glory*, which played with great success in all markets. Nevertheless, there were cases in which it became an issue. Films would play in rural districts or conservative sections of large towns which would never have been exposed to the latest *succès de scandale* on Broadway. Writing in 1926 about the film *Far Cry*, an exhibitor in Melville, Louisiana complained to the *Moving Picture World*: "Here it was a case of 'another lemon from the First National orchard.' These pictures may have gone over big in the large cities, but the average country patron does not enjoy eight reels of a cigarette smoking heroine, who makes unchaperoned visits to the hero's studio."[19] The industry was clearly aware of the disparities in taste with which it had to deal. The Formula, one of the first self-regulatory policies adopted by the film industry in 1924, states that: "The members of the Motion Picture Producers and Distributors of America, Inc. in their continuing effort 'to establish and maintain the highest possible moral and artistic standards of motion picture production' are engaged in a special effort to prevent the prevalent type of book and play from becoming the prevalent type of picture. . . ."[20] But it was not simply a matter of ignoring current literary and dramatic productions: Broadway was a more important market than Melville, Louisiana, and it was in such major metropolitan centers that the vertically integrated producer-distributors owned their theaters. Film producers thus had to steer a course between the minority whose tastes might be epitomized by the hip and irreverent *Smart Set* and the vast majority who remained loyal to Norman Rockwell and the *Saturday Evening Post*.

The divide between "Broadway and Main Street" becomes all the more apparent when we consider two genres: stories of rural life, which were generally considered a hard sell on Broadway, and sophisticated comedy, which was generally considered a hard sell west of the Hudson.

Rural life

Joseph De Grasse's *The Old Swimmin' Hole* (First National, 1921), starring Charles Ray, is an unusual example of a story of rural life, an experiment in a reduced "slice of life" plot told without intertitles. Ray had been very successful in a series of 1910s program features produced by Thomas Ince. He played juvenile male leads, in some ways the male equivalent of Pickford's girl-woman roles, except that Ray was more consistently associated with rural types than was Pickford.[21] In its review of *The Old Swimmin' Hole, Wid's* (20 February 1921: 2) noted that Ray was "best known and liked for his 'Rube' country boy portrayals." The press uniformly praised the film, and related the absence of intertitles to the simplicity of the plot. *Wid's* advised exhibitors: "Maybe you won't believe it but there are no sub-titles in it. The pictures tell the story and you understand it perfectly. There isn't any plot or 'intrikut' business. It's just a series of incidents in the life of a country boy but they're important enough to keep you interested all the time." *Variety* (25 February 1921: 42) judged *The Old Swimmin' Hole* to be well-made if inexpensive, and also thought that the lack of plot made it easy to produce a film without intertitles: "In some respects it is hardly a fair test, inasmuch as there is little or no story to the picture, merely a series of incidents in the life of a bucolic youth." The *New York Times* (28 February 1921: 16:2) defended the experiment: "It may be pointed out that a story more complex in plot and characterization would not lend itself to wordless treatment as 'The Old Swimmin' Hole' does. And this is all true . . . But there's more to the matter than this. The photoplay remains significant and promising. For, in the first place, no matter how easily its story may be told without words, 999 directors out of every 1,000 would have done the usual thing and used words."

Despite its experimental narrative, and a critical reception that suggested the film might well have appealed to "sophisticated" audiences, the film received minimal playing time in the metropolitan theaters, probably due to its star's "rube" persona. It opened on Broadway in late February 1921 at the Mark Strand, a 2750-seat Broadway theater, where it played for a week billed alongside the two-reel "juvenile comedy" *Edgar the Explorer*. The following week, the film moved to the Brooklyn Strand, where it played with *My Barefoot Boy*. At the end of April, it played at the Plaza as part of the "Junior Cinema Club." It opened in Chicago prior to its New York release, playing at the 750-seat Boston Theater in early February for one week. After a significant gap, it reappeared on Chicago screens in early April, when it played in two large B&K neighborhood houses. It moved into a number of other neighborhood houses in April and reappeared for a sub-sub run screening in July, at the 20th Century. In Syracuse it played for three days at the Strand in early June with a one-day sub-sub run the following year. It arrived in Bradford and Fayetteville in August. Further research on rural and especially Western and Southern markets might reveal that this film played more extensively in those areas, as Glancy and Sedgwick indicate was the case with films of a similar rural appeal, starring Will Rogers, in the 1930s.[22]

Laddie (James Leo Meehan for FBO, 1926), an adaptation of Gene Stratton Porter's best selling novel of 1913, concerned the relationship between two Indiana farm families who are initially at odds. The courtships of the older siblings are enabled by Little Sister, whom *Film Daily* (22 August 1926: 7) described as "a Pollyanna sort of youngster whose philosophy of happiness is contagious." *Film Daily* thought that the film was "likely to prove especially appealing to the small town audience," while *Variety* (11 March 1926: 17) judged that "it is a certainty that the picture will make its best score in the neighborhoods and add further to its lustre in the theatres outside the big cities." All the evidence suggests that this film simply died in the urban markets. It appears to have reached Davenport first among the cities in our sample. Originally intended for a four-day run, it played there for a week in mid-September 1926, supposedly drawing "record crowds." It did not have a major downtown Broadway release, just one day at the Plaza in mid-October 1926. Nor did it open in a major Loop theater. Rather it played neighborhood houses in Chicago during a two-week span in late November and early December 1927. In October, the film showed at the Empire in Syracuse (with a sub-run showing in June 1927). It arrived in Fayetteville in mid-November (though the print was late, shaving a day off its intended 3-day run and leaving the Ozark theater dark). It played in Bradford for three days in mid-July 1927, with a film called *Naughty Boy*.

It was not the case, however, that rural dramas inevitably failed in urban markets. The most obvious exception to this rule in our period was D. W. Griffith's *Way Down East* (United Artists, 1920), which was based on the play by Lottie Blair Parker, a combination melodrama that enjoyed great success at the turn of the century in a production by William Brady. Contemporary press commentary on the film emphasized that the source was old-fashioned but that it was redeemed by Griffith's spectacular staging of the rescue of the heroine as an icy river breaks up beneath her. *Variety* (9 October 1920: 36), extremely enthusiastic about the film ("it would be sacrilege to cut a single foot"), saw Griffith's role as that of transforming an old warhorse: "'D.W.' has taken a simple, elemental, old-fashioned, bucolic melodrama and 'milked' it for 12 reels of absorbing entertainment." For many intellectuals, however, the story Griffith had chosen to tell simply overwhelmed his treatment of it: they could see the appeal of the film, especially of its last minute rescue over the ice, but they still could not take it seriously. Writing anonymously in the *New York Times* (4 September 1920: 7), Alexander Woolcott quipped:

> Anna Moore, the wronged heroine of *Way Down East*, was turned out into the snowstorm again last evening, but it was such a blizzard as she had never been turned out into in all the days since Lottie Blair Parker first told her woes nearly twenty-five years ago. For this was the screen version of that prime old New England romance, and the audience that sat in rapture at the Forty-fourth Street Theatre to watch its first unfolding here realized finally why it was that D. W. Griffith has selected it for a picture. It was not for its fame. Nor for its heroine. Not for the wrong done her. It was for the snowstorm.[23]

Nonetheless, the high-budget, spectacular showmanship was clearly sufficient to put the film over in all markets. It opened in September 1920, and according to the *American Film Institute Catalog* "was originally released on a road show basis with twenty companies, including symphonic orchestras and effects. The film was shown in two parts with an intermission."

Other producers took note. One suspects that Harry Millarde's *Over the Hill to the Poorhouse* (Fox, 1920), based on poems by Will Carleton published in his *Farm Ballads* in 1873, was made to compete with Griffith's production. It opened in the same month (September

1920), and it followed *Way Down East* in a number of theaters. However, *Over the Hill* was not a big-budget spectacular along the lines of Griffith's film. The film concerns a long suffering mother who is neglected by her grown, respectable children and finally rescued from the poorhouse by the return of one of her sons, the black sheep of the family. All the reviewers praised Mary Carr's performance as Ma Benton. *Motion Picture News* (2 October 1920: 2703) wrote: "Her role is a crystallization of the ideals of mother-love, mother-devotion, and mother-understanding, and her acting, especially in the second act, is the most beautiful, the most exquisite, and the most touching feature of the production." *Exhibitor's Trade Review* (2 October 1920: 1919) was similarly impressed: "Mary Carr does honor to *your* mother, and to *mine*, and to the others, everywhere." So great was Carr's success that she was typecast: although middle-aged she played mother and grandmother roles, often in small parts, throughout the 1920s and early 1930s.

Variety (24 September 1920: 44) reported that the film initially did scanty business at the Astor in mid-September, but business built on Broadway due to word-of-mouth.[24] In a later review for a comparable mother-love story with the same star, *Variety* (review of *Drusilla with a Million*, 27 May 1925: 40) described Fox's distribution strategy for *Over the Hill* as "a forced Broadway showing," a process of keeping a film in the major Broadway theaters and allowing demand to build on the assumption that urban audiences would eventually flock to it. The evidence we have uncovered suggests that this was in fact the case.

The film played at the Astor from 17 September to 10 October 1920. Then it began to move around Broadway: first to Nora Bays (October 11–October 17), then the Central (18 October–7 November), then the Lyric (8 November–27 December), then the Broadhurst (27 December 1920–29 April 1921), then the Park (30 April–2 August). After a brief stint at the Central, it "approached Broadway again" at the end of September, arriving at the Tivoli. Screening announcements in the *Times* were for weeks accompanied by lengthy prose pieces (attributed to William Fox himself) that vigorously championed the film, including an excerpt from the maudlin letter of a deaf fan and complaints about the "landlords" that forced the film to keep moving venues. Near the end of its Broadway run, in September of 1921, the film began a roadshow release in Chicago. It played from 3 September to 8 November 1921 at the Woods Theater, a legitimate house, at high ticket prices ($1.50 except for the last six rows of the balcony according to one review).[25] After a six-week clearance, *Over the Hill* began showing at the Randolph, a small Loop theater (where it replaced *Way Down East*); it stayed for approximately three weeks. Following a briefer clearance (1–2 weeks), it hit a number of neighborhood houses in early February 1922. It re-opened in other neighborhood theaters in mid-May. The film arrived in Bradford and Fayetteville in the first and second weeks of January 1922.

While most of the trade press commentary on *Over the Hill* lauds it as what *Variety* called a "hymn to mother love," there are several hints that even as early as 1920 this kind of plot was considered old-fashioned and a tough sell in urban venues. *Variety* thought the film itself was too long, and was worried about the lack of a love interest. *Wid's Daily* (26 September 1920: 4) praised the film wildly, but noted that the "sob stuff" was laid on a bit thick at times. The reviewer concluded "everyone is going to enjoy this picture except the snobs and you don't have to bother about them." The *New York Times* (20 September 1920: 21), apparently the newspaper of record for snobs, was much less enthusiastic: "Seldom has a motion picture been so deliberately sentimental as this one. Its assault upon the emotions is undisguised and sweeping. It is ruthless in its mass attack. Its capacity for tear-water and gallery cheers is unlimited.... Without subtlety, without restraint, it pours out its appeal—its towering demand—for the laughs and the smiles and the sobs of the susceptible spectators—of whom, as its makers know, there are many—and many more." In a more ironic vein, Robert Sherwood, writing in the pages of the New York humor magazine *Life*, called *Over the Hill* "the sort of rural sob drama that warms the cockles of your heart, provided you have such outlandish features."[26] The examples of *Way Down East* and *Over the Hill* are important in that they demonstrate that the divisions between rural and metropolitan markets were not absolute, and that given a sufficiently spectacular presentation, and/or a compelling melodrama with a producer willing and able to hold a film on Broadway, it was possible to conquer the urban markets with rural fare. Jacobs' research on the maternal melodrama indicates that one continues to find examples of the successful placement of film genres usually aligned with rural or old-fashioned taste in major metropolitan markets throughout the decade and into the early 1930s.[27]

Sophisticated comedy

The distinction between sophisticated and naïve taste is one of the most far-reaching in the trade press discourse, as Maltby notes above. Presumably the stability of this idea derives from the fact that it referred back to the fundamental distinction between the metropolitan theaters, which were largely owned by the major producer-distributors and which determined access to the distribution system, and the rest. Thus the notion of sophistication is applied much more broadly than to the narrow subset of films dubbed "sophisticated comedy" which arose in the 1920s, following the model of Charlie Chaplin's *A Woman of Paris* (United Artists, 1923) and Ernst Lubitsch's *The Marriage Circle* (Warners, 1924). However, while it would be an error to reduce the idea of sophisticated taste as it appears in the trade press to the genre of sophisticated comedy, the study of the distribution of such films does provide a telling counterexample to the rural dramas discussed above.

Sophisticated comedies were thought to be understated (especially in relation to slapstick) and unusually risqué, thus posing the dual problems of a lack of appeal in the nabes and potential action by state and local censors. *Variety* (20 July 1927: 19) writes of *The World at Her Feet*, an adaptation of *Maître Bolbec et son mari* by Georges Berr and Louis Verneuil:

> These French triangle stories never seem to bull's-eye at the box office, though this one is a first-rate sophisticated comedy with excellent wise humor and a lot of sparkle. It has much elegance of atmosphere and a brisk play of wit.
>
> The answer seems to be that the fans run to either low comedy in domestic stories or high intense drama, and the graduations between the extremes don't register. There have been a score of suave comedies of this sort on the Broadway screen, but not one of them sticks in memory as a commercial success. . . .
>
> It's all very smooth and casual, without theatrical parade, and perhaps the screen public wants its dramatic punch delivered with more force than grace. . . . The French are a discriminating, fastidious people, sipping their pastimes like old wine. This American people gulp their screen and stage sensations like straight redeye.

Throughout the 1920s, films of this type garnered positive comments in the press but experienced relatively short runs. Moreover there was not much co-ordination between runs in New York and Chicago (a sign that the distributors were not attempting to build word of mouth in a systematic way, as Fox had done for *Over the Hill*). Upon its release, *The Marriage Circle* received good reviews in the trade press, the *New York Times*, the *Spectator* and *Exceptional Photoplays*. It was almost always compared to *A Woman of Paris* which was similarly praised for its innovative technique.[28] In a typical review, *The Moving Picture World* (16 February 1924: 581) compared these films as follows:

> With a technique as revolutionary as Chaplin's in "A Woman of Paris" and resembling it in its subtlety, he [Lubitsch] has handled a rather daring and sensational theme with simplicity and directness; concentration of action, incident and even sets being always evident. For instance, he confines his scenes to the particular portion of the set in which the action occurs, puts over his points with a minimum of footage, having his characters portray whole situations in a gesture, a look and even by absolute inaction at times. It is an excellent example of finely handled pantomime; there is a minimum of subtitles, but few are needed, for the situations are so deftly handled as to render them unnecessary.

Despite glowing reviews such as this, *The Marriage Circle* had a lackluster first run. It opened in New York for one week at the Mark Strand in early February 1924. Six weeks later, it played at the Loew's New Lexington for one week and two weeks later it showed at the Plaza. It opened in Chicago on March 8 at the Orpheum (a small Loop theater, part of a local chain) where it played for two weeks. After a clearance of approximately six weeks, it moved into a large L&T neighborhood house, then to the Stratford in mid-May. After a few days clearance, it went to play in several L&T and other neighborhood houses, moving quickly from one theater to the next and wrapping up its cycle at the end of the month. It played at the Empire in Syracuse for a week (30 March–5 April), with a three-day sub-run at the Regent a month later and it arrived in Davenport (at the Capitol) in the first week in June. As far as we can determine, it never played in Bradford or Fayetteville.

Sidney Franklin's charming *Her Sister from Paris* (First National, 1925) followed a similar pattern, though perhaps because of Constance Talmadge's star appeal, it opened at a better theater on Broadway, the Capitol, where it played for a week in August. After a week's clearance, it moved to Loew's New Lexington for a week and two weeks later spent seven days at the Plaza. Its opening in Chicago was delayed, perhaps because it experienced difficulties with local censorship.[29] It eventually opened on October 12 for a week in the Loop at B&K's Chicago. After a clearance of approximately three weeks, it went to two large B&K neighborhood houses, the Uptown and the Tivoli (it stayed at the latter for two weeks and the former for five days). It then went to two large L&T houses, the Senate and the Harding, and an Orpheum circuit house, the Diversey, for a three-day run. It entered a number of other neighborhood houses in December, and wrapped up its screenings by Christmas. It went on to play at the Syracuse Strand for a week-long run in mid-July the following year, followed by a 3-day sub-run at the Regent in late September. We have no record of it playing in Davenport, Bradford or Fayetteville.

By the time of the February release of Malcolm St. Clair's *The Grand Duchess and the Waiter* (Paramount, 1926), sophisticated comedy was an established trend and Adolphe Menjou, who had had only a small part in *The Marriage Circle*, was acknowledged as the predominant star for this type of film. Both *Variety* (10 February 1926: 40) and *Film Daily* (21 February 1926: 6) assumed that the film would be successful and that, unlike many others in the genre, it would not pose censorship problems. *Variety* also noted that Menjou went through his role "like Sherman through Georgia." Nonetheless, the film did not even have the benefit of an initial New York run. It opened first in the Loop at B&K's McVickers, where it ran for a week from 30 November to 6 December 1925. After a two-week clearance, it entered the large neighborhood theaters: the Uptown and the Woodlawn, where it played for a week, and the Stratford, where it played for three days. It opened the newly renovated, 900-seat Metro theater on New Year's Eve, then began moving around the neighborhood theaters with scattered screenings through the end of January.

After it had already gone through all its Chicago runs, the film opened at the Mark Strand in New York in the second week of February 1926. Menjou's growing association with the genre is evident in the fact that he was advertised as making a personal appearance. Despite this publicity push, and despite a good review from Mordaunt Hall in the *New York Times* (9 February 1926: 22), the film moved on to the Brooklyn Strand after only a week. Mordaunt Hall complained about this fact in a subsequent feature: "At the Mark Strand last week there was a sparkling farce comedy . . . an unusually clever piece of sophisticated fun. There were no slapstick barrages, no chasing people around the passages of hotels, no fast automobile race for the wind-up chapter, but just clever comedy. . . . 'The Grand Duchess and the Waiter' is such a fine picture that it is regrettable that it was not kept on Broadway for more than a week."[30] Perhaps in response to these and other glowing comments, *The Grand Duchess and the Waiter* played for a week at the Cameo in early April, sponsored by Symon Gould's International Film Arts Guild (a group which held repertory screenings of older American classics and European films).[31] It also had short Broadway runs at the Loew's Lexington, Sheridan and Plaza through the end of May.

The film did not do well outside of New York. It played a Sunday-to-Wednesday run in late March in Davenport (concurrently at a downtown theater and at the nearby military base), where local newspapers repeatedly referred to it as a "French comedy." It went on to play a week in Syracuse in early April, followed by a brief second run in early May. It spent two days in Fayetteville in late May and finally reached Bradford in August, playing for three days on a double bill with *Hell-Bent for Heaven*.

Throughout the 1920s, then, sophisticated comedies consistently had short runs in the major metropolitan theaters, even though they were described as appealing primarily to the supposedly sophisticated audiences in those markets. This apparent anomaly may be explained by the fact that the major downtown theaters on Broadway and in the Loop were considered places to build up the reputation of films in order to improve their performance in other cities and in the nabes and subsequent-run markets. The whole roadshow strategy was based on this principle: the distribution pattern did not simply license higher-than-normal ticket prices for the period in which a film was in limited release, it also helped build demand for the time when it went into general release. The same principle applies to the "forced Broadway showing" of a film like *Over the Hill*: the goal was to build up prices and audiences down the line.

Industry wisdom on how to handle sophisticated comedy went just the other way: these films, made with more grace than force, were considered unlikely to have broad appeal in the neighborhoods and small towns. In 1926, Sidney Kent, then Vice-President and General Manager of Famous Players–Lasky, told a reporter for the *Times* that: "'The Grand Duchess and the Waiter,' an excellent picture, had met with success in the larger cities, but he explained that it was too sophisticated or high-brow for the smaller places."[32] Given the assumption by industry executives (and by *Variety* reviewers) that sophisticated comedies had limited appeal, it was not considered worthwhile to keep them on Broadway for long, even when they generated good press. Thus, somewhat paradoxically, the films were not accorded long runs in the metropolitan markets for which they were thought to be best-suited (which fact all but guaranteed that they would do poorly in the nabes and smaller towns). One presumes that sophisticated comedies continued to be produced in the late 1920s because they brought studios prestige and good reviews, because they had a reliable market niche, and because they were relatively cheap to make (providing one already had under contract the services of brilliant directors like Ernst Lubitsch and subtle comedians like Menjou).

Conclusions

This pilot study has a number of limitations. Expanding the number of markets considered, and in particular extending the geographical boundaries of the research would make it possible to consider the less populated rural areas of the South and West, and thus take "hick taste" more fully into account. Gender remains a difficult category to deal with given the method proposed here. While gender is certainly an important means of differentiating the audience within the trade press discourse, it is difficult to correlate with any given type of theater.[33] The films thought to be preferred by the "flaps" or "jazz babies" in the cities do not necessarily tally with those thought to be preferred by the suburban matrons in the nabes. In addition, irrespective of whether films were thought to be preferred by male, by female or by mixed audiences, they played in the same sorts of theaters, and followed very similar distribution patterns. Thus the distribution of Henry King's *Stella Dallas* (United Artists, 1925), dubbed a "woman's picture" by *Variety* (18 November 1925: 42), looks much like *7th Heaven* which, as noted above, *Variety* thought would be appealing for "man, woman or child."[34] *Stella Dallas* was held for a run in New York city at special prices for six months (*7th Heaven* was held for five months). Then, like Borzage's film, it moved into a major theater in the Loop at popular prices, then fanned out into the neighborhoods and smaller towns for extended runs.

Despite the limitations of this study, its findings suggest that the hypothesis of a newly formed "mass audience" for the movies in the 1920s is not tenable. Indeed, the distribution system that was constituted as vertical integration took hold in this period was predicated on refined and far-reaching differentiations of the audience. These differentiations were marked by such factors as the length of roadshows or initial major metropolitan runs; the choice of theaters for opening runs; the price of the tickets for opening runs; the timing of movements from Broadway to other metropolitan centers and from metropolitan centers to smaller cities and towns; the timing and number of screenings in the nabes and sub-run markets; and the nature of the double bills or accompanying live acts (if any) with which films were screened. Moreover, these differences in distribution patterns were reinforced by the ways in which the trade press discussed particular films and understood audience preferences in various markets. The distinctions apparent in the trade press discourse between the metropolitan centers and the nabes, and between sophisticated and naïve taste, provide an important context for understanding the logic of the distribution strategies we have sought to highlight here, and help to explain the extent to which assumptions about where certain types of films normally played, and at what prices, were already institutionalized.

Notes

1 See, for example, Miriam Hansen, "Early Silent Cinema—Whose Public Sphere?" *New German Critique* 29, The Origins of Mass Culture: The Case of Imperial Germany (1871–1918) (Spring–Summer 1983): 147–84; Ben Singer, "Manhattan Nickelodeons: New Data on Audiences and Exhibitors," *Cinema Journal* 34, no. 3 (Spring 1995): 5–35. It should be noted that Singer's position has been contested by Robert C. Allen, "Manhattan Myopia; Or, Oh! Iowa! Robert C. Allen on Ben Singer's 'Manhattan Nickelodeons: New Data on Audiences and Exhibitors,' *Cinema Journal* 35, no. 3 (Spring 1996): 75–103.

2 Gregory A. Waller, *Main Street Amusements: Movies and Commercial Entertainment in a Southern City, 1896–1930* (Washington: Smithsonian Institution Press, 1995); Gregory Waller, ed., *Moviegoing in America* (Oxford: Blackwell Publishers, 2002); Kathryn H. Fuller, *At the Picture Show: Small-Town Audiences and the Creation of Movie Fan Culture* (Charlottesville: University Press of Virginia, 2001); Kathryn H. Fuller-Selley, ed., *Hollywood in the Neighborhood: Historical Case Studies of Local Moviegoing* (Berkeley: University of California Press, 2008); George Peter Potamianos, "Hollywood in the Hinterlands: Mass Culture in Two California Communities, 1896–1936," Ph.D. Diss., University of Southern California, 1998.

3 Maltby, "Sticks, Hicks and Flaps: Classical Hollywood's generic conception of its audiences," in *Identifying Hollywood's Audiences: Cultural Identity and the Movies*, ed. Melvyn Stokes and Maltby (London: BFI, 1999), 25.

4 Mark Glancy and John Sedgwick, "Cinemagoing in the United States in the mid-1930s: A Study Based on the *Variety* Dataset," in *Going to the movies: Hollywood and the social experience of cinema*, ed. Richard Maltby, Melvyn Stokes and Robert C. Allen (Exeter: University of Exeter Press, 2007), 173.

5 Maltby, 28; Lea Jacobs, *The Decline of Sentiment: American Film in the 1920s* (Berkeley: University of California Press, 2008).

6 Maltby, 23–24.

7 See, for example, the reviews of *Man Crazy* in *Variety* (21 December 1927: 22) and *Souls for Sables* (16 September 1925: 40). All subsequent references to *Variety* are given in the text.

8 Sheldon Hall and Steve Neale, *Epics, Spectacles and Blockbusters: A Hollywood History* (Detroit: Wayne State University Press, 2010): 44, note that by the late 1910s and 1920s, the idea of the "roadshow" had shifted from a highly individualized mode of distribution for feature films within a market dominated by variety nickelodeon programs to one which differentiated among features with "long runs in movie theaters and the showing of films at high prices in premiere venues prior to a later conventional release."

9 We accessed the *New York Times* and the *Chicago Tribune* through Proquest Historical Newspapers for which the general URL is http://www.proquest.com/en-US/catalogs/databases/detail/pq-hist-news.shtml (we actually accessed the site through our University library). We utilized http://www.newspaperarchive.com to access newspapers for Syracuse: *Syracuse Herald, The Post Standard*; for Bradford: *The Bradford Era*; for Davenport: *Davenport Democrat and Leader, Davenport Daily Leader, Davenport Democrat*; for Fayetteville: *Fayetteville Daily Democrat.*

10 Douglas Gomery, *The Hollywood Studio System* (London: British Film Institute, 1986), 28.

11 John Driscoll, entry on "Laurence Stallings," Source Database: *Dictionary of Literary Biography*, vol. 44: *American Screenwriters*, Second Series, ed. Randall Clark (A Bruccoli Clark Layman Book, The Gale Group: 1986): 357–63.

12 There were two reviews of the film in *Variety*, one which reported on the print screened at Grauman's Egyptian Theater in Hollywood on November 9 (11 November 1925: 36) and one which reported on the print screened at the Astor in New York City on November 19 (2 December 1925: 40). The latter review notes some censorship cuts made between the Los Angeles and New York screenings, but both prints ran 130 minutes.

13 The Astor is listed as a Loew's theater in the *Film Year Book* for 1926 (New York: Wid's Films and Film Folks), 625.

14 Lubliner & Trinz is listed as a subsidiary of Balaban & Katz in the *Film Year Book* for 1927 (New York: Wid's Films and Film Folks), 649, but they may not have been affiliated prior to that date.

15 See http://cinematreasures.org/theater/15178/(accessed 27 November 2010).

16 These include *Why Change Your Wife, The Sheik, The Marriage Circle, Her Sister from Paris, Mantrap, Three Hours*, and *Sorrell and Son*.

17 See *The Decline of Sentiment*, 79–90.

18 The relationship between the play *What Price Glory*, the various film versions, and vanguard literary tastes is discussed in detail in *The Decline of Sentiment*, chapter 4.

19 H. H. Hedberg, Melville, Louisiana, "Straight from the Shoulder Reports," *Moving Picture World* (13 November 1926): 106.

20 Richard Maltby, "'To Prevent the Prevalent Type of Book': Censorship and Adaptation in Hollywood, 1924–34," *American Quarterly*, vol. 44, no. 4 (December 1992): 554–82.

21 Pickford certainly played rural types, most prominently in *Rebecca of Sunnybrook Farm* (1917), *Pollyanna* (1920), and the Western *M'liss* (1918), but some of her most successful comedic roles were in an urban setting: *Amarilly of Clothes-Line Alley* (1918), *The Hoodlum* (1919), and *Suds* (1920).

22 Glancy and Sedgwick, 160.

23 For some reason this review does not appear in the anthology *The New York Times Film Reviews 1913–1931*; it is reprinted in George Pratt, *Spellbound in Darkness: A History of the Silent Film* (Greenwich, Connecticut: New York Graphic Society, rev. edition 1973): 252. For more on the reception of *Way Down East*, see *The Decline of Sentiment*, 185–87.

24 In *Variety*'s review (24 November 1931: 17) of the 1931 remake of *Over the Hill*, the paper noted the film "was a furore in 1920, starting slowly and growing into a country-wide sensation."

25 "A Mother Picture with a Great Mother," *Chicago Tribune* (5 September 1921: 14).

26 The quote is from Sherwood's encapsulated reviews in the "Recent Developments" section of his column, *Life* (17 March 1921): 396.

27 Lea Jacobs, "Unsophisticated Ladies: the Vicissitudes of the Maternal Melodrama in Hollywood," *Modernism/Modernity* 16, no. 1 (January 2009): 123–40.

28 For more on the critical reception of the two films, see *Decline of Sentiment*, 92–96 and 99–101.

29 Censorship difficulties are alluded to in two reviews: "Here is Connie Twice as Good as Ever Before," *Chicago Tribune* (12 October 1925: 25); "Blue Ribbon Films Viewed Last Month," *Chicago Tribune* (1 November 1925: H1).

30 "The Beautiful Duchess and the Hopeful Waiter," *New York Times* (14 February 1926: X4).

31 For examples of other repertory screenings organized by Symon Gould see the *New York Times*, "The Screen" (4 June 1926: 26); "Finding Unusual Films" (5 December 1926: X11); "The Screen" (12 October 1929: 18).

32 "How Pictures Are Sold Explained by Executive," *New York Times* (19 September 1926: X5); see also Sidney Kent "Distributing the Product," in *The Story of the films; as told by leaders of the industry to the students of the Graduate School of Business Administration, Harvard University*, ed. Joseph P. Kennedy (Chicago: A.W. Shaw, 1927), 203–32.

33 For more on the complex ways in which films are differentiated by gender in the trade press discourse, see *The Decline of Sentiment*, chapters 4 and 6.

34 In its second review of *The Big Parade, Variety* (2 December 25) dubbed it a "man's picture": "Coming in the same week as 'Stella Dallas,' the contrast was striking, the former being very much of a woman's film and 'Parade' having a strong masculine appeal."

PART II

Sound and the studio system, 1926–46

Introduction

This section of the Reader begins with the advent of a number of new sound technologies in the mid- to late 1920s, the policies that governed their adoption, and their effects on the structure and practices of Hollywood, both short term and long. It goes on to consider the organisation of production and distribution, the impact of the Great Depression, the role or roles played by independents, cinema-going and cinema programming, the industrial role played by film stars, the impact of World War Two, and the involvement of the US government in the industry's policies and practices.

As has now become a cliché, 'silent' films were nearly always accompanied by live sound performances of one kind or combination or another: 'a spoken lecture or commentary; a live musical performance, ranging from a single instrumentalist to a live ensemble or orchestra; sound effects corresponding to, and generated in approximate synchronisation with the action in the film; and, more rarely (in the European and North American film industries at least), a theatrical performance by actors' either in full view of the audience or else hidden behind the screen.[1] It is a further cliché that some of the earliest films produced by Thomas Edison in the US were designed as visual accompaniments to phonograph recordings and that the silent era was peppered with sound systems (some using cylinders, others using discs), which often served as feature attractions in their own right. However, as Douglas Gomery explains in 'The Coming of Sound' (Chapter 7), nearly all these systems suffered from two major problems: inadequate amplification and unreliable synchronisation. As he also explains, these problems were largely overcome by the mid-1920s, thanks in the main to electronic technology, and thanks in part to the efforts of small-scale technological entrepreneurs such as Lee De Forest, Earl I. Sponable and Theodore W. Case, and in part to the research programmes of large-scale radio and telecommunications corporations such as the American Telephone & Telegraph Corporation and its subsidiary, Western Electric, and General Electric and Westinghouse and its subsidiary, the Radio Corporation of America (RCA). The result was a number of viable systems, some based on the recording and reproduction of sounds on discs, others on strips of film.[2]

Gomery's account goes on to stress three key aspects of the adoption of sound in the US. The first is the extent to which smaller companies such as Warner Bros. and Fox pioneered sound as part of a programme of expansion. The second is the extent to which the sound-on-disc system adopted by Warner Bros. (and indeed the Phonophone sound-on-film system devised by De Forest and Case) was initially used as a means of providing pre-recorded orchestral scores and pre-recorded prologues, presentations and high-class vaudeville acts at a time when their live equivalents were standard attractions in premier venues but not in ordinary theatres.[3] (The same is at least partly true of Fox, which used its sound-on-film system to record vaudeville acts and musical performances prior to its use as a means of enhancing its newsreels. The success of part-talking and all-talking films such as *The Jazz Singer* [1927] and *The Singing Fool* [1928] came as something of a surprise.)[4] The third is the extent to which five of the then major companies – MGM, Universal, First National, Paramount and Producers Distributing Corporation – monitored the systems adopted by Warner Bros. and Fox in an attempt to agree on a system that would suit everyone. Complications arose when RCA launched Photophone, another sound-on-film system, in 1927. By then Western Electric had devised a sound-on-film system too. The majors opted for the latter. In retaliation, RCA acquired a production company, national and international distribution facilities and a chain of theatres, and set up RKO, a new vertically integrated major. Along with Phonophone, Photophone continued to be available throughout the 1930s, proving particularly popular with smaller independent theatres and chains. In the meantime, since 'an enormous potential for profits existed' (and since large amounts of money would need to be spent on production facilities, retraining and new personnel), 'it was incumbent on the majors to make the switchover as rapidly as possible'. By the end of 1929, Paramount, MGM, Fox, RKO, Universal, United Artists and Warner Bros. had all 'completed their transitions'. First National had been acquired by Warner Bros. and the Producers Distribution Company by Pathé in 1928. Pathé was acquired by RKO in turn in 1931. By the end of 1932, Gomery concludes, most of the remaining unconverted independent production companies and most of the remaining unwired theatres had been equipped with sound technology.

Some have challenged the idea that the transition to sound in the US was as orderly as Gomery suggests. Focusing on small-scale independents such as Liberty, Gotham, Sono-Art and Tiffany-Stahl, often known collectively as 'Poverty Row', Paul Seale has drawn attention not only to their importance as suppliers of films for double bills (the showing of two feature films for the price of one), but also to their role as consumers of small-scale sound production systems, as beneficiaries of a price war between RCA and Western Electric, and as suppliers of sound shorts and sound and silent Westerns (the latter for unwired rural theatres) 'against the dominant industry trends'.[5] Either way, however orderly or disorderly the transition to sound may have been in the US itself, the advent and impact of new sound technologies in Europe and elsewhere abroad led to a series of experiments, crises and decisions that were to have a considerable impact on the nature of film production and distribution and on the import and export of films in either the short term or the long term, or both.

Following the successful introduction of sound in the US, Europe witnessed the formation of Tonbild-Syndikat ('Sound Film Syndicate') and a rival company known as Klangfilm. These companies merged to form Tobis-Klangfilm in 1929. Tobis-Klangfilm threatened the international hegemony of US companies and US sound systems. In response, the MPPDA and the firms allied to Western Electric agreed to establish 'an international cartel that would divide up the world market. Tobis-Klangfilm received exclusive rights to sell sound equipment in Germany, Scandinavia, most of eastern and central Europe, and a few other countries. American manufacturers gained control of Canada, Australia, India, the USSR, and other regions. Some countries, such as France, remained open to free competition'.[6] However, while the availability of sound technology was internationalised, the advent of talking pictures created international language barriers that were harder to circumvent. By translating their inter-titles, silent films had proven relatively cheap and easy to adapt to foreign-language markets. Talking pictures were a different matter. Because it was initially difficult to remix dialogue and other forms of sound in post-production, dubbing different sound tracks in different languages was either expensive or clumsy or both. Expedients such as adding subtitles or eliminating spoken dialogue and substituting inter-titles or foreign-language commentaries were tried but rejected. By '1929, many producers decided that the only way to preserve foreign markets was to reshoot additional versions of each film, with actors speaking different languages in each'.[7]

As Ginette Vincendeau explains in 'Hollywood Babel' (Chapter 8), MGM produced multiple language versions (MLVs) in Hollywood, Paramount in Joinville near Paris.[8] As she goes on to point out, MLVs were usually filmed simultaneously 'in one of several fashions'. Films shot in two or three languages would often be directed by the same person, but if more languages and versions were involved the number of directors was likely to increase. In most cases, different sets of actors would be cast for different versions, but polyglot stars like Claudette Colbert and Maurice Chevalier would feature in more than one version, while other members of the cast with speaking parts would be changed. In addition, whenever MLVs were based on the same source material but remade in different languages at different times, the mix of variables tended to increase. As Vincendeau goes on to point out, MLVs exemplified the tensions between film as an industry and as an art form, film as a vector of national and international values, and film as a commodity designed for national and international markets. The production and variable reception of MLVs seemed to highlight the limitations as well as the range of Hollywood's power. With the advent of the Depression and of nationalist governments and policies, Film Europe died away, national quotas were reinforced and Hollywood's exports declined in value. However, although markets in Germany and the USSR 'became less accessible to American films, the remaining foreign countries lost any particular coherence in resisting American competition'.[9] US exports picked up again in 1935 then fell off in the late 1930s and the early years of World War Two.[10] In the meantime, as political developments in Europe swelled a further influx of overseas actors, directors and other production personnel,[11] the introduction of a new Moviola in 1930 made it possible to mix separate sound tracks after shooting.[12] As a result, original music and sound effects could be combined with new and more easily synchronised voices. The practice of dubbing increased and MLVs were eventually abandoned.[13]

The technological and stylistic aspects of the advent of sound and its impact on the nature of films, on film production, and on the careers of actors, performers and other production personnel have been detailed at length in a number of publications.[14] To a lesser extent, the same is true of the ways in which the advent of sound established, cemented or extended links between the film industry, the radio and music industries, and the legitimate, musical and variety theatres.[15] From an industrial and organisational perspective, the advent of sound led, on the one hand, to a reconfiguration of a number its practices and departmental structures and, on the other, to a newly reconfigured oligopoly. A detailed snapshot of a range of departmental structures, activities and practices in the 1929–30 season is provided in 'Organization' by Howard T. Lewis (Chapter 9). An overview of the structures, practices and history of the new oligopoly between the late 1920s and the mid-1940s is provided in 'Hollywood: The Triumph of the Studio System' by Thomas Schatz (Chapter 10).[16]

Lewis looks in turn at the organisation of production at Paramount, distribution at Pathé and United Artists, and exhibition, distribution and the management of theatres at Fox. Worthy of particular note are the sheer scale of these operations, the number of activities, functions and personnel they each involved, and the figures, costs and other details that Lewis provides.[17] Worthy of note as well is that Paramount planned its productions 'about the first of April', that the 'general manager in charge of distribution' rather than production 'states the number of pictures desired for the coming year', that 'Paramount pictures are released on an average of one each week, with two released during one week in every four', that 'the best-quality feature pictures' were rarely released in the 'late spring, summer, and early fall', and that the sales department at Fox 'often withheld the sale of some Fox pictures to certain Fox theatres because it was possible to derive a greater profit from a sale to a rival theatre-operating company'. However, with the exception of large first-run theatres, whose films were booked individually, and with the exception of United Artists, whose films were rented individually, Paramount, Pathé, Fox and most of the other major producer-distributors generally sought to rent their films in blocks.[18]

Schatz begins his overview with the coming of sound and the delayed but devastating impact of the Great Depression. Because of the Depression, theatre admissions and industry revenues fell drastically, thousands of theatres closed their doors, and Paramount, Fox and RKO 'suffered financial collapse' in the early 1930s.[19] However, the members of the MPPDA were able to restore their fortunes by perpetuating their 'run-zone-clearance' distribution system and practices such as blind bidding and block booking thanks largely to President Roosevelt's National Industrial Recovery Act (NIRA), which sought 'to promote recovery by condoning certain monopoly practices by US industries – including motion pictures'.[20] In order to 'mitigate the potential for worker exploitation and abuse', the NIRA also 'authorized labour organization and collective bargaining'. As a result (though not without resistance from a number of company managers), the 1930s and early 1940s witnessed the foundation of Screen Actors Guild (SAG), the Screen Directors Guild (SDG) and the Screen Writers Guild (SWG) as 'Hollywood evolved from an essentially "open shop" to a "union town"'.[21] Following an account of management practices and structures, double billing, Poverty Row and the key role played by stars and the 'A-class feature', Schatz goes on to detail production policies, categories and trends within and across a number of different companies before turning to two key developments in 1940–41.[22]

The first of these developments was the signing of a consent decree by the Big Five vertically integrated majors in response to the filing of an anti-trust suit on behalf of the nation's independent theatre owners by the US Justice Department in 1938. The NIRA had been declared unconstitutional in 1935, and in signing the decree ('essentially a plea of no contest') these companies 'agreed to limit block booking to groups of no more than five pictures, to end blind bidding by holding trade screenings of all features and to modify their run-zone-clearance policies'. At this point, charges of collusion by the Little Three were dropped, though neither the independent exhibitors nor the Justice Department were fully satisfied and the 'Paramount case', as it was called, was resurrected – with a number of major consequences – in 1945.

The second development was 'the outbreak and escalation of war overseas', which devastated Hollywood's overseas trade with the Axis countries and occupied Europe, and which had a number of important economic, industrial, political and demographic effects on the US film industry both before and after the bombing of Pearl Harbor in 1941. Among the former were increases in domestic taxation, uncertain forays into war-themed productions, increasing demands for 'top features' as a result of the consent decree, increasing employment and an 'overheated first-run market' in cities associated with the defence built-up, and an increase in the authority granted to both freelance and in-house contract talent. Among the latter was a 'five-year war boom'. Restrictions were placed on production materials and film stock, but there was an increasing rapport with government, which 'now saw the "national cinema" as an ideal source of diversion, information and propaganda for citizens and soldiers alike', which mandated genres such as the combat film and home-front drama, which welcomed the war-related re-tooling of many of its musicals, and which helped promote films with appeal to Hollywood's remaining foreign markets in the UK and Latin America.[23] Reducing their overall output, the major companies were able to concentrate their output on '"bigger" pictures which played longer runs and enjoyed steadily increasing revenues'.[24] Fuelled by returning service personnel, these revenues were to reach a peak in 1946.

The remaining contributions to this section of the Reader deal with cinema-going and with the industrial role played by stars in the 1930s. Before detailing their contents, one or two additional points should be made about production supervision, independent production and double billing. Schatz argues that a 'central producer system, wherein one or two studio executives supervised all production, typified the mass-production mentality and factory-oriented operations of the studio system' in the early 1930s, that this approach 'gradually gave way to a more flexible "unit" system', and that this shift 'was fairly pervasive by the late 1930s'. However, as he himself notes, the producer-unit system (in which relatively small groups of films were supervised by a producer or producer-director) was adopted by nearly all the major companies in 1931. Prompted, as Matthew Bernstein has pointed out, by the need 'for greater cost control and for diversity of output in a period of box-office decline' as well as by the growing demand for double bills, the 1930s witnessed the advent of B-unit production at the majors and a growth in A-scale production by independents such as Walter Wanger and David O. Selznick (whom Bernstein views as independent equivalents to in-house unit producers) as well as in A-scale unit production by the majors.[25] In other words, unit production existed in many guises, forms and contexts and may have been in place even earlier than Schatz suggests. Whatever the case may be, the move toward unit production was not universal, for, as Schatz points out, Darryl Zanuck 'personally supervised all of Fox's A-class pictures' from the point at which he and Joe Schenck merged Fox and Twentieth Century Pictures in 1935. He might have added that Harry Cohn personally supervised Columbia's features too, and that while Wanger was a relatively hands-off producer, Selznick tended to micromanage the films he produced from start to finish.[26] Wanger and Selznick both supplied films for United Artists in the 1930s and 1940s, and Schatz notes that 'by 1940–41 UA's strategy of distributing major independent productions had been adopted by four other studios: RKO, Warners, Universal and Columbia'. He also points out that 'all the studios also saw a marked increase in the number of "hyphenates" under contract ... notably writers and directors who attained producer status'.

As has already been noted, the practice of double billing became particularly widespread in the 1930s as a means of attracting audiences during the Depression. As Schatz points out, most of the nation's theatres 'began showing two features per programme, and changed programmes two or three times per week. The increased product demand was met largely via B movies – that is, quickly and cheaply made formula fare, usually Western or action pictures, which ran about sixty minutes and were designed to play in subsequent-run theatres outside the major urban markets'. This is largely true. However, it should be pointed out that double billing, which originated in New England in the late 1910s, was practised in the 1920s too.[27] It should also be pointed out that the distinction between A films and B films was by no means always clear cut. As Lea Jacobs has noted, inexpensive features such as *The Informer* (1935) and B productions such as *Pacific Liner* (1938) could find themselves treated as A films, and costly A features such as *Room Service* (1938) could find themselves relegated to the second half of double bills. Moreover, there were incentives for studios 'to sell attractive B films to exhibitors if they could. This was especially the case for so-called "intermediates", which cost between $250,000 and $500,000. The cost of "intermediates" precluded the studio making a profit unless some exhibitors could be persuaded (or coerced!) to pay a percentage of the gross rather than the flat rental fee usual for Westerns and other B pictures'.[28]

Along with stage shows, competitions and other forms of live entertainment, double billing, single billing and the programming of shorts and newsreels feature extensively in 'Cinemagoing in the United States in the mid-1930s' by Mark Glancy and John Sedgwick (Chapter 11). Drawing on the weekly reports on the exhibition market published in *Variety*, Glancy and Sedgwick note that 'the standard running time for a cinema programme was just under three hours' and that 'programmes could include one or two feature films, a live stage performance, cartoons, newsreels, short films, as well as "bank nights" or "giveaways"'.[29] Thus they underline the extent to which 'the cinemagoing experience was seldom if ever limited to a single film ... Rather, the cinema offered a diverse range of attractions and entertainment forms, in which a single feature film might be a secondary or even an incidental consideration'. In addition, they stress the heterogeneity of the exhibition market, noting that 'there was a remarkable gap in regional tastes; a gap between "classy" metropolitan audiences, who paid "top price" for biopics, adaptations of Shakespeare and exotic melodramas, and the more provincial audiences, who preferred their entertainment to be decidedly less expensive and less exotic', though they also draw attention to the fact that the 'top-earning films' in the mid-1930s were either costume dramas or musicals (or both). Finally, they draw attention, too, to the ways in which double billing and the consequent increase in the demand for films benefited not just Poverty Row producers, but producers and importers of foreign films as well. Of these, by far the most numerous were imports from Britain.

The popularity of films was governed by a number of factors, one of which was stars. Based initially on theatrical precedents, stardom in the cinema became a major phenomenon in the 1910s. By the mid-1910s, stars were distinguished from other performers. Along with their literal and metaphorical images, their names (whether real or invented) were circulated in the media, used to promote the films in which they appeared, and used to sell those films to exhibitors and in turn to the general public. Some of them set up their own production companies and units, a practice that continued into the 1920s.[30] By then, as Tino Balio points out in 'Selling Stars' (Chapter 12), onscreen performers were categorised as supporting players (or extras), stock players (or character actors), featured players or stars, and as such paid at different rates and under different contractual arrangement.[31] As he also points out, some stars worked on a freelance basis, others under 'option contracts that lasted as long as seven years'. These contracts could be terminated at any time. They enabled the studios to insist on the appearance of stars in specific films and to loan them out to other studios. However, along with a number of directors and writers, stars were often represented by agents, who would often act as advisors to the studios as well as to their clients and who would often mediate between conflicting parties.[32] In order to minimise the influence of stars and their agents, the major producers sought to introduce a 50 per cent pay cut in 1933. Led by Eddie Cantor, a number of actors campaigned against the cuts and helped establish the SAG.[33]

Cantor himself was one of a number of stars whose backgrounds were in vaudeville, theatre or radio. As such, while radio in particular acted as an additional medium of promotion for the studios, he and they were central to a number of new production trends that drew on these forms. In some instances, this simply meant transferring ready-made and well-known public personas and performance skills to the screen. In others and in general, however, performance skills had to be taught, physical characteristics altered, and personas built by trial and error through different 'star-genre formulations' or 'star-formula combinations', as Schatz has aptly called them.[34] Along with their names and images, these personas were circulated in fan magazines and other organs of studio publicity and used to advertise the industry's films. As such the value of stars was economic as well as aesthetic; as such it was measured in box-office returns as well in polls and other indices of popularity; and as such, stars themselves functioned as 'brands' for individual films, for the studios that made them and, indeed, for Hollywood itself.[35]

Notes

1 Leo Enticknap, *Moving Image Technology: From Zoetrope to Digital* (London: Wallflower Press, 2005), 102. For more on sound and music in silent era cinema, see Richard Abel and Rick Altman (eds), *The Sounds of Early Cinema* (Bloomington: Indiana University Press, 2002); Rick Altman, *Silent Film Sound* (New York: Columbia University Press, 2004); Gillian

Anderson, *Music for the Silent Films, 1894–1929* (Washington, DC: Library of Congress, 1988); Mervyn Cooke, *A History of Film Music* (Cambridge: Cambridge University Press, 2008), 1–30; Kathryn Kalinak, *Settling the Score: Music and the Classical Hollywood Film* (Madison: University of Wisconsin Press, 1992), 40–65; Martin Miller Marks, *Music and the Silent Film: Contexts and Case Studies, 1895–1924* (New York: Oxford University Press, 1997), Roy M. Prendergast, *Film Music: A Neglected Art* (New York: W.W. Norton, 1992 edition), 3–18. It is important to point out that a number of popular songs were written for or otherwise tied into the promotion of a number of silent films. See Jeff Smith, *The Sounds of Commerce: Marketing Popular Film Music* (New York: Columbia University Press, 1998), 27–29.

2 These were all analogue systems in which patterns of vibration emitted from a sound source were recorded as a continuously modulated groove on a disc or cylinder or as a continuously modulated pattern of light on a strip of film. These patterns were reproduced either by placing a needle in the groove and moving the cylinder or disc at the requisite speed or by moving the strip of film past a photoelectric cell, a device that converts patterns of light back into sonic vibrations.

3 Hall and Neale, *Epics, Spectacles and Blockbusters*, 78; Douglas Gomery, 'The Economics of U.S. Exhibition Policy and Practice', *Cine-Tracts*, vol. 3 no. 4 (Winter 1981), 37. Prologues and presentations were live performances whose ingredients were either linked to the subject matter of the films they preceded (in the case of the former) or not (in the case of the latter).

4 In his article, Gomery focuses on the success of *The Jazz Singer*. But in his recent book, also entitled *The Coming of Sound* (New York: Routledge, 2005), 61, he underlines the importance and success of *The Singing Fool*. According to Donald Crafton, *The Talkies: American Cinema's Transition to Sound, 1926–1931* (New York: Scribners, 1997), 273 and 549, *The Singing Fool* cost $388,000 and grossed $3,821,000 domestically and a staggering $5,916,000 worldwide. For this reason alone, writes Gomery: 'We should abandon the myth of *The Jazz Singer* and replace it with economic reality of *The Singing Fool*. This is what the industry saw from the fall of 1928 through the spring of 1929. *The Singing Fool* convinced all doubters that talkies were not just a fad. They represented – by the public embracement of this film by paying their dollars – the turning point of a new era'.

5 Paul Seale, '"A Host of Others": Toward a Nonlinear History of Poverty Row and the Coming of Sound', *Wide Angle*, vol. 13 no. 1 (January 1991), 94. For more on Poverty Row, see Michael R. Pitts, *Poverty Row Studios, 1929–1940* (Jefferson, NC: McFarland, 1997), and Brian Taves, 'The B Film: Hollywood's Other Half', in Balio, *Grand Design*, 313–50.

6 Thompson and Bordwell, *Film History*, 201.

7 Thompson and Bordwell, *Film History*, 210. As Thompson points out in *Exporting Entertainment*, 162, RKO used the Dunning process, a process in which 'the backgrounds and extras were filmed in American studios, with these shots being sent abroad; foreign producers could then film local artists in their native language against the Hollywood scenes in back projection'. It is important to note that Hollywood's MLVs were not just produced for foreign consumption. As Colin Gunckel has shown in 'The War of the Accents: Spanish Language Hollywood Films in Mexican Los Angeles', *Film History*, vol. 20 no. 3 (2008), 325–43, Spanish-language MLVs were made for Spanish-speaking audiences in Los Angeles as well as for export to Spain and Latin America. Finally, as Richard Maltby and Ruth Vasey have pointed out in 'The International Language Problem: European Reactions to Hollywood's Conversion to Sound', in David W. Ellwood and Rob Kroes (eds), *Hollywood in Europe: Experiences of a Cultural Hegemony* (Amsterdam: VU University Press, 1994), 88, it should be noted that subtitled versions found 'a satisfactory response in several markets, particularly in Latin America, where subtitled movies were often preferred to those dubbed, or indeed produced, in standard Castilian Spanish. Conversely, in Portugal, audiences found the Brazilian Portuguese of the casts employed by American companies to be unendurable, and pleaded for subtitles instead. As these instances illustrate, subtitles not only solved the problems of translation, they also managed to ameliorate some of the problems produced by the cultural specificity that characterised sound production, re-introducing some of the advantages of semantic flexibility provided by silent intertitles'.

8 For a detailed listing of the films produced by Paramount at Joinville, see Harry Waldman, *Paramount in Paris: 300 Films Produced at the Joinville Studios, 1930–1933, with Credits and Biographies* (Lanham, MD: Scarecrow Press, 1998).

9 Thompson, *Exporting Entertainment*, 169–70. It should, however, be noted that sound served or continued to boost the film industries in France, Eastern Europe, India and Latin America, helping to increase the proportion of domestically produced films shown in their own exhibition markets. See Thompson and Bordwell, *Film History*, 210. For detailed studies of the distribution and exhibition of imported and domestically produced sound films in Australia and in the Netherlands in the mid-1930s, see John Sedgwick, 'Patterns in First-Run and Suburban Filmgoing in Sydney in the mid-1930s', Mike Walsh, 'From Hollywood to the Garden Suburb (and Back to Hollywood): Exhibition and Distribution in Australia', and Clara Pafort-Overduin 'Distribution and Exhibition in the Netherlands, 1934–36', in Maltby, Biltereyst and Meers (eds), *Explorations in New Cinema History*, 140–58, 159–70 and 125–39, respectively.

10 According to John Sedgwick and Mike Pokorny, 'Hollywood's Foreign Earnings During the 1930s', *Transnational Cinemas*, vol. 1 no. 1 (2010), 83, 'the idea that Hollywood garnered its profits from overseas, while meeting production costs at home, is too simplistic'. On average, Hollywood releases generated twice as much income from the domestic market than they did from markets abroad. The 'most successful films overseas were big budget films that needed to be highly popular with domestic audiences if they were to be profitable'. It should be noted here that Hollywood's 'domestic markets' included Canada.

11 Helmut G. Asper and Jan-Christopher Horak, 'Three Smart Guys: How a Few Penniless German Émigrés Saved Universal Studios', *Film History*, vol. 11 no. 2 (1999), 134–53; Jan-Christopher Horak, 'German Exile Cinema', *Film History*, vol. 8 no. 4 (1996), 373–89; John Russell Taylor, *Strangers in Paradise: The Hollywood Émigrés, 1933–1950* (New York: Holt, Rinehart and Winston, 1983).

12 A Moviola was a machine that enabled exposed film stock to be viewed frame by frame by those involved in editing. The new Moviola enabled various optical sound takes, including those in different languages, to be cut in correlation with the image track.

13 For more on sound, Europe and overseas markets, and MLVs and foreign-language imports, see Crafton, *The Talkies*, 418–41.

14 In addition to the contribution to this Reader by Helen Hanson and Steve Neale (Chapter 14), these publications include Bordwell, Staiger and Thompson, *The Classical Hollywood Cinema*, 294–308; Evan William Cameron (ed.), *Sound and the Cinema: The Coming of Sound to American Film* (New York: Redgrave Publishing Company, 1980); Crafton, *The Talkies*; Scott Eyman, *The Speed of Sound: Hollywood and the Talkie Revolution* (New York: Simon Schuster, 1997); Hall and Neale, *Epics, Spectacles and Blockbusters*, 78–84; Hampton, *History of the American Film Industry*, 388–405; Lewis, *The Motion Picture Industry*, 121–23, Salt, *Film Style and Technology*, 187–90, 212–13.

15 Crafton, *The Talkies*, 106–7, 195; Michele Hilmes, *Hollywood and Broadcasting: From Radio to Cable* (Urbana: Illinois University Press, 1990), 26–48; Richard B. Jewell, 'Hollywood and Radio: Competition and Partnership in the 1930s', *Historical Journal of Film, Radio and Television*, vol. 4 no. 2 (1984), 125–41; Robert McLaughlin, *Broadway and Hollywood: A History of Economic Interaction* (New York: Arno Press, 1974), 88–131; Ron Mottram, 'American Sound Films, 1926–30', in Elisabeth Weis and John Belton (eds), *Film Sound: Theory and Practice* (New York: Columbia University Press, 1985), 228–31; Katherine Spring, 'Pop Go the Warner Bros. et al.: Marketing Film Songs During the Coming of Sound', *Cinema Journal*, vol. 48 no. 1 (Fall 2008), 68–89; John Tibbetts, *The American Theatrical Film: Stages in Development* (Bowling Green, OH: Bowling Green State University Popular Press, 1985), 112–45. As these and other publications make clear, the US film industry drew on or set up links with the music and theatre industries throughout the silent era too.

16 For a vivid account of Hollywood and the studio system, see Ronald L. Davis, *The Glamour Factory: Inside Hollywood's Studio System* (Dallas: Southern Methodist University Press, 1993).

17 Among these operations and functions were the physical storage, repair and transportation of prints, as Lewis notes. Prints of films were highly flammable as well as being easily susceptible to wear and tear. Along with the other tasks and functions they performed, film exchanges and distribution centres were therefore generally housed in large brick buildings well away from residential areas and were usually clustered together in what was known colloquially as 'Film Row'.

18 As Lewis points out, and as F. Andrew Hanssen has pointed out more recently, block-booking was much more flexible in practice than most accounts suggest. The majority of films were booked for a range rather than a fixed number of days, which meant that runs of popular films could be extended and those of unpopular ones reduced. Others could be cancelled altogether. Given the move to 'revenue-sharing' (paying the distributor a percentage of the box-office take rather than a flat fee, as had been the norm in the silent era), flexible booking was as advantageous to distributors as it was to those who booked their films. See F. Andrew Hanssen, 'The Block Booking of Films Re-Examined' and 'Revenue Sharing and the Coming of Sound', in John Sedgwick and Michael Pokorny (eds), *An Economic History of Film* (London: Routledge, 2005), 121–50 and 86–120, respectively.

19 For further details, see Balio, *Grand Design*, 13–30; Crafton, *The Talkies*, 263–65.

20 According to Balio, *Grand Design*, 20, the run-zone-clearance system worked as follows: 'the majors had divided the country into thirty markets, with each market subdivided into zones. Theatres within each zone were classified by run. Located in the downtowns of the largest cities, first-run theatres seated thousands and charged the highest admission prices. Second-run houses were typically located in neighbourhood business districts and charged lower prices. Subsequent-run theatres, going down the scale to fifth-, sixth-, seventh-run, and more, were located in outlying communities and charged still less. A film would move from zone to zone like clockwork, with each zone separated by a [period of] clearance ranging from fourteen days to forty-two days or more ... Since the value of a motion picture to an exhibitor depended on its novelty, the granting of excessive clearance to prior-run houses had the effect of increasing their drawing power and keeping patronage in subsequent-run houses at low levels'. It also favoured the majors' theatres, most of which were first or second run. (See also Richard Maltby, *Hollywood Cinema* [Oxford: Blackwell, 2003 edition], 121–23.) It should be noted here that Schatz also appears to attribute the recovery to Wall Street as well as to the NIRA. However, Balio argues in *Grand Design*, 23, that Wall Street bankers and financiers 'proved singularly inept in managing motion-picture businesses' and that movie business veterans and the NIRA were largely responsible for restoring the industry's fortunes. For more on the NIRA and Hollywood, see Douglas Gomery, 'Hollywood, the National Recovery Administration, and the Question of Monopoly Power', in Gorham Kindem (ed.), *The American Movie Industry: The Business of Motion Pictures* (Carbondale: Southern Illinois University Press, 1982), 205–14.

21 For more on unions and unionization, see Balio, 'Part III: A Mature Oligopoly, 1930–48', in Balio (ed.), *The American Film Industry* (1985), 271–79; Balio, *Grand Design*, 79–82 and 82–85; Gerald Horne, *Class Struggle in Hollywood: Moguls, Mobsters, Stars, Reds, and Trade Unions* (Austin: University of Texas Press, 2001); Mike Nielsen and Gene Miles, *Hollywood's Other Blacklist: Union Struggles in the Studio System* (London: BFI, 1995); Tom Stempel, *Framework: A History of Screenwriting in the American Film* (New York: Continuum, 1988), 136–43. The formation of the SAG is further discussed by Balio in 'Selling Stars' (Chapter 12).

22 For additional overviews of trends in production in the 1930s, see Balio, *Grand Design*, 179–312, and Hall and Neale, *Epics, Spectacles and Blockbusters*, 88–111.

23 For more on the period between 1940 and 1945, see Thomas Schatz, *Boom and Bust: American Cinema in the 1940s* (New York: Scribners, 1997), 11–281. For more on Hollywood and World War Two, see Bernard Dick, *The Star-Spangled Screen: The American World War II Film* (Lexington: University of Kentucky Press, 1985); Thomas Doherty, *Projections of War: Hollywood, American Culture and World War II* (New York: Columbia University Press, 1993); Clayton R. Koppes and Gregory D. Black, *Hollywood Goes to War: How Politics, Profits and Propaganda Shaped World War II Movies* (New York: The Free Press, 1987). For further details on Hollywood and Britain, see H. Mark Glancy, *When Hollywood Loved Britain: The Hollywood 'British' Film, 1939–45* (Manchester: Manchester University Press, 1999). For further details on Hollywood and Latin America, see Kerry Segrave, *American Films Abroad: Hollywood's Domination of the World's Movie Screens* (Jefferson, NC: McFarland, 1997), 122–26.

24 For more on 'bigger' pictures in the 1940s, see Hall and Neale, *Epics, Spectacles and Blockbusters*, 112–34.

25 Bernstein, 'Hollywood's Semi-Independent Production', 45, 48–49.

26 For a detailed account of Wanger's career as a producer, see Matthew Bernstein, *Walter Wanger: Hollywood Independent* (Berkeley: University of California Press, 1994). For detailed accounts and evidence of Selznick's practices, see Rudy Behlmer (ed.), *Memo from Darryl F. Zanuck: The Golden Years at Twentieth Century-Fox* (New York: Grove Press, 1993) and *Memo from David O. Selznick* (Los Angeles: Samuel French, 1972); Hall and Neale, *Epics, Spectacles and Blockbusters*, 113–17, 129–32; Leonard J. Leff, *Hitchcock and Selznick* (New York: Weidenfeld, 1987); Alan David Vertrees, *Selznick's Vision: Gone with the Wind and Hollywood Filmmaking* (Austin: University of Texas Press, 1997). For case studies of supervision and production management from the late 1920s to the late 1950s, see Schatz, *The Genius of the System*, 58–481. For a detailed account of the Freed Unit at MGM, see Hugh Fordin, *The Movies' Greatest Musicals Produced in Hollywood USA by the Freed Unit* (New York: Ungar, 1975).

27 Hall and Neale , *Epics, Spectacles and Blockbusters*, 95; Seale, 'A Host of Others', 74–75.

28 Lea Jacobs, 'The B Film and the Problem of Cultural Distinction', *Screen*, vol. 33 no. 1 (Spring 1992), 3. It should also be noted that modestly budgeted A features were sometimes rented for flat fees. According to *Variety*, 10 January 1947, 7, examples included *My Friend Flicka* (1943) and *Laura* (1944). Despite numerous attempts to eliminate the practice, double billing remained in force throughout the 1940s and, indeed, the 1950s and 1960s too. For a detailed discussion of attempts to end double billing in the 1930s and 1940s, see Gary D. Rhodes, '"The Double Feature Evil": Efforts to Eliminate the American Dual Bill', *Film History*, vol. 23 no. 1 (2011), 57–74.

29 Along with the selling of popcorn, candy, soft drinks, the spread of smoking in audititoria and other facets of cinema-going in the 1930s, the topic of cinema programming and the role played by newsreels and shorts is also discussed by Thomas Doherty in 'This Is Where We Came in: The Audible Screen and the Voluble Audience of Early Sound Cinema', in Melvyn Stokes and Richard Maltby (eds), *American Movie Audiences: From the Turn of the Century to the Early Sound Era* (London: BFI, 1999), 143–63. For additional discussion of shorts and cartoons, see Michael Barrier, *Hollywood Cartoons: American Animation in Its Golden Age* (New York: Oxford University Press, 1999); Crafton, *The Talkies*, 381–401; Leonard Maltin, *Selected Short Subjects: From Spanky to the Three Stooges* (New York: Da Capo Press, 1972) and *Of Mice and Magic: A History of American Animated Cartoons* (New York: Plume, 1987 edition). For additional discussion of newsreels, see Thomas Doherty, 'Documenting the 1940s', in Schatz, *Boom and Bust*, 398–405; Raymond Fielding, *The American Newsreel, 1911–1967* (Norman: University of Oklahoma Press, 1972) and *The March of Time, 1935–1951* (New York: Oxford University Press, 1978). It should be noted that there is very little written about serials, which were part and parcel of 'kiddie matinees', beyond the listing of titles and credits and the provision of synopses – a comprehensive example is Buck Rainey, *Serials and Series: A World Filmography, 1912–1956* (Jefferson, NC: McFarland, 1999). For a relatively detailed study of sound serials and their aesthetic characteristics, see William C. Cline, *In the Nick of Time: Motion Picture Sound Serials* (Jefferson, NC: McFarland, 1984). For a discussion of the 'balanced program' and its constituents prior to the advent of sound, see Koszarski, *An Evening's Entertainment*, esp. 163–90. For serials in the silent era, see Kalton C. Lahue, *Continued Next Week: A History of the Motion Picture Serial* (Norman: University of Oklahoma Press, 1964). For further discussion of kiddie matinees, see Gomery, *Shared Pleasures*, 138–39, and Stones, *America Goes to the Movies*, 125–30.

30 Bordwell, Staiger and Thompson, *The Classical Hollywood Cinema*, 101–02; Bowser, *The Transformation of Cinema 1907–1915*, 103–19; Richard deCordova, *Picture Personalities; The Emergence of the Star System in America* (Urbana: University of Illinois Press, 1990); Hampton, *History of the American Film Industry*, 146–69, 190–207, 216–18, 230–34; Koszarski, *An Evening's Entertainment: The Age of the Silent Feature Picture* (New York: Scribners, 1990), 259–314.

31 In addition to Balio, see Phil Friedman, 'The Players Are Cast', in Nancy Naumberg (ed.), *We Make the Movies* (New York: W.W. Norton, 1937), 106–16; Lewis, *The Motion Picture Industry*, 117–24; and Milton Stills, 'The Actor's Part', in Kennedy (ed.), *The Story of the Films*, 175–202. It is worth pointing out that a tiny handful of stars secured contracts that paid them a percentage of the box-office gross in the 1930s. They included John Barrymore and Al Jolson (in 1930) and Mae West and James Cagney (in 1939). See Mark Weinstein,

'Profit-Sharing Contracts in Hollywood: Evolution and Analysis', *Journal of Legal Studies*, vol. 27 no. 1 (January 1998), 81. As we shall see, profit-sharing became much more common in the postwar era.

32 For an extensive account of the roles played by agents in the 1930s and 1940, see Tom Kemper, *Hidden Talent: The Emergence of Hollywood Agents* (Berkeley: University of California Press, 2010).

33 For a history of the unionization of actors prior to the 1930s, and for details of other pay cuts, see Sean P. Holmes, 'And the Villain Still Pursued Her: The Actors' Equity Association in Hollywood, 1919–29', *Historical Journal of Film, Radio and Television*, vol. 25 no. 1 (March 2005), 27–50. For more on stars, labour and unions, see Danae Clark, 'Acting in Hollywood's Best Interests: Representations of Actors' Labor During the National Recovery Administration', *Journal of Film and Video*, vol. 42 no. 4 (Winter 1990), 3–19, and *Negotiating Hollywood: The Cultural Politics of Actors' Labor* (Minneapolis: University of Minnesota Press, 1995); Sean P. Holmes, 'The Hollywood Star System and the Regulation of Actors' Labour, 1916–34', *Film History*, vol. 12 no. 1 (2000), 97–114.

34 Schatz, *The Genius of the System*. As Schatz points out, stars could on occasion be 'off-cast', cast in roles that complicated or appeared to contradict their established personas. The personas of stars could also change over time. Little has been written about the training of stars, thus tending to reinforce the idea they possessed natural talent, that they merely played themselves, and thus that they did not really act. For a rare exception, see Sharon Marie Carnicke, 'Crafting Film Performances: Acting in the Hollywood Studio Era', in Alan Lovell and Peter Krämer (eds), *Screen Acting* (London: Routledge, 1999), 31–45.

35 Gerben Bakker, 'Stars and Stories: How Films Became Branded Products', in Sedgwick and Pokorny, *An Economic History of Film*, 48–85. For an economically oriented account of stars at Warner Bros. in the 1930s, see Michael Pokorny and John Sedgwick, 'Warner Bros. in the Inter-War Years: Strategic Responses to the Risk Environment of Filmmaking', in Sedgwick and Pokorny, *An Economic History of Film*, 151–85. For detailed studies of stars and stardom in the 1930s and 1940s, see Jeanine Basinger, *The Star Machine* (New York: Vintage Books, 2007); Sean Griffin, *What Dreams Were Made of: Movie Stars of the 1940s* (New Brunswick, NJ: Rutgers University Press, 2011); Adrienne J. McLean (ed.), *Glamour in a Golden Age: Movie Stars of the 1930s* (New Brunswick, NJ: Rutgers University Press, 2011).

Chapter 7

Douglas Gomery

THE COMING OF SOUND: TECHNOLOGICAL CHANGE IN THE AMERICAN FILM INDUSTRY

THE COMING OF SOUND DURING THE LATE 1920s climaxed a decade of significant change within the American industry. Following the lead of the innovators—Warner Bros. Pictures, Inc., and the Fox Film Corporation—all companies moved, virtually en masse, to convert to sound. By the autumn of 1930, Hollywood produced only talkies. The speed of conversion surprised almost everyone. Within twenty-four months a myriad of technical problems were surmounted, stages soundproofed, and theaters wired. Engineers invaded studios to coordinate sight with sound. Playwrights (from the East) replaced title writers; actors without stage experience rushed to sign up for voice lessons. At the time, chaos seemed to reign supreme. However, with some historical distance, we know that, although the switch-over to talkies seemed to come "overnight," no major company toppled. Indeed the coming of sound produced one of the more lucrative eras in U.S. movie history. Speed of transformation must not be mistaken for disorder or confusion. On the contrary, the major film corporations—Paramount and Loew's (MGM)—were joined by Fox, Warner, and RKO in a surge of profits, instituting a grip on the marketplace which continues to the present day.

Moreover, sound films did not spring Minerva-like onto the movie screens of twenties America. Their antecedents reached back to the founding of the industry. We need a framework to structure this important thirty-year transformation. Here the neoclassical economic theory of technical change proves very useful. An enterprise introduces a new product (or process of production) in order to increase profits. Simplified somewhat, three distinct phases are involved: invention, innovation, and diffusion. Although many small-inventory entrepreneurs attempted to marry motion pictures and sound, it took two corporate giants, the American Telephone & Telegraph Corporation (AT&T), and the Radio Corporation of America (RCA), to develop the necessary technology. AT&T desired to make better phone equipment; RCA sought to improve its radio capabilities. As a secondary effect of such research, each perfected sound recording and reproduction equipment. With the inventions ready, two movie companies, Warner and Fox, adapted telephone and radio research for practical use. That is, they innovated sound movies. Each developed techniques to produce, distribute, and exhibit sound motion pictures. The final phase, diffusion, occurs when the product or process is adopted for widespread use. Initially, the movie industry giants hesitated to follow the lead of Warner and Fox but, after elaborate planning, decided to convert.

All others followed. Because of the enormous economic power of the major firms, the diffusion proceeded quickly and smoothly. During each of the three phases, the movie studios and their suppliers of sound equipment formulated business decisions with a view toward maximizing long-run profits. This motivation propelled the American motion picture industry (as it had other industries) into a new era of growth and prosperity.

Invention

Attempts to link sound to motion pictures originated in the 1890s. Entrepreneurs experimented with mechanical means to combine the phonograph and motion pictures. For example, in 1895 Thomas Alva Edison introduced such a device, his Kinetophone. He did not try to synchronize sound and image; the Kinetophone merely supplied a musical accompaniment to which a customer listened as he or she viewed a "peep show." Edison's crude novelty met with public indifference. Yet, at the same time, many other inventors attempted to better Edison's effort. One of these, Léon Gaumont, demonstrated his Chronophone before the French Photographic Society in 1902. Gaumont's system linked a single projector to two phonographs by means of a series of cables. A dial adjustment synchronized the phonograph and motion picture. In an attempt to profit by his system, showman Gaumont filmed variety (vaudeville) acts. The premiere came in 1907 at the London Hippodrome. Impressed, the American monopoly, the Motion Picture Patents Company, licensed Chronophone for the United States. Within one year Gaumont's repertoire included opera, recitations, and even dramatic sketches. Despite initially bright prospects, Chronophone failed to secure a niche in the marketplace because the system, relatively expensive to install, produced only coarse sounds, lacked the necessary amplification, and rarely remained synchronized for long. In 1913, Gaumont returned to the United States for a second try with what he claimed was an improved synchronizing mechanism and an advanced compressed air system for amplification. Exhibitors remembered Chronophone's earlier lackluster performance and ignored all advertised claims, and Gaumont moved on to other projects.

Gaumont and Edison did not represent the only phonograph sound systems on the market. More than a dozen others, all introduced between 1909 and 1913, shared common systems and problems. The only major rival was the Cameraphone, the invention of E. E. Norton, a former mechanical engineer with the American Gramophone Company. Even though in design the Cameraphone nearly replicated Gaumont's apparatus, Norton succeeded in installing his system in a handful of theaters. But like others who preceded him, he never solved three fundamental problems: (1) the apparatus was expensive; (2) the amplification could not reach all persons in a large hall; and (3) synchronization could not be maintained for long periods of time. In addition, since the Cameraphone system required a porous screen, the image retained a dingy gray quality. Therefore it was not surprising that Cameraphone (or Cinephone, Vivaphone, Synchroscope) was never successful.

It remained for one significant failure to eradicate any further commercial attempt to marry the motion picture and the phonograph. In 1913, Thomas Edison announced the second coming of the Kinetophone. This time, the Wizard of Menlo Park argued, he had perfected the talking motion picture! Edison's demonstration on January 4, 1913, impressed all present. The press noted that this system seemed more advanced than all predecessors. Its sensitive microphone obviated traditional lip-sync difficulties for actors. An oversized phonograph supplied the maximum mechanical amplification. Finally, an intricate system of belts and pulleys erected between the projection booth and the stage could precisely coordinate the speed of the phonograph with the motion picture projector.

Because of the success of the demonstration, Edison was able to persuade vaudeville magnates John J. Murdock and Martin Beck to install the Kinetophone in four Keith-Orpheum theaters in New York. The commercial premiere took place on February 13, 1913, at Keith's Colonial. A curious audience viewed and listened to a lecturer who praised Edison's latest marvel. To provide dramatic evidence for his glowing tribute, the lecturer then smashed a plate, played the violin, and had his dog bark. After several music acts (recorded on the Kinetophone), a choral rendition of "The Star-Spangled Banner" stirringly closed the show. An enthusiastic audience stood and applauded for ten minutes. The wizard, Tom Edison, had done it again!

Unfortunately this initial performance would rank as the zenith for Kinetophone. For a majority of later presentations, the system functioned badly—for a variety of technical reasons. For example, at Keith's Union Square theater, the sound lost synchronization by as much as ten to twelve seconds. The audience booed the picture off the screen. By 1914, the Kinetophone had established a record so spotty that Murdock and Beck paid off their contract with Edison. Moreover, during that same year, fire destroyed Edison's West Orange factory. Although he quickly rebuilt, Edison chose not to reactivate the Kinetophone operation. The West Orange fire not only marked the end of the Kinetophone, but signaled the demise of all serious efforts to mechanically unite the phonograph with motion pictures. (The later disc system would use electronic connections.)

American moviegoers had to wait nine years for another workable sound system to emerge—and when it did, it was based on the principle of sound on film, not on discs. On April 4, 1923, noted electronics inventor Lee De Forest successfully exhibited his Phonofilm system to the New York Electrical Society. De Forest asserted that his system simply photographed the voice onto an ordinary film. In truth, Phonofilm's highly sophisticated design represented a major advance in electronics, begun when De Forest had patented the Audion amplifier tube in 1907. Two weeks later Phonofilm reached the public at large at New York's Rivoli theater. The program consisted of three shorts: a ballerina performing a "swan dance," a string quartet, and another dance number. Since the musical accompaniment for each was *non*-synchronous, De Forest, whose brilliance shone in the laboratory rather than in showmanship or business, generated little interest. A *New York Times* reporter described a lukewarm audience response. No movie mogul saw enough of an advancement, given the repeated previous failures, to express more than a mild curiosity.

In fact, De Forest never wanted to work directly through a going motion picture concern, but go it alone. Consequently, legal and financial roadblocks continually hindered substantial progress. De Forest tried but could not establish anywhere near an adequate organization to market films or apparatus. Movie entrepreneurs feared, correctly, that the Phonofilm Corporation controlled too few patents ever to guarantee indemnity. Still, De Forest's greatest difficulties came when he attempted to generate financial backing. This brilliant individualist failed ever to master the intricacies of the world of modern finance. Between 1923 and 1925, Phonofilm, Inc., wired only thirty-four theaters in the United States, Europe, South Africa, Australia, and Japan. De Forest struggled on, but in September 1928, when he sold out to a group of South African businessmen, only three Phonofilm installations remained, all in the United States.

It took AT&T, the world's largest company, to succeed where others had failed. In 1912, AT&T's manufacturing subsidiary, Western Electric, secured the rights to De Forest's Audion tube to construct amplification repeaters for long-distance telephone transmission. In order to test such equipment the Western Electric Engineering Department, under Frank Jewett, needed a better method to test sound quality. After a brief interruption because of World War I, Jewett and his scientists plunged ahead, concentrating on improving the disc method. Within three months of the armistice, one essential element for a sound system was ready, the loudspeaker. The loudspeaker was first used in the "Victory Day" parade on Park Avenue in 1919, but national notoriety came during the 1920 Republican and Democratic national conventions. A year later, by connecting this technology to its long-distance telephone network, AT&T broadcast President Harding's address at the burial of the Unknown Soldier simultaneously to overflowing crowds in New York's Madison Square Garden and San Francisco's Auditorium. Clear transmissions to large indoor audiences had become a reality. Other necessary components quickly flowed off Western Electric's research assembly line. The disc apparatus was improved by creating a single-drive shaft turntable using 33⅓ revolutions per minute. Ready in 1924, the complete new disc system included a high-quality microphone, a nondistortive amplifier, an electrically operated recorder and turntable, a high-quality loudspeaker, and a synchronizing system free from speed variation.

In 1922, in the midst of these developments, Western Electric began to consider commercial applications. Western Electric did advertise and sell the microphones, vacuum tubes, and loudspeakers in the radio field, but Jewett's assistant, Edward Craft, argued that more lucrative markets existed in "improved" phonographs and sound movies. Employing the sound-on-disc method, Craft produced the first motion picture using Western Electric's sound system. To *Audion*, an animated cartoon originally created as a silent public relations film, he added a synchronized score. Craft premiered *Audion* in Yale University's Woolsey Hall on October 27, 1922. He followed this first effort with more experiments. On February 13, 1924, at a dinner at New York's Astor Hotel, Craft presented *Hawthorne.* This public relations film showing Western Electric's plant in Chicago employed a perfectly synchronized sound track. By the fall of 1924, the sound-on-disc system seemed ready to market.

Laboratory success did not constitute the only criterion which distinguished Western Electric's efforts from those of De Forest and other inventors. Most important, Western Electric had almost unlimited financial muscle. In 1925, parent company AT&T ranked with U.S. Steel as the largest private corporation in the world. Total assets numbered over $2.9 billion; revenues exceeded $800 million. At this time America's national income was only $76 billion, and government receipts totaled only $3.6 billion. Western Electric, although technically an AT&T subsidiary, ranked as a corporate giant in its own right with assets of $188 million and sales of $263 million, far in excess of even Paramount, the largest force in the motion picture industry at the time. If absolute economic power formed the greatest advantage, patent monopoly certainly added another. AT&T spent enormous sums to create basic patents in order to maintain its monopoly position in the telephone field. Moreover, AT&T's management actively encouraged the development of nontelephone patents to use for bargaining with competitors. For example, between 1920 and 1926 AT&T protected itself by cross-licensing its broadcasting-related patents with RCA. In turn, RCA and its allies agreed not to threaten AT&T's monopoly for wire communication. In particular, in the cross-licensing agreement of 1926, AT&T and RCA contracted to exchange information on sound motion pictures, if and when required. Thus by 1926, AT&T had control over its own patents, as well as any RCA created.

Using its economic power and patent position, Western Electric moved to reap large rewards for its sound-recording technology. As early as 1925, it had interested and licensed the key phonograph and record manufacturers Victor and Columbia. Movie executives proved more stubborn, so Western Electric hired an intermediary, Walter J. Rich. On May 27, 1925, Rich inked an agreement under which he agreed to commercially exploit the AT&T system for nine months.

Innovation

Warner Bros.

Warner Bros. would eventually be the company to innovate sound motion pictures. However, in 1925, Warner ranked low in the economic pecking order in the American film industry. Certainly brothers Harry, Albert, Sam, and Jack had come a long way since their days as nickelodeon operators in Ohio some two decades earlier. Yet in the mid 1920s, their future seemed severely constrained. Warner neither controlled an international system for distribution, nor owned a chain of first-run theaters. The brothers' most formidable rivals, Famous Players (soon to be renamed Paramount), Loew's, and First National, did. Eldest brother Harry Warner remained optimistic and sought help.

In time, Harry Warner met Waddill Catchings, a financier with Wall Street's Goldman, Sachs. Catchings, boldest of the "New Era" Wall Street investors, agreed to take a flyer with this fledgling enterprise in the most speculative of entertainment fields. Catchings correctly reasoned that the consumer-oriented 1920s economy would provide a fertile atmosphere for boundless growth in the movie field. And Warner seemed progressive. The four brothers maintained strict cost accounting and budget controls, and seemed to have attracted more than competent managerial talent. Catchings agreed to finance Warner, only if it followed his master plan. The four brothers, sensing they would find no better alternative, readily agreed.

During the spring of 1925, Harry Warner, president of the firm, formally appointed Waddill Catchings to the board of directors, and elevated him to chairman of the finance committee. Catchings immediately established a $3 million revolving credit account through New York's National Bank of Commerce. Although this bank had never loaned a dollar to a motion picture company, not even the mighty Paramount, Catchings possessed enough clout to convince president James S. Alexander that Warner would be a good risk. Overnight Warner had acquired a permanent source for financing future productions. Simultaneously Warner took over the struggling Vitagraph Corporation, complete with its network of fifty distribution exchanges throughout the world. In this deal Warner also gained the pioneer company's two small studios, processing laboratory, and extensive film library. Finally, with four million more dollars that Catchings raised through bonds, Warner strengthened its distribution system, and even launched a ten-theater chain. Certainly by mid 1925 Warner was becoming a force to be reckoned with in the American movie business.

Warner's expansionary activities set the stage for the coming of sound. At the urging of Sam Warner, who was an electronics enthusiast, the company established radio station KFWB in Hollywood to promote Warner films. The equipment was secured from Western Electric. Soon Sam Warner and Nathan Levinson, Western's Los Angeles representative, became fast friends. Until then, Walter J. Rich had located no takers for Western Electric's sound inventions. Past failures had made a lasting and negative impression on the industry leaders, a belief shared by Harry Warner. Consequently, Sam had to trick his older brother into even attending a demonstration. That screening, in May 1925, included a recording of a five-piece jazz band. Quickly Harry and other Warner executives reasoned that if the company could equip their newly acquired theaters with sound and present vaudeville acts as part of their programs, they could successfully challenge the Big Three. Then, even Warner's smallest house could offer (1) famous vaudeville acts (on film); (2) silent features; and (3) the finest orchestral accompaniments (on disc). Warner, at this point, never considered feature-length talking pictures, only singing and musical films.

Catchings endorsed such reasoning and gave the go-ahead to open negotiations with Walter J. Rich. On June 25, 1925, Warner signed a letter of agreement with Western Electric calling for a period of joint experimentation. Western Electric would supply the engineers and sound equipment; Warner the camera operators, editors, and the supervisory talent of Sam Warner. Work commenced in September 1925 at the old Vitagraph studio in Brooklyn. Meanwhile, Warner continued to expand under Waddill Catchings' careful guidance. Although feature film output was reduced, more money was spent on each picture. In the spring of 1926, Warner opened a second radio station and an additional film-processing laboratory, and further expanded its foreign operations. As a result of this rapid growth, the firm expected a $1 million loss on its annual income statement issued in March 1926.

By December 1925, experiments were going so well that Rich proposed forming a permanent sound motion picture corporation. The contracts were prepared and the parties readied to sign, but negotiations ground to a halt as Western Electric underwent a management shuffle. Western placed John E. Otterson, an Annapolis graduate and career navy officer, in charge of exploiting nontelephone inventions. Otterson possessed nothing but contempt for Warner. He wanted to secure contracts with industry giants Paramount and Loew's, and then take direct control himself. Hitherto, Western Electric seemed content to function as a supplier of equipment. Catchings saw this dictatorial stance as typical of a man with a military background unable to adjust to the world of give-and-take in modern business and finance. Unfortunately for Warner, AT&T's corporate muscle backed Otterson's demands.

Only by going over Otterson's head to Western Electric's president, Edgar S. Bloom, was Catchings able to protect Warner interests and secure a reasonable contract. In April 1926, Warner, Walter J. Rich, and Western Electric formed the Vitaphone Corporation to develop sound motion pictures further. Warner and Rich furnished the capital. Western Electric granted Vitaphone an exclusive license to record and reproduce sound movies on its equipment. In return, Vitaphone agreed to lease a minimum number of sound systems each year and pay a royalty fee of 8 percent of gross revenues from sound motion pictures. Vitaphone's total equipment commitment became twenty-four hundred systems in four years.

As *Variety* and the other trade papers announced the formation of the alliance, Vitaphone began its assault on the marketplace. Its first goal was to acquire talent. Vitaphone contracted with the Victor Talking Machine Company for the right to bargain with its popular musical artists. A similar agreement was reached with the Metropolitan Opera Company. Vitaphone dealt directly with Vaudeville stars. In a few short months it had contracted for the talent to produce the musical short subjects Harry Warner had envisioned. So confident was Vitaphone's management that the firm engaged the services of the New York Philharmonic Orchestra. Throughout the summer of 1926, Sam Warner and his crew labored feverishly to ready a Vitaphone program for a fall premiere, while the Warner publicity apparatus cranked out thousands of column inches for the nation's press.

Vitaphone unveiled its marvel on August 6, 1926, at the Warners' Theatre in New York. The first-nighters who packed the house paid up to $10 for tickets. The program began with eight "Vitaphone Preludes." In the first, Will Hays congratulated the brothers Warner and Western Electric for their pioneering efforts. At the end, to create the illusion of a stage appearance, Hays bowed to the audience, anticipating their applause. Next, conductor Henry Hadley led the New York Philharmonic in the Overture to *Tannhäuser.* He too bowed. The acts that followed consisted primarily of operatic and concert performances: tenor Giovanni Martinelli offered an aria from *I Pagliacci*, violinist Mischa Elman played "Humaresque," and soprano Anna Case sang, supported by the Metropolitan Opera Chorus. Only one "prelude" broke the serious tone of the evening and that featured Roy Smeck, a popular vaudeville comic-musician. Warner, playing it close to the vest, sought approval from all bodies of respectable critical opinion. The silent feature *Don Juan* followed a brief intermission. The musical accompaniment (sound-on-disc) caused no great stir because it "simply replaced" an absent live orchestra. All in all, Vitaphone, properly marketed, seemed to have a bright future.

That autumn, the *Don Juan* package played in Atlantic City, Chicago, and St. Louis. Quickly Vitaphone organized a second program, but this time aimed at popular palaces. The feature, *The Better 'Ole*, starred Charlie Chaplin's brother, Sydney. The shorts featured vaudeville "headliners" George Jessel, Irving Berlin, Elsie Janis, and Al Jolson. These performers would have charged more than any single theater owner could have afforded, if presented live. The trade press now began to see bright prospects for the invention that could place so much high-priced talent in towns like Akron, Ohio, and Richmond, Va. By the time Vitaphone's third program opened in February 1927. Warner had recorded fifty more acts.

As a result of the growing popularity of Vitaphone presentations, the company succeeded in installing nearly a hundred systems by the end of 1926. Most of these were located in the East. The installation in March 1927 of apparatus in the new Roxy theater and the attendant publicity served to spur business even more. Consequently, Warner's financial health showed signs of improvement. The corporation had invested over $3 million in Vitaphone alone, yet its quarterly losses had declined from about $334,000 in 1925 to less than $110,000 in 1926. It appeared that Catchings' master plan was working.

John Otterson remained unsatisfied. He sought to take control of Vitaphone so that Western Electric could deal directly with Paramount and Loew's. To accomplish this he initiated a harassment campaign by raising prices on Vitaphone equipment fourfold, and demanding a greater share of the revenues. By December 1926, Western Electric and Warner had broken off relations. Simultaneously, Otterson organized a special Western Electric subsidiary called Electrical Research Products, Inc. (ERPI), to conduct the company's nontelephone business—over 90 percent of which concerned motion picture sound equipment.

Realistically Warner, even with Catchings' assistance, could not prevent Otterson from talking with other companies—even though exclusive rights were contractually held by Warner. However, only Fox would initial an agreement. The majors adopted a wait-and-see stance. In fact, the five most important companies—Loew's (MGM), Universal, First National, Paramount, and Producers Distributing Corporation—signed an accord in February 1927 to act together in regard to sound. The "Big Five Agreement," as it was called, recognized that since there were several sound systems on the market, inability to interchange this equipment could hinder wide distribution of pictures and therefore limit potential profits. These companies agreed to jointly adopt only the system that their specially appointed committee would certify, after one year of study, was the "best" for the industry. As further protection, they would employ no system unless it was made available to all producers, distributors, and exhibitors on "reasonable" terms.

Otterson needed to wrest away Warner's exclusive rights if he ever hoped to strike a deal with the Big Five. To this end, he threatened to declare Warner in default of its contractual obligations. Catchings, knowing such public statements would undermine his relations with the banks, persuaded Warner to accede to Otterson's wishes. In April 1927, ERPI paid Vitaphone $1,322,306 to terminate the old agreement. In May, after the two signed the so-called New License Agreement, Vitaphone, like Fox, became merely a licensee of ERPI. Warner had given up the exclusive franchise to exploit ERPI sound equipment and lost its share of a potential fortune in licensing fees.

Now on its own, Warner immediately moved all production to several new sound stages in Hollywood. While the parent company continued with its production program of silent features, Vitaphone regularly turned out five shorts a week, which became known in the industry as "canned vaudeville." Bryan Foy, an ex-vaudevillian and silent film director, now worked under Sam Warner to supervise the sound short subject unit. At this juncture, Vitaphone's most significant problem lay in a dearth of exhibition outlets for movies with sound. By the fall of 1927, six months since the signing of the New License Agreement, ERPI had installed only forty-four sound systems. ERPI was holding back on its sales campaign until the majors made a decision. Warner would later charge that ERPI had not used its best efforts to market the equipment and had itself defaulted. This accusation and others were brought to arbitration and, in a 1934 settlement, ERPI was forced to pay Vitaphone $5 million.

As the 1927–28 season opened, Vitaphone began to add new forms of sound films to its program. Though *The Jazz Singer* premiered on October 6, 1927, to lukewarm reviews, its four Vitaphoned segments of Al Jolson's songs proved very popular. Vitaphone contracted with Jolson immediately to make three more films for $100,000. (The four Warner brothers did not attend *The Jazz Singer*'s New York premiere because Sam Warner died in Los Angeles on October 5. Jack Warner took over Sam's position as head of Vitaphone production.) Bryan Foy pushed his unit to create four new shorts each week, becoming more bold in programming strategies. On December 4, 1928, Vitaphone released the short *My Wife's Gone Away*, a ten-minute, all-talking comedy based on a vaudeville playlet developed by William Demarest. Critics praised this short; audiences flocked to it. Thus Foy, under Jack Warner's supervision, began to borrow even more from available vaudeville acts and "playlets" to create all-talking shorts. During Christmas week, 1927, Vitaphone released a twenty-minute, all-talking drama, *Solomon's Children*. Again revenues were high, and in January 1928, Foy moved to schedule production of two all-talking shorts per week.

Warner had begun to experiment with alternative types of shorts as a cheap way to maintain the novelty value of Vitaphone entertainment. Moreover, with such shorts it could develop talent, innovate new equipment, and create an audience for feature-length, all-sound films. In the spring of 1928, with the increased popularity of these shorts, Warner began to change its feature film offerings. On March 14, 1928, it released *Tenderloin*—an ordinary mystery that contained five segments in which the actors spoke all their lines (for twelve of the film's eighty-five minutes). More part-talkies soon followed that spring.

Harry Warner and Waddill Catchings knew the investment in sound was a success by April 1928. By then it had become clear that the *The Jazz Singer* show had become the most popular entertainment offering of the 1927–28 season. In cities that rarely held films for more than one week *The Jazz Singer* package set records for length of run: for example, five-week runs in Charlotte, N.C., Reading, Pa., Seattle, and Baltimore. By mid February 1928, *The Jazz Singer* and the shorts were in a (record) eighth week in Columbus, Ohio, St. Louis, and Detroit, and a (record) seventh week in Seattle, Portland, Ore., and Los Angeles. The Roxy even booked *The Jazz Singer* package for an unprecedented second run in April 1928, where it grossed in excess of $100,000 each week, among that theater's best grosses for that season. Perhaps more important, all these first-run showings did not demand the usual expenses of a stage show and orchestra. It took Warner only until the fall of 1928 to convert to the complete production of talkies—both features and shorts. Catchings and Harry Warner had laid the foundation for this maximum exploitation of profit with their slow, steady expansion in production and distribution. In 1929, Warner would become the most profitable of any American motion picture company.

The Fox-Case Corporation

As noted above, only the Fox Film Corporation had also shown any interest in sound movies. Its chief, William Fox, had investigated the sound-on-film system developed by Theodore W. Case and Earl I. Sponable and found it to be potentially a great improvement over the cumbersome Western Electric disc system. Theodore Case and Earl Sponable were two recluse scientists. In 1913, the independently wealthy, Yale-trained physicist Case established a private laboratory in his hometown of Auburn, New York, a small city near Syracuse.

Spurred on by recent breakthroughs in the telephone and radio fields, Case and his assistant, Sponable, sought to better the Audion tube. In 1917 they perfected the Thalofide cell, a highly improved vacuum tube, and began to integrate this invention into a system for recording sounds. As part of this work, Case met Lee De Forest. For personal reasons—envy perhaps—Case turned all his laboratory's efforts to besting De Forest. Within eighteen months, Case labs produced an improved sound-on-film system, based on the Thalofide cell. Naively, De Forest had openly shared with Case all his knowledge of sound-on-film technology. So as De Forest unsuccessfully attempted to market his Phonofilm system, Case quietly constructed—with his own funds—a complete sound studio and projection room adjacent to his laboratory.

In 1925, Case determined he was ready to try to market his inventions. Edward Craft of Western Electric journeyed to Auburn, and saw and heard a demonstration film. Craft left quite impressed. But after careful consideration, he and Frank Jewett decided that Case's patents added no substantial improvement to the Western Electric sound-on-disc system, then under exclusive contract to Warner. Rebuffed, Case decided to directly solicit a show business entrepreneur. He first approached John J. Murdock, the long-time general manager of the Keith-Albee vaudeville circuit. Case argued that his sound system could be used to record musical and comedy acts—the same idea Harry Warner had conceived six months earlier. Murdock blanched. He had been burned by Edison's hyperbole only a decade earlier, and De Forest a mere twenty-four months before. Keith-Albee would never be interested in talking movies! Executives from all the "Big Three" motion picture corporations, Paramount, Loew's (MGM), and First National, echoed Murdock's response. None saw the slightest benefit in this latest version of sight and sound.

Case moved to the second tier of the U.S. film industry—Producers Distributing Company (PDC), Film Booking Office (FBO), Warner, Fox, and Universal. In 1926, Case signed with Fox because Courtland Smith, president of Fox Newsreels, reasoned that sound newsreels could push that branch of Fox Film to the forefront of the industry. In June 1926, Smith arranged a demonstration for company owner, founder, and president William Fox. The boss was pleased, and within a month helped create the Fox-Case Corporation to produce sound motion pictures. Case turned all patents over to the new corporation, and retired to his laboratory in upstate New York.

Initially, William Fox's approval of experiments with the Case technology constituted only a small portion of a comprehensive plan to thrust Fox Film into a preeminent position in the motion picture industry. Fox and his advisors had initiated an expansion campaign in 1925. By floating $6 million of common stock, they increased budgets for feature films and enlarged the newsreel division. (Courtland Smith was hired at this point.) Simultaneously Fox began building a chain of motion picture theaters. At that time Fox Film controlled only twenty small neighborhood houses in the New York City environs. By 1927, the Fox chain included houses in Philadelphia, Washington, D.C., Brooklyn, New York City, St. Louis, Detroit, Newark, Milwaukee, and a score of cities west of the Rockies. To finance these sizable investments, William Fox developed close ties to Harold Stuart, president of the Chicago investment house of Halsey, Stuart. Meanwhile, Courtland Smith had assumed control of Fox-Case, and, in 1926, initiated the innovation of the Case sound-on-film technology. At first all he could oversee were defensive actions designed to protect Fox-Case's patent position. In September 1926, exactly two months after incorporation, Fox-Case successfully thwarted claims by Lee De Forest, and a German concern, Tri-Ergon. For the latter, Fox-Case advanced $50,000 to check the future court action.

At last, Fox-Case could assault the marketplace. Although Smith pushed for immediate experimentation with sound newsreels, William Fox conservatively ordered Fox-Case to imitate the innovation strategy of Warner and film popular vaudeville acts. On February 24, 1927, Fox executives felt confident enough to stage a widely publicized demonstration of the newly christened Movietone system. At ten o'clock in the morning, fifty reporters entered the Fox studio near Times Square and were filmed by the miracle of Movietone. Four hours later these representatives of the press corps saw and heard themselves as part of a private screening. In addition, Fox-Case presented several vaudeville sound shorts: a banjo and piano act, a comedy sketch, and three songs by the then-popular cabaret performer Raquel Mueller. The strategy worked. Unanimous favorable commentary issued forth; the future seemed bright. Consequently, William Fox ordered sound systems for twenty-six of Fox's largest first-run theaters, including the recently acquired Roxy.

However, by this time Warner had signed nearly all popular entertainers to exclusive contracts. Smith pressed William Fox to again consider newsreels with sound. Then, Smith argued, Fox Film could offer a unique, economically viable alternative to Warner's presentations, and move into a heretofore unoccupied portion of the market for motion picture entertainment. Furthermore, sound newsreels would provide a logical method by which Fox-Case could gradually perfect necessary new techniques of camerawork and editing. Convinced, William Fox ordered Smith to adopt this course for technological innovation. This decision would prove more successful for Fox Film's overall goal of corporate growth than either William Fox or Courtland Smith imagined at the time.

Smith moved quickly. The sound newsreel premiere came on April 30, 1927, at the Roxy in the form of a four-minute record of marching West Point cadets. And despite the lack of any buildup, this newsreel elicited an enthusiastic response from the trade press and New York-based motion picture reviewers. Quickly Smith seized upon one of the most important symbolic news events of the 1920s. At eight in the morning, on May 20, 1927, Charles Lindbergh departed for Paris. That evening Fox Movietone News presented footage of the takeoff—with sound—to a packed house at the Roxy. Six thousand persons stood and cheered for nearly ten minutes. The press saluted this new motion picture marvel and noted how it had brought alive the heroics of the "Lone Eagle." In June, when Lindbergh returned to a tumultuous welcome in New York City and Washington, D.C., Movietone News cameramen also recorded portions of those celebrations. Both William Fox and Courtland Smith were now satisfied that the Fox-Case system had been launched on a propitious path.

That summer, Smith dispatched camera operators to all parts of the globe. They recorded the further heroics of aviators, beauty contests, and sporting events, as well as produced the earliest filmic records of statements by Benito Mussolini and Alfred Smith. Newspaper columnists, educators, and other opinion leaders lauded these latter short subjects for their didactic value. Fox Film's principal constraint now became a paucity of exhibition outlets. During the fall of 1927, Fox Film did make Movietone newsreels the standard in all Fox-owned theaters, but that represented less than 3 percent of the potential market. More extensive profits would come as Fox Film formed a larger chain of first-run theaters. In the meantime, Courtland Smith established a regular pattern for release of Movietone newsreels, one ten-minute reel per week. He also increased the permanent staff and established a worldwide network of stringers.

In addition, Smith and William Fox decided again to try to produce vaudeville shorts and silent feature films accompanied by synchronized music on disc. Before 1928, Fox-Case released only one scored feature, *Sunrise*. The two executives moved quickly. By January 1928, Fox had filmed ten vaudeville shorts and a part-talkie feature, *Blossom Time*. During the spring of 1928, these efforts, Fox's newsreels, and Warner's shorts and part-talkies proved to be the hits of the season. Thus in May 1928, William Fox declared that 100 percent of the upcoming production schedule would be "Movietoned." Simultaneously Fox Film continued to wire, as quickly as possible, all the houses in its ever-expanding chain and draw up plans for an all-sound Hollywood-based studio. Fox's innovation of sound neared completion; colossal profits loomed on the horizon.

The rise of RKO

Only RCA offered Warner, Western Electric, or Fox any serious competition. In 1919, General Electric and Westinghouse had created RCA to control America's patents for radio broadcasting. Like rival AT&T, GE conducted fundamental research in radio technology. The necessary inventions for what would become RCA's Photophone sound-on-film system originated when, during World War I, the U.S. Navy sought a high-speed recorder of radio signals. GE scientist Charles A. Hoxie perfected such a device. After the war, Hoxie pressed to extend his work. Within three years, having incorporated a photoelectric cell and a vibrating mirror, he could record a wide variety of complex sounds. In December 1921, GE executives labeled the new invention the Pallo-Photophone.

To test it, Willis R. Whitney, head of the GE Research Laboratory, successfully recorded speeches by Vice-President Calvin Coolidge and several Harding Administration cabinet members. At this point, GE executives conceived of the Pallo-Photophone as a marketable substitute for the phonograph. During 1922 and 1923, Hoxie and his assistants continued to perfect the invention. For example, they discovered that the recording band need not be 35 millimeters wide. A track as narrow as 1.5 millimeters proved sufficient, and thus freed sound to accompany a motion picture image. Simultaneously other GE scientists, Chester W. Rise and Edward W. Kellogg, developed a new type of loudspeaker to improve reception for the radio sets General Electric manufactured for RCA. Late in 1922, Whitney learned of Lee De Forest's efforts to record sound on film. Not to be outdone, Whitney ordered Hoxie and his research team to develop a sound reproducer that could be attached to a standard motion picture projector. In November 1923, Hoxie demonstrated such a system for GE's top executives in an almost perfect state. However, by that time, Whitney and his superiors sensed that De Forest's failure to innovate sound motion pictures proved there existed no market for Hoxie's invention. Whitney promptly transferred all efforts toward the development of a marketable all-electric phonograph. GE successfully placed its new phonograph before the public during the summer of 1925.

One year later, because of Warner's success, Whitney reactivated the sound movie experiments. At this point he christened the system "Photophone." By the end of that year, 1926, GE's publicity department had created several experimental short subjects. Quickly GE executives pondered how to approach a sales campaign. However before they could institute any action, Fox sought a license in order to utilize GE's amplification patents. Contemplating the request, David Sarnoff, RCA's general manager, convinced his superiors at GE that RCA should go out on its own, sign up the large movie producers Paramount and Loew's, and not worry about Fox. The GE high brass agreed and assigned Sarnoff the task of commercially exploiting GE's sound movie patents.

Sarnoff easily convinced Paramount and Loew executives to seriously consider RCA's alternative to Western Electric's then monopoly, even though RCA had yet to publicly demonstrate Photophone. Presently the "Big Five Agreement" was signed. Sarnoff immediately went public. On February 2, 1927, Sarnoff demonstrated Photophone for invited guests and the press at the State theater in GE's home city of Schenectady, New York. Musical short subjects featuring a hundred-piece orchestra impressed all present. Nine days later Sarnoff recreated the event for more reporters at New York's Rivoli theater. Here two reels of MGM's *Flesh and the Devil* were accompanied by a Photophone recording of the Capital theater orchestra. Then three shorts featured the Van Curler Hotel Orchestra of Schenectady, an unnamed baritone, and a quartet of singers recruited from General Electric employees. A *New York Times* reporter praised the synchronization, volume, and tone. Sarnoff, in turn, lauded Photophone's ease of installation and simplicity of operation.

In private, Sarnoff tried to convince the producers' committee of his company's technical and financial advantages. The producers had established three specific criteria for selection: (1) the equipment had to be technically adequate; (2) the manufacturer had to control all required patents; and (3) the manufacturer had to have substantial resources and financial strength. Only two systems qualified: RCA's Photophone and Western Electric's Vitaphone. At first, the producers favored RCA because it had not licensed any movie concern, whereas Western Electric had formal links to Warner and Fox. In October 1927, Sarnoff proposed an agreement which called for a holding company, one-half owned by RCA and one-half by the five motion picture producers. All of GE's sound patents would be vested in this one corporation. Sarnoff demanded 8 percent of all gross revenues from sound movies as a royalty. The producers countered. They sought individual licenses and fees set at $500 per reel. For a typical eight-reel film (90 minutes) with gross revenues of $500,000, the 8 percent royalty would be $40,000; at the new rate the amount came to $4,000, a savings of $36,000.

Sarnoff reluctantly acceded to the per reel method of royalty calculation, but stubbornly refused to grant individual licenses. On the other hand, the motion picture corporations held fast to their belief that they should play no role in the manufacture of the apparatus. They wanted a license only to produce and distribute sound films. For two months the two parties stalemated over this issue. Late in November 1927, John Otterson of Western Electric stepped forward and offered individual licenses. Western Electric's engineers had made great progress with their sound-on-film system, and there no longer existed exclusive ties to Warner. Consequently, in March 1928, the movie producers, with all the relevant information in hand, selected Western Electric. Each producer—Paramount, United Artists, Loew's, and First National—secured an individual license and would pay $500 per reel of negative footage. All four signed on May 11, 1928. Universal, Columbia and other companies quickly followed. The movie producers had adroitly played the two electrical giants off each other and secured reasonably favorable terms.

Sarnoff reacted quickly as the tide turned toward Western Electric. First General Electric purchased (for nearly $500,000) 14,000 shares of stock of the Film Booking Office from a syndicate headed by Joseph P. Kennedy. This acquisition guaranteed Photophone a studio outlet. FBO was the only producer with national distribution which was not linked in talks with Western Electric. Next Sarnoff formed RCA Photophone, Inc. Sarnoff now controlled production facilities and the necessary sound technology. To generate significant profits, RCA needed a chain of theaters.

In 1928, the Keith-Albee vaudeville chain controlled such a chain. Faced with declining business in vaudeville, Keith-Albee executives developed two approaches. First they took over the Orpheum vaudeville chain, and thus merged all major American vaudeville under one umbrella. The new Keith-Albee-Orpheum controlled two hundred large downtown theaters. Second, Keith-Albee acquired a small movie company, Pathé, just to hedge its bets. When Sarnoff approached the owners of the new Keith-Albee-Orpheum they were more than ready to sell. Sarnoff quickly moved to consolidate his empire. FBO and Pathé formally acquired licenses for Photophone. FBO and Pathé executives supervised the addition of music-on-film to three features, *King of Kings*, *The Godless Girl*, and *The Perfect Crime*. Upcoming sound newsreels and vaudeville shorts were promised. However, these films would be useless unless Sarnoff could wire the Keith-Albee-Orpheum theaters with Photophone equipment. Warner, Fox, and Western Electric had taken almost two years to eliminate all the problems of presenting clear sounds of sufficient volume in large movie palaces. As of this point Photophone had yet to be tested in a commercial situation. And Sarnoff and his staff would need at least six months to iron out technological problems. Promised first in April, then July, commercial installations commenced in October 1928. In the meantime, Sarnoff used a low installation price and sweeping prognostications of future greatness to persuade a shrinking number of prospective clients to wait for Photophone equipment.

That October, Sarnoff legally consolidated RCA's motion picture interests by creating a holding company, Radio-Keith-Orpheum (RKO). Sarnoff became president of the film industry's newest vertically integrated combine. The merger united theaters (Keith-Albee-Orpheum), radio (NBC), and motion pictures (FBO and Pathé, subsequently renamed Radio Pictures in May 1929). Although late on the scene, RCA had established a secure place in the motion picture industry. RKO released its first talkies in the spring of 1929, and Photophone could battle Western Electric for contracts with the remaining unwired houses. Gradually during the 1930s, RCA Photophone would become as widely accepted as Western Electric's system.

Diffusion

The widespread adoption of sound—its diffusion—took place quickly and smoothly, principally because of the extensive planning of the producers' committee. Since an enormous potential for profits existed, it was incumbent on the majors to make the switchover as rapidly as possible. Paramount released its first films with musical accompaniment in August 1928; by September its pictures contained talking sequences; and by January 1929, it sent out its first all-talking production. By May, one year after signing with ERPI, Paramount produced talkies exclusively and was operating on a level with Warner and Fox. In September 1929, MGM, Fox, RKO, Universal, and United Artists completed their transitions. Those independent production companies which survived took, on average, one year longer.

Elaborate plans had been laid by the industry to facilitate diffusion. In Hollywood, the Academy of Motion Picture Arts and Sciences was designated as a clearing house for information relating to production problems. The local film boards of trade handled changes in distribution trade practices. And a special lawyers' committee representing the major producers was appointed to handle disputes and contractual matters with equipment manufacturers. For example, when ERPI announced a royalty hike, the committee initiated a protest, seeking lower rates. Unions presented no difficulties. The American Federation of Musicians unsuccessfully tried to prevent the wholesale firing of theatrical musicians; Actors' Equity, now that professionals from the Broadway stage began to flock west, failed to establish a union shop in the studios. All problems were resolved within a single year; the industry never left an even keel.

ERPI's task all the while was to keep up with the demand for apparatus. It wired the large, first-run theaters first and then, as equipment became available, subsequent-run houses. Installations were made usually from midnight to nine in the morning. For example, in January 1930, ERPI installed more than nine systems each day. To facilitate the switchover, Western Electric expanded its Hawthorne, Illinois, plant, and ERPI established training schools for projectionists in seventeen cities and opened fifty district offices to service and repair equipment. Many smaller theaters, especially in the South and Southwest, could not afford ERPI's prices and signed with RCA, or De Forest. As late as July 1930, fully 22 percent of all U.S. theaters still presented silent versions of talkies. That figure neared zero two years later.

The public's infatuation with sound ushered in another boom period for the industry. Paramount's profits jumped $7 million between 1928 and 1929, Fox's $3.5 million, and Loew's $3 million. Warner, however, set the record; its profits increased $12 million, from a base of only $2 million. A 600 percent leap! Conditions were ripe for consolidation, and Warner, with its early start in sound, set the pace. It began by acquiring the Stanley Company, which owned a chain of three hundred theaters along the East Coast, and First National. In 1925, when Waddill Catchings joined the Warner board of directors, the company's assets were valued at a little over $5 million; in 1930 they totaled $230 million. In five short years, Warner had become one of the largest and most profitable companies in the American film industry.

Not content merely to establish RKO, David Sarnoff of RCA set out to sever all connections with General Electric and Westinghouse and acquire sound manufacturing facilities of his own. The first step in this direction was the acquisition in March 1929 of Victor Talking Machine Company and its huge plant in Camden, New Jersey. In the process, RCA secured Victor's exclusive contracts with many of the biggest stars in the musical world. By December 1929, Sarnoff had reached his goal. RCA was now a powerful independent entertainment giant with holdings in the broadcasting, vaudeville, phonograph, and motion picture industries.

William Fox had the most grandiose plan of all. In March 1929, he acquired controlling interest in Loew's, the parent company of MGM. Founder Marcus Loew had died in 1927 and left his widow and sons one-third of the company's stock. Nicholas Schenck, the new president, pooled his stock and that belonging to corporate officers with the family's and sold out to Fox at 25 percent above the market price. The new Fox-Loew's merger created the largest motion picture complex in the world. Its assets totaled more than $200 million and an annual earning potential existed of $20 million. Fox assumed a substantial short-term debt obligation in the process, but during the bull market of the late twenties he could simply float more stock and bonds to meet his needs.

Adolph Zukor of Paramount, meanwhile, added more theaters, bringing Paramount's total to almost one thousand. He also acquired a 49 percent interest in the Columbia Broadcasting System. Then, in the fall of 1929, he proposed a merger with Warner that would create a motion picture and entertainment complex larger than Fox-Loew's and RCA combined. Catchings and Harry Warner were agreeable, but the new U.S. attorney general, William D. Mitchell, raised the red flag. If that merger went through, the industry would be dominated by three firms. As it happened, though, it was to be dominated by five. After the stock market crash, William Fox was unable to meet his short-term debts and had to relinquish ownership of Loew's. The oligopolistic structure of the industry, now formed by Warner, Paramount, Fox, Loew's, and RKO, would continue to operate well into the 1950s. The coming of sound had produced important forces for industry consolidation, immediately prior to the motion picture industry's first crisis of retrenchment—the Great Depression.

Bibliographical note

This article is based on a series of articles I wrote between 1976 and 1980. The innovative activities of Warner Bros. are analyzed in "Writing the History of the American Film Industry," *Screen* 17 (Spring 1976): 40–53. For a separate discussion of the experiences of Fox and RCA, see "Problems in Film History: How Fox Innovated Sound," *Quarterly Review of Film Studies* 2 (August 1976): 315–30, and "Failure and Success: Vocafilm and RCA Innovate Sound," *Film Reader 2* (January 1977): 213–21.

The reaction of the industry at large can be seen in two distinct stages. I treat the initial reluctance and waiting in "The Warner-Vitaphone Peril: The American Film Industry Reacts to the Innovation of Sound," *Journal of the University Film Association* 28 (Winter 1976): 11–19. For general acceptance and prosperity, see "Hollywood Converts to Sound: Chaos or Order?" in Evan W. Cameron, ed., *Sound and the Cinema* (Pleasantville, N.Y.: Redgrave, 1980), pp. 24–37.

A complete list of my publications on the coming of sound can be found in Claudia Gorbman, "Bibliography on Sound in Film," *Yale French Studies* 60 (Winter 1980): 276–77.

Chapter 8

Ginette Vincendeau

HOLLYWOOD BABEL: THE COMING OF SOUND AND THE MULTIPLE LANGUAGE VERSION

IN HISTORIES OF THE CINEMA THE QUESTION of multiple language versions (hereafter MLVs) figures generally as a negligible episode worthy of a line or two, at most a paragraph, which goes as follows: with the coming of sound, films are no longer automatically exportable. Hollywood studios find themselves obliged to produce films adapted to national markets (at least the most important ones linguistically and financially, i.e., the Spanish, German, French and Swedish) in order to satisfy the demand for films in European languages as well as to dodge import quotas then imposed by most European countries. One of two strategies is usually adopted: importing directors, scriptwriters and actors from each country to Hollywood (the MGM solution) or setting up production centres in Europe (the Paramount method). Both solutions, against the background of the Depression, prove equally costly and are rapidly dropped in favour of dubbing, or (more, rarely) subtitling, as we know them today.

MLVs have remained unexplored for two main reasons. First of all, in terms of industrial practice, the phenomenon is overshadowed by the sound patents struggle between the USA and Europe for the domination of European markets. Secondly, in terms of an aesthetic history of world cinema, or of the national cinemas concerned, multi-language films, and particularly those produced by Paramount in Paris, are considered worthless – the universally recognised exceptions (*Marius*, *The Threepenny Opera*) being attributed entirely to the talent of their (European) *auteur.*

My intention is not to attack this interpretation of film history in order to replace it with my own version, although even a quick scan through *Variety* of that period reveals facts and positions much more complex than is usually recognised. Nor is the point to unearth forgotten 'masterpieces'. My intention is, rather more modestly, to shift the terms of the debate and concentrate on what is generally regarded as a *weakness.* MLVs interest me because they failed: aesthetically these films were 'terrible', and financially they turned out to be a disaster. However, it seems to me that they can be of use to the historian as they are located at the point of contact between the aesthetic (this term being used here rather loosely to cover cultural, thematic and generic constructs) and the industrial dimensions of cinema. In this sense they can become precious instruments of knowledge for a crucial period of the history of cinema. Beyond the debate on European resistance *versus* American cultural imperialism, the different solutions applied to linguistic and cultural barriers raise generic, theoretical and ideological questions that are also important for an exploration of the notion, apparently self-evident but rarely challenged, of national cinemas.

I

MLVs were usually films shot *simultaneously* in different languages, in one of several fashions:

- Where a film was shot in two or three languages only, the director was often the same; Pabst shot the German and French versions of *The Threepenny Opera* in Berlin, Lubitsch the French and US versions of *One Hour With You* in Hollywood, Jean de Limur the French and US versions of *The Parisian* in Paris, etc. This formula tended to be used, although not exclusively, in European studios.
- For higher numbers of versions (up to 14), particularly at Paramount, each version could have a different director, generally of the nationality corresponding to the language used. *The Lady Lies* (Paramount), for instance, had as many directors as versions (six). Sometimes the same director shot two or three versions. Charles de Rochefort in his memoirs remembers having shot the Czech and Romanian versions of *Paramount on Parade* (and Raymond Chirat's catalogue attributes the Italian version to him too).
- For actors the permutations were as numerous. When polyglot actors were used, the star of the film could remain the same while the rest of the cast (except extras) changed. Claudette Colbert and Maurice Chevalier are the stars of the US and French versions of *The Big Pond*; Brigitte Helm stars in the French and German versions of *Gloria*, *Gold* and *Die schönen Tage von Aranjuez*; Adolphe Menjou in the French and US versions of *The Parisian*, etc. In other cases, the whole cast was different, the individual combinations on the whole being dictated by the coincidence of linguistic competence and script requirements. In *Baroud* (directed by Rex Ingram), Pierre Batcheff, a Russian émigré living in Paris at the time, plays the part of an Arab in both the French and US versions of the film.

There is a second type of MLV: films made from the same source material, but with a short time gap. This category is more difficult to define and catalogue, as the filmographies available adopt different criteria. Variations can be considerable: E.A. Dupont shot the French version of *Atlantic* in January 1930 after he had directed the German and English versions in July 1929 in London. Two years after he had directed *L'Equipage* in Paris, Anatole Litvak remade his own film under the title *The Woman I Love* in Hollywood. Should *Atlantic* be considered a trilingual film and *The Woman I Love* a remake? Herbert Mason shot *First Offence* (also known as *Bad Blood*) in London, based on *Mauvaise Graine* which Billy Wilder had shot in Paris a year before. The country of origin, the studio, the cast and the director were all different. Yet the production of *First Offence* attempted to copy *Mauvaise Graine* to the letter: careful duplication of outdoor locations, meticulous research to retrieve as many as possible of the original costumes; some actors were even imported for character parts (such as Maupi, out of the Pagnol stable, who plays the same part as in *Mauvaise Graine*, speaking execrable English). The two films are remarkably similar, except for the superior quality of the sound in the English version.

The two main categories above still have enough in common to be considered as part of the same phenomenon. In both cases the intention is to adapt the same text for different audiences, within a short period of time. Remakes as we know them now tend to copy an older text, appealing to the cinematic memory of the spectator. The relationship between the two versions (e.g., of *Scarface*, *A Star is Born*, *Breathless*) is diachronic, in MLVs it is synchronic. The distinction is important as one of the main characteristics (and limitations) of the MLV is that only one version is given to a particular audience who in most cases has not (and more importantly must not have) any knowledge of the others.[1]

Finally a third, small, category must be mentioned: the polyglot film, in which each actor speaks his or her own language, such as Pabst's *Kamaradschaft* and Duvivier's *Allo? Berlin? Ici Paris.* Unlike other attempts like *Camp Volant* (Max Reichmann) or *Les Nuits de Port Said* (Léo Mittler), these two films were very successful, which clearly has to do with the fact that they integrate diegetically the inter-lingual apparatus which is their industrial *raison d'être.*

II

Seen from Europe, the coming of sound only reinforced US hegemony. Nino Frank, in his essay on the Paramount Studio in Paris entitled 'Babel-on-Seine', summarises the situation thus: 'One would have thought that the "100% talkies" by establishing cinematic national borders, would demolish the American penetration of our studios. Well, rather the opposite: we are the new Eldorado. The Americans are upon us, loaded with millions of dollars, and they merrily start reorganising French production'.[2] On the other side of the Atlantic, however, the Society of Motion Picture Engineers saw the situation in a different light: 'The demand of people for films in their own language plus the difference in technique between the silent and sound picture is operating to increase competition for American films *abroad*'.[3]

Nino Frank's error was to call Europe the *new* Eldorado. As the Society of Motion Picture Engineers put it, 'Europe, [. . .] after all, remains our principal revenue market'[4] (though, as a matter of fact, the profitability of US films for foreign markets was hotly debated within the studios). By ascribing the reorganisation of the French industry caused by the coming of sound only to the US, Frank was also ignoring the 'helpful' role played by French personnel such as Bernard Natan (of Pathé-Natan), who was responsible for installing the RCA sound system in most French studios, over and above the interests of European patents.

MLVs, as a film industry phenomenon, are situated along a rather blurred line of division between the two types of discourse mentioned above: that of a European resistance (however ephemeral, disorganised and doomed to failure) to US hegemony, and that of a continued expansion by Hollywood in the face of a sudden increase in foreign competition. For, despite the instant success of *The Jazz Singer* and other early Hollywood sound productions, the news of hostile reactions – sometimes going as far as riots[5] – started flooding in from all over Europe and South America, from audiences outraged at the shoddy adaptation of US films (with the addition of sub-titles, music and dubbed passages) they were presented with. At the same time, the popular success of German and French speaking films in Berlin and Paris was becoming increasingly evident.

The first MLV was an Anglo-German production, *Atlantic*, shot by E.A. Dupont for British International Pictures in these languages at Elstree, and which came out in November 1929 (with a third, French, version shot at the beginning of 1930). Then came another Anglo-German film also shot at Elstree, *The Hate Ship*, followed by American MLVs shot in Europe – United Artists produced *Knowing Men* (Anglo-German) and Warner Brothers *At the Villa Rose* (Anglo-French), both made in London. (Meanwhile some US studios briefly produced a few films in foreign languages only, such as *The Royal Box*, shot by Warner Brothers entirely in German.[6]) The first film with a French version shot in Hollywood was *The Unholy Night*, directed by Jacques Feyder for MGM. These had been preceded by a few shorts in foreign languages produced by various US studios, notably Paramount. Purely European MLVs continued to be produced through the 1930s (although their production slowed down after 1933), but the vast majority of MLVs were made under the impulse of Hollywood studios. It is on these that I will concentrate.[7]

On October 2, 1929, *Variety* announced that 'while Paramount is in the lead in foreign tongue shorts production, Warner Brothers is ahead on features. These two companies, of all in the field, so far as could be ascertained, are the only companies indulging in this foreign stuff. What was first considered an indulgence soon became a necessity. A month later, the same paper declared that, apart from projects by Paramount to shoot *The Big Pond* in French, Radio (RKO) was making its first steps towards the 'invasion' of foreign lands (with the dubbing of *Rio Rita* in Spanish and German without, however, any illusions about the efficiency of the dubbing system), that MGM was about to make foreign versions of practically all its films (to start with, a German version of *Sunkissed* directed by Victor Sjöström and a Spanish version of *The Singer of Seville* by Ramon Novarro) and that Fox had decided not to waste any time with dubbing but to start straight away with MLVs produced in Europe.

Although they are generally seen as manifestations of the efficiency of the Hollywood machine, US-generated MLVs are, under close inspection, symptomatic of a great deal of disorganisation, despite the creation, in February 1930, of a special commission of the Academy of Motion Picture Arts and Sciences, responsible for consultation between major studios and with the aim of reaching a standardisation of foreign language films.[8] The reason for the failure to reach such standardisation for many years can be understood if MLVs are seen as symptoms of one of the basic characteristics of the film industry (as of all capitalist industries), the constant tension between the necessity for standardisation to increase profitability on the one hand, and on the other the need for differentiation to ensure the renewal of demand. MLVs were, on the whole, too standardised to satisfy the cultural diversity of their target audience, but too expensively differentiated to be profitable.

Despite assertions to the contrary, dubbed films did not come after MLVs but preceded them, their low cost making them desirable despite the recognised technical deficiencies of the dubbing process. Articles in *Variety* over the period 1929–32 reveal an incredible confusion on the subject – strategies varied between studios, and changes of policy within a studio were common. On April 9, 1930, *Variety* indicated that 'All foreign picture managers [. . .] have reached different conclusions as to the solution which at the moment is the leading trade question. Most are in the air and stalling while they wonder if the potentialities of foreign markets are worth the cost'. On January 14, 1931, MGM 'is still uncertain whether dubbing or direct shooting (in a foreign language) is the best policy'.[9] So, in spite of the self-satisfied Hollywood discourse on its superior competence in terms of planning, it was precisely the lack of long-term strategies which engendered this confusion. The fiasco of Spanish versions in Latin America offers a good example.

As soon as the question of foreign versions came up in Hollywood, the Latin American Spanish-language market was clearly the most attractive, in terms both of audience and number of theatres. All studios immediately launched into Spanish versions, facilitated by the presence of Spanish-speaking personnel in Los Angeles. Two years later, most of that production had to be declared redundant; not only did many linguistic blunders hinder distribution, but they lacked outlets, as many cinemas in South America had not converted to sound. Hesitation was the order of things, policy within studios changing from day to day, following the success or failure of individual films.

Equally, once the principle of MLVs was accepted, the debate shifted to the location of production. There, too, the standardisation/differentiation dialectic operated. Policies switched between production in Hollywood, more cost-effective and conducive to standardisation, and production in Europe, economically less favourable, but more apt to adaptation to local needs, thus to product differentiation. In May 1929, Jesse Lasky confirmed Paramount's intention to make films in foreign languages in France, while Warner Brothers opened studios in Germany and Britain. In March/April 1930, Paramount and MGM were openly in favour of films made in Europe, to ensure better choice of foreign actors at a lower cost, except for Spanish-language versions. Warner Brothers acquired 20 per cent of Tobis-Klangfilm and signed an agreement with Nero for the production of ten films in June 1930. By September 1931, all major US studios had established a presence (in terms of production, that is, since most of them were already present as distributors) in Europe: Warner Brothers, Universal, RKO, Paramount, United Artists and MGM in London; Paramount, United Artists and Fox in Paris; Fox and United Artists in Berlin.

The high water mark for European-made MLVs was reached in July 1930, when *Variety* announced that 75 per cent of all foreign language films would be produced in Europe over the next six months, soon to be followed by the entire production, within a year. A major exception was MGM, Arthur Loew having declared that foreign language versions had to be made in Hollywood 'in order to combine the genius of Hollywood with the European mentality'.[10]

Whatever the location of production, however, multi-language films meant an enormous increase in costs. In December 1930, for instance, MGM had more than 60 foreign actors, scriptwriters and directors under contract in Hollywood, at a total estimated cost to the studio of 40,000 dollars a week (most of the foreigners were repatriated in February 1631). Productions in Europe, however, were not much more profitable, and both types of solution were thus soon revealed to be equally inadequate. By late 1930/early 1931, most Hollywood studios concluded that their main error had been importing foreign personnel to Hollywood, while a US official report on foreign trade strongly advised the film industry not to set up production units in Europe. Their conclusions seem to have been based on the performance of Paramount's Joinville studio, near Paris, which cost 1.5 million dollars in fixed investments and had an annual salary bill of six million dollars.[11] Later, Paramount would admit that the Joinville studio only became profitable after two years of operation, by which time it had become a simple dubbing laboratory.

III

The activities of the Paramount studios in Joinville have attracted enough attention for it to be unnecessary to dwell on the more picturesque details – crews of all nationalities sharing canteen tables, sound stages working 24 hours a day, and the like. Observers such as Ilya Ehrenburg and Marcel Pagnol (who also satirised it in his 1938 film *Le Schpountz*) have left lively accounts of this particular episode.[12] But the myth which has developed around Joinville needs re-examining on a couple of points. First, it is as well to remember that this ostensibly American production-line method (so apparently incompatible with 'European sensitivity') actually originated in London and Berlin. As *Variety* of May 7, 1930, put it, 'Elstree, London's Hollywood, started that idea of working different companies day and night, leaving the sets standing'. Secondly, though legend has it that Bob Kane and his team at Paramount were blind to the gap between its films and their local reception, constant adjustments were made. The material (often plays) on which the films were based changed from predominantly North American to local texts. In January 1931, Kane also decided to reduce the number of versions to French, German, Spanish and Swedish, in order to 'give their production for each one of these countries a more individual treatment, such as is now necessary on account of local improvement in production',[13] adding that 'There's a better chance of bringing up the quality of the Paris made pictures than of lowering the cost of the versions here'.[14] These adjustments did not prove sufficient. The Depression, as well as internal difficulties within Paramount (as Dudley Andrew and Douglas Gomery have shown),[15] contributed to the end of the Joinville studio as a production unit in July 1932. In other parts of Europe, as in Hollywood, studios reverted to dubbing, more economical (about one third of the cost of a MLV) and by now somewhat improved.

Finally, another solution was applied, which consisted in selling the rights to a script as soon as a film had been distributed in a country, a method rightly considered detrimental to European cinema. P.A. Harlé, the editor of the main French trade journal *La Cinématographie française*, warned his readers that: 'The sale of the story of a French film can ruin its career abroad. It is a new method. Instead of selling *Pépé le Moko* in America, its subject was sold. Not only will the Gabin film not be shown on the other side of the Atlantic, but even in French-speaking countries, *Algiers*, with Charles Boyer in the Gabin part, will be shown instead of the French original!'.[16]

IV

The relevance of MLVs to a history of the cinema goes beyond economic and industrial factors, though. Other variables, notably cultural and technical, come into play. The main period of MLVs (1929–32) more or less corresponds to the time lapse necessary for Hollywood to establish the sound cinema and for film-makers to integrate sound to their practice acquired during silent cinema so completely that, according to David Bordwell, 'By 1933, to make a sound film corresponded exactly to shooting a silent film only with sound'.[17] The struggle to improve and impose sound technology was combined with a struggle to improve and impose its *credibility* and given the primacy of the human body and thus the human voice, that of dialogue.

Much attention was paid, at the beginning of sound cinema, to the relationship between sound and the 'reality' of hearing. Linguistic barriers added another dimension to this problem. However, at the same time, there was also concern with the audibility of dialogue.[18] This tension between the desire for a certain auditory 'realism' and the necessity for audible dialogue explains why dubbing was not immediately accepted by either audiences or critics. The coming of sound caused not just technological problems, but altered fundamentally the relation of spectator to film.

In one issue of the *Journal of the Society of Motion Picture Engineers*, a sound engineer asked the following question: 'As the chief objection to [dubbing] as straight dialogue is the fact that it shows actors talking perfectly in a language of which obviously they have no knowledge [. . .] I wonder if any American producer has ever considered saying quite frankly to his foreign audience by means of an explanatory title, that while the actors do not speak the language in question, it was considered fair, in the interests of realism, to employ voice doubles'.[19] Dubbing, a process which underlines the separation between body and voice, disconcerted audiences of the late 1920s/early 1930s. Speaking of the dubbing of Anny Ondra by Joan Barry in Hitchcock's *Blackmail*, *Film Weekly* asked: 'What if the voice, tired and resentful of anonymity, clamours for publicity, to be featured as itself? [. . .] like the Siamese twins, Face and Voice are inseparable, the death of one implying the death of both.'[20] Dubbing upset the feeling of unity, of plenitude, of the character, and thus the spectator position. Moreover, it produced in the contemporary audience a feeling of being duped. A trade paper announced in June 1930: 'Dubbed films are easily recognized as such by audiences who daily get more sophisticated'.[21] Dubbing was on the whole accepted only because of its novelty, and even then it was considered that it would 'go for a while on the novelty angle'[22] but would soon be found unsatisfactory on account of poor synchronisation.

The very naïvety of these reactions draws attention to a problem which is not so much repressed as transparent. The millions of non English-speaking contemporary spectators of *Dallas* know that Sue Ellen does not speak their language, but they do not care very much. In a similar situation, the spectators of 1930 would have perceived an incongruity. It is as if, at its beginning, sound film having just offered a new sense of completeness to the spectator in reconciling body and voice, then immediately upset it by the lack of credibility of dubbing or the sound/image dislocation of sub-titles. MLVs tried to remedy this by the extreme solution of dubbing the body of the actor – until the new norm of dubbed films (or marginally sub-titled films) finally managed to incorporate all these changes and disturbances, thus making MLVs redundant. The linguistic problems, if they underline some of the factors relating to the role of sound in the constitution of the spectator as subject, do not, however, modify them fundamentally. Reasons other than economic for the failure of MLVs are to be found elsewhere – in the field of cultural specificity, as well as casting and narrative patterns.

V

The notion that it is possible to make several versions of the same film, like a piece of clothing in different colours, was/still is abhorrent to the critic, historian or *auteur.* All commentators on MLVs, whether politically to the right or to the left, were unanimous in their condemnation. 'On the seven stages', said Ilya Ehrenburg, 'work is going on day and night. There are shifts for directors, concierges, caterers',[23] while Charles de Rochefort remembered that 'from 8am to 7pm, a Serbian version; from 8pm to 7am, a Romanian version. . . .',[24] and Nino Frank added, 'They shoot while they eat, shoot while they sleep, shoot while they wash, shoot while they talk'.[25] One critic from the ultra-right journal *Action Française* declared: 'We are fed up with these counterfeiters who should be punishable by law, with this cheat which consists of putting the name of a famous *auteur* on some miserable "ersatz" by a third-rate director'.[26]

The important point here is that the criticism of MLVs always went hand in hand with a phobia of the cinema in its industrial as opposed to artistic dimension. Directors of MLVs or 'remakes' were no more flattering. René Clair, who shot *Break the News* in Britain (the English-language version of *Le Mort en fuite*, directed by André Berthomieu), declared 'it ought to be burned [. . .] it was based on an idea which had been used already for another film; and that kills all inspiration'.[27] As Clair's opinion exemplifies, the main objection to MLVs from European 'art' directors was their standardising attitude to film-making, a conception of cinema tainted by its lack of concern for originality and creativity. On the other hand, the necessity of showing films in the language of their country of exhibition confronted Hollywood with the ethnic, linguistic and cultural diversity of its audience. Suddenly studios were aware that Latin American audiences did not appreciate films in Castillian accents, that British accents provoked mirth in the Midwest, and that in the Midlands Yankee voices seemed equally funny. Hollywood was also alerted to a large ethnic range on its home territory.[28]

Such cultural diversity ran counter to the need to rationalise production costs, making MLVs an exemplary meeting point of the economic and the cultural. The irreducible nature of cultural difference even affected the attempt to standardise shooting schedules. It became the practice early on to divide shooting time into as many arbitrary units as there were different languages (for a simple bilingual version, one team worked days, the other nights; in the case of three languages, each team worked for eight hours, etc.), since the more logical break-down, by scene, proved impossible. It was found that the same scene required widely different screening times according to the country it was destined for.[29]

Not only did cultural difference upset the rationalisation of the shooting process; it could also undermine success based largely on technical superiority. *The Unholy Night*, an MGM prestige production, from a novel by Ben Hecht and directed by Jacques Feyder in Hollywood, scored very disappointingly at the box-office when it came out in its French version in Paris. *The Parisian*, on the other hand, a formulaic filmed play about the return of the illegitimate son of the hero (Menjou) at the age of 20, produced by Pathé-Natan and directed by Jean de Limur, was a triumph. Clearly the major element in box-office appeal was not production values, but the audience's familiarity with certain narrative patterns, as this contemporary commentary made clear: 'Story would be simply impossible in Hollywood [. . .] Mere glance at the idea furnishes ample proof that the Americans are disqualified by temperament from picking themes for this market. Such a story wouldn't get a second glance in California. Here it is accepted quite placidly'.[30] What is brought out here is the crucial importance of intertextual familiarity with genres and narrative patterns in the source material – in this case the boulevard play and its archetypical plot revolving around illegitimacy and adultery – for audience appeal and identification. It could be objected that this analysis disregards the importance of the director. The example of *L'Equipage* shows the auteurist factor to be minimal in this case.

Though shot with a time gap, *L'Equipage* (Pathé-Natan) and *The Woman I Love* (RKO) are typical MLVs in that they share the same source material (a novel by Joseph Kessel), the same director (Anatole Litvak), the same music (Arthur Honneger) and some scenes re-used wholesale – a staple technique of MLVs where crowd and street scenes were generally presented in long shot and used as a basis for all versions. Despite being virtually the same film, *L'Equipage* was much more successful with audiences in France than *The Woman I Love* in the USA, a difference hard to impute to the actors, Miriam Hopkins being at least as considerable an actress as Annabella, and Paul Muni (who had just received an Academy Award) as Charles Vanel. Contemporary reviews of both films show that the US version does not 'work' as well as the French one. Two reasons can be found: the credibility of the plot, and the question of morals.

Changes to the film show that an attempt was made to adapt the story to fit American censorship and moral codes. The French version begins on a platform at the Gare de l'Est in Paris in 1918: the hero, Herbillon (Aumont) is going to the front, saying his farewell to both his family and a young woman, Denise (Annabella). The couple exchange vows of eternal love. The young man is going to join the other member of his air crew, who is, unknown to all characters, none other than Denise's husband (Maury/Charles Vanel). The US version adds a long scene prior to the station platform *adieux* which aims to anchor the film in the myth of Paris for a North American audience (the couple meet at an operetta called *Love in Paris*). More importantly, it shows Denise as seduced by Herbillon against her will. Later on, when Herbillon confronts her with the 'horror' of her adultery (for he has come to like and respect her husband), reference is made to the beginning of the film and she reminds him that he initially pursued her. In the French version Denise is left to take full responsibility – and guilt – for her desire for Herbillon. Despite this effort at adaptation, however, the character played by Hopkins was considered by several contemporary critics as 'unsympathetic'. The relative failure of the US version is easier to understand if we consider the narrative, and the fact that a considerable number of French films in the '30s privileged the symbolic or real father–daughter axis to the detriment of other Oedipal relationships (see most films with Raimu, Jules Berry, Victor Francen, Harry Baur and, indeed, Charles Vanel). The relationship between Annabella and Charles Vanel in *L'Equipage*, and the fact that the lover of her own age loses out to the older man, was perfectly acceptable to the French audience of 1935, but less so to North Americans accustomed to different models, among them the victory of the young pretender.[31] Equally, the accent that the French version puts on male bonding, also culturally typical, runs against the stronger drive in contemporary Hollywood films toward the romantic heterosexual couple.

The contrast between acceptance in one culture and rejection in another is characteristic of US-produced MLVs. When contemporary reviews in the country of exhibition can be traced, they constantly show such discrepancies. *Une Femme á menti*, a Paramount production which triumphed in Paris, outraged Italian audiences (under the title *Perché No!*). The Hungarian version of *The Doctor's Secret* was a resounding flop domestically, despite – but also because of – the fact that its crew and cast included great Central European names. The production was cheap: the same decors were used for the eight versions of the film, totalling 49,000 dollars (whereas the contemporary French and Spanish language versions of *Olympia* cost 100,000 dollars each). The same costumes were used in all versions, and the Hungarian public were apparently shocked to see their national star Gizi Bajor so poorly dressed. The reception of the Polish version of *The Doctor's Secret* in Poland is equally telling: there, the fact that the version had been filmed by a Polish director of repute (Ryszard Ordynski) provoked negative reactions on account of the perceived discrepancy between the status of that director and the mediocrity of the film.[32]

Besides, Hollywood's vision of a country rarely coincided with the idea that country had of itself. Hence the hostile reactions in European countries to films based on North American texts. As *Variety* put it: 'We have tried to understand the Spanish psychology, given them pictures about things of which they know more than we do and have naturally laid ourselves open to ridicule'.[33] Not surprisingly, closer cultural understanding between European countries promoted better reception of European produced MLVs like *Le Tunnel* and *The Threepenny Opera*.

VI

Actors in MLVs are another locus of conflict between cultural and economic factors, as well as a focal point for the contradiction inherent in the tendency towards both standardisation and differentiation. In principle, the advantage of MLVs was that they allowed the possibility of using local stars, but this turned out also to be their drawback. One of the major reasons for the success of Hollywood films with non-American audiences was the attraction of stars who had become international. Conversely, one of the main obstacles to the export of European films to the USA was their actors' relative lack of celebrity. MLVs underlined and reinforced this situation. As *Film Weekly* remarked: 'No longer will British actors be able to increase their following by appearing in foreign cinemas; no longer shall we see German and French artists on our screens. If multi-lingual talkies succeed, as I think they are bound to, stars, instead of being international, will be purely local'.[34]

Against this potential obstacle, Hollywood had recourse to polyglot stars: Adolphe Menjou, Maurice Chevalier, Brigitte Helm, Greta Garbo. But here again, interesting contradictions emerged. The foreign accent, for example, had to be diegetically integrated. That being so, it had to be light enough to be acceptable, but strong enough to be picturesque (allegedly Chevalier was forbidden by Paramount to take English lessons). But for the majority of those who could act in only one language, other problems surfaced.

Beyond their impact in the film itself, the commodity value of actors was greatly curtailed by MLVs. For instance, foreign actors posted in Hollywood were encouraged to limit their press statements to the press of their country of origin, as their publicity potential in the USA was nil. Too Americanised for their compatriots, but condemned to remain foreigners in the USA, they were relegated to a sort of media 'no man's land' which uncannily reflects the fate of the MLVs themselves. On the whole, the film industry of their own country was not too happy about their presence on US territory, and in some cases went as far as taking sanctions against them (concurrently, the French actors' union tried to stop its members accepting dubbing work). But, more importantly, it became evident that too much publicity for actors working in MLVs, in Europe as in the United States, inevitably attracted attention to the 'production-line' aspect of these films, and provoked possibly detrimental comparisons – for example, if stars in another version were of higher international status. This goes some way towards accounting for the striking lack of publicity given MLVs at the time.

Notes

1 Except in very rare cases where some art cinemas experimented with showing several versions of the same film (but then always *auteur* films) – for instance, the Studio des Ursulines in Paris programmed French and German versions.
2 Nino Frank, *Petit cinéma sentimental*, Paris, La Nouvelle Edition, 1950.
3 C.J. North and N.D. Golden, 'Meeting Sound Film Competition Abroad', *Journal of the Society of Motion Picture Engineers*, December 1930, p. 750.
4 Ibid., p. 752.
5 For example, in Poland where audiences rebelled (June 12, 1929) against German sub-titles and in France (December 9, 1929), where seats were torn out at a screening of *Les Innocents de Paris* in Nice.
6 *Variety*, November 6, 1929.
7 On Franco-Italian versions, see one of the few articles on the subject by Rémy Pithon, 'Présences françaises dans le cinéma italien pendant les dernières années du régime mussollinien (1935–43)'; in *Risorgimento*, 1981–82/3, pp 181–95.
8 *Variety*, February 12, 1930.
9 Ibid., January 14, 1931.
10 Ibid., April 9, 1930.
11 Ibid., January 14, 1931.
12 Nino Frank, *Petit cinéma sentimental*, op. cit.; Henri Jeanson, 'Cinq semaines à la Paramount, choses vécues', *Le Craponillot*, numéro spécial, November 1932; Ilya Ehrenburg, *Usine de Rêves*, Paris, Gallimard, 1936; Charles de Rochefort, *Le Film de mes souvenirs*, Paris, Société Parisienne d'Edition, 1943. Marcel Pagnol, interview in *Cahiers du cinéma*, no. 173, December 1965.
13 *Variety*, January 14, 1931.
14 Ibid., January 21, 1931.
15 Dudley Andrew, 'Sound in France: the Origins of a Native School', in *Yale French Studies*, no. 60, 1980, pp 94–114; Douglas Gomery, 'Economic Struggle and Hollywood Imperialism: Europe Converts to Sound', *Yale French Studies*, no. 60, pp 90–93.
16 *La Cinématographie française*, September 23, 1938.
17 David Bordwell, Janet Staiger, Kristin Thompson, *The Classical Hollywood Cinema*, London, Routledge and Kegan Paul, 1985.
18 Théophile Pathé, 'Doublage du film étranger', in *Le Cinéma*, Paris, Corréa Editeur, 1942. Rick Airman, in a paper delivered at the 1985 Cerisy conference on Film History, noted a similar concern expressed by US sound engineers at the coming of sound.
19 C.J. North and N.D. Golden, 'Meeting Sound Film Competition Abroad', op. cit., p. 757.
20 *Film Weekly*, September 30, 1929.
21 *Variety*, June 18, 1930.
22 Ibid., November 6, 1929.
23 Ilya Ehrenburg, *Usine de Rêves*, op. cit., p. 117.
24 Charles de Rochefort, *Le Film de mes souvenirs*, op. cit., p. 212.
25 Nino Frank, *Petit cinéma sentimental*, op. cit., p. 66.
26 *L'Action française*, October 24, 1930.
27 René Clair, interviewed by John Gillet, *Focus on Film* no. 12, Winter 1972, p. 41.
28 *Variety*, May 7, 1930.
29 As *Variety*, May 7, 1930 put it, 'The French thought they'd top it by having all companies and making scene by scene, only to find what might merit expansion and building-up for French edification did not go for another audience'. As the example of *L'Equipage/The Woman I Love* below suggests, this scene by scene discrepancy is another characteristic of MLVs.
30 *Variety*, May 7, 1930.
31 See Martha Wolfenstein and Nathan Leites, *Movies, a Psychological Study*, Glencoe, Illinois, The Free Press, 1950, and Ginette Vincendeau, 'The Eye of the Father, Oedipal Narratives in French Films of the 1930s', *Iris*, forthcoming.
32 Marek Halberda, 'Polskie filmy made of Paramount', *Kino* (Poland), May 1983, pp. 22–25.
33 *Variety*, January 7, 1931.
34 *Film Weekly*, November 25, 1929.

Chapter 9

Howard T. Lewis

ORGANIZATION

ROUGHLY SPEAKING, THE INDUSTRY MAY BE divided into the three major aspects of production, distribution, and exhibition. Production may be said to cover all the steps up to and including the completion of the requisite number of positive prints. Distribution activities relate to the rental of films to exhibitors, the "dating in" of the pictures, the physical distribution of the films, and the collection of the amounts due. Exhibition obviously relates to the problem of securing films and to the various other problems of theater management.

According to the United States Bureau of the Census (Census of Distribution), there were in 1929 approximately 142 motion picture establishments in the United States as compared with 127 in 1921 and 132 in 1925. There were 40.84% of these establishments, producing 70.28% of the total output, located in California; 21.12% of all plants, producing 23.83% of the total production, were located in New York State. These 142 establishments produced films "valued, on a production cost basis, at $184,102,419. These producing establishments, commonly known as studios, carried on their pay rolls 19,639 people, of whom 37 were proprietors and firm members, 8,818 were salaried officers and employees (including actors) on the pay rolls on or about December 14, 1929, and 10,784 was the average number of wage earners for the year. Of the total pay roll for the year, the salaried officers and employees received $60,167,520, and the wage earners received $24,860,092. Thus, salaries constituted 32.68% of the total cost, the largest single item of expense; wages were 13.5 per cent of cost; and materials, fuel, and purchased electric energy amounted to 20.88 per cent of the total cost of production." Of the total cost value "$128,496,710 consisted of negative films, the remainder comprising unfinished productions, the development of positive films, receipts for laboratory work done for others, receipts for use of studio facilities, and other work done for others. No less than 92.39 per cent of the value of all negative films was in theatrical pictures (both feature pictures and short subjects), in which but 13.06 per cent was in silent pictures."

The organization of the production department for a typical company may be seen by reference to Exhibit 9.1, which indicates the organization for production of the Paramount Publix Corporation as of the first part of 1929.[1] Smaller companies obviously have a much less elaborate organization, but usually the same general outline is followed.

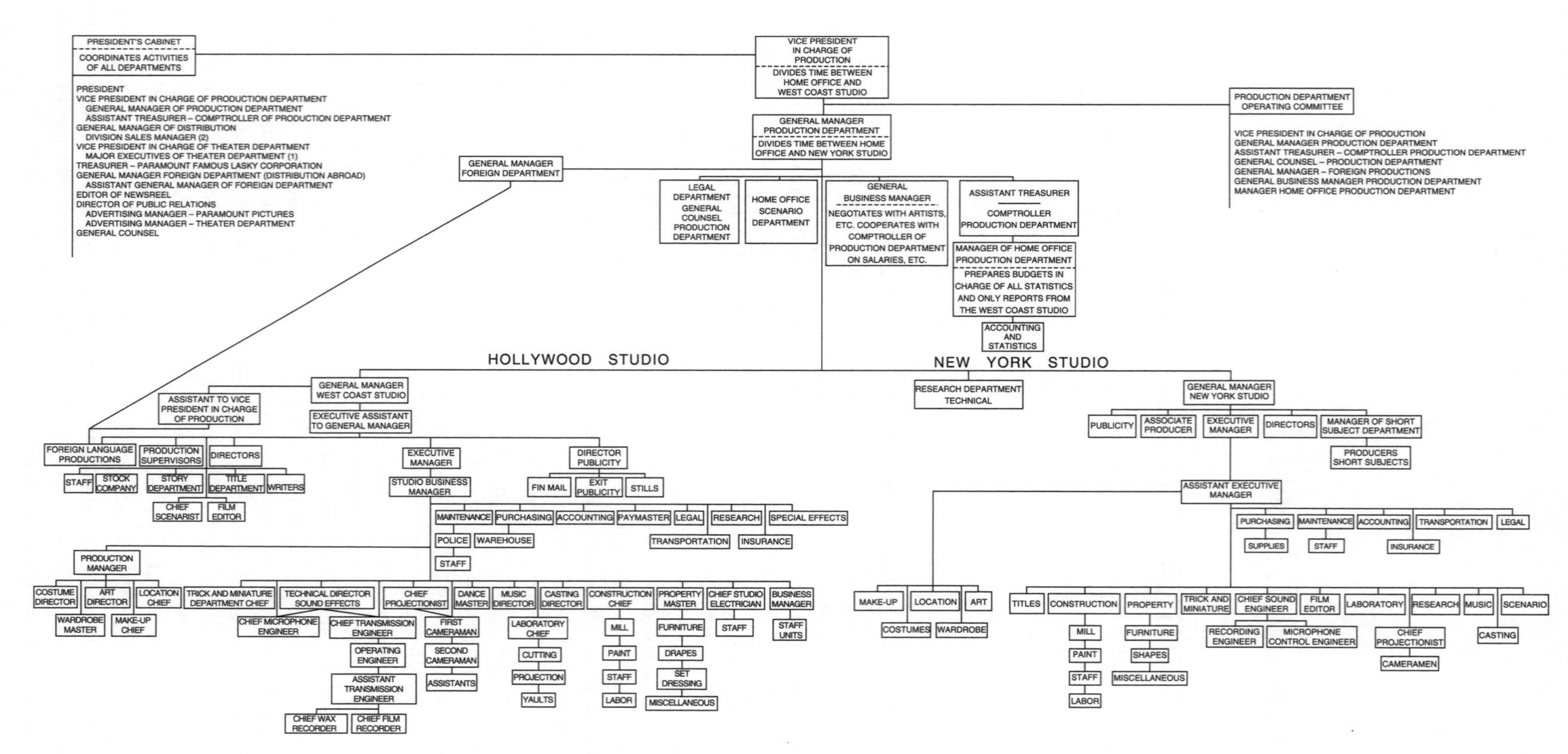

Exhibit 9.1 Production Department, Paramount Famous Lasky Corporation, 1929

The production organization of the Paramount Publix Corporation is designed to effect the maximum amount of flexibility. Directors, as a rule, are free to proceed unmolested with their respective assignments. Supervision by associate producers is of an advisory nature; it is of little importance in productions assigned to outstanding directors. Studio cabinets, consisting of the studio production manager, assistant production manager, directors, supervisors, and studio executive manager, hold weekly meetings to discuss the current problems of the various productions. In this manner each picture benefits from the ideas of all production officials. As a rule, the vice president in charge of production is present each week at either the studio cabinet meeting on Long Island or the one in Hollywood. It is his duty to coordinate this production function with the other major functions such as finance, distribution, and exhibition.

Story departments are maintained both at the home office in New York City and at the Hollywood studio. They are operated as separate units, although the former, because of its proximity to the theatrical, book publishing, and style center is considered first in line of authority.

The story departments obtain information and maintain records of current, past, and future stories, magazine articles, plays, poems, and all other material that might be used as plots for motion pictures. The department managers make contacts with publishers, writers, playwrights, and composers and endeavor at all times to have current knowledge of matters pertaining to the literary world. To facilitate the proper selection of material, it is necessary for them to understand the story requirements of motion pictures, the censorship laws of the several stages, and in particular the type of plot best suited to each Paramount actor, actress, and director.

The story department maintains separate readers to glean material for motion picture plots from magazine stories. Generally magazine articles are made available to motion picture producers in advance of publication; to be of value as a motion picture plot it is necessary that a serial story be made available long before the last episode appears in a magazine. Additional material is secured from "fan" mail, actors, actresses, directors company employees, especially those holding important positions in the production departments, and from numerous other voluntary sources.

The two units of the story department submit briefs of all reviewed material to the studios once each week. It is also customary, because of the lack of literary facilities in Hollywood, for the New York division to supply the California studio with the reviewed books and articles. If the production heads at either studio are interested in the briefs, the story is studied in detail and a rough script, or motion picture version, is prepared by an adapter. If the script is approved, it is edited and usually submitted to the Motion Picture Producers and Distributors of America, or some similar organization, for censorship. In some cases censorship is not considered until after the story rights have been purchased. After approval of a story a price limitation based on the executives' opinion of the motion picture value of the story is set, and the legal department is instructed to complete the purchase. In acquiring story rights, careful consideration is given to copyrights and to pictorial, publishing, dialogue and musical right. Finally, the script is either assigned for production or catalogued for future use.

All catalogue materials are frequently revaluated by the committee in charge. Amortization is based upon the various executives' opinion of each script's worth as a motion picture plot. Scripts of questionable production value are inventoried as $1 each.

The home office is supplied with a summary record of all players working on flat contracts. All stock players are retained in Hollywood, but are transferred temporarily upon request of the Long Island studio. Stock players, as a rule, are well-known character actors.

A separate record is kept for high-salaried stars not classified as stock players. That record contains information on an actor's present occupation, on the time of his availability for the next production in which he will take part, and on the title and starting date of that production. Ail assignments are made approximately one month in advance of the date designated for starting actual production. This precaution is taken to prevent delays arising from changes in assignments caused by such uncontrollable factors as outbursts of temperament on the part of actors.

The report on directors under contract serves the same purpose as did the Stock Company List and the report on high-salaried stars. It is revised every 10 days because, as a rule, changes in directorial assignments are numerous, These changes, however, rarely occur after a production is in actual process. All directorial assignments are made at least 30 days prior to production.[2] Directors' salaries, in 1929, ranged from $100 to $750 a week and upward.[3]

Like its competitors, the Paramount Publix Corporation uses various contracts for actors, actresses, and directors. Some contracts stipulate a certain salary for one picture, others a salary for a series of pictures, and still others a stated weekly salary to be paid on the basis of the ratio of the number of hours in which the particular actor is actually employed in a week to the total number of working hours in the week. All stock actors are paid on a flat salary basis.

Aside from contracts for single pictures, those of the Paramount Publix Corporation are in general for six months' duration with subsequent option at the expiration of each six months' period for five years. Usually, five-year contracts stipulate definite salary increases operative at each renewal; that clause, however, is not an essential requisite. If for any reason the company should not elect to take up its option, the contract automatically becomes inoperative and the actor becomes a free agent. The actor does not have a similar privilege of rejection. The longest straight contract is for twelve months.

In general all salaries, regardless of position, are paid weekly. Extras, temporary electricians, and others not under contract are paid daily.

Production executives and other executives holding important positions are placed under irrevocable five-year contracts. To insure against their resignation within that period of time, their salaries for the full five-year period are placed in escrow. This unusual precaution is taken because of a realization of the importance of executive positions in the motion picture industry and especially because of an appreciation of the value of assured loyalty.

Included among the more tangible materials of production are such items as sets, costumes, general studio supplies, raw film stock, cameras, sound equipment, etc. In general the purchase and maintenance of these materials are supervised by the studio executive manager, whose duties pertain principally to the business aspects of production.

The art department, which operates in close conjunction with the directorial staff, has charge of the making of all sets necessary for the production of pictures. It is equipped and prepared to design an unlimited variety and number of structures. The functions performed are carpentering, painting, plastering, model-making, and wall-papering.

The property department performs the function of supplying all decorative furniture and a large supply of heterogeneous small articles. A corps of experienced employees is maintained to purchase and to construct the articles required. They are assisted by a research division, which maintains a library of volumes dealing with types of furniture, costumes, and designs of all periods of history.

Mechanisms required for illuminating purposes in both indoor and outdoor photography are maintained in the electrical department. A great many scenes are taken under artificial lights whether the work is done by day or by night.

The costume department functions much in the same manner as does the art department. The studio wardrobe requires a large and ever-changing supply of clothing of all kinds and of jewelry.

The photography department maintains a number of expert photographers and keeps in stock cameras of all types.

Raw film stock, unless in natural color, is purchased along with regular supplies. Natural color film and its accessory camera equipment, as well as sound recording equipment, require special contract agreements.

As a general rule the Paramount Publix Corporation plans its production about the first of April. The general manager in charge of distribution states the number of pictures desired for the coming year. He bases his estimates on his own and his associates' opinions in regard to the number of pictures which can be distributed profitably by the company.

Upon determination of the distribution department's requirements, the studio cabinet operative committee, which comprises the chief production executives, maps out the production program, its classification and total cost, and the material content and cost of each picture. These data are submitted to the president for recommendations and approval. While neither the manager in charge of distribution nor the president of the Publix Theaters Corporation, the company's subsidiary theater operating company, is entitled to vote on final decisions, their opinions regarding the submitted program and their recommended changes are influential in the drawing up of the final program.

Feature motion pictures in general are either program or special pictures. Specials are generally the more important productions, some even warranting roadshowing. Either a program picture or a special may or may not be designated as a starring production, according to whether or not it is cast with a star of great box office value. A separate group at one time contained silent pictures which had a few talking sequences. By 1929, however, because such pictures had practically disappeared from the market, this subclassification was not used. Exhibit 9.2 shows how the Paramount Publix Corporation, then called Paramount Famous Lasky Corporation, presented its 1929–30 program for distribution.

To facilitate production and release control, the company's 1929–30 program was segregated into four divisions known as Personality Pictures, Commander Specials, Leader Specials, and New Show World Specials. Personality Pictures were the least costly. Their individual budgets were based on a statistical knowledge of the starring artist's value as a box office attraction. Where more than one Personality production was assigned to a star, a total budget was computed from the individual picture estimates. In such cases, the control department was interested only in the total amount, and as a result some pictures profited at the expense of others. There was a wide range in star salaries; it was estimated in 1929 that the average star received $2,500 a week while working.[4] Three or possibly four Personality Pictures represented the maximum number assigned to any one star. Stories for these pictures were selected currently, and in many cases were written by the scenario department especially for the assigned artist.

Commander Specials ranked third in quality and cost of production. As a rule the stories for these pictures were secured before the beginning of the production season. The casts of Commander Specials did not contain high-salaried stars; generally they were made up of well-known featured players, none of whom was advertised as the particular attraction. The average featured player earned $750 per week while working.[5]

Leader Specials were the company's highest-quality regular feature pictures and often were given extended-run exhibition. All stories for Leader Specials were selected and assigned, and in some cases were in process, prior to the production season. Sometimes such important players as George Bancroft and Nancy Carroll were featured in conjunction with an all-star cast. In others, the title of the story predominated, and the cast of stars was featured

Paramount Productions (36)	Paramount Personalities (29)	Paramount Short Features
Harold Lloyd*	3 Charles "Buddy" Rogers	52 Paramount Sound News
Moran and Mack	2 Maurice Chevalier	104 Paramount Silent News
The Marx Brothers	3 Gary Cooper	*Two-reel Shorts*
The Dance of Life	1 Richard Dix	24 Christie Talking Plays
Mysterious Dr. Fu Manchu	2 Jeanne Eagels	6 Paramount Talking Comedies
The Vagabond King	3 Richard Arlen	—
The Love Parade	3 Nancy Carroll	30
Glorifying the American Girl	4 William Powell	*One-reel Shorts*
The Four Feathers	4 Evelyn Brent	32 Paramount Talking Singing Acts
Illusion	2 Ruth Chatterton	12 Screen Songs
The Virginian	1 Hungarian Rhapsody	6 Talkartoons
Applause	1 Soul of France	—
The Children		50
Greene Murder Case		
Sweetie		
4 Clara Bow		
4 George Bancroft		
Pointed Heels		
Return of Sherlock Holmes		
Escape		
The Lost God		
Woman Trap		
Charming Sinners		
The Lady Lies		
Behind the Make-up		
Youth Has Its Fling		
Fast Company		
Kibitzer		
Sarah and Son		
The Gay Lady		

* Produced by Harold Lloyd Corporation. A Paramount release.

Exhibit 9.2 Production program of Paramount Famous Lasky Corporation for 1929–30, as classified for distribution

as of secondary importance. Because of the impossibility of accurately measuring public opinion, Leader Specials were known to have exceeded more costly pictures in total box office receipts.

New Show World Specials were the Paramount Publix Corporation's extended-run pictures that commanded $1.50 and $2 admission prices when prereleased for roadshow purposes. "Glorifying the American Girl," a Ziegfeld production, assigned to a high-salaried director, cast with all-star players, and staged on an elaborate basis, was typical of this type of picture. In that particular case, the use of Ziegfeld's name represented a large cost in itself. Because New Show World Specials required extensive preparations, as a general rule their production was started before the opening of the regular production season.

Immediately following the approval of the production program, the studio cabinet prepares a tentative release schedule for the first six months of the new theatrical year. Such a schedule is limited to six months because of the following considerations: current changes in public demand; unavoidable production substitutions; the possibility of acquiring rights to some current story or stage play success for immediate production; the general consensus of opinion that a more extended schedule would be unwise; and finally, the probable introduction of new devices for the production of motion pictures.

Tentative release schedules are used to inform the general production manager of changes in release dates as they occur from time to time throughout the year. These enforced changes, which are not uncommon, are the result of unavoidable production delays.

Approved, tentative release schedules are put into final form after all changes have been definitely decided upon. As a rule, changes in the final release schedule occur in one out of ten pictures. Since these changes affect the entire mechanism of distribution and exhibition, it is necessary for the production department to fill in vacancies by substitution. Substitute pictures are always available. They are derived from productions finished ahead of schedule, pictures filmed before the production season, and roadshows not yet released for regular distribution.

Final release schedules are sent immediately to members of the distribution department, to production executives, and in some cases to managers of Publix theaters and other theaters.

Short subjects [generally] require less than one week for production. Consequently, their release dates are determined in final form without recourse to a tentative schedule.

Paramount pictures are released on an average of one each week, with two released during one week in every four. This system does not always operate as planned, but as a general rule changes are infrequent. During the 1929–30 season every fifth picture released by the company was a costly special.

This large number of expensive pictures was a direct result of the advent of talking pictures and the subsequent availability of stage stars, stage plays, musical comedies, and operettas. To insure its position in the motion picture industry, the Paramount Famous Lasky Corporation, as well as its competitors, purchased numerous stage successes and procured famous stage actors at high salaries. In the opinion of many of the company's executives, the production of a large number of costly specials is unwise. They believe that the market cannot absorb so many costly cinemas, and that as a result none of them will return the maximum possible income. At the present time, therefore, the number of this type of production has been reduced. Musical comedies and operettas, which are expensive to produce, are being used sparingly.

Because of the time element in musical pictures, immediate production and release are necessary for them. The time element is not so important in the case of dialogue plays; since the stage actors are under contract, however, the company cannot afford to delay production. It might withhold release of its dialogue pictures until a more opportune time, but in so doing it would risk the possibility of obsolescence due to the introduction of new devices. Furthermore, it would tie up large sums of working capital.

In addition to the release schedules for the first six months or first two quarters of the theatrical year, the studio cabinet prepares a tentative release summary for each of the four quarters. This schedule is based on the entire year's production and is concerned primarily with the total number of pictures allocated to the various seasons. Because of the inadvisability of releasing the best-quality feature pictures during late spring, summer, and early fall, however, consideration is given also to the type and classification of each picture.

Careful allocation of such prerelease productions as New Show World Specials is not necessary. The company controls a sufficient number of extended-run theaters in the New York theatrical district and in Los Angeles, California, to provide all necessary theater facilities for these pictures, even though their exhibition dates should overlap.

As a general rule, box office receipts vary during the four seasons of the theatrical year. The first quarter, including September, October, and November, is fair, gaining momentum in late October. In the winter season, receipts are greatest. In the early part of the spring season, receipts equal those of late fall; in the late spring, they are approximately the same as in September. The summer season usually is the least profitable. In 1929, however, the introduction of talking pictures, a decline in the popularity of legitimate plays except in New York City, and the widespread use of air-cooling in theaters forecast a leveling out of box office receipts. With this fact in mind, production officials of the Paramount Famous Lasky Corporation diverged from the usual practice of varying the number of releases according to the season of the year. At the present time the tendency is to level out the releases over the year. Tentative quarterly release schedules can be changed without seriously affecting the mechanism of distribution and exhibition.

The production process of a picture usually begins with the directorial assignment about 30 days in advance of the date set for starting actual production. Upon assignment the director, in conjunction with the casting, art, and sets departments, selects respectively the cast, locations, and sets. When possible, Paramount contract actors are given first preference in casting. At the same time a scenarist prepares the script in final form making all necessary additions and corrections, determining the number of sets to be used, the amount and text of the dialogue, the number of titles, the musical scores, and the general pictorial effects.

Since the introduction of sound, all Paramount actors, regardless of their experience, have had to submit to a thorough voice diagnosis, known as the "screen-voice test." This process is expensive but necessary, in order that the company's production officials may have a complete knowledge of all prominent players' voices, as well as of their camera possibilities.

Subsequent to the directorial assignment for a picture, the studio production manager appoints an associate producer to supervise the making of the picture, and the studio executive manager appoints a business manager to take charge of the business aspects of his assigned pictures. These two, together with the director, decide upon a definite production schedule and draw up detailed budgets, the total of which shall not exceed the amount allocated to that specific production.

Dialogue parts are usually assigned immediately after the cast has been selected. Rehearsals, which begin one week in advance of the date set for starting production, are held again on the sets just before the shots are taken. Rehearsals were not the customary practice before the introduction of synchronization.

In arranging the schedule, plans are made at the same time for taking both indoor and outdoor shots. This precaution is taken to insure progress of the picture against inclement weather. Schedule delays cost between $5,000 and $10,000 a day, the average cost being $8,000. The more difficult scenes are assigned to the last stages of production. To avoid waste in time, the scenes in which high-salaried actors appear are arranged in sequence.

In the production of a motion picture, pictures of each scene are made from two or more perspectives. Often different types of cameras are used in order to reduce the risks due to errors in the judgment of the directors, photographers, or actors. After each day's work the negative is taken to the laboratory for development. After the negative has been developed and dried, it is carefully polished by hand and inspected for defects. A print is made from the negative as soon as possible, in order that changes suggested by the director, supervisor, or other members of the production department may be made before the sets and other effects have been dismantled.

Upon the completion of a production, the director and his assistant, in conjunction with the cutting department, assemble the negatives in continuous form in approximately the order in which they finally are to be arranged, trimming them to the desired length and, if necessary, inserting temporary titles. The negative is then "light tested" and one copy taken off, which is known as the "answer copy." The answer copy is projected, and upon approval of it by the director, his assistant, the supervisor, and other members of the production department, the negative is pronounced ready for positive printing and subsequent distribution.

Dating from assignment and continuing through production, all directors and supervisors attend the weekly studio cabinet meetings. Each director reports on the progress of his production, the difficulties faced, the coordination of the production schedule with the budget, and any contemplated changes, as well as on various other details. The problems are discussed by the group as a whole, and ideas are exchanged among the several directors and supervisors. Detailed reports of these meetings are retained by the studio production managers for their respective groups and by the general production manager for both studios.

In addition to the weekly report, daily wires are sent from the California studio to the home office, and vice versa. As a rule they are night letters of from 100 to 1,000 words. They contain information concerning stories, script, artists, costs, transfers, release dates, substitution, and all information and requests which should be made known immediately to the designated official. The general production manager retains copies dated back two weeks. Inasmuch as the general production manager visits the Long Island studio frequently, daily reports from that studio are not required.

In order that the company's executives may have a bird's-eye view of the current financial status of production, they are furnished with weekly summary reports giving cash disbursements for each feature picture, and total expenditures for all feature pictures. Such reports are sent from the Hollywood studio and from the Long Island studio. Separate expenditure reports are submitted for short subjects, which are produced only at the East Coast studio.

An estimated final cost schedule presents a detailed report of expenditures of individual productions as contrasted with their respective established budgets.

Half-year summaries list the actual results without reference to the expected results. Half-year summaries for the previous four years are made available for current use by the general production manager.

According to the United States Bureau of the Census in 1929, there were 143 establishments, including both studios and laboratories, producing motion pictures in this country. Such of these production units as were engaged in making pictures for the theatrical trade appear to have been operated by 46 producing companies.

Substantially the larger proportion of the pictures produced up to the present have been made at Hollywood. Around this major task of producing pictures, there have grown up also certain affiliated organizations. Thus there are approximately five distributors[6] of raw film stock, nearly a hundred casting agencies, a substantial group of insurance brokers supplying various forms of motion picture insurance, laboratories,[7] play and story brokers, storage companies, and numerous other companies such as title studios, costumers, camera dealers, film laboratories, and lighting specialists.

As has already been indicated, distribution relates to the rental of films to the exhibitors, dating in of the pictures, physical distribution of the films, and the collection of the amounts due. The organization created for the proper performance of these functions is naturally extensive. Considering the character of the work to be done, however, it is not unduly complex. No better understanding of what is involved in the task of distributing pictures can be obtained than by describing the organization of a particular company. For this purpose the sales organization of the Pathé Exchange, Incorporated, as of 1929, may be cited. While it is true that this company was acquired in January, 1931, by the Radio-Keith-Orpheum Corporation, this transfer of control does not affect materially the value of this illustration for our present purpose.

The Pathé Exchange, Incorporated, operated a national exchange system with branch offices, called exchanges, in 31 key cities of the United States. Foreign distribution was carried on by subsidiaries of the company, except in Canada, where Regal Films, Limited, distributed Pathé pictures. The offices of the sales department were located at the executive headquarters in New York City.

In addition to its own product, the company distributed films of other producing companies on a percentage arrangement, usually receiving 30% to 40% of the gross revenue.[8] The company was also the largest American distributor of films of educational value to the nontheatrical market. The company's gross sales, in 1928, were approximately $18,000,000.

The organization of the sales department of the Pathé Exchange, Incorporated, was as is shown in Exhibit 9.3. The general sales manager, located at the home office, had as functional assistants an assistant sales manager, a manager of exchange operations, and a director of advertising. The assistant sales manager exercised control over the routine selling activities, including the analysis of contract applications, and was available for field work. The manager of exchange operations had direct control of the nonselling activities of the exchanges. In financial matters affecting exchange operations, he cooperated with the comptroller and the treasurer. Three auditors, who were responsible to the manager of exchange operations, visited the exchanges for the purpose of checking the activities of the bankers and the cashiers. The advertising director maintained contact with the exchanges through the manager of exchange operations.

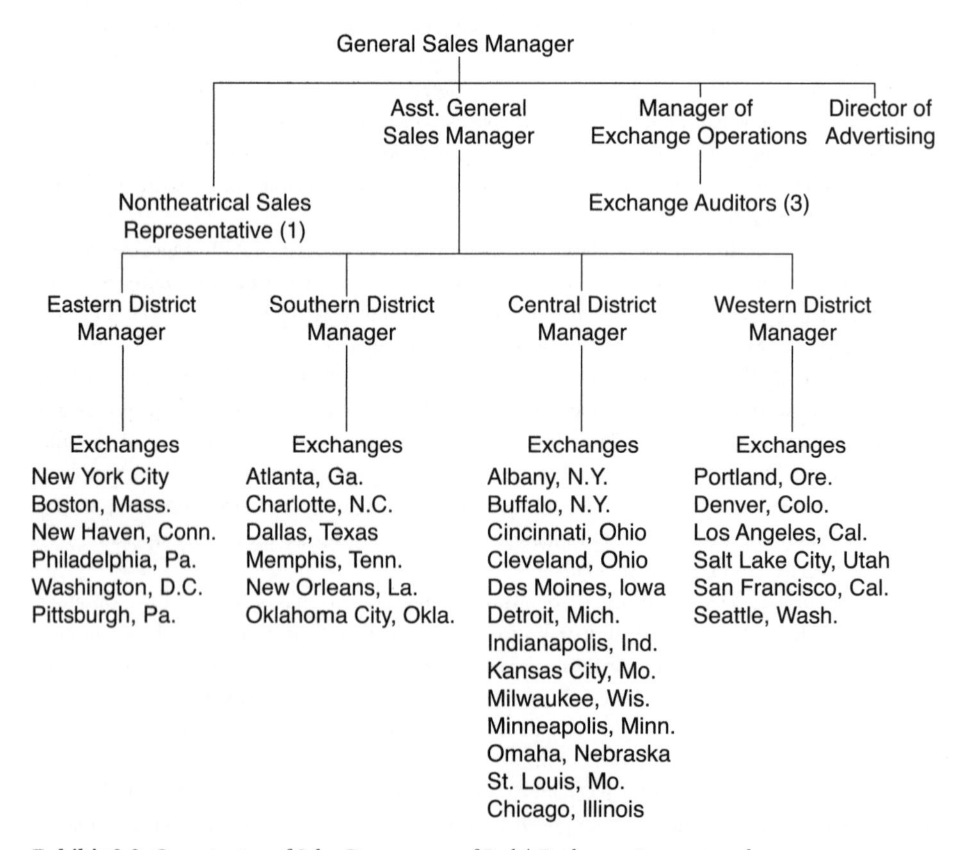

Exhibit 9.3 Organization of Sales Department of Pathé Exchange, Incorporated

The company's advertising activities included the advertising and exploitation of individual pictures and the general advertising and publicity of the company. The work was divided into such departments as the following: advertising—paid space, outdoor advertising, exhibitor helps, and advertising on theater screens; publicity—news stories and other forms of free publicity; and exploitation—stunts, contests, cooperation of merchants, and tie-ups with music distributors. The company employed an advertising agency.

For purposes of selling, the company had divided the country into four divisions: eastern, southern, central, and western. Four divisional managers located at the home office were responsible for all the selling activities, except those involved in the sale of nontheatrical product, in their respective territories. These division managers were directly responsible to the assistant general sales manager. The sale of nontheatrical films was in charge of a nontheatrical sales representative who, while under the supervision of the general sales manager, was somewhat detached from the organization because a majority of the sales of nontheatrical products were made by special salesmen and by correspondence. Educational films, while sold largely to nontheatrical accounts, were considered a part of the company's regular line and were available for theatrical exhibition.

Physical distribution was done from the branch exchanges, which obtained positive prints from the company's laboratories in New Jersey. The activities of an exchange included selling pictures and advertising accessories, booking pictures for exhibition, shipping, inspecting, and servicing films, billing exhibitors, and making collections. These activities were organized into three departments, as shown in Exhibit 9.4.

Each branch exchange manager was responsible directly to the manager of the division in which his exchange was located. The branch manager was largely concerned with the selling activities of the exchange; in addition to supervising the salesmen, he sold films to the large accounts in the exchange territory. The number of salesmen varied with each exchange. The company maintained a force of about 130 salesmen during the selling season, reducing that number somewhat during the remainder of the year.

At the beginning of the selling season,[9] a convention was held in Chicago at which sales quotas were assigned by the general sales manager to the several branch exchanges. These branch quotas were broken down by the branch managers into blocks. In this sense a block constituted a certain percentage of a territory and should not be confused with a block of pictures for block booking. A block was assigned to each salesman at an exchange. Each salesman except those selling nontheatrical pictures sold the entire line of product in the block assigned to him.[10]

Each salesman at an exchange was furnished with a sales manual and an announcement of the product for the season. A price classification of product and price schedules for posters

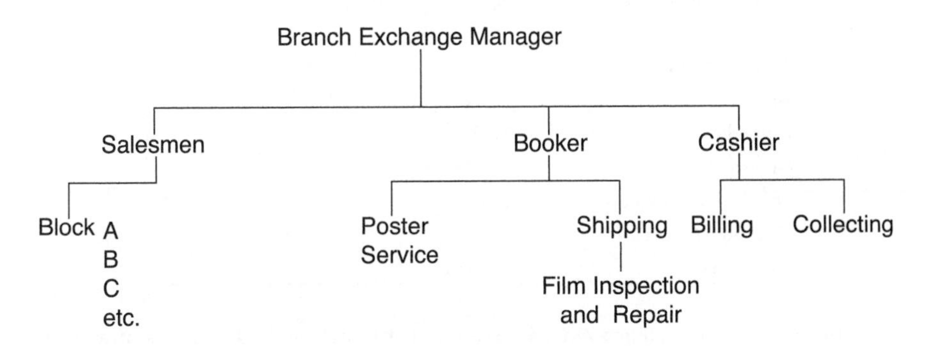

Exhibit 9.4 Exchange Organization of Pathé Exchange, Incorporated

and accessories were available to the salesmen at the exchange. With this information and a work sheet for each theater, a salesman visited exhibitors to solicit contracts for picture rentals. It was extremely advisable for a salesman to secure satisfactory play dates in addition to selling the pictures. No contract signed by an exhibitor became effective until it had been approved by the general sales manager at the home office of the Pathé Exchange, Incorporated. Accordingly, each contract obtained by a salesman, accompanied with a contract enclosure embodying the recommendations of the branch manager as to the contract, was forwarded by the manager of the exchange to the home office. The exchange maintained a record of sales for each theater, giving information as to the pictures sold to that theater, the prices secured, and the dates of showing.

Educational films were sold on order blanks which specified the entire product of the department with an agreement to execute a uniform exhibition contract upon preparation and delivery. Advertising accessories also were sold on order with payment on delivery. From nontheatrical account records maintained at each exchange, each salesman received information relative to such accounts in his territory.

Each salesman submitted a daily report on the results of interviews and a weekly traveling expense report for the approval of the branch manager. Salesmen were paid on a straight salary-and-expense basis.

Pathé pictures were sold in groups of varying numbers, depending upon the requirements of the exhibitors and the sales ability of the salesmen. The company endeavored, by all fair and reasonable inducements, to persuade exhibitors to contract for the entire number in the group. While the company never refused to sell an exhibitor less than an entire group of films, it did not permit an exhibitor to make his choice of one or more of the pictures offered in a group at a group average price. In the case of Pathé News, 104 issues, divided into 2 series, comprised the output for a year.

Of the 15,000 or so theaters considered as potential accounts, satisfactory distribution commonly constituted from 6,000 to 8,000 contracts on an individual picture. In presenting the company's product, the salesmen placed short subjects first, because the Pathé Exchange, Incorporated, was recognized as the leader in this field and because the widely diversified output of these pictures made them adaptable for sale to nearly every theater in the country. After an agreement had been reached on the short-subject product, other groups were presented in turn. It was the policy of the company to sell an entire program of available product wherever possible. Sales of the several classes of product during the 1928–29 season, in relation to total sales, were approximately as follows:

Superspecials		16.3% of total sales
Programs and Specials	3.4.0	
Westerns	1.5	
Total Features	–	51.8%
Serials	3.0%	
Comedies	10.2%	
Newsreels	20.4%	
Other Short Subjects	14.6%	
Total Short Subjects	–	48.2%
Total Sales	100.0%	

Pictures were sold on a percentage basis or as flat rentals; in both cases, contracts were made between the exhibitor and the distributor. Sales on a percentage split necessitated daily and weekly reports on box office receipts, with a summary on each contract. The price at which Pathé pictures, except certain short subjects, were offered was established by the adoption of a national quota for each picture, which was in turn broken up into percentages assigned to each branch exchange and further broken up into percentages thereof assigned to each sales district within the territory covered by a branch exchange. From the quota figures, and from the sales manager's, the branch manager's, and the salesman's knowledge of prices, which had been paid and thereby had become more or less established for given theaters for the different classes of product, the rentals to be derived from the several theaters in a territory and zone were determined. Such prices necessarily remained flexible, because in the final analysis the price at which a group of pictures was sold was determined as a result of negotiations between the salesman and the exhibitor in arriving at an average price acceptable for the entire number of pictures taken. Large first-run theaters did not, generally speaking, take pictures in groups. It was the salesman's responsibility to sell as much of the company's product to as many exhibitors as possible at the best prices obtainable. The company's records showed a great variance between the number of contracts taken on the various pictures comprising a group. For example, a popular picture in one group had received approximately 10,000 contracts, whereas a less popular picture had netted only 3,000 contracts.

The work of making arrangements for the playing of the pictures, together with the directing of the shipping and other work in connection with the physical handling of the positive prints, was performed by the booker, who in most instances acted as office manager as well. His department was also responsible for the storing and issuing of advertising accessories such as posters. Sales contracts, with specified dates of exhibition or arrangements to play so many pictures per week, month, or season, constituted his authority for booking a print. A record of bookings contained the necessary information for the shipment of prints. Each of several prints was scheduled for continuous use, with allowance for delivery and return time. It was the booker's duty to keep the salesmen informed as to unspecified play dates in order that the Pathé product might secure an early showing. A delay in exhibition caused cancellations and difficulties in selling new releases. Protection rights of the several theaters in a zone were carefully guarded by the booker. The necessary information for determining such rights was maintained in customers' service records for the product before the current year.

The control system which the company used provided standardized forms to be used by the booker in each exchange. At regular intervals the booker sent to each exhibitor a notice of availability which listed the current productions contracted for by the exhibitor and the dates available for choice in exhibition. At the same time the exhibitor was reminded that, according to a provision of the Standard Exhibition Contract, failure to select exhibition dates within 14 days of the notice gave the exchange authority to designate such dates. In the event that an exhibitor failed to exercise his privilege of selection, a designation of play dates was sent to him. Applications for play dates received from exhibitors were confirmed on a special form. Notice of substitution or cancellation was furnished the home office and the exhibitor with the reasons therefore. A copy of each of these two forms was forwarded to the home office, where a record of bookings in each exchange was prepared. In addition, the booker forwarded a weekly report of spot bookings. A special report was required by the home office for immediate information on the play dates booked for a particular release.

As custodian of prints the booker requisitioned new prints, returned films to the laboratory, and maintained vault record cards for each print in storage. A monthly report of inventory was sent to the home office. The inspector prepared a daily inspection report which was submitted to the manager but not forwarded. The shipper prepared a daily report of late returns on films, which was turned over to the manager for decision as to the action to be taken by the booker, and a special report on refused shipments.

In the poster-room of the exchange was maintained a record (by theater) of exhibitors' advertising orders for current product. A monthly report of accessory inventory was submitted by the booking department to the branch manager for his approval; the original was then forwarded to the home office. Requisitions for accessories were submitted to the manager for approval.

In addition to his duties in connection with billing the exhibitors and making collections, the exchange cashier had charge of the accounting work of the exchange. As part of the company's system of control the cashier was required to submit reports to aid the traveling auditors in checking his work. These reports were specified to be as follows: (1) a weekly statement of business, which included a report of billings, collections, and ledger entries, prepared by the cashier and certified by the branch manager; (2) a daily cash sheet, showing the amounts received on picture rentals with subcolumns for rentals on particular releases; (3) a record of departmental disbursements accompanied by signed vouchers submitted weekly; (4) an employees' salary sheet substantiated by vouchers; and (5) a monthly report of customers' balances prepared from the ledger. All reports were to be approved by the manager. The cashier's department was to prepare also a daily shipping sheet which reported all rentals received and shipments made. All supplies exceeding $10 in value were requisitioned from the home office; requisitions were approved by the manager.

The company used a system of sales control which made possible close home office supervision of the exchange activities. This central supervision resulted in a close control over expense with a realization of economies in exchange operation. Large savings were made because of centralized purchasing of supplies. A close control of credit adjustments minimized the number and amount of such credits.

A close control of sales was possible as a result of standardized home office reports. With these reports, the central sales organization was in touch with the field and could direct activities during the sales season; inefficiency was noticeable, and corrective measures could be taken.

In setting up a control system, the executives of the Pathé Exchange, Incorporated, were interested primarily in the realization of sales volume; the routing of salesmen and the control of sales expense remained largely with the branch managers, who were considered to be in a better position to supervise the visits of the salesmen. The executives desired up-to-date information, however, on contracts, bookings, and play dates which would enable them to instruct and direct the activities of the branches.

The sales in key cities were particularly important in the early season and were watched carefully because of their effect upon subsequent runs. There were estimated to be between 700 and 800 accounts in key cities, cities of more than 25,000 population, in which the initial exhibition of a feature motion picture was given and in which a showing had an additional value to the comparative from the advertising of the picture. The comparative value of different theaters in a key city varied according to such factors as the location, the size, the entertainment policy, and the management. Generally speaking, the longer the period of exhibition in an important first-run theater in a key city, the greater was the value of such exhibition to the Pathé Exchange, Incorporated.

Under the system of control, a special book register was prepared for recording the contracts, bookings, rentals, and play dates in the key cities. The cities were listed by branch territories, and first-run contracts were recorded as obtained. The record for each city showed the theaters operating, the classification as to run, the seating capacity, past rentals, and other data on the competitive situation. With this record the sales executives were enabled to direct the efforts of the division managers, branch managers, and salesmen towards the strengthening of the company's position in this important part of the market.

Potential accounts in each territory were maintained in card form. Similar records were maintained at each branch. Contracts, as approved, with data relative to rentals, runs, and playing arrangements, were posted to the card accounts as received. This record disclosed sales possibilities, thereby furnishing a basis for instructing the branch manager as to "open spots" in the territory. The card records and the special register were kept up to date by a force of clerks working under the supervision of the assistant sales manager.

Taking into consideration a few slight modifications in individual companies, the distribution organization which has just been described may be regarded as typical of the system used by all the larger distributors except one, the United Artists Corporation. The general organization of this company is such as to make expedient the use of a distribution procedure quite different from that followed generally in the industry.

The United Artists Corporation is made up of a number of independent producing units, each of which delivers its completed pictures to the company for distribution. The total number of pictures produced by these units is somewhat smaller than the number distributed by any one of the other major companies. In 1930, for example, when the average number of productions distributed by the larger companies was about 35, the number distributed by the United Artists Corporation was only 16.

The United Artists Corporation receives, for the performance of the services of a distributor, a stipulated percentage of the gross earnings of a picture. It makes no advances to the producing units against their pictures in process but makes payment to them only out of actual receipts. In return for its commission the United Artists Corporation performs all the services of distribution including selling, advertising, booking of the pictures for definite exhibition dates, shipment of the films to exhibitors, inspection and repair of films, and collection of payments from exhibitors. The payment received from an exhibitor, less the company's commission, is paid to the producer three weeks after the exhibition of the pictures. The producers furnish the company a negative of each picture and 130 positive prints ready for use.

The total amount spent on advertising is determined at the beginning of the year by the company. This sum of money is expended in equal amounts on all the pictures to be distributed. If any producer desires to obtain more advertising for his pictures than can be afforded by the budget assigned to them, he is at liberty to expend his own money for advertising. The company offers the services of its advertising department to any of the producers who wish to do additional advertising at their own expense.

Because of the limited number of productions which the company distributes, it sells its pictures individually. One of the principal reasons for this is that only this method of sale permits an equitable division of the receipts from the sales of the pictures. By selling the pictures individually the company knows the exact amount earned by each picture, thus obviating the necessity for an arbitrary division of the receipts in order to sell the pictures in groups.

Another reason which the company gives for selling individually is that such a method permits special stress on the merits of the individual pictures and enables the company to obtain higher prices for pictures of a given quality than it otherwise might obtain. Inasmuch as the company makes only a few pictures each year and attempts to maintain a high quality, it considers this factor to be of considerable importance.

Thus the United Artists Corporation has a distribution organization which is designed to meet the peculiar problems arising out of an unusual situation. It is able to use a method of individual selling, whereas for other companies the same system might prove quite unsuccessful.

The relationship between the distributor and the theater exhibiting the films naturally raises a series of very important and crucial issues. These problems center around selection of pictures, price, terms of sale, and dating. Important as these problems are, a consideration of them is not pertinent at the present moment, since we are now interested only in gaining a picture of the structure of the industry itself. Neither is it important here to consider certain distinctly theater management problems, such as those relating to labor, local advertising, building restrictions and requirements, and the like. It is necessary only to indicate briefly the typical theater organization, in order that some idea may be gained as to the requirements to be met in the successful management of a theater. In general these may be classified in two main groups: those relating to the service and business end of the theater, and those relating to what may be termed the back of the house, including those factors pertaining to the entertainment, such as musicians, stage hands, projectionists, and performers. A typical large theater may operate with as many as 10 departments, namely, manager's office, service department, maintenance, housekeeping, engineering, production, projection, musical, advertising, and accounting.

The organization chart, Exhibit 9.5, indicates somewhat definitely the relationship between these various departments.

Ever since the development of chains of theaters, the problems involving the degree of centralization of control have caused a good deal of concern. Here again the degree of centralization of authority is a very real issue that has not always been met in the same way by various companies, nor has the same company always followed the same procedure. However, an illustration of the procedure actually followed is necessary.

In 1929, the Fox Theaters Corporation established a buyer in New York City to purchase motion pictures for all Fox theaters. Before that time, the function had been decentralized, in some cases being delegated to the various division officers of the Fox Theaters Corporation and in others to the unit theater managements.

The Fox Theaters Corporation divided its theaters into five divisions, each group operating under the control of a division manager. The five divisions were: West Coast, Midwest, Southern, New England, and Metropolitan New York. Individual theater managers were to operate under the supervision of district managers,[11] who in turn would report directly to their respective division managers.

The company did not maintain a general theater manager as did a majority of its chief competitors; instead, the activities of the five division managers were coordinated by two regional directors, whose territories were divided theoretically by the Mississippi River. The eastern director had headquarters in New York City; the western director, in Los Angeles. To provide all the benefits of a highly centralized organization, the company rotated the directors as between the two territories.

In general, before the appointment of a general film buyer for Fox theaters, films were purchased by the unit theater managers or their buyers, by division buyers, and in some cases through the cooperation of both. When films were purchased by the individual theater, the transaction from purchase to delivery, although governed in part by predetermined budgets, and to a lesser extent by general company policy, was not unlike the purchases made by individual independent exhibitors. Purchases made for a group or a division were negotiated by a division buyer or a division manager. The division buyer, because of his personal contact with the theater managers in his territory, was well acquainted with the situation of each theater and was therefore able to judge the suitability of each producer's product for specific theaters.

Division sales managers of the various distributing companies usually negotiated with the division buyers for the sale of their pictures to Fox theaters. Unless he was offering his company's product in units, the distributor-representative endeavored to sell to the buyer as many of his pictures for as many of the Fox theaters in his district as possible, and to secure for them a maximum total price. The buyer, on the other hand, usually selected certain pictures and endeavored to purchase them, for a minimum price. He was guided by an already

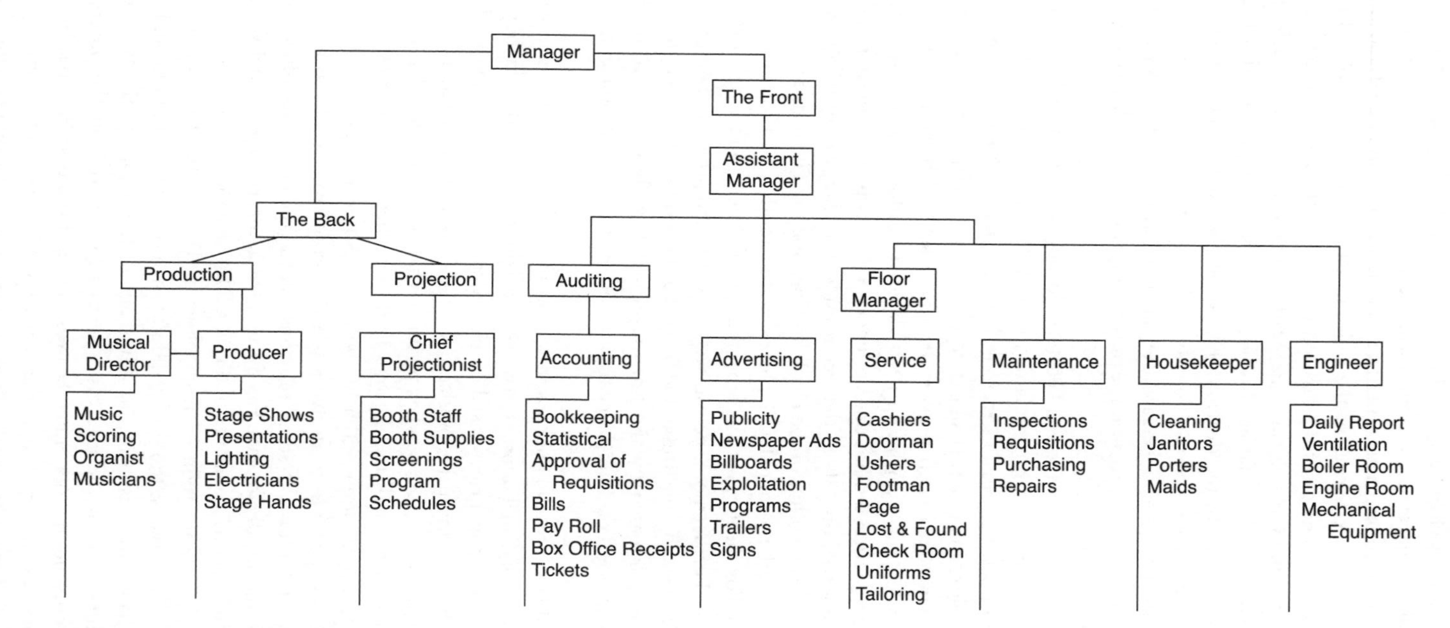

Exhibit 9.5 Theater organization chart

Source: *Motion Picture Theater Management* by Harold B. Franklin. New York: Doubleday, Doran and Company, 1927.

established theater budget. Although prices were estimated for each theater, both buyer and seller were interested primarily in the aggregate totals. Pricing was made complicated by the use of flat rentals and the resultant bargaining process.

If an agreement was reached between the salesman and the buyer, the seller submitted the proposal to his sales manager, and the buyer submitted the proposal to the theater division manager. If satisfactory to both parties, the transaction was consummated and the project turned over to the various exchanges of the seller and to the division booker for the Fox Theaters Corporation. In some cases the division booker for the Fox Theaters Corporation cooperated with the distribution division manager of the company making the sale and in that way scheduled all play dates, in so far as possible, for the pictures purchased. Copies of the booking schedules were sent to the distributor's exchanges from which deliveries were to be made, and to the Fox theaters booked to exhibit the pictures. In other cases each theater district office maintained its own booker, and the individual theater managers were permitted to express their dating preferences to him. The purchasing by district bookers was coordinated by a division booker or buyer.

The purchase of Fox films by the Fox Theaters Corporation, whether by division buyers or by individual theater buyers, was negotiated in much the same manner as the purchase of a rival producer's product. For obvious reasons, however, the problems of negotiating sales were much less complex, and to a certain extent were governed by company policy.

Although a decentralized control over film purchases had been adequate during the period of early expansion, the company, in 1929, decided to adopt a centralized policy. Since the Fox Theaters Corporation did hot operate under a duly appointed theater operator, the general film buyer reported directly to the vice president in charge of distribution, who was in active charge of the home office of the Fox Film Corporation. The man appointed to serve as film buyer understood thoroughly all phases of the motion picture industry. He was acquainted with each Fox theater, with its box office possibilities, and to some extent with its community problems. Furthermore, he was well versed in the product of all producers and the ingredients which usually constituted the basis of a successful motion picture.

Motion pictures purchased by the film buyer for Fox theaters were derived from three general sources: the Fox Film Corporation, other American producers, and foreign producers.

The purchase of Fox films did not constitute a problem.[12] A committee comprised of the vice president in charge of distribution, the two regional directors, and the film buyer selected pictures, established equitable prices based on individual theater quotas, and made all other arrangements which were mutually satisfactory to the parties involved. Division theater managers had the right to reject certain Fox pictures which they considered inappropriate for exhibition in any one of or in all the theaters in their respective territories. On the other hand, the sales department often withheld the sale of some Fox pictures to certain Fox theaters because it was possible to derive a greater profit from a sale to a rival theater-operating company or because of some previous sales agreement. Fox films, in so far as possible, were booked centrally for each Fox theater. The actual routine and delivery, however, were delegated to the various exchanges of the Fox Film Corporation.

The competing American producers from whom the Fox Theaters Corporation purchased films were of two general types: those who did not own or control theaters, and those who operated theater chains. By reason of this difference in the nature of the two groups, the purchasing procedure used in the two cases necessarily differed. In the purchase of the product of those companies not operating theaters, such as Columbia Pictures, Incorporated, and Tiffany Productions, Incorporated, the procedure followed was relatively simple. A representative or the distributor, usually either a division manager or a ranking sales official, approached the Fox buyer in New York.

The sales representative endeavored to sell all his company's product to all the Fox theaters in his territory. The buyer for Fox theaters, in turn, made his selections, if any, and discussed prices, run, and all factors usually connected with the sale of motion pictures. When an agreement of some sort had been reached, the distribution's representative returned a copy of the agreement to his general sales manager and the Fox buyer returned one to the vice president in charge of distribution. The various theater division managers of the Fox Theaters Corporation were notified of the prospective purchase and instructed to state their approval or objections.[13] If all division managers, the regional directors, and the vice president—and, on the other hand, the chief sales executives of the distributing company—were agreed, the deal was completed. If any complications arose, they usually were adjusted in subsequent discussions; otherwise the transaction was rejected.

In the event of a purchase, the individual pictures were booked for each Fox theater as far in advance as possible. Duplicate copies of the booking sheets and sales arrangements were then sent by the distributor to its exchanges, which were to make the deliveries during the course of the theatrical year. The Fox Theaters Corporation sent similar schedules to its various division bookers, who in turn notified the district bookers; the latter notified those theaters operating under their jurisdiction which were booked to exhibit these pictures. Although the booking process appeared comparatively simple, in actual practice it was extremely complicated. In considering the process one should bear in mind the necessity of scheduling all purchases made by Fox theaters from the several producers. Naturally, booking is not the duty of the general film buyer, although for obvious reasons it becomes a part of every purchase negotiated by him.

The procedure followed in making purchases from producers operating theater chains is much more involved than in making any others. The purchase of films by the Fox Theaters Corporation from the Paramount Publix Corporation may be taken as an exemplary case. In this instance, a division manager of the Paramount Publix Corporation approached the buyer for Fox theaters, an agreement was reached, and both parties involved submitted their projects for approval. Likewise a sales representative of the Fox Film Corporation negotiated an agreement with the film buyer for Paramount Publix theaters. Before being carried further, negotiations for all divisions of the country had to proceed to the same stage of development.

When the negotiations for the entire country had been concluded both as to the purchase of Paramount pictures for Fox theaters and Fox films for Paramount Publix theaters, the two companies withdrew for intracompany consultation. The Fox film buyer and the vice president in charge of distribution for the Fox Film Corporation considered the prospective purchase of Paramount pictures in the light of the proposed sale of Fox films to Paramount Publix theaters. In the meantime, division managers of the Fox Film Corporation had been notified of the proposed purchase. In the majority of cases these officials remained in New York during the selling season or were represented there by the buyer or booker for their respective divisions.

If the proposed purchase and the prospective sale were satisfactory to both corporations, the transaction was completed and booking arrangements were then made in the manner already described.

In actual practice consummation of a deal of this sort was a long-drawn-out process. Usually innumerable difficulties arose from all sources involved. For example, division managers might object to certain pictures, knowing from actual experience that these pictures would not succeed in certain theaters located in their territories. Then, too, frequently it was difficult to arrive at an agreeable trade as between the sales departments of both companies involved. Consequently, as a general rule, three and sometimes four weeks or more were required to complete the purchase, for example, of Paramount pictures for Fox theaters.

The trading process did not involve coercion. On the contrary, it was considered both ethical and good business practice for the Fox Film Corporation to insist on an equitable trade with competing producing companies operating theaters. Furthermore, the process often was advantageous to both companies involved, since no one theater operating company possessed enough theaters to insure its affiliated producing company adequate exhibition in all sections of the country; on the other hand, no one producing company produced a sufficient number of pictures to supply all its affiliated theaters.

The purchase of pictures made by foreign producers was principally a matter of company policy based almost entirely on exchange quotas established by the several foreign countries. For example, at one time France required the purchase of one French film by an American company for every four films exported from the United States for sale in France. The buyer, to the extent to which he was acquainted with foreign films, took some part in the selection. Foreign films were sometimes purchased outright. They were more often purchased on flat-rental and percentage agreements. The booking of these pictures and their distribution to the various Fox theaters usually were performed by the Fox Film Corporation through its regular channels of distribution.

Notes

1 The material dealing with the Paramount Publix Corporation is taken from *Motion Pictures* by Howard T. Lewis, Harvard Business Reports, Vol. 8 (McGraw-Hill Book Co., 1930). The present production organization does not correspond exactly with the description here given.

2 Immediately following the introduction of sound, there was considerable discussion in some of the producing companies as to whether or not two directors, one a dialogue director and the other a pictorial director, should be used instead of one director for dialogue pictures.

3 See *Motion Picture Almanac*, 1929 edition, p. 111.

4 See *Motion Picture Almanac*, 1929 edition, p. 111. In general, the salaries paid stars are less than they were in 1929; see p. 118.

5 See *Motion Picture Almanac*, 1929 edition, p. 111.

6 Including such concerns as the Eastman Kodak Company, the DuPont-Pathé Film Manufacturing Corporation, and the Agfa Ansco Corporation.

7 Including such a concern as Consolidated Film Industries, Incorporated.

8 The product of one outside producer was not commingled with that of another outside producer or with Pathé product, but was sold independently as individual pictures or as groups of a given series.

9 There has been a good deal of agitation directed toward securing an agreement by which distributors would agree to launch the selling season in September rather than during the early summer. A few years ago, this was the common practice, but gradually the time has been put back to an earlier date under the stress of competition. Up to the present, no results have come from this effort.

10 Before 1928, salesmen had specialized in the sale of features and in the sale of short subjects, comedies, and newsreels in certain territories.

11 See Lewis, Howard T., *op. cit.*

12 Several competitors of the Fox Theaters Corporation have experienced difficulties in making purchases from their affiliated distributing companies. Prices, quantity of product, and numerous other factors have been advanced as fundamental causes for this recognized difficulty.

13 Division bookers and buyers, through contact with their district managers and unit theater managers, were in a position to relay to the division manager all important opinions regarding the purchase of product. The company had retained its division buyers to purchase films for spot bookings and other films not on the regular schedule. These buyers also negotiated with the various exchanges for all bookings not scheduled in New York.

Chapter 10

Thomas Schatz

HOLLYWOOD: THE TRIUMPH OF THE STUDIO SYSTEM

The American film industry in the 1930s

The 1920s had been a decade of tremendous growth and prosperity for the American motion picture industry, with all phases of production, distribution, and exhibition expanding rapidly as movie-going became the nation's – and indeed much of the world's – preferred form of entertainment. Following the industry's conversion to sound films in 1927–28, the so-called 'talkie boom' capped this halcyon period, providing an additional market surge at the decade's end and further solidifying the dominant position of Hollywood's major studio powers. The talkie boom was so strong, in fact, that Hollywood was touting itself as 'Depression-proof' in the wake of Wall Street's momentous collapse in October 1929, and the American movie industry enjoyed its best year ever in 1930 as theatre admissions, gross revenues, and studio profits reached record levels.

The Depression caught up with the movie industry in 1931, however, and its delayed impact was devastating. Between 1930 and 1933, theatre admissions fell from 90 million per week to only 60 million, gross industry revenues fell from $730 million to about $480 million, and combined studio profits of $52 million became net losses of some $55 million. Thousands of the nation's 23,000 theatres closed their doors in the early 1930s, leaving only about 15,300 in operation by 1935. Among the Hollywood powers, the Depression hit the Big Five integrated major studios especially hard because of the massive debt service on their theatre chains. Three of the Big Five – Paramount, Fox, and RKO – suffered financial collapse in the early 1930s, and Warner Bros. survived only by siphoning off roughly one-quarter of its assets. MGM, meanwhile, not only survived but prospered during the Depression due to its relatively limited chain of first-class metropolitan theatres, the deep pockets of powerful parent company Loew's Inc., and the quality of product turned out by its Culver City studio.

Hollywood's three 'major minor' studios – Columbia, Universal, and United Artists (UA)–fared somewhat better in the early 1930s. These companies produced top product and had their own nation-wide and overseas distribution operations, like the Big Five, but they did not own theatre chains. While this had been a tremendous disadvantage during the 1920s, it proved to be a blessing during the Depression, since these studios avoided the related mortgage commitments. The major minors also adjusted their production and market strategies more effectively than the integrated majors. UA, which was essentially a releasing company for the A-class productions of its active founders (Charlie Chaplin, Mary Pickford, and Douglas Fairbanks) and for major independent producers like Sam Goldwyn and Joe Schenck, simply limited its output to about a dozen A-class pictures per annum. Columbia and Universal pursued a very different course, gearing their factories to low-cost, low-risk features which fell into a new and significant 1930s product category: the 'B movie'.

The rapid rise of the B movie and the 'double feature' on a national scale was a direct result of the Depression. To attract patrons in those troubled economic times, most of the nation's theatres began showing two features per programme, and changed programmes two or three times per week. The increased product demand was met largely via B movies – that is, quickly and cheaply made formula fare, usually Westerns or action pictures, which ran about sixty minutes and were designed to play on double bills in subsequent-run theatres outside the major urban markets. [M]any companies specialized in B pictures, some of them notably Monogram and Republic, not only survived the Depression but became relatively important companies by the late 1930s.

B-movie production was scarcely confined to the 'poverty row' outfits, but in fact became an important element in the studio system at large. All of the integrated majors produced Bs during the 1930s, with up to half of the output of Warners, RKO, and Fox falling into that category as the decade wore on. While most of the major studios' revenues came from A-class features, the production of B movies enabled them to keep their studio operations running smoothly and their contract personnel working regularly, to develop new talent and try new genres, and to ensure a regular supply of product.

While Bs and double bills helped the studios and exhibitors weather the early Depression years, the real key to Hollywood's survival – as well as its tremendous late-1930s success – was the intervention of both Wall Street and Washington, DC. In fact, one might argue that Hollywood's classical era could not have occurred without the financial support of Wall Street and the government's economic recovery programme. New York financiers and banking firms had been involved with the movie industry's development since the early 1920s, particularly in the studios' theatre chain expansion and conversion to sound. Wall Street's involvement increased significantly during the early 1930s, as various firms engineered and financed the reorganization of foundering studios and became more directly involved in their management and operations as well.

The federal government, meanwhile, initiated an economic recovery programme during the Depression which enabled the major Hollywood powers to solidify their control of the industry. The crucial factor here was the election of Franklin D. Roosevelt to the presidency in late 1932, and the impact of FDR's National Industrial Recovery Act (NIRA), which went into effect in June 1933. The NIRA strategy, basically, was to promote recovery by condoning certain monopoly practices by major US industries – including motion pictures. The studios now enjoyed what Tino Balio (1985) has termed 'government sanction for the trade practices that [the majors] had spent ten years developing through informal collusion'. Hollywood's own trade association, the Motion Picture Producers and Distributors Association (MPPDA), drafted the Code of Fair Competition required by the NIRA. This formally codified trade practices such as block booking, blind bidding, and a run-zone clearance, which minimized the studios' financial risks and maximized the profitability of their top features.

Another significant effect of the NIRA on the industry, and particularly on studio production, was the rise of organized labour. To mitigate the potential for worker exploitation and abuse inherent in the recovery programme, the NIRA authorized labour organizing and collective bargaining – an effort reinforced by Congress via the Wagner Act in 1935 and the creation of the National Labor Relations Board. Thus Hollywood evolved from essentially an 'open shop' to a 'union town' in the 1930s, with the division and specialization of film-making labour now mandated by the government and codified by the various unions and guilds.

Beyond the government-mandated regulation of the industry, the 1930s saw increasingly heavy regulation of movie content by the MPPDA. [T]he Production Code Administration (PCA), an MPPDA agency, regulated movie content. Beginning in 1934, PCA approval was required on all scripts before production and then on the finished film, which was assigned a PCA seal prior to release.

The extensive codification and regulation of the movie industry during the 1930s enhanced economic recovery, stabilized film-making operations, and solidified the major studios' control of virtually every phase of the industry. By the late 1930s the eight major studios produced about 75 per cent of the features released in the USA, which generated 90 per cent of all box-office revenues. As distributors they took in 95 per cent of all rental receipts. Hollywood films also accounted for an estimated 65 per cent of all films exhibited world-wide, and roughly one-third of studio revenues throughout the 1930s came from foreign markets – predominantly Britain, which supplied nearly half of Hollywood's overseas revenues.

The five integrated majors, meanwhile, had solidified their dominant position via command of the crucial first-run market. Taken together, the Big Five owned or controlled about 2,600 'affiliated' theatres in 1939; this was only 15 per cent of the nation's total, but it comprised over 80 per cent of metropolitan first-run theatres – that is, the de luxe town-centre houses in the largest cities with seating in the thousands which ran only top features, operated day and night, and generated the lion's share of industry revenues. While the majors' theatre holdings had been a severe financial burden during the darkest Depression years, by the late 1930s they again were keying the Big Five's utter domination of the movie market-place. As Mae Huettig stated (1944) in her study of industry economics in the late 1930s: 'As a control device, the development of strategic first-run theaters as the showcase of the industry proved remarkably effective. Ownership of these relatively few theaters gave [the integrated major studios] control over access to the market.'

Studio management in the 1930s

Given the changing economic fortunes of the studios during the 1930s, as well as the changes in industry trade practices, there were significant changes in management at virtually all levels: the executive control of the motion picture companies themselves, the operation of the studio-factories in Los Angeles, and the supervision of actual film production. The Depression era collapse of five of the Big Eight studios gave several Wall Street firms the opportunity, usually through the trustees or boards of directors of these salvaged companies, directly to control particular studios and to impose their own notions of efficiency and sound business practices. These efforts had relatively little real impact, however, which is instructive in terms of studio ownership and management. As film historian Robert Sklar has pointed out in his analysis of studio control during the 1930s:

> The ultimate issue is not who owns the movie companies but who manages them. When several studios began to founder in the early Depression years, the outside interests who took them over were determined to play a direct role in running them, to demonstrate how practical business and financial minds could make money where movie men could not. By the end of the 1930s, however, all the studios were back under the management, if not the ownership, of men experienced in the world of entertainment.
>
> (Sklar 1976)

This latter distinction is crucial, as has been demonstrated by film historians applying the tenets of neoclassical economic theory. Balio, for example, argues that the ownership–management split which took hold in the 1930s indicated that the studios had developed into modern business enterprises. 'As they grew in size', writes Balio, the studios 'became managerial, which is to say, they rationalized and organized operations into autonomous departments headed by a professional manager.' The studio founders themselves either became 'full-time career managers' – as with the Cohns (at Columbia) and the Warners – or, as was more often the case, they relinquished direct control to salaried executives. And as Balio notes, most of the chief executives appointed during the 1930s who successfully managed the studios had backgrounds in either distribution or exhibition; management of actual film-making operations was invariably left to salaried executives with direct production experience.

These refinements in top management of the motion picture companies during the 1930s were accompanied by equally important adjustments in production supervision. During the early studio era, executive management of the major motion picture companies involved a classic 'top-down' procedure which well indicated the commercial imperatives of the vertically integrated industry. The New York office was the site of ultimate power and authority; there the chief executives managed the direction of capital, marketing and sales, and, for the Big Five, theatre operations. The New York office also set the annual budget and determined the general production requirements of the studio-factory. The Hollywood plant, in turn, was managed by one or two corporate vice-presidents who were responsible for day-to-day studio operations and the overall output of pictures.

Generally, the West Coast studios were run by a two-man team comprised of a 'studio boss' and a 'production chief' – Louis B. Mayer and Irving Thalberg at MGM, Jack Warner and Darryl Zanuck at Warner Bros., Jesse Lasky and B. P. Schulberg at Paramount, and so on. And in every case, there was a clear chain of command extending from the New York office, through the 'front office' at the studio, and into the production arena itself – usually via 'supervisors' who oversaw production and acted on behalf of the higher corporate executives.

As Janet Staiger has noted (Bordwell *et al.* 1985), 'In 1931 the film industry moved away from the central producer management system to a management organization in which a group of men supervised six to eight films per year, usually each producer concentrating on a particular type of film'. This shift to a 'unit-producer system' was fairly pervasive by the late 1930s. By then the only widespread remnant of the factory-oriented, assembly-line approach to production was in the B-picture realm, where production was still managed by a foreman: J. J. Cohn at MGM, Bryan Foy at Warners, Sol Wurtzel at Fox, Harold Hurley at Paramount, and Lee Marcus at RKO.

While the shift to unit production involved a variation of sorts on the United Artists model, UA itself faced increasingly severe management problems in the 1930s and early 1940s. These problems invariably centred on conflicts between UA's board of directors and its top management executive – that is, independent producers Joe Schenck, Sam Goldwyn, and David Selznick, who, in succession, managed UA's production and release programme during this period. Each was crucial to UA's feature output as well as its overall operations, and each left UA due to difficulties with its still-active founding partners, Chaplin and Pickford. Thus while UA's film-makers enjoyed more autonomy as unit producers than their counterparts at the major studios, the company itself was in almost constant turmoil at the management level – and particularly in 1940–41 during Goldwyn's bitter battle (and much-publicized lawsuit) with Chaplin and Pickford, which resulted in Goldwyn releasing through RKO.

One notable exception to this shift to a production-unit system in the 1930s was 20th Century-Fox. In 1933 independent producer Joe Schenck convinced Warners' production chief, Darryl Zanuck, to leave the studio and join him in creating Twentieth Century Pictures. Twentieth released through UA and provided roughly half its output in 1934–35, but Schenck and Zanuck broke with UA in 1935 when Chaplin and Pickford refused to make them full partners. Twentieth then merged with the (previously bankrupt) Fox Film Corporation, with Schenck and Zanuck forming a two-man studio management team and Zanuck reverting to the role he had played at Warners of central producer.

In fact Zanuck, having risen through the writers' ranks at Warners, was the only 'creative executive' running a studio in the late 1930s, and the only top studio executive with an active hand in actual production – a practice he would continue well into the 1950s. Zanuck personally supervised all of Fox's A-class pictures, from such revenue-generating 'hokum' (Zanuck's own term) as the Tyrone Power adventure-romances to its acclaimed 'John Ford pictures' with contract star Henry Fonda: *Drums along the Mohawk* (1938), *Young Mr. Lincoln* (1939), and *The Grapes of Wrath* (1940). At Warner Bros., meanwhile, Zanuck had been replaced by Hal Wallis, who along with Jack Warner orchestrated a steady shift to unit production in the late 1930s, although at Warners top directors like Michael Curtiz and William Dieterle (and later John Huston and Howard Hawks) rather than producers were the key unit personnel.

Columbia Pictures was also a notable exception in terms of studio management in the 1930s, although here the distinction involved the owners and top executives. Until the early 1930s Columbia's ownership and executive control operated much like Warners, in that the company was owned and operated by brothers – Jack and Harry Cohn (along with founding partner Joe Brandt) – and the siblings' executive roles reflected the relations of power within the industry. As with Warners, the older brother (in this case, Jack Cohn, along with Joe Brandt) ran the New York office as the senior executive, while vice-president and younger sibling Harry Cohn ran the studio. When Brandt retired in 1931, however, a power struggle ensued; Harry prevailing, thanks largely to the support of A. H. Giannini of the Bank of America. Thus Columbia, with Harry Cohn as president and Jack as vice-president handling marketing and sales out of New York, was the only major producer-distributor whose studio boss was also the chief executive of the company.

Studio film-making in the 1930s

While the 'creative control' and administrative authority of studio film-making in the 1930s steadily dispersed throughout the producers' ranks, the industry remained as market-driven and commercially motivated as ever. Indeed, in that era of well-regulated block booking, blind bidding, and run-zone clearance, each studio's film-making set-up-was attuned to market conditions and to the industry's vertically integrated structure. The co-ordination of market strategies, management structure, and production operations was in fact the essential feature of the Hollywood studio system – it was the basis for what Tino Balio has termed the 'grand design' of 1930s Hollywood, and for what André Bazin has aptly termed the 'genius of the system' of classical Hollywood film-making.

The primary studio product was the feature film, of course, which accounted for over 90 per cent of the $150 million invested in Hollywood film production in 1939. Feature production at the eight major studios included both A-class pictures and B movies, with the proportion of As to Bs dependent on the company's resources, theatre holdings (or lack thereof), and general market strategy. The majors also turned out occasional 'prestige pictures' – that is, bigger and more expensive features, usually released on a 'roadshow' basis in select first-run theatres with increased admission prices and reserved seating. As the economic recovery took hold and as market conditions improved in the late 1930s, prestige output increased – best exemplified by major independent productions like Disney's *Snow White and the Seven Dwarfs* (1938) and Selznick's *Gone with the Wind* (1939), along with ambitious studio productions such as MGM's *The Wizard of Oz* (1939) and Paramount's Cecil B. DeMille epics like *Union Pacific* (1939) and *North West Mounted Police* (1940).

Actually, these occasional high-cost, high-risk pictures were fundamentally at odds with the market structure and trade practices of the studio era. The most crucial commodity in classical Hollywood, without question, was the A-class feature, and particularly the routine 'star vehicle'. Each studio's 'star stable' was its most visible and valuable resource, an inventory of genre variations which drove its entire operation. Indeed, there was a direct correlation between a studio's assets and profits, its stable of contract stars, and the number of star-genre formulas in its repertoire – ranging from the talent-laden MGM, which boasted 'all the stars in the heavens', to companies like RKO, Columbia, and Universal, which had only one or two top stars under contract and turned out only a half-dozen or so A-class pictures per year. Each company constructed its entire production and marketing strategy around its established star-genre formulations, which dominated the first-run market and generated the majority of studio revenues. A-class star vehicles also provided veritable insurance policies not only with the public but also with the unaffiliated theatres that were subject to block booking, in that a company's top features 'carried' its entire annual programme of pictures.

Each studio's A-class star-genre formulations also keyed its distinctive 'personality' or 'house style' during the classical era. The Warners style in the 1930s, for instance, coalesced around its steady output of crime dramas and gangster films with James Cagney and Edward G. Robinson, crusading bio-pics with Paul Muni, backstage musicals with Dick Powell and Ruby Keeler, epic swashbucklers with Errol Flynn and Olivia de Havilland, and, in a curious counter to its male ethos, a succession of 'women's pictures' starring Bette Davis. These pictures were the key markers in Warners' house style, the organizing principles for its entire operation from the New York office to the studio-factory. They were a means of stabilizing marketing and sales, of bringing efficiency and economy into the production of some fifty features per annum, and of distinguishing Warners' collective output from that of its competitors.

The contract talent and resources necessary to sustain these formulas extended well beyond a studio's star stable, of course. In fact, with the development of the unit-production trend during the 1930s, the studios tended to form collaborative 'units' around these star-genre formulas. And while a studio's stars were crucial to the formulation of its house style, the chief architects of that style were its production executives and top producers. As Leo Rosten noted in his landmark 1941 study:

> Each studio has a personality; each studio's product shows special emphases and values. And, in the final analysis, the sum total of a studio's personality, the aggregate pattern of its choices and its tastes, may be traced to its producers. For it is the producers who establish the preferences, the prejudices, and the predispositions of the organization and, therefore, of the movie which it turns out.

In some cases, production units were informal and fluid, changing somewhat from one star-genre formulation to the next, while in other cases the units were remarkably consistent, particularly in terms of directors, writers, cinematographers, composers, and other key creative roles. Not surprisingly, studio-based production units (and the pictures they turned out) tended to be most consistent in the realm of low-budget film-making, especially on the series pictures which comprised roughly 10 per cent of Hollywood's total feature output in the 1930s and early 1940s. These were produced with ruthless, factory-like efficiency, invariably under the supervision of a particular producer.

Some A-class star-genre formulas attained series or quasi-series status in the 1930s, as well, and also were turned out by relatively consistent units – Paramount's Marlene Dietrich films via the Josef von Sternberg unit, for instance, or Universal's Deanna Durbin musicals via its Joe Pasternak unit. But while unit production brought economy and efficiency even to high-end studio output, it was not necessarily a rigid or mechanical process – an important concern regarding A-class pictures, given the demand for product differentiation. Indeed, under the right circumstances and given adequate resources, unit production proved to be remarkably flexible, and quite responsive to changes in both audience tastes and studio personnel.

For an illuminating example of studio-based unit production and star-genre formulation at the A-class feature level, consider the late-1930s emergence at MGM of the so-called 'Freed unit' and of a new musical cycle featuring two emerging contract stars, Mickey Rooney and Judy Garland. Rooney's star was already in rapid ascent at the time, due largely to yet another unit-based star-genre formula, MGM's Hardy Family pictures. In 1937 Rooney was cast as the son of a middle-class, middle-American couple (Lionel Barrymore and Spring Byington) in a modest comedy-drama, *A Family Affair*. Audiences responded and exhibitors clamoured for more, so Mayer replaced Barrymore and Byington with two lesser stars, Lewis Stone and Fay Holden, and assigned J. J. Cohn's low-budget unit to develop a series. Cohn assembled a specialized unit around associate producer Lou Ostrow, director George B. Seitz, writer Kay Van Riper, cinematographer Lew White, *et al.*, which turned out a new Hardy instalment every few months from 1938 to 1941. The Hardy pictures qualified as As in terms of budget, running time, and release schedule, and they were by far the most profitable pictures on MGM's schedule.

With each Hardy instalment, Rooney's character took greater command of the series. By 1939 every picture carried the name 'Andy Hardy' in the title, and Rooney was the nation's top star. While an ideal vehicle for Rooney, the series also provided a context for trying out new talent, particularly contract *ingénues* who could serve as a friend or 'love interest' for Andy. Besides Ann Rutherford, who became a regular in 1938, the series enjoyed guest appearances from such emerging Metro stars as Lana Turner, Ruth Hussey, Donna Reed, Kathryn Grayson, and most significantly (and most frequently) Judy Garland. While these *ingénues* had an obvious dramatic value in terms of Andy's male adolescent rites of passage, Garland also introduced a musical dimension to the series and thus brought out quite another facet of Rooney's character and talent.

The first real Hardy musical was *Love Finds Andy Hardy* in 1938. A surprisingly strong box-office hit, even for a Hardy film, it suggested that MGM might have a new musical team on its roster. Garland's success in the Hardy musical also confirmed Mayer's decision to give her the lead in *The Wizard of Oz* (in a role initially conceived for Shirley Temple, whom MGM planned to borrow from Fox). *Oz* was MGM's biggest risk and most expensive picture of the decade, costing $2.77 million to produce and tying up five sound stages and hundreds of personnel for twenty-two weeks of shooting. But remarkably enough, considering the film's subsequent success as a perennial television event in the USA, *Oz* was not a big commercial hit on its initial release. The picture did gross over $3 million, but due to its high production and marketing costs it actually lost about $1 million.

The disappointing revenues on *Oz* were offset by its prestige value and by the emergence of Judy Garland as a bona fide star, and also by the ascent of Arthur Freed to the producer ranks at MGM. Freed had worked uncredited on *Oz* as an assistant to producer Mervyn LeRoy on the assurance that he could produce a musical of his own once *Oz* was completed. Freed convinced Mayer to purchase the rights to a 1937 Rodgers and Hart stage hit, *Babes in Arms*, and to bring Busby Berkeley over from Warner Bros. to direct and choreograph the film. He also convinced Mayer to let him team Rooney and Garland as co-stars. Freed hoped to combine the energy and appeal of the Hardy pictures with the 'backstage musical' formula that Berkeley had refined at Warners in pictures like *42nd Street* and *Gold Diggers of 1933* (both 1933). The Rodgers and Hart musical was one of those 'Hey kids, let's put on a show!' types, which Freed and Berkeley redesigned as a showcase for Rooney and Garland. Kay Van Riper rewrote the script to add some of the Hardy series flavour, while musical director Roger Edens revamped the song-and-dance numbers.

Shot in only ten weeks for just under $750,000, a remarkably low figure for a major musical, *Babes in Arms* grossed well over $3 million. One of MGM's biggest hits in 1939, it surpassed not only the high-stakes *Wizard of Oz* but a number of other more reliable star vehicles as well. Indeed, Mayer could gamble on pictures like *Oz* or *Babes in Arms* knowing that MGM's release schedule included such bankable products as *Ninotchka* with Greta Garbo, *The Marx Brothers at the Circus*, *Another Thin Man* with William Powell and Myrna Loy, *The Women* with Joan Crawford, Norma Shearer, and Rosalind Russell, *Northwest Passage* with Spencer Tracy, and three more Hardy Family pictures.

While it may have lacked the critical cachet of *Oz* or other top features, *Babes in Arms* also underscored what the studio system did best: it was an economical, efficiently produced star vehicle, an A-class genre amalgam with just enough novelty to satisfy audiences and ensure its success in the lucrative first-run market, and a prime candidate for reformulation. Immediately after *Babes in Arms*, Freed, Berkeley, Roger Edens, *et al.* set to work on a cycle of Rooney–Garland show musicals – *Strike up the Band* (1940), *Babes on Broadway* (1942), *Girl Crazy* (1943), etc. – marking the birth of MGM's Freed unit, which produced many of Hollywood's greatest musicals during the 1940s and 1950s.

Into the 1940s: the approach of war

MGM's 1939 output contributed heavily to what many critics and film historians consider Hollywood's greatest year ever, a view underscored by the Oscar contenders for best picture of 1939: *Mr. Smith Goes to Washington*, *The Wizard of Oz*, *Stagecoach*, *Dark Victory*, *Love Affair*, *Goodbye, Mr. Chips*, *Of Mice and Men*, *Ninotchka*, *Wuthering Heights*, and *Gone with the Wind*. The view of 1939 as Hollywood's peak year was reinforced in quite another way by the industry's rapid transformation and decline in the 1940s, a decade-long process that began in 1940–41 during the odd, intense interval between the Depression and America's entry into the Second World War.

While a range of factors and events were involved in that process, two were of principal importance in 1940–41: the US government's anti-trust campaign against the studios, and the outbreak of war overseas. The antitrust campaign actually had been brewing for several years, beginning in earnest in July 1938 when the US Justice Department filed suit against the Big Eight studios on behalf of the nation's independent theatre owners for monopolizing the movie industry. The government's objectives in the so-called 'Paramount case' (named after the first defendant cited) were extensive trade reforms and the 'divorcement' of the integrated majors' theatre chains. The studios managed to keep the US attorneys at bay until early 1940, when the case was ordered to trial. At that point the Big Five (MGM, Warners, Paramount, Fox, and RKO) signed a consent decree – essentially a plea of no contest and a compromise with the government. The studios agreed to limit block booking to groups of no more than five pictures, to end blind bidding by holding trade screenings of all features, and to modify their run-zone-clearance policies. The 1940 Consent Decree scarcely satisfied either the independent exhibitors or the Justice Department; both vowed to keep the Paramount case alive through appeals, and to continue the anti-trust campaign.

Meanwhile, the outbreak and escalation of war overseas threatened to devastate Hollywood's vital overseas trade. The studios' exports to the Axis nations – principally Germany, Italy, and Japan – had declined to almost nil in 1937–38, but still Hollywood derived roughly one-third of its total revenues from overseas markets. The primary overseas client was Europe, which supplied about 75 per cent of the studios' foreign income in 1939, and the United Kingdom in particular, alone providing 45 per cent. Those markets were severely disrupted by the outbreak of war in September 1939, and they went into free fall as the fighting on the Continent intensified. The Nazi blitzkrieg across Europe eliminated one key foreign market after another in 1940, culminating in the fall of France in June. By late 1940 Britain stood virtually alone against Germany in the west – and it stood alone as Hollywood's only significant remaining overseas market as well.

The war in Europe induced the US government to initiate a massive military and defence buildup, which had tremendous impact on the movie industry. While the buildup promised an eventual boost to Hollywood's domestic market, the short-term effects were primarily negative. The government financed its defence programme mainly via tax increases, which hit the industry particularly hard, while pressure from the White House to support the buildup also put Hollywood in a delicate position politically. President Roosevelt lobbied for industry support in 1940, even though the USA was still officially neutral and public sentiment was split into isolationist and interventionist camps. Not surprisingly, Hollywood played it safe, producing a few war-related features in 1940 but confining its treatment of the war mainly to newsreels and documentary shorts.

By 1941 both Hollywood and the nation had shifted into a more pronounced 'pre-war' mode, as Axis aggression solidified US support of Britain and rendered American intervention against Germany and Japan all but inevitable. On the domestic front, the defence buildup went into high gear, signalling the definitive end of the Depression and the onset of a five-year 'war boom' for the American movie industry. Hollywood stepped up its war-related production in 1941, including features with clear interventionist and 'preparedness' themes. The public was clearly buying; five of the top ten box-office hits in 1941 were war related, topped by *Sergeant York* (Warner Bros., 1941) and Charlie Chaplin's late-1940 UA release *The Great Dictator.* Hollywood's war-related efforts provoked America's dwindling isolationist contingent, including several US Senators who demanded an inquiry into the movies' pro-war, anti-German bias. The Senate 'propaganda hearings' of September 1941 became a major media event, with public and press support swinging solidly behind the movie industry.

While the anti-trust campaign and the war involved external forces well beyond the industry's direct control, the Hollywood studio powers faced a number of serious internal threats as well in 1940–41. Many of these were related to the unprecedented demand for top product, which resulted from both the 1940 Consent Decree (with its blocks-of-five and advance-screening provisions) and the defence buildup (with its overheated first-run markets).

To meet the increased demand for top features, the studios either turned to independent producers, whose ranks grew rapidly in the early 1940s, or granted their own contract talent greater freedom and authority over their productions. This not only accelerated the unit-production trend already under way, but gave it a strong independent impetus as well. Indeed, by 1940–41 UA's strategy of distributing major independent productions had been adopted by four other studios: RKO, Warners, Universal, and Columbia.

Actually, each of these studios developed a significant variation on the UA model, depending on whether production facilities and financing were provided as well as distribution. The number of these 'outside' productions also varied considerably. Warners, for instance, was quite tentative in its foray into independent production, entering only two such deals in 1940: Frank Capra's *Meet John Doe* and Jesse Lasky's *Sergeant York* (directed by freelancer Howard Hawks). The most aggressive of the four companies was RKO, which had over a dozen independent unit productions under way in 1940 and was relying on outside units for virtually all of its A-class product.

Meanwhile, all of the studios also saw a marked increase in the number of 'hyphenates' under contract in 1940–41 – notably writers and directors who attained producer status. The most aggressive studio in this regard was Paramount, which elevated director Mitchell Leisen to producer-director and writer Preston Sturges to writer-director, while the writing team of Billy Wilder and Charles Brackett was recast as a writer-producer (Brackett) and writer-director (Wilder) unit. And in perhaps the most significant – and most forward-looking – move along these lines, Paramount not only granted Cecil B. DeMille the status of an 'in-house independent' producer, but gave him a profit-participation deal on his pictures as well.

The growing power of independent film-makers and top contract talent in the early 1940s was reinforced by the concurrent rise of the talent guilds – that is, the Screen Actors Guild (formally recognized by the studios in 1938), the Screen Directors Guild (recognized in 1939), and the Screen Writers Guild (recognized in 1941). The top talent guilds represented yet another serious challenge to studio control, particularly in terms of film-makers' authority over their work. Moreover, top contract talent was 'going freelance' in unprecedented numbers in 1940–41 due to the demand for A-class product and the shift to independent production, as well as the considerable tax advantages involved in avoiding salaried income and creating independent production companies. This further undermined the long-standing contract system, a crucial factor in studio hegemony and in efficient studio production.

Thus Hollywood was on the verge of a sea-change in late 1941, with the studios' control of both production and exhibition facing serious challenges. The crisis mentality was underscored by sociologist Leo Rosten in *Hollywood: The Movie Colony*, an influential, in-depth industry analysis published in November 1941. Lamenting what he termed the 'end of Hollywood's lush and profligate period', Rosten surveyed the range of concerns and crises facing the studios at the time:

> Other businesses have experienced onslaughts against their profits and hegemony; but the drive against Hollywood is just beginning. No moving picture leader can be sanguine before the steady challenge of unionism, collective bargaining, the consent decree (which brought the Department of Justice suit to a temporary armistice), the revolt of the independent theater owners, the trend toward increased taxation, the strangulation of the foreign market, and a score of frontal attacks on the citadels of the screen.

War boom

Rosten's assessment is notable for two principal reasons: first, its accurate inventory of the deepening industry crises in 1940–41, and second, how completely those crisis conditions would change within only weeks of the book's publication, following the US entry into the Second World War. With America suddenly engaged in a global war, Hollywood's social, economic, and industrial fortunes changed virtually overnight. Among the more acute ironies of the period – and there were many – was Hollywood's sudden rapport with a government that now saw the 'national cinema' as an ideal source of diversion, in formation, and propaganda for citizens and soldiers alike. In the months after the USA entered the war, Hollywood's film-making operations, from its studio-factories to its popular genres and cinematic forms, were effectively retooled for war production. Within a year of Pearl Harbor nearly one-third of Hollywood's feature films were war related, as were the vast majority of its newsreels and documentaries.

Thus the war was a period of massive paradox for the movie industry, and especially for the Hollywood studios. Granted a reprieve of sorts by the Justice Department 'for the duration', the studios reasserted their control of the movie market-place during the war and enjoyed record revenues, while playing a vital role in the US war effort. The nation's wholesale conversion to war production sustained Hollywood's 'war boom' for five years, with millions of defence plant workers concentrated in major urban-industrial centres pushing weekly theatre attendance to its record pre-Depression levels. Hollywood's overseas trade was limited primarily to the UK and Latin America, but with Britain undergoing an industrial and movie-going boom of its own during the war – and relying more than ever on Hollywood product, given the severe cutbacks of film production in Britain – Hollywood's foreign revenues also reached record levels. Indeed, the degree of integration between the US and British markets and film-making efforts during the Second World War was not only unprecedented but altogether unique in motion picture annals, and was crucial to the war effort in each nation.

Hollywood's studio production system saw some modification during the war, as the unit-production trend continued and as A-class output was increased to accommodate the overheated first-run market. The Big Five radically reduced their overall output (from an average of fifty films per year for each studio to about thirty), concentrating on 'bigger' pictures which played longer runs and enjoyed steadily increasing revenues. The studios readily adapted to changing social conditions, developing two government-mandated genres, the combat film and home-front melodrama, and a new crop of wartime stars – notably Bob Hope, Betty Grable, Greer Garson, Abbott and Costello, and Humphrey Bogart. Traditional genres (particularly the musical) were converted to the war effort, while Hollywood's long-standing bias for 'love stories' was adjusted to favour couples who separated at film's end to perform their respective patriotic duties. There were significant stylistic adjustments during the war as well. Hollywood's war-related features, particularly the combat film, employed quasi-documentary techniques which were altogether distinctive for American cinema. And in something of a stylistic and thematic counter to the realism and enforced optimism of the war films, Hollywood also cultivated a darker and more 'anti-social' vision in urban crime films and 'female Gothic' melodramas steeped in a style which post-war critics would term film noir.

The war boom continued through 1945 and into 1946, fuelled by returning servicemen, the lifting of wartime restrictions, and a wide-open global market-place starved of Hollywood product. But as box-office revenues and studio profits reached staggering levels in 1946, the 'drive against Hollywood' that Rosten had described in 1941 resumed with a vengeance. In fact, the mood in Hollywood grew decidedly downbeat in 1946 due to deepening labour strife, the renewed anti-trust campaign, and the threats of 'protectionist' policies in major foreign markets. Moreover, the studios faced a host of distinctly post-war woes at home, from suburban migration and the rise of commercial television, to the Cold War and anti-Communist crusades. With each passing month in the post-war era it became more evident that the Hollywood studio system and the classical cinema that it had engendered were rapidly coming to an end.

Bibliography

Balio, Tino (1976), *United Artists.*

— (ed.) (1985), *The American Film Industry.*

— (1993), *Grand Design: Hollywood as a Modern Business Enterprise, 1930–1939.*

Bordwell, David, Staiger, Janet, and Thompson, Kristin (1985), *The Classical Hollywood Cinema.*

Finler, Joel (1988), *The Hollywood Story.*
Gomery, Douglas (1986), *The Hollywood Studio System.*
Huettig, Mae D. (1944), *Economic Control of the Motion Picture Industry.*
Jacobs, Lea (1991), *The Wages of Sin.*
Jowett, Garth (1976), *Film: The Democratic Art.*
Kindem, Gorham (ed.) (1982), *The American Movie Industry.*
Rosten, Leo (1941), *Hollywood: The Movie Colony.*
Schatz, Thomas (1988), *The Genius of the System.*
Sklar, Robert (1976), *Movie-Made America.*
Staiger, Janet (1983), 'Individualism versus Collectivism'.

Chapter 11

Mark Glancy and John Sedgwick

CINEMAGOING IN THE UNITED STATES IN THE MID-1930s: A STUDY BASED ON THE *VARIETY* DATASET

THROUGHOUT THE 1930s, THE ENTERTAINMENT industry trade journal *Variety* published extensive information on the North American exhibition market in each of its weekly issues. The 'Pictures' section of *Variety* included weekly box-office reports from as many as 200 cinemas in 30 cities.[1] The cinemas included were mainly 'first-run' venues with a large seating capacity, but there were also some smaller, second-run and specialty cinemas. The reports were organized by city, and they provided general comments on the trading conditions, audience preferences and promotional strategies seen in each city during the previous week, as well as more specific reports from a selection of the city's cinemas, indicating the box-office results for the week, and listing the items on each cinema's programme. In the 1930s, the standard running time for a cinema programme was just under three hours, and programmes could include one or two feature films, a live stage performance, cartoons, newsreels, short films, as well as 'bank nights' or 'giveaways'.[2] *Variety*'s comments on these programmes would consider if one half of a double feature was regarded as a greater attraction than the other, whether the stage show interested audiences more than the film, or if in fact it was the bank night or a newsreel that drew the crowds. The overwhelming impression given by the reports is that the cinemagoing experience was seldom if ever limited to a single film in this era. Rather, the cinema offered a diverse range of attractions and entertainment forms, in which a single feature film might be a secondary or even an incidental consideration.

Altogether, the array of information available in *Variety* constitutes a valuable historical source, and one that offers a highly revealing and often fascinating glimpse into both the film industry and its audiences during the Depression era. This paper has been designed as a pilot study of the *Variety* data, demonstrating the range of information available and how it can be used. However, even within our designated time frame, October 1934 to October 1936, it is not within the scope of the study to utilize the reports from all of the 200 cinemas and all of the 30 cities. Instead, we have drawn data from tables that were published monthly in *Variety*, which presented a selection of the data from the previous four weeks. These tables offer a condensed and more easily digested overview of each month's box-office activity, and they include reports from 100 first-run cinemas-in 23 cities across the United States plus another four cinemas in Montreal, Canada.[3] In addition to the weekly box-office grosses, the *Variety* tables also list each cinema's seating capacity, admission price range and its record box-office 'high' and 'low'. These cinemas and attendant statistics are listed in the Appendix (Table 11.6). This condensed sample of the data involves a substantial body of information: the 104 cinemas represented in the tables played 967 films on 8,694 separate programmes during the twenty-five month period.[4] In addition to the tables of data, we have also drawn upon the weekly textual reports for further information on exhibition conditions and practices, at least insofar as they relate to the data sample, in order to offer a fuller picture than that provided by the box-office grosses alone.

The data will be used here to investigate four aspects of cinemagoing during the Depression. The first centres on audience preferences and film popularity, although it will be argued that, even with extensive box-office data, judgements in this area often require careful qualification. The second is a consideration of the diversity of films shown in these mainstream cinemas, with particular attention to the presence of foreign films and the extent to which the major Hollywood studios dominated this sector of the exhibition market. It should be noted that the specific time period chosen for this study, October 1934 through October 1936, coincides with a concerted campaign by one foreign producer, Gaumont-British, to market its films in the United States, and the data sample offers one means of assessing the success of that campaign. The third concerns the practice of 'double billing', which proliferated in the early 1930s. Of the 8,694 programmes recorded in the dataset, 6,384 are single bills and 2,310 are double bills, and we will consider which films played on double bills and the circumstances in which they were paired.[5] The fourth aspect is the use of live stage shows to accompany feature films in cinemas. In the mid-1930s, this remained a common feature of cinemagoing in the first-run cinemas of major cities. Approximately 18 per cent of all programmes in the sample combined a feature film with a live stage show, and the *Variety* data offers an opportunity to examine the nature and success of this now largely forgotten aspect of Depression-era exhibition practices.[6]

The validity of the data

Given that there were approximately 15,000 cinemas operating in the U.S. during the mid-1930s, and that most films received between 2,000 and 10,000 bookings, it is clear that the dataset-sample for this pilot study is a highly select one.[7] If includes reports from just 104 cinemas, and only 14 per cent of the films (134 films) in the sample have more than twenty bookings, while 55 per cent of the films (533 films) have ten or fewer bookings. Hence, questions as to the validity of the sample should be addressed. On what basis can a survey of cinemagoing practices and preferences at such a small sample of cinemas be said to represent the cinemagoing experiences of Americans more generally? The answer lies in the fact that the cinemas in the sample represent a large proportion of the first-run sector of the exhibition market.[8] The Motion Picture Producers and Distributors of America, Inc. calculated that in 1941 there were approximately 450 first-run cinemas in cities with over 100,000 inhabitants, and so the sample represents approximately one-quarter of the cinemas in this sector.[9] These cinemas were at the top of the exhibition hierarchy. They tended to be located in the most populous areas, typically on the main streets of city centres, and they tended to be the largest cinemas. They played films first, they played them exclusively (within their area), and they charged the highest admission charges. As this profile suggests, they were not entirely typical as cinemas and they had some distinctive exhibition practices; for example, they were more likely to have live entertainment and less likely to have double bills. However, the size and scale of this market, and the large number of patrons that it drew, make it an important one to examine and understand. In regard to cinemagoing preferences, the fact that *Variety* focused its reports on this sector of the market is itself revealing. These showcase cinemas provided the first test of a film's popularity before it went on to subsequent runs (the 'second-run', 'third-run' 'and so on) in cinemas that tended to have fewer seats, to charge less for an admission, and to be located away from the main streets and in outlying neighbourhoods or small towns.[10]

The pattern of popularity experienced in the set of first-run cinemas and reported in *Variety* would not necessarily be replicated exactly in the lower orders of cinemas. Sedgwick's study of cinemagoing in Britain during the 1930s noted distinctive patterns of preferences between metropolitan and regional audiences, and in particular between audiences attending cinemas in London's West End and those attending cinemas in Brighton and in Bolton. It was noted, too, that the lower-order cinemas exhibited films not seen in the higher-order tiers.[11] Maltby's study of American cinema audiences also suggests that the notion of a single undifferentiated audience is a misleading one, and that audience preferences were fragmented along class, gender and geographical lines. The divide between rural and metropolitan tastes is said to have been particularly pronounced, with a wide gap existing between the preferences of metropolitan audiences attending first-run cinemas on Broadway and those of provincial American audiences.[12]

Yet the reports in *Variety* do not promote the notion of a single, homogenous audience, and they came from much further afield than Broadway and New York City. The cities in the data sample include the five largest cities in the United States (New York, Chicago, Philadelphia, Detroit and Los Angeles) and also much smaller cities such as Birmingham, Indianapolis, New Haven, St Louis and Tacoma. All regions of the country are represented, and in fact divergent tastes are readily apparent in the data. Nevertheless, the fact of local, regional and other distinctions in taste is not sufficient to explain the overall level of popularity of a film. Or, to put it differently, films that were the 'hits' of their day needed to perform extremely well across the first-run sector, and it is apparent that, for the greater part, those films that did so were also relatively popular amongst audiences attending lower-order cinemas. This can be demonstrated with reference to the studio ledgers unearthed by Glancy and Jewell.[13] Among the figures reported in the Eddie Mannix (MGM), William Schaefer (Warner Bros.) and C.J. Trevlin (RKO) ledgers are 'domestic earnings figures' that represent the sum of all exhibition revenues received by these studios for each of their films. These revenue figures included all earnings accrued by the studios from every stage of exhibition and throughout all regions of the United States and Canada. An analysis comparing the *Variety* dataset grosses and the ledger grosses for all MGM, RKO and Warner Bros. films released in this period, indicates a very strong correlation between the: two sets of figures, from which it is possible to conclude that, in the vast majority of cases, films that proved popular in the first-run market were similarly popular across all exhibition sectors.

The issue of the validity of what was reported in *Variety* should also be addressed. Were the figures reported each week truly an accurate measure of the gross earnings of films? That question simply cannot be answered with absolute certainty. However, there is an apparent emphasis on accuracy in the reports: projected grosses were reported in each weekly issue and then followed by confirmed grosses in the following week's issue. Furthermore, and as argued elsewhere, some confidence can be taken by the fact that the trade treated *Variety* with respect. It told a story about the relative and absolute popularity of films that accorded with the experience of those whose livelihood was bound up in the film business. This is most important, because without such veracity it seems highly unlikely that *Variety* would have continued to serve as the industry's principal trade publication.[14]

Film preferences

The weekly box-office reports in *Variety* offer a rich source of information for reception studies centred on the tastes and interests of local audiences, as well as studies centred on the reception of stars, genres or individual films across different locales. The value of the source lies not only in the financial figures themselves, but in the textual reports that accompany them. It is these reports that reveal the influences behind the figures, and the comments regularly discuss local pricing strategies, promotional campaigns, critical views expressed in the local press, and the age, class and gender of those attending a film. For example, the success of *The Gay Divorcee* in Cincinnati was said to stem, at least in its first week, from the excellent reviews the film received in the local papers.[15] In Brooklyn, *The Crusades* was reported to have benefited from endorsements by local ministers, who had been given a special preview screening of the film and then urged their congregations to see it.[16] *A Midsummer Night's Dream* was reported to have had limited success in St Louis because ticket prices were set as high as $1.50 and 'locals won't pay top price'.[17] In Kansas City, *The Story of Louis Pasteur* did not attract a sizeable audience because the biopic was 'too classy for this town'.[18] In Minneapolis, *The Painted Veil* was a hit because audiences considered its star, Greta Garbo, to be a 'Scandinavian luminary', but in Birmingham the same picture failed because 'Garbo doesn't mean anything here and this film means less'.[19] As some of these comments indicate, there was a remarkable gap in regional tastes; a gap between 'classy' metropolitan audiences, who paid 'top price' for biopics, adaptations of Shakespeare and exotic melodramas, and the more provincial audiences, who preferred their entertainment to be decidedly less expensive and less exotic. This divide is particularly apparent in the reception given to the populist dramas of film star Will Rogers, whose homespun values and common sense celebrated the virtues of small town, middle-America. His films were merely routine releases in eastern cities such as New York, Boston, Philadelphia and Washington, D.C. Yet the further west and south they travelled, the more their fortunes improved. In Birmingham, Rogers was actually the city's top star, and his *In Old Kentucky* and *Steamboat Round the Bend* were by far the two top-earning films of the period in this city.[20] He had a similar status in cities such as Indianapolis, St Louis and Tacoma. In these cities, as in Birmingham, the preference for Rogers' films was accompanied by a predilection for Westerns, the family entertainment of Shirley Temple's films and the small-town dramas *Ah, Wilderness* and *The Country Doctor*.[21] More sophisticated fare, meanwhile, did not last long, and metropolitan favourites such as Chaplin's *Modern Times* and the costume drama *The Barretts of Wimpole Street* had a much more limited drawing power in these smaller, provincial cities.[22]

A comparison of local and national film preferences is one of the interesting opportunities the data offers. However, the comments that accompany the box-office reports consistently indicate that levels of attendance were governed by an array of factors, some of which had no relation to a single film's individual qualities or popularity. The notion of 'audience preferences' thus becomes problematic, or at least in need of careful scrutiny and qualification. Box-office takings in individual cities were often reported to have been affected by competition from alternative entertainment forms such as circuses, sporting events and state fairs. Games, giveaways, bank nights and raffles, on the other hand, were said to attract audiences who otherwise were reluctant to attend, not least because the first-run houses offered prizes as substantial as a new car and, in one instance at least, a fully furnished four-bedroom house.[23] Good and bad weather was frequently reported to have affected box-office either positively or negatively. And establishing audience preferences is further complicated by the fact that a single feature film was often only one item in a programme that had other attractions. Even minor items such as short films were sometimes said to be the primary attraction for audiences. The most notable example of this was a film made of a boxing match between heavyweights Max Schmeling and Joe Louis. The match took on a particular significance because Schmeling was representing Nazi Germany while Louis was a black American, and the short film, which circulated in the week after the match, was reportedly a more important draw than the feature films it preceded.[24] In some instances newsreels were also cited as significant attractions for audiences. A *March of Time* segment on the Nazi persecution of the Jews, for example, was said to have garnered considerable interest, and in some cinemas more interest than the feature films it accompanied.[25] It was exceptional, though, for such short items to be cited as a more significant attraction than the feature film. Much more frequently, the issue of a single film's popularity was clouded by its inclusion on a programme with another feature film or with live entertainment.

On their own, most feature films simply could not fill these capacious theatres for several showings a day throughout an entire week-long engagement, and so they were often paired with other features or with live entertainment. However, there was a distinct minority of films that needed little or no support. These were the top-earning films, and, as Table 11.1 vindicates, it was common for these films to enjoy runs of several weeks in a single city. They rarely appeared on double bills during their first-runs, and when they did it was usually in one of the smaller cities. In the larger cities, an additional feature simply was not necessary to

Table 11.1 The fifty top earning films in the data sample

A ranking of the fifty top earning films, based on all of the earnings recorded in the data sample, with indications of how many engagements were recorded, the total number weeks that each film played, and the number of weeks that it appeared on a double bill and the number of weeks it appeared with a stage show.

	Film (studio, director, year)	*Sum of all box-office earnings*	*No. of cities played*	*No. of weeks played*	*Weeks on a double bill*	*With a stage show*	*Top-billed stars*
1	*San Francisco* (MGM, Van Dyke, 1936)	1,147,650	22	79	3	0	Clark Gable Jeanette MacDonald
2	*Top Hat* (RKO, Sandrich, 1935)	1,132,550	20	54	1	3	Fred Astaire Ginger Rogers
3	*The Great Ziegfeld* (MGM, Leonard, 1936)	966,700	22	66	0	0	William Powell Myrna Loy
4	*Swing Time* (RKO, Stevens, 1936)	964,650	21	48	6	8	Fred Astaire Ginger Rogers
5	*Mutiny on the Bounty* (MGM, Lloyd, 1935)	939,100	21	53.5	0	0	Clark Gable Charles Laughton
6	*Roberta* (RKO, Seiter, 1935)	873,650	19	51	0	10	Irene Dunne Fred Astaire
7	*Follow the Fleet* (RKO, Sandrich, 1936)	806,600	20	47	0	0	Fred Astaire Ginger Rogers
8	*Anthony Adverse* (WB, Leroy, 1936)	793,000	20	47.5	0	1	Fredric March Olivia De Havilland
9	*David Copperfield* (MGM, Cukor, 1935)	738,450	23	48	2	2	W.C. Fields Lionel Barrymore
10	*Love Me Forever* (Col, Schertzinger, 1935)	731,900	21	43	3	9	Grace Moore Leo Carrillo

(continued)

Table 11.1 The fifty top earning films in the data sample *(continued)*

Film (studio, director, year)	*Sum of all box-office earnings*	*No. of cities played*	*No. of weeks played*	*Weeks on a double bill*	*With a stage show*	*Top-billed stars*
11 *One Night of Love* (Col, Schertzinger, 1934)	711,300	21	47.5	5	8	Grace Moore Tullio Carminati
12 *The Gay Divorcee* (RKO, Sandrich, 1934)	661,500	21	41.5	2	3	Fred Astaire Ginger Rogers
13 *China Seas* (MGM, Garnett, 1935)	644,900	23	45.5	0	0	Clark Gable Jean Harlow
14 *Rose Marie* (MGM, Van Dyke, 1936)	634,100	22	44	0	0	Jeanette MacDonald Nelson Eddy
15 *The Barretts of Wimpole St* (MGM, Franklin, 1934)	623,700	21	40	0	5	Norma Shearer Fredric March
16 *Broadway Bill* (Col, Capra, 1934)	620,850	22	41	8	6	Warner Baxter Myrna Loy
17 *G-Men* (WB, Keighley, 1935)	613,650	23	39	3	8	James Cagney Ann Dvorak
18 *Modern Times* (Chaplin, Chaplin, 1936)	608,270	18	40.5	5	0	Charles Chaplin Paulette Goddard
19 *Lives of a Bengal Lancer* (Par, Hathaway, 1935)	605,200	22	37	1	6	Gary Cooper Franchot Tone
20 *My Man Godfrey* (Uni, La Cava, 1936)	605,050	22	45	12	1	William Powell Carole Lombard
21 *The Gorgeous Hussy* (MGM, Brown, 1936)	598,450	23	48	9	4	Joan Crawford Robert Taylor
22 *Mr Deeds Goes to Town* (Col, Capra, 1936)	591,150	21	48.5	10	2	Gary Cooper Jean Arthur
23 *Mary of Scotland* (RKO, Ford, 1936)	589,800	21	34	7	6	Katharine Hepburn Fredric March
24 *The Big Broadcast of 1937* (Par, Leisen, 1936)	567,950	20	32	5	5	Jack Benny George Burns
25 *Under Two Flags* (TCF, Lloyd, 1936)	567,150	21	33	1	3	Ronald Colman Claudette Colbert
26 *Belle of the Nineties* (Par, McCarey, 1934)	564,825	19	31.5	1	3	Mae West Roger Pryor
27 *The Broadway Melody of 1936* (MGM, 1935)	530,400	22	43	0	0	Jack Benny Eleanor Powell
28 *Wife vs. Secretary* (MGM, Brown, 1936)	529,300	23	39	2	2	Clark Gable Jean Harlow
29 *The Littlest Rebel* (TCF, Butler, 1935)	527,800	21	29.5	7	4	Shirley Temple John Boles
30 *Libeled Lady* (MGM, Conway, 1936)	517,275	19	43	13	0	Jean Harlow Myrna Loy
31 *Poor Little Rich Girl* (TCF, Cummings, 1936)	512,450	22	35	9	7	Shirley Temple Alice Faye

Film (studio, director, year)	*Sum of all box-office earnings*	*No. of cities played*	*No. of weeks played*	*Weeks on a double bill*	*With a stage show*	*Top-billed stars*
32 *The Green Pastures* (WB, Connelly & Keighley, 1936)	509,350	21	34	5	0	Rex Ingram Oscar Polk
33 *The Little Colonel* (Fox, Butler, 1935)	507,800	19	27	1	5	Shirley Temple Lionel Barrymore
34 *Show Boat* (Uni, Whale, 1936)	505,950	20	45	4	1	Irene Dunne Allan Jones
35 *Forsaking All Others* (MGM, Van Dyke, 1934)	503,145	21	39	2	2	Joan Crawford Clark Gable
36 *The Little Minister* (RKO, Wallace, 1934)	500,500	19	30.5	2	11	Katharine Hepburn John Beal
37 *His Brother's Wife* (MGM, Van Dyke, 1936)	492,850	23	34	8	1	Robert Taylor Barbara Stanwyck
38 *Becky Sharp* (Pioneer, Mamoulian, 1935)	490,550	20	33	1	2	Miriam Hopkins Frances Dee
39 *Naughty Marietta* (MGM, Van Dyke, 1935)	488,450	20	44	3	6	Jeanette MacDonald Nelson Eddy
40 *The Country Doctor* (TCF, King, 1936)	486,350	22	32.5	2	4	Dionne Quintuplets Jean Hersholt
41 *Anna Karenina* (MGM, Brown, 1935)	485,400	21	34	1	1	Greta Garbo Fredric March
42 *A Tale of Two Cities* (MGM, Conway, 1936)	481,200	21	33	1	0	Ronald Colman Elizabeth Allan
43 *Little Lord Fauntleroy* (Selznick, Cromwell, 1936)	476,300	21	31	6	2	Freddie Bartholomew Dolores Barrymore
44 *Curly Top* (Fox, Cummings, 1935)	475,500	21	33	3	0	Shirley Temple John Boles
45 *No More Ladies* (MGM, Griffith, 1935)	474,050	22	39	4	6	Joan Crawford Robert Montgomery
46 *Strike Me Pink* (Goldwyn, Taurog, 1936)	466,600	20	37	2	0	Eddie Cantor Ethel Merman
47 *Trail of the Lonesome Pine* (Par, Hathaway, 1936)	450,250	21	36	5	5	Sylvia Sidney Henry Fonda
48 *Les Miserables* (TCF, Boleslawski, 1935)	447,600	15	30	2	3	Fredric March Charles Laughton
49 *The Bride Comes Home* (Par, Ruggles, 1936)	446,650	21	26	6	7	Claudette Colt Fred MacMun
50 *Captain Blood* (WB, Curtiz, 1935)	444,650	22	34	2	1	Errol Flynn Olivia De Havilland

draw audiences. Paradoxically, these top-earning films were actually more likely to appear on programmes with live entertainment. This is because they played in the largest cinemas of New York City and Chicago, and many of these venues regularly featured live entertainment, regardless of the film that was showing. Outside of these cities and that type of venue, they played without the support of a live show. *San Francisco,* for example, had engagements recorded in twenty-two of the twenty-four sample cities for a total of seventy-nine weeks. It never appeared with live support and it only played on a double bill during an unusually long three-week engagement in Denver. Similarly, *Top Hat* had engagements recorded in twenty cities for a total of fifty-four weeks. It appeared with live entertainment only during its three-week engagement at New York's Radio City Music Hall, where the stage show was a regular feature, and it played on a double bill only once, in Tacoma. The exceptional popularity of such films is evident in the fact that a small number of films took a disproportionately large amount of the box-office. The ten top earning films, for example, represent only 1.0 per cent of the films in the sample, but they account for 7.5 per cent of the box-office gross for the period; and the fifty top earning films represent just 5.2 per cent of the films in the sample but they took 25.3 per cent of the total earnings.

The most popular films in the sample have some striking similarities. All of the ten top-earning films are either costume dramas or musicals, and the film that earned more than any other, *San Francisco*, belongs to both categories. The popularity of costume dramas was not reliant upon any single studio or set of stars. Those that rank among the fifty top-earning films of the period, as listed in Table 11.1, include a wide array of performers and were produced by several different studios. If there is a dominant strand of the genre, it is the tasteful or 'culturally elevated' strand, represented by such films as *Mutiny on the Bounty*, *Anthony Adverse*, *David Copperfield*, *Lives of a Bengal Lancer*, *The Gorgeous Hussy*, *The Barretts of Wimpole Street*, *Becky Sharp*, *Anna Karenina* and *A Tale of Two Cities*. These were films that were set within a relatively modern historical period (the eighteenth or nineteenth centuries), and films that were either adapted from canonical literature or centred on the lives of key historical figures. Another type of costume drama, centred on medieval or ancient times, and favouring spectacle over literary values, was far less prominent, as is evident in the poorer performance of films such as *Cleopatra*, *The Crusades* and *The Last Days of Pompeii*.[26] In contrast to the costume drama, a large measure of the popularity of the musical can be attributed to the phenomenal success of a single star team. Fred Astaire and Ginger Rogers made five films at RKO during this time period. Four of their films (*Top Hat*, *Swing Time*, *Roberta* and *Follow the Fleet*) appear among the ten top-earning films in the sample, and the other (*The Gay Divorcee*) is not far behind. These distinctly contemporary, witty and sophisticated musicals, which combined singing and dancing, were broadly popular across audiences and in all regions. MGM's three-hour-extravaganza, *The Great Ziegfeld*, was another musical, in-costume that was exceptionally popular. The operettas of Grace Moore, and Jeanette MacDonald and Nelson Eddy were also successful, but none was in quite the same league as Astaire and Rogers.

Only two other stars came close to matching Astaire and Rogers. One was Clark Gable. Altogether, Gable's films earned more than those of any other star in the sample period, but this was partly the result of the sheer number of films that he had in cinemas over these twenty-five months. They include the highly successful costume dramas *San Francisco* and *Mutiny on the Bounty*, the romantic dramas in which he starred with Jean Harlow (*China Seas*, *Wife Versus Secretary*) and Joan Crawford (*Forsaking All Others*, *Chained*), and a few new releases that must have been considered commercial disappointments (*After Office Hours*, *Cain and Mabel*, *The Call of the Wild*). They also include re-releases of three of his earlier films (*Dancing Lady*, *It Happened One Night* and *Men in White*). The latter are particularly noteworthy as there were only twenty-five re-releases in the data sample, and the fact that three of Gable's films were chosen for re-release offers another sign of his box-office standing. The other leading star was Shirley Temple, and she too made an impact at least partly as a result of the sheer volume of her films. Over the twenty-five months, she starred in no fewer than eight new releases, some of which were markedly more successful than others. This was an era in which audiences were able to see their favourite stars in several films each year, and they were apparently willing to pick and choose among the films.

With costume dramas and musicals as the most consistently popular genres, it appears that escapism was a key aspect of cinemagoing for most people in this period. Indeed, it is notable how very few of the most successful films hear any significant traces of the harsh economic climate of the 1930s. Among the 50 top earning films, those that come the closest are Frank Capra's *Broadway Bill* and *Mr. Deeds Goes to Town*, the Warner Bros. crime drama *G-Men* and the comedies *Modern Times* and *My Man Godfrey*, but their engagement with contemporary concerns is oblique at best. Nevertheless, the decade is often characterized as one in which a significant number of films engaged with social issues and problems, and did so in a realist manner.[27] Many of the films chosen by historians to demonstrate this view were notably poor performers within the data sample. The grosses for both *Fury*, the story of small town prejudice and injustice, and *Black Fury*, the story of a labour dispute, fell short of $250,000. So, too, did the gross for *Bullets or Ballots*, another Warner Bros. crime film. *The Informer*, John Ford's account of 'the troubles' in Ireland, and the political drama *The President Vanishes* fared even worse, with grosses below $200,000. *Our Daily Bread*, meanwhile, was almost uniquely unpopular and grossed only $51,250. This film, directed by King Vidor, centres on an impoverished young couple who struggle to work a farm as a collective. It was shown throughout the country, but it had the dubious distinction of ranking among the lowest grossing films in almost every city it played. It is apparent that such films were perhaps the least representative of audience preferences in the period, as social relevance and box-office failure invariably went hand in hand.

Film diversity

In some respects the data sample may seem to cover a narrow range of films. Not only does a disproportionately large amount of the earnings go to a minority of very popular films, but, as Table 11.2 indicates, the vast majority of earnings are attributed to films made by the major Hollywood studios. Altogether, the sample includes 741 films made by MGM, Paramount, Warner Bros., RKO, Twentieth Century-Fox, Fox, Columbia, Universal and the independent American companies that produced for these studios and for United Artists, and these films account for 94.6 per cent of all of the earnings recorded in the sample. However, there is greater diversity than this figure would seem to suggest, because the remaining 5.4 per cent of earnings was derived from 226 films made by a farther 52 production companies. Double billing undoubtedly facilitated this diversity. As Taves has argued, the proliferation of double billing in the early 1930s created a demand for feature films that the major studios could not fulfil, and that demand was met by smaller studios.[28] This is borne out by the data sample, which includes films produced by Republic (twenty-five films), Monogram (twenty-four), Chesterfield (fifteen), Mascot (thirteen), Invincible (eleven), Liberty (ten), Atherton (eight), Buck Jones (four), Majestic (three), and an additional twenty-five companies that had only one film listed. These films rarely appeared on single bills. For the most part, their entrance into the first-run market was a result of the need for 'second features'. This was particularly true in the smaller cities, where double billing was much more common and engagements were shorter, thus intensifying the demand for additional films.

Foreign films also benefited from double billing. The Department of Commerce reported that 190 foreign films were released in the United States during 1935 alone, and that German

Table 11.2 The leading production companies

A ranking of the leading production companies in the data sample, based on the sum of all box-office earnings.

Rank	*Studio/production company*	*Box-office total ($)*	*Percentage of total box-office earnings*	*Number of releases*	*Average box-office per film ($)*
1	MGM	23,179,352	19.14	107	216,629
2	Paramount	19,631,855	16.21	131	149,861
3	Warner Bros.	17,526,849	14.47	123	142,495
4	RKO	13,715,798	11.32	82	167,266
5	Twentieth Century-Fox	8,250,675	6.81	45	183,348
6	Fox	8,210,000	6.78	60	136,833
7	Columbia	7,875,432	6.50	89	88,488
8	Universal	6,623,519	5.47	69	95,993
9	Goldwyn (UA)	3,121,066	2.58	9	346,785
10	20th Century (UA)	2,677,500	2.21	8	334,688
11	Gaumont-British	1,805,935	1.49	24	75,247
12	Reliance (UA)	1,201,950	0.99	6	200,325
13	London Films (UA)	1,021,925	0.84	8	127,741
14	Republic	908,355	0.75	25	36,334
15	Chaplin (UA)	608,270	0.50	1	608,270
16	Pioneer (RKO)	582,175	0.48	2	291,088
17	Selznick (UA)	476,300	0.39	1	476,300
18	Pickford-Lasky (UA)	395,450	0.33	2	197,725
19	Principal	366,850	0.30	5	73,370
20	Monogram	311,690	0.26	24	12,987
21	Gainsborough	272,190	0.22	8	34,024
22	Mascot	270,750	0.22	13	20,827
23	Roach (MGM)	266,175	0.22	3	88,725
24	British & Dominions (UA)	252,300	0.21	7	36,043
25	Hecht-MacArthur (Paramount)	195,767	0.16	3	65,256
26	Liberty	143,985	0.12	10	14,399
27	Atherton	122,850	0.10	8	15,356
28	Invincible	93,700	0.08	11	8,518
29	Chesterfield	79,650	0.07	15	5,310
30	British International	77,000	0.06	12	6,417
31	Van Beuren	73,800	0.06	2	36,900
32	Select	68,966	0.06	3	22,989
33	Mr & Mrs Martin Johnson	68,750	0.06	1	68,750

Rank	*Studio/production company*	*Box-office total ($)*	*Percentage of total box-office earnings*	*Number of releases*	*Average box-office per film ($)*
34	Lianofilm	67,085	0.06	1	67,085
35	Howard B. Franklin	64,450	0.05	1	64,450
	35 other companies	503,550	0.42	48	10,491
	TOTAL	*121,111,913*	*100.00*	*967*	*125,245*

and 'Spanish' films were the leading imports.[29] Very few of the films from non-English speaking countries appear in the data sample, but when they do it is always as one feature within a double bill. The Swiss travel documentary *Wings Over Ethiopia* was released in the month that Italy invaded Ethiopia, and curiosity about the conflict enabled the film to garner engagements on double bills in eleven cities. The intriguingly titled *Legong: Dance of the Virgins* was actually an anthropological documentary from the Dutch East Indies, and it played on double bills in three cities. The German film *Mädchen in Uniform* and the Mexican film *She-Devil Island* were in the sample only by virtue of (separate) one-week engagements on double bills. Of course, a distinct market for foreign films and for 'poverty row' films existed beyond the first-run cinemas, and their appearance in the data sample is only sporadic and marginal. Yet it demonstrates the increased exhibition opportunities offered by double billing, which appears to have opened at least some of the first-run sector's cinemas to a wide range of films.

British films benefited from double billing to a much greater extent than other foreign films. Of the seventy-five British films in the data sample, thirty-one played only on double bills, and in fact many had their original running time cut to facilitate this.[30] Other British films were much more significant releases, which played widely and mainly as single features. London Films was the most successful of all British companies, and *The Scarlet Pimpernel*, with data sample earnings of nearly $335,000, was by far the top earning foreign film of the period, followed by the same company's *The Ghost Goes West* (with $255,000) and *Things to Come* ($205,000).[31] It is notable, though, that these films took an unusually large share of their earnings from their highly successful New York City engagements. *The Scarlet Pimpernel* earned a staggering $162,500 in the two weeks that it played in Radio City Music Hall, but that level of success was not maintained throughout the country. In fact, those earnings represented 49 per cent of the film's earnings throughout the country, whereas the average film took 21 per cent of its earnings from New York. *The Ghost Goes West* took 44 per cent of its gross from four weeks at New York's Rivoli Theatre, and *Things to Come* took 37 per cent of its gross from three weeks at the Rivoli. It is also notable that, beyond New York City, these films were most likely to be held over for additional weeks in large cities such as Boston, Chicago, Philadelphia and Washington, DC, while engagements elsewhere did not last for more than a single week and yielded comparatively small sums. In the divide between metropolitan and provincial tastes, British films landed firmly on the metropolitan side.

London Films held a relatively privileged place among the British production companies seeking access to American audiences. Its films were released through United Artists, a major American distribution firm, and that ensured that they received a high profile and extensive playing dates. Yet the company produced only eight films within this period, and so its impact was limited. Gaumont-British had much more ambitious plans. It intended to release most, if not all, of its British-made films in North America, and in 1934 it established its own American distribution company to facilitate this. Some success in this endeavour is apparent: no fewer than twenty-four Gaumont-British productions appear in the data sample, and many of these played in one of New York City's largest first-run houses, the Roxy Theatre. There, in one of the country's largest and most lucrative venues, films such as the Alfred Hitchcock thrillers *The 39 Steps* and *The Secret Agent*, the musical *It's Love Again*, the futuristic drama *Transatlantic Tunnel*, the historical adventure film *Rhodes of Africa* and the melodrama *Little Friend* were popular enough to be held over for a second week, and each earned between $60,000 and $80,000 in this venue alone. These notable successes undoubtedly helped the company to gain playing dates further afield, and the films all showed in major venues throughout the country. A familiar pattern emerged, however, in which the box-office takings dropped markedly outside of New York City and particularly in the smaller and more provincial cities. The company's other films, meanwhile, did not enjoy such wide releases and were more likely to be seen on double bills wherever they did play. This intermittent success was clearly not enough for Gaumont-British, which ceased film production in 1937.[32] Nevertheless, it is apparent that for a time in the mid-1930s the company was able to release more British films in the United States than any other company, and to distribute them widely. In doing so, Gaumont-British came close to matching Hollywood's Columbia and Universal studios (in both total revenues and per-film averages), and achieved a status that was far above the 'poverty-row' level. The same cannot be said for the other British companies, including major concerns such as British International and British and Dominions, which, by any measure, languished at the bottom of the rankings.[33]

Double billing

The question of how to attribute the earnings of a programme that included two feature films is a difficult one. One option would be to attribute the earnings to both films, on the grounds that audiences saw both, and earnings levels are meant to reflect the size of the audience. However this method would double the revenue recorded for some programmes and thereby distort the true levels of earnings within the data sample. A second option rests on the notion that double bills were formed by combining a popular 'A' film with a more obscure 'B' film, and that, because audiences were primarily drawn to the 'A' films, all or most the earnings should be attributed to the 'A' film. This combination of 'A' and 'B' films is evident in some double bills. For example, the only double billing recorded in the data sample for MGM's *San Francisco* was in Denver, where it was accompanied by another MGM film, *The Three Godfathers*, which had a much lower level of earnings and played most of its engagements as a double feature. It therefore seems reasonable to assume that audiences in Denver were drawn primarily to see *San Francisco* and that the earnings should be attributed accordingly. However, in many other instances the 'A' and 'B' divide is not so clear. For example, *The Three Godfathers* also played on double features with RKO's *The Witness Chair*, Columbia's *Meet Nero Wolfe* and Warner Bros.' *The Case of the Velvet Claws*, and it is impossible to determine which might be the 'A' or 'B' film in these cases. All of these films played most but not all of their engagements on double bills, and they were usually paired with similarly low-profile films. The majority of double billings, in fact, seem to be various combinations of what Taves refers to as 'shaky A films': 'programmers' from major studios, and 'B' films from smaller studios.[34] That is, most double bills did not have a strong and weak component, but were combinations of relatively weak films. Hence, the third option, to divide the earnings evenly between the two films, seems to be the fairest and most appropriate method of dealing with double bills, and it is the one that we have adopted here.[35]

By 1934, the number of cinemas that regularly showed double features was estimated to be between 50 per cent and 75 per cent.[36] Within the data sample, it is apparent that practices varied widely and were set according to local conditions. In the smaller cities, such as New

Haven and Tacoma, double billing was the norm within the first-run sector, and single features were a rarity. In larger cities, double billing was used selectively and to support films that were not perceived to fit clearly into either the 'A' or 'B' category. The fact that *Variety* often reported on double bills by commenting on which film was the greater attraction is indicative of this uncertainty.[37] Many of the very largest cities, meanwhile, did not have double bills at all within the first-run sector. Exhibitors in Chicago, Minneapolis, New York, Philadelphia and Washington, DC, held to mutual agreements that banned double billing in these venues. One reason for this was that the practice increased costs for the exhibitor, who had to pay for two films rather than one, but another and perhaps a more important reason was to protect the special quality of the first-run cinemas, which were meant to be movie palaces rather than bargain cinemas. Audiences paid a higher admission fee to attend these cinemas and they may have welcomed the change of pace that a single-feature programme provided. This is evident in the fact that, although double bills were appreciated as a good value, there were also many objections to them.

A survey conducted by Warner Bros. in 1936 indicated that 78 per cent of those polled preferred single bills, and an array of complaints about double bills explained this preference.[38] Among them was the length of programmes, some of which were extended to four hours in order to include a second feature. Conversely, there were complaints that some programmes were kept within three hours by cutting the films to the point where they became noticeably 'jerky'. It was also said that 'a good picture is invariably coupled with a bad one'. This complaint did not necessarily refer to the aforementioned divide between 'A' films and 'B' films; but could also reflect the often haphazard method of pairing films. Audiences complained that the films on a double feature did not 'match' one another.[39] This complaint was voiced separately by exhibitors, who pointed out that films were not planned as double bills at either the production or distribution stage, to ensure that they would complement one another on a single programme.[40] Instead, they were often combined simply on the basis of availability. Thus, odd pairings abounded. Within the data sample, these include double bills that combined *The Informer* with the society farce *Going Highbrow*; Fritz Lang's *Fury* and the children's musical *Let's Sing Again*; Robert Flaherty's documentary *Man of Aran* with the crime serial *Charlie Chan in Paris*; an adaptation of Dickens' *Great Expectations* and the dating agency comedy *Bachelor Bait*; and the German film *Mädchen in Uniform* with *Dealers in Death*, a documentary exposé of the munitions industry in the First World War. As intriguingly odd as some of these combinations appear to be it is plain to see that an audience paying to see one of the films was unlikely to find the other as appealing. Of course, the Warner Bros. survey also found a minority of 22 per cent who favoured double bills, and one commonly cited reason for preferring them was that second features replaced vaudeville stage acts in many venues. Or, as one patron put it, 'If you only have only one feature at the movies you'll have vaudeville, and vaudeville is lousy'.[41]

Live acts

Many of the theatres in the data sample had been built as vaudeville houses rather than as cinemas, and in earlier decades films had been only one item on programmes dominated by vaudeville acts. Films then gradually pushed the vaudeville acts off the programmes. By the mid-1930s the decline of vaudeville was frequently discussed in the trade papers, and it was termed the 'orphan child of show business' by *Variety*.[42] One major factor was the expense of live entertainment, which many venues could not afford in the Depression. Another factor was said to be that vaudeville had relied upon the same material and formulae for too long, and the familiar range of acts, including acrobats, magicians, comedy sketches, dancers, vocalists, impersonators and animal acts, was said to have lost its appeal.[43] The most significant factor, however, was undoubtedly the advent of talking pictures, which allowed films to take up vaudeville's staple entertainments: spoken comedy, tap-dancing, and music in a variety of forms. Films could not duplicate the excitement of a good live performance, of course, but they could record the best performances of some of the world's greatest performers. They could be distributed more widely than even the most extensive stage tour, and they could do all of this at less cost to the exhibitor. Many of the most popular films of this period borrowed from vaudeville even as they replaced it. Fred Astaire and Ginger Rogers, for example, began their careers (separately) in vaudeville, and their films typically foreground their status as stage entertainers. While the films do not adhere to a 'revue' format, they nonetheless offer a variety of entertainment forms, including a range of singing and dancing styles as well as comedy in the form of both verbal 'gags' and situational sketches. Moreover, the stage is frequently foregrounded within their films. *Swing Time*, for example, begins with a shot of the proscenium arch in a large theatre. On stage, a team of dancers can be seen, but it is a long shot from the perspective of the theatre audience, and so it is difficult to discern the individual performers, or even to see that Astaire himself is leading the team. This opening shot thus offers only the limited vantage point of an audience watching a stage performance. The shots that follow demonstrate the advantages of cinema. They offer not only much closer views of Astaire, but they also allow the audience to see the spaces backstage and in the dressing rooms, and they reveal musical performances that are private (or 'integrated' into the film's story). Furthermore, the subsequent stage performances are freed from the constraints of the proscenium, once again demonstrating the privileged position available to the cinema audience. The only blatant reference to vaudeville comes in the very first lines of dialogue in which an ageing magician stands just off-stage complaining that his act has been cut from the show because it was 'too old-fashioned'. He attempts to perform a card trick for a stage hand, but the stage hand is too captivated by his close view of Astaire dancing on stage to notice the old-timer's routine. It is a brief scene, but the camera's backstage view, the stage hand's privileged position, and the notion of new and old entertainment forms can serve as a succinct demonstration of the decline of vaudeville and the triumph of musical cinema.

Another key film of this period, the musical *The Great Ziegfeld*, demonstrates a different type of self-consciousness toward vaudeville. The film seems designed to satiate the audience's appetite for staged entertainment. As a 'biopic' of the showman Florenz Ziegfeld, it offers a veritable history of popular staged entertainments, at least within its subject's lifetime. It begins with carnival attractions and culminates on Broadway with the spectacular Ziegfeld Follies stage show, and throughout it features all manner of acts and performances. It is essentially a very long and varied 'revue' programme, and one that is linked together rather thinly by the biographical storyline. Its extraordinary 170-minute running time may have been a canny strategy on the part of MGM, which produced the film. Very little time was left for anything else on the programme. An abbreviated newsreel, a cartoon or a trailer may have been possible, but there certainly could not have been a stage programme or second feature. And that, of course, meant that the box-office earnings would not have to be shared with any other attraction. It is clear that audiences did not feel cheated, though, as *The Great Ziegfeld* earned one of the top grosses of the period, and in fact its fifteen-week engagement at the Carthay Circle in Los Angeles was the longest recorded in the data sample.

Even if vaudeville was being eclipsed in the mid-1930s, it was not yet time to sound the death knell for live entertainment in combination with films. One of the most striking aspects of the data sample is the extent to which live entertainment was still used as an accompaniment to film screenings in the 1930s. Of the 104 cinemas represented in the data sample, 45 were what *Variety* referred to as 'combination houses', or theatres that combined live and film entertainment on the same bill. For the most part these were the largest theatres in the largest cities. The two largest, New York's Radio City Music Hall and the Roxy, had approximately 6,000 seats each, and each regularly combined film and stage shows. The stage shows were a key part of their identity and appeal, and they were also crucial to the theatres' ability to draw in thousands of patrons at advanced prices and for several daily shows. Of the other New York venues in the sample, the Capitol (5,486 seats) and Paramount (3,664) had live shows in most but not all weeks, while the Center (3,700), Strand (2,758), Rivoli (2,092) and Rialto (750) theatres never did. A similar pattern held in other large and medium-sized cities, with the larger venues more likely to offer live entertainment and the smaller venues very rarely offering it. In the smaller cities 'combination houses' were rare, and cinemas offered a 'double bill' of two feature films rather than combining stage and screen entertainments. Even in these cities, however, the occasional live performance could be extraordinarily successful. Throughout the country and in cities large and small, the appeal of live entertainment in combination with films was often remarkably strong.

In some rare instances, live performances were tailored to fit with or to celebrate a film. This was reserved for the most important releases, such as *Top Hat*, and it was done only in the grandest venues, such as Radio City Music Hall. There, *Top Hat* was preceded by a stage show that featured a projected backdrop of the film's title designs, a recreation on stage of the film's ballroom set, dancers in top hats and tails, and an all-male choir with twenty-four members singing the film's signature song, 'The Piccolino'.[44] In most other instances, if a strong 'A' film was accompanied by a live attraction, it would not be a particularly prominent one. When MGM's *Wife Versus Secretary* was shown at the Loew's State in New York City, for example, *Variety* commented that the 'strong screen fare' meant that the theatre was offering only a 'routine' live programme.[45] It was the weaker 'A' films and 'B' films playing as single features that needed to be coupled with a strong live programme in order to improve or maintain audience numbers. One particularly notable example of this concerns Paramount's *The Scarlet Empress.* Directed by Josef Von Sternberg and starring Marlene Dietrich, *The Scarlet Empress* is now considered a classic film, but when it was first released in the autumn of 1934 it met with dismal box-office returns. It was not held over for a second week in any of the thirteen engagements represented in the data sample, and it earned its four highest box-office grosses in the cities where it was accompanied by a stage show. In one of those cities, Pittsburgh, the Stanley Theatre postponed its engagement of the film for several weeks, waiting until it could book a stage show that should 'carry' the film through its one week engagement. Hence, *The Scarlet Empress* was screened on a programme that also included a variety show led by the popular Fred Waring Orchestra – and the box-office gross for the week was a remarkably high $26,000. This would make it seem as though *The Scarlet Empress* was one of the leading attractions in Pittsburgh during the 1934–36-period, and yet according to *Variety*, it was the live act that drew audiences to the theatre that week and the live act that compensated for the '100 minutes of dull celluloid'.[46]

While traditional vaudeville was seldom seen as a key attraction in the mid-1930s, stage shows led by a prominent 'headline name' often drew remarkable results. These appearances could add considerably to the venue's overhead costs. Top-rated performers such as Jack Benny, Eddie Cantor and Milton Berle were able to earn between $7,500 and $15,000 for a one-week engagement, as well as garnering a percentage of the box-office takings.[47] Exhibitors were willing to pay such sums only during the weeks in which the main feature film was thought to be a weak attraction. Pittsburgh offers further examples of the effect that prominent live acts could have in cinemas, as well as examples of the way in which 'combination houses' operated outside of New York City. The two largest venues in Pittsburgh, the Penn and Stanley theatres, regularly used headline acts to support minor feature films. As Table 11.3 indicates, this strategy was at times highly successful. Films such as *Behold My Wife*, *Exclusive Story*, *Hide Out*, *Dangerous*, *O'Shaughnessy's Boy*, *Sequoia* and *Hands Across the Table*, which had a

Table 11.3 The top attractions in Pittsburgh

The top attractions in Pittsburgh, ranked by local earnings and with a comparison to national ranking.

Film/studio	*Local earnings ($)*	*Local ranking*	*National ranking*	*Local cinema*	*Weeks played*	*Live artist/ headline name*
G-Men (WB)	$52,500	1	17	Stanley	2	Folies Bergere
Anthony Adverse (WB)	$47,500	2	8	Stanley	2	none
Mutiny on the Bounty (MGM)	$44,200	3	5	Penn Warner	1	none
Green Pastures (WB)	$35,300	4	32	Penn Warner	1	none
Behold My Wife (Par)	$34,500	5	217	Stanley	1	Jack Benny
The Gorgeous Hussy (MGM)	$33,000	6=	21	Penn	2	none
San Francisco (MGM)	$33,000	6=	1	Penn Warner	1	none
Top Hat (RKO)	$33,000	6=	2	Penn	2	none
Rose Marie (MGM)	$32,700	9	14	Penn Warner	1	none
The Great Ziegfeld (MGM)	$32,500	10	3	Penn	2	none
Broadway Melody of 1936 (MGM)	$31,500	11	27	Penn Warner	1	none
Exclusive Story (MGM)	$31,000	12=	215	Stanley	1	Jack Benny
Hide Out (MGM)	$31,000	12=	210	Penn	1	Ted Lewis Orchestra
China Seas (MGM)	$30,500	14	13	Penn Warner	1	None
Dangerous (WB)	$30,000	15=	176	Stanley	1	Major Bowes
O'Shaughnessy's Boy (MGM)	$30,000	15=	242	Stanley	1	Major Bowes
Devil Dogs of the Air (WB)	$28,500	17	106	Stanley	2	none
Sequoia (MGM)	$28,000	18	251	Penn	1	Eddie Cantor
Hands Across the Table (Par)	$27,600	19	114	Stanley	1	Guy Lombardo
Swing Time (RKO)	$127,000	20=	4	Stanley	1	none
				Penn	1	none
Follow the Fleet (RKO)	$27,000	20=	7	Stanley	1	none
				Warner	1	none

much lower profile in other cities, were among the twenty top-earning attractions in Pittsburgh during this period, and in each case a live performance was credited with drawing the crowds. When more prominent films played in these theatres, however, stage shows were not usually offered and there was no need for shows with an expensive 'headline' act. A similar pattern is also seen in other major cities, and Table 11.4 and Table 11.5 demonstrate the impact that live performances had in Detroit and Minneapolis, respectively. Live appearances by performers such as Amos 'n' Andy, John Boles, George Burns and Gracie Allen, the Marx Brothers, Stepin Fetchit, Ed Sullivan, and the orchestras led by Cab Calloway,

Table 11.4 The top attractions in Detroit

The top attractions in Detroit, ranked by local earnings and with a comparison to national ranking. All of the films played as the single feature film.

Film/studio	*Local earnings ($)*	*Local ranking*	*National ranking*	*Local cinema*	*Weeks played*	*Live artist/ headline name*
The Littlest Rebel (TCF)	$60,000	1	29	Fox	2	Molasses 'n' January
Roberta (RKO)	$54,000	2	6	Fox	2	none
Curly Top (Fox)	$53,000	3	44	Fox	2	none
In Old Kentucky (Fox)	$50,000	4	81	Fox	2	vaudeville
Life Begins at Forty (Fox)	$49,500	5	97	Fox	2	Dorsey Brothers
The Country Doctor (TCF)	$49,000	6	40	Fox	2	Phil Baker
Love Me Forever (Col)	$47,900	7	10	Fox	2	none
San Francisco (MGM)	$45,000	8	1	United Artists	4	none
Mutiny on the Bounty (MGM)	$44,500	9=	5	United Artists	3	none
Private Number (TCF)	$44,500	9=	58	Fox	2	Eddie Duchin
King of Burlesque (TCF)	$40,000	11	123	Fox	2	Clyde Beatty
Poor Little Rich Girl (TCF)	$39,500	12	31	Fox	2	Ed Sullivan Unit
His Brother's Wife (MGM)	$39,000	13	37	Michigan	1	NBC Radio Unit
Modern Times (Chaplin)	$38,800	14	18	United Artists	3	none
Libeled Lady (MGM)	$37,500	15	30	United Artists	3	none
Charlie Chan in Shanghai (Fox)	$36,000	16	262	Fox	1	Cab Calloway
Anthony Adverse (WB)	$35,500	17	8	United Artists	3	none
The Little Minister (RKD)	$35,000	18	36	Fox	2	vaudeville
The Bride Walks Out (RKO)	$34,000	19=	168	Michigan	1	Major Bowes' Amateurs
Rhythm on the Range (Par)	$34,000	19=	53	Michigan	1	Bob Ripley Unit

Eddie Duchin and Guy Lombardo, were able to draw audiences to see films that did not fare nearly so well on their own.

Many of these performers had begun in vaudeville, but by the mid-1930s they were known primarily for their work in radio and films. That they could return to the stage as vital support for struggling film exhibitors is one indication of the cross-fertilization that allowed vaudeville, radio and film to intermix and evolve. At this point, radio was of course much more important to their drawing power than vaudeville. Radio ownership doubled over the course of the decade, and although there was initially some suspicion and hostility in

Table 11.5 The top attractions in Minneapolis

The top attractions in Minneapolis, ranked by local earnings and with a comparison to national ranking. All of the films played as the single feature film.

Film studio	*Local earnings ($)*	*Local ranking*	*National ranking*	*Local cinema(s)*	*Weeks played*	*Live artist/ headline name*
Roberta (RKO)	$34,000	1	6	Orpheum	2	none
The Bride Comes Home (Par)	$32,000	2	49	Minnesota	1	Burns & Allen
One Night of Love (Col)	$30,500	3	11	Orpheum	2	none
San Francisco (MGM)	$29,000	4	1	Minnesota	2	none
Libeled Lady (MGM)	$28,000	5	30	Lyric	2	none
				Minnesota	1	none
Top Hat (RKO)	$27,860	6	2	Orpheum	2	none
Broadway Bill (Col)	$27,500	7	16	Orpheum	2	none
My American Wife (Par)	$27,000	8=	170	Minnesota	1	Eddie Duchin
Trail of the Lonesome Pine (Par)	$27,000	8=	47	State	2	none
				Minnesota	1	none
Follow the Fleet (RKO)	$26,500	10	7	Orpheum	2	none
Goose and Gander (WB)	$26,000	11	221	Orpheum	1	Folies Bergere
Swing Time (RKO)	$24,500	12	4	Orpheum	2	none
Mr Deeds Goes to Town (Col)	$23,500	13 =	22	Orpheum	2	none
My Man Godfrey (Uni)	$23,500	13=	20	Orpheum	2	none
Cain and Mabel (WB)	$23,000	15	178	Minnesota	1	John Boles
The Great Ziegfeld (MGM)	$22,500	16	3	Minnesota	1	none
				State	1	none
Wife vs Secretary (MGM)	$22,000	17	28	Minnesota	1	none
				State	1	none
The First Baby (TCF)	$21,000	18	522	Minnesota	1	The Marx Brothers
The Gay Divorcee (RKO)	$20,000	19=	12	Orpheum	2	none
Two in the Dark (RKO)	$20,000	19=	370	Orpheum	1	Wayne King Orchestra
Walking on Air (RKO)	$20,000	19=	216	Orpheum	1	Folies Parisienne

Hollywood toward the new medium, by the middle of the decade the relationship between film and radio had become a mutually beneficial one.[48] Nationally syndicated radio programmes promoted and publicized film stars and the latest film releases. Films such as Paramount's *The Big Broadcast of 1936* and *The Big Broadcast of 1937* portrayed the backstage operations of a radio station and offered the dramatic rationale for a succession of comedy and musical acts. Crucially, they also gave audiences the opportunity to see the performers they normally only heard. There was also at least one instance of a syndicated radio programme forming the basis of a touring stage show that played in cinemas. This was the 'Major Bowes Amateur Show', an amateur talent contest hosted on the radio by Major Bowes. The winners of the radio competition were placed in Bowes' stage revues, which toured the country playing in combination houses. These stage shows were essentially amateur vaudeville hours, featuring the usual mix of singers, tap dancers, magicians and comedians, and yet they drew audiences into cinemas in a manner that standard vaudeville no longer could. They could even support 'B' films playing as a single feature in a major venue.[49] Their popularity was so remarkable that in 1935 the shows were also filmed as 'short subjects' so that they could be distributed more widely and to smaller venues.

The persistence of vaudeville – albeit in various guises – is one of the most notable characteristics of this period of cinemagoing, but the combination houses were always searching for new and different acts. Among the many other stage acts that preceded film screenings in this period were acrobats, roller-skaters and, for a brief time, some venues even tried staging badminton or basketball games before showing a film.[50] These were short-lived phenomena. Nevertheless, they offer further evidence that notions of what constituted an evening's entertainment at the cinema continued to include a surprisingly wide array of different attractions in the 1930s.

Conclusion

Today, *Variety* continues to offer its readers information on box-office grosses in North America, but its reports focus entirely on the national level and little (if any) attention is given to individual cities or to specific cinemas. This is undoubtedly appropriate in the contemporary context. Audiences today usually go to the cinema to see a single film rather than a programme of attractions. The majority of cinemas lack individual character and a distinct identity. And films are promoted through national rather than local campaigns. In the 1930s, by contrast, there was apparently little interest among *Variety*'s readers in national box-office grosses; the box-office earnings that were reported so carefully each week were not added together to form a national gross. The national level simply was not important. Rather, it was important to report what was happening in individual cities and cinemas, and how audiences were responding to specific programmes, promotions and pricing strategies. The cinemagoing experience itself varied widely even at the local level, and the task of *Variety* was to report on the success or failure of any number of variations. Thus, the reports allow film historians to observe changes in industry and exhibition practices, to study the context in which films were shown and how that changed over time, to assess audience preferences and compare the tastes of different cities or regions, and, of course, to examine the journey of a single film as it makes its way across the country and through various exhibition contexts.

Notes

1. The number of cinemas varied slightly because some of the smaller venues reported irregularly. Also, cities came and went. For example, New Orleans was included in 1934 but then dropped from the reports in 1935.
2. 'Bank nights' were lotteries that an audience member entered by buying an admission ticket. 'Giveaways' typically offered customers a piece of crockery for the price of admission.
3. The cities reported in the tables were Birmingham, Boston, Brooklyn, Buffalo, Chicago, Cincinnati, Denver, Detroit, Indianapolis, Kansas City (Missouri), Los Angeles, Minneapolis, Montreal, New Haven, New York, Philadelphia, Pittsburgh, Portland, Providence, St Louis, San Francisco, Seattle,

Tacoma, and Washington, D.C. Reports from a further six cities are included in the text of *Variety*, but not in the monthly tables, and so these cities are not included here. They are Baltimore, Cleveland, Lincoln, Louisville, Newark and Omaha.

4 Films included in the sample are those whose principal billing, as reported in *Variety*, was during the twenty-five months between 4 October 1934 and 29 October 1936. The records of films released before 4 October 1934 but exhibited predominantly during and after this month will be included. Likewise included are the records of films released during October 1936 and receiving subsequent exhibitions in November and December 1936.

5 A 'single' bill could include a live stage show. The 'single' aspect indicates the presence of only one feature film on the programme.

6 The figure of 18 per cent is likely to be an underestimate. This is because the tables in some instances do not report on live acts, and it has not been possible to review all of the weekly text reports. Future studies, in this respect, will need to make more extensive use of the text reports.

7 *The International Motion Picture Almanac, 1936–37* (New York, 1937), p. 992.

8 Ibid.

9 *International Motion Picture Almanac,* 1946–47 (New York, 1947). The populations of the cities from which the sample set of cinemas is drawn sum to just under 26 million, out of a total U.S. population of 128 million for the mid-1930s. U.S. Department of Commerce, Bureau of the Census, *U.S. Historical Statistics: Colonial Times to 1970* (Washington, DC, 1975), Appendix One.

10 Gomery states that in some large cities there could be as many as eleven runs. Douglas Gomery, *The Hollywood Studio System* (London: Macmillan, 1986), p. 17.

11 John Sedgwick, *Popular Filmgoing in 1930s Britain: A Choice of Pleasures* (Exeter: University of Exeter Press, 2000).

12 Richard Maltby, 'Sticks, Hicks and Flaps: Classical Hollywood's Generic Conception of its Audiences', in Melvyn Stokes and Richard Maltby (eds), *Identifying Hollywood's. Audiences: Cultural Identification and the Movies* (London: BF1, 1999), pp. 25–29.

13 Mark Glancy, 'MGM Film Grosses, 1924–48: The Eddie Mannix Ledger', *The Historical Journal of Film, Radio and Television*, 12 (1992), pp. 127–44; Mark Glancy, 'Warner Bros. Film Grosses, 1921–51: The William Schaefer Ledger', *The Historical Journal of Film, Radio and Television,* 15 (1995), pp. 55–74; Richard Jewell, 'RKO Film Grosses, 1929–51: The C.J. Tevlin Ledger', *The Historical Journal of Film Radio and Television,* 14 (1994), pp. 37–51.

14 John Sedgwick, 'Product Differentiation at the Movies: Hollywood, 1945–65', *Journal of Economic History,* 62 (2002), pp. 682–83.

15 *Variety*, 25 October 1934, p. 8.

16 *Variety*, 30 October 1935, p. 8.

17 *Variety*, 2 April 1936, p. 8.

18 *Variety*, 31 October 1935, p. 12.

19 *Variety*, 11 December 1934, p. 11; and *Variety*, 18 December 1934, p. 11.

20 Birmingham was the only city in the deep South that *Variety* covered. This has been attributed to the poor box-office returns of the region. See Thomas Cripps, 'The Myth of the Southern Box-Office: A Factor in Racial Stereotyping in American Movies, 1920–40', in J.C. Curtis and L.L. Lewis (eds), *The Black Experience in America: Selected Essays* (Austin, TX and London, 1970), pp. 116–44.

21 In Birmingham, for example, Temple's *Dimples, The Littlest Rebel* and *The Little Colonel* placed among the 20 top earning films during this period; and one of the city's leading cinemas, the Strand, regularly offered week-long engagements to low budget Westerns of stars such as Richard Dix, George O'Brien and Randolph Scott.

22 *The Barretts of Wimpole Street* was declared to be 'too snooty', 'too stylish' and 'too highbrow' for Birmingham audiences, and it lasted only one week at the Alabama Theatre, where it earned $6,500. The week before, Will Rogers' *Judge Priest* had earned $8,500 in the same venue. *Variety*, 30 October 1934, p. 8. *Modern Times* earned a remarkable $230,500 during the six weeks it played New York's Rivoli Theatre, but this proved to be 38 per cent of its total earnings, indicating that its success elsewhere was not so great.

23 The house, said to be worth $16,000, also included home insurance and groceries for a full year. The lottery took place in Denver and four first-run cinemas participated. Denver's Orpheum, which is included in the sample, was one participant. *Variety*, 4 September 1935, p. 4.

24 The Louis–Schmeling fight film was reported to be a significant attraction in Boston, Denver, Detroit, Indianapolis, Los Angeles, Minneapolis, Montreal, Portland, San Francisco and St Louis. See *Variety*, 25 June 1936.

25 *Variety*, 23 October 1935, p. 9.

26 Earnings for *Cleopatra* reached a moderately successful $415,500, but this was below the level of earnings reached by other costume dramas. The grosses for *The Crusades* and *The Last Days of Pompeii* were much lower, at $212,900 and $186,400, respectively.

27 See, for example, Nick Roddick, *A New Deal in Entertainment:Warner Brothers in the 1930s* (London: BFI, 1983); Peter Roffman and James Purdy, *The Hollywood Social Problem Film: Madness, Despair and Politics from the Depression to the 1950s* (Bloomington, IN: Indiana University Press, 1981); and Colin Shindler, *Hollywood in Crisis: American Cinema and Society, 1929–39* (London: Routledge, 1996).

28 Brian Taves, 'The B Film: Hollywood's Other Half', in Tino Balio (ed.), *Grand Design: Hollywood as Modern Business Enterprise 1930–1939* (New York: Scribner's, 1993), p. 321.

29 The 'Spanish' films were actually Spanish-language films, and many of these came from Mexico and South American countries, but the report does not categorize them in this way. See *Variety*, 1 January 1936, p. 43.

30 The discrepancy in running times is between the time listed for the original British release and the (shorter) time listed for the American release. For example, *Evergreen* was cut from 92 to 82 minutes; *First a Girl* from 93 to 78 minutes, *Things to Come* from 110 to 96 minutes, *Scrooge* from 78 to 72 minutes, and *Man of Aran* from 80 to 70 minutes.

31 For further analysis of London Films and the American market, see Sarah Street, *Transatlantic Crossings. British Feature Films in the USA* (London: Continuum, 2002), chap. two.

32 For further analysis of Gaumont British's efforts in the USA, see Sedgwick, *Popular Filmgoing*, chap, ten; the American release of *The 39 Steps* is considered in Glancy, *The 39 Steps: A British Film Guide* (London: I.B. Tauris, 2002).

33 One exception to this was the British and Dominions film *Escape Me Never*, which earned $189,950, but the company's other films had very few engagements and earnings levels far below this.

34 Taves, 'The B Film', pp. 318–20.

35 This method does under represent those very strong 'A' films, such as *San Francisco*, that occasionally played on double bills with much less popular films. This is an unavoidable problem, but also a slight one. As we have seen, the major 'A' films were actually the least likely to appear on double bills.

36 *Variety*, 25 September 1934, p. 9.

37 To give one example, when MGM's *It's in the Air* was paired with Universal's *Fighting Youth* at the Broadway in Portland, the report commented that audiences were coming chiefly for *Air Variety*, 30 October 1935, p. 9.

38 Polling was reported to have included 725,824 people. *Variety*, 12 August 1936, p. 5.

39 For example, it was said that while one of the films might be 'suitable for children, the second feature generally is not'. *Variety*, 12 August, 1936, p. 34.

40 *Variety*, 8 July 1936, p. 5.

41 *Variety*, 12 August 1936, p. 34.

42 *Variety*, 16 October 1934, p. 49.

43 *Variety*, 1 January 1935, p. 115.

44 *Variety*, 4 September 1935, p. 17.

45 *Variety*, 1 April 1936, p. 20.

46 *Variety*, 13 November 1934, p. 17. Of course, such reports suggest that the earnings of *The Scarlet Empress* and other films that played with prominent live acts should perhaps be altered to take account of another significant attraction on the programme. However, there is no clear and obvious method of doing this. Instead, we have chosen to draw attention to the presence of live acts when reporting box-office grosses. See Tables 11.1, 11.3, 11.4 and 11.5 and the Appendix for examples of this.

47 *Variety*, 22 January 1935, p. 49.

48 Gregory Waller, 'Hillbilly Music and Will Rogers: Small Town Picture Shows in the 1930s', in Melvyn Stokes and Richard Maltby (eds), *American Movie Audiences: From the Turn of the Century to the Early Sound Era* (London: BFI, 1999), p. 171. See also Richard B. Jewell, 'Hollywood and Radio: Competition and Partnership in the 1930s', *The Historical Journal of Film, Radio and Television*, 4 (1984), pp. 125–51.

49 The Roxy Theatre, for example, paid $6,000 for the Bowes stage show to support the Republic film *Laughing Irish Eyes*. *Variety* declared that this was a 'fair-gamble' given the 'weak picture.' The week's earnings, at $24,000, were twice as high as the film earned in any other engagement. *Variety*, 8 April 1936, p. 19.

50 *Variety*, 1 April 1936, p. 6.

Table 11.6 Appendix: the sample cinema set for the period October 1934 to October 1936

Cinema	*City*	*Affiliation*	*Seats*	*Price range (cents)*	*Box-office over the period ($)*[a]	*Regular live acts? (yes/no)*	*Films screened*	*Best week's film/act*	*Best week's box-office ($)*	*Worst week's box-office ($)*	*Best run film/act*	*Best run box-office ($)*	*Best run weeks*
Alabama	Birmingham	Publix	2,800	30–40	743,950	No	117	*Steamboat Round The Bend*	8,700	3,000	*Steamboat Round The Bend*	8,700	1
Empire	Birmingham	Acme	1,100	25 only	285,100	No	112	*Devil Dogs of the Air*	5,500	1,300	*Mr Deeds Goes To Town*	7,300	2
Strand	Birmingham	Publix	800	25 only	187,000	No	135	*In Old Kentucky / Keeper Of The Bees*	3,000	500	*In Old Kentucky*	4,250	1.5
Boston	Boston	RKO	2,900	25–50	1,690,625	Yes	144	Folies Bergeres Unit / *Hot Tip*	38,000	3,100	Folie Parisienne Unit / *Walking On Air*	58,500	2
Memorial	Boston	RKO	3,212	25–55	1,509,180	Occ.	89	*Top Hat*	40,000	5,000	*Top Hat*	111,200	5
Metropolitan	Boston	Publix	4,331	35–65	2,803,200	Yes	112	Jack Benny / *Private Worlds*	49,000	2,500	Paul Lukas / *Trail Of The Lonesome Pine*	58,000	2
Orpheum	Boston	Loew's	2,900	25–55	1,529,450	No	126	*Mutiny On The Bounty*	24,000	6,000	*San Francisco*	67,900	4
State	Boston	Loew's	3,700	30–55	1,501,550	No	151	*Barretts Of Wimpole Street*	24,500	4,000	*San Francisco*	55,000	4
Albee	Brooklyn	RKO	3,245	25–50	1,301,500	No	138	*Swing Time*	25,000	2,500	*Swing Time*	43,000	2
Fox	Brooklyn	Independent	4,075	25–50	1,479,100	Occ.	141	stage show / *One Night Of Love*	29,000	8,900	stage show / *One Night Of Love*	72,500	3
Metropolitan	Brooklyn	Loew's	3,618	25–50	1,658,500	Occ.	115	Eddie Cantor / *Transatlantic Lines*	36,000	13,000	*San Francisco*	85,000	6

Paramount	Brooklyn	Publix	4,156	25–65	1,581,500	No	102	*Captain Blood*	40,000	5,600	*Captain Blood*	65,000	2
Strand	Brooklyn	WB	2,870	25–50	565,200	No	198	*Show No Mercy / $1000 A Minute*	13,000	2,500	*Show No Mercy $1000 A Minute*	31,000	3
Buffalo	Buffalo	Publix	3,489	30–65	1,566,000	Occ.	113	Ted Lewis Orch/ *Ladies In Love*	25,000	5,700	Ted Lewis Orch/ *Ladies In Love*	25,000	1
Century	Buffalo	Publix	3,076	25 only	643,350	No	204	*Don't Turn 'em Loose / Old Hutch*	10,000	3,200	*Robin Hood Of El Dorado / Widow From Monte Carlo*	14,300	2
Hippodrome	Buffalo	Publix	2,089	25–40	742,500	No	146	*David Copperfield*	22,000	3,100	*David Copperfield*	13,500	2
Chicago	Chicago	Publix	3,861	35–75	3,652,000	Yes	97	Veloz and Yolanda / *Bride Comes Home*	59000	14,000	stage show / *Belle Of The Nineties*	95,900	2
Oriental	Chicago	Publix	3,217	25–40	1,805,600	Yes	111	vaudeville/ *One-Way Ticket*	27,600	10,300	vaudeville/ *One-Way Ticket*	27,600	1
Palace	Chicago	RKO	2,500	25–55	2,280,300	Occ.	80	*Swing Time*	34,700	5,900	*Top Hat*	141300	6
State Lake	Chicago	Jones	2,734	20–35	1,375,400	Yes	107	vaudeville/ *Iron Man*	19,200	9,800	vaudeville/ *Iron Man*	19,200	1
UA	Chicago	Publix	1,696	35–65	1,523,500	No	47	*Mutiny On The Bounty*	28500	7,000	*Great Ziegfeld*	95900	5
Albee	Cincinnati	RKO	3,317	35–42	1,377,950	No	108	*Follow The Fleet*	26,000	5,500	*Libeled Lady*	32,000	2
Keith's	Cincinnati	Libson	1,500	30–42	547,300	No	98	*Flirtation Walk*	10,500	2,100	*Flirtation Walk*	15,000	2
Lyric	Cincinnati	RKO	1,432	35–42	503,900	No	107	*Night At The Opera*	16,000	1,800	*Night At The Opera*	16,000	1
Palace	Cincinnati	RKO	2,614	35–42	1,171,900	No	103	*San Francisco*	22,000	3,750	*San Francisco*	36,000	2
Denham	Denver	Cooper	1,392	25–50	659,500	Occ.	110	*Belle Of The Nineties*	16,000	1,000	*Cleopatra*	22,500	2

(continued)

Table 11.6 Appendix: the sample cinema set for the period October 1934 to October 1936 *(continued)*

Cinema	*City*	*Affiliation*	*Seats*	*Price range (cents)*	*Box-office over the period ($)*[a]	*Regular live acts? (yes/no)*	*Films screened*	*Best week's film/act*	*Best week's box-office ($)*	*Worst week's box-office ($)*	*Best run film/act*	*Best run box-office ($)*	*Best run weeks*
Denver	Denver	RKO	2,525	25–50	899,900	No	115	*Mutiny On The Bounty*	15,000	4,000	*Mutiny On The Bounty*	15,000	1
Orpheum	Denver	RKO	2,600	25–50	732,400	Occ.	118	Ben Bernie Orch/*Romance In The Rain*	16,000	2,000	*San Francisco*	31,500	3
Paramount	Denver	RKO	2,096	25–40	319,925	No	160	*Bride Of Frankenstein*	7,000	1,000	*Bride Of Frankenstein*	9,000	1.5
Fisher	Detroit	Publix	2,975	30–40	85,350	No	37	*David Copperfield*	6,20(3	3J100	*David Copperfield*	6,200	1
Fox	Detroit	Independent	5,500	25–55	1,929,600	Yes	78	Cab Calloway Orch/*Charlie Chan In S'pore*	36,000	12,000	*Littlest Rebel*	60,000	2
Michigan	Detroit	Publix	4,038	25–55	1,852,200	Yes	86	NBC Radio Unit/*His Brother's Wife*	39,000	10,000	NBC Radio *Unit/ His Brother's Wife*	39,000	1
UA	Detroit	Publix	2,070 .	25–55	656,800	No	45	*Mutiny On The Bounty*	20,000	3,500	*San Francisco*	45,000	4
Apollo	Indianapolis	Fourth Avenue	1,171	25–40	327,950	No	59	*Steamboat Round The Bend*	9,800	1,300	*Steamboat Round The Bend*	19,900	3.5
Circle	Indianapolis	Monarch	2,638	25–40	318,350	No	99	*Swing Time*	10,500	1,900	*Swing Time*	14,700	2
Loew's	Indianapolis	Loew's	2,431	25–40	517,050	No	112	*Mutiny On The Bounty*	14,000	2,500	*Mutiny On The Bounty*	19,600	2

Lyric	Indianapolis	Olson	1,896	25–40	724,800	Yes	83	Major Bowes' Amateurs/*Pepper*	14,000	5,000	Major Bowes' Amateurs/Pepper	14,000	1
Main Street	Kansas City	RKO	2,500	25–40	1,046,000	Yes	99	Folies Bergere Unit/*Case Of Lucky Legs*	25,000	3,000	*Top Hat*	38,000	3
Midland	Kansas City	Loew's	4,000	25–40	1,222,345	No	101	*China Seas*	24,000	2,400	*San Francisco*	47,500	3
Newman	Kansas City	Publix	1,800	25–40	800,100	No	108	*Belle Of The Nineties*	18,000	2,700	*Both Belle Of The Nineties* and *Goin' To Town*	25,000	2
Uptown	Kansas City	Fox	2,045	25–40	460,375	No	108	*Steamboat* or *Poor Little Rich Girl*	11,000	1,200	*Steamboat Round The Bend*	19,400	3
Carthay Circle	Los Angeles	Independent	1,518	55–165	298,800	No	2	*Great Ziegfeld*	19,500	8,100	*Great Ziegfeld*	201,600	15
Chinese	Los Angeles	Fox	2,020	30–55	982,080	No	129	*Modern Times*	26,230	4,200	*Modern Times* i	37,530	2
Down Town Los Angeles	Los Angeles	WB	2,500	30–40	655,800	No	131	*Captain Blood*	14,500	2,200	*Captain Blood*	27,500	3
Hollywood	Los Angeles	WB	2,758	30–55	768,760	No	100	*Roberta*	15,000	2,200	*Dodsworth/Case Of The Velvet Claws*	33,000	3
Panatges	Los Angeles	RKO	2,812	25–40	669,250	No	134	*My Man Godfrey/Yellowstone*	23,700	1,500	*My Man Godfrey/Yellowstone*	45,200	3
Paramount	Los Angeles	Patmar	3,347	30–55	1,922,160	Occ.	105	Eddie Cantor/*Paris In Spring*	33,860	8,400	*Big Broadcast Of 1937*	56,200	3
RKO	Los Angeles	RKO	2,916	25–55	807,700	No	99	*My Man Godfrey/Yellowstone*	20,300	2,600	*Top Hat*	42,000	4
State	Los Angeles	Fox	2,422	30–55	1,311,700	No	139	*Mutiny On The Bounty*	24,300	5,100	*Mutiny On The Bounty*	38,900	2

(continued)

Table 11.6 Appendix: the sample cinema set for the period October 1934 to October 1936 *(continued)*

Cinema	*City*	*Affiliation*	*Seats*	*Price range (cents)*	*Box-office over the period ($)*[a]	*Regular live acts? (yes/no)*	*Films screened*	*Best week's film/act*	*Best week's box-office ($)*	*Worst week's box-office ($)*	*Best run film/act*	*Best run box-office ($)*	*Best run weeks*
Capitol	Manhattan	Independent	5,486	35–110	3,665,300	Occ.	64	*Mutiny On The Bounty*	75,300	7,000	*David Copperfield*	235,000	5
Center	Manhattan	Independent	3,700	25–110	564,500	No	21	*Ah, Wilderness!*	37000	6,000	*Thanks A Million*	86,000	4
Paramount	Manhattan	Publix	3,664	35–85	3,264,900	Yes	60	*Cleopatra*	68,000	8,500	*Big Broadcast Of 1937*	156,600	4
Radio City Music Hall	Manhattan	Independent	6,200	40–165	8,952,500	Occ.	78	*Top Hat*	134,000	48,000	*Top Hat*	348,000	3
Rialto	Manhattan	Publix	750	25–65	411,200	No	26	*Lives Of A Bengal Lancer*	21,800	5,000	*Lives Of A Bengal Lancer*	39,200	2
Rivoli	Manhattan	Independent	2,092	35–99	2,692,800	No	39	*Modern Times*	74,500	10,200	*Modern Times*	230,500	6
Roxy	Manhattan	Independent	6,000	25–65	3,492,910	Yes	90	Stage show/*If You Could Only Cook*	62,500	16,000	*Sing Baby Sing*	141,800	3
Strand	Manhattan	WB	2,758	35–85	2,366,900	No	57	*G–Men*	61,300	6,500	*Anthony Adverse*	200,000	5
Isyric	Minneapolis	Publix	1,126	20–25	224,400	No	112	*Kelly The Second*	7,000	900	*Libeled lady*	10,000	2
Minnesota	Minneapolis	Publix	4,024	25–55	477,000	Occ.	37	Burns and Allen/*Bride Comes Home*	32,000	6,500	*San Francisco*	29,000	2
Orpheum	Minneapolis	Singer	2,600	25–40	1,095,000	Occ.	100	Folies Bergere Unit/*Goose And The Gander*	26,000	2,200	*Roberta*	34,000	2
State	Minneapolis	Publix	2,290	25–40	761,900	Occ.	110	Major Bowes' Amateurs/*Redheads On Parade*	17,000	2,500	*China Seas*	19,800	2

Capitol	Montreal	Famous Players	2,603	50 only	841,950	No	200	*Lives Of A Bengal Lancer*	18,000	3,000	*Lives Of A Bengal Lancer*	28,000	2
Loew's	Montreal	Loew's	3,200	50 only	995,100	Occ.	149	John Boles/*Public Enemy's Wife*	20,000	6,000	John *Boles/Public Enemy's Wife*	20,000	1
Palace	Montreal	Famous Players	2,582	50 only	929.80C	No	105	*Mutiny On The Bounty* or *Roberta*	16,000	6,000	*San Francisco*	60,200	7
Princess	Montreal	CT	2,200	50 only	788,800	No	160	*Modern Times/Guard That Girl*	15,000	3,000	*Modern Times/Guard That Girl*	34,500	3
Paramount	New Haven	Publix	2,373	35–50	695,600	Occ.	178	*Belle Of The Nineties*	11,000	2,400	*Belle Of The Nineties*	14,500	2 i
Poli's	New Haven	Loew's	3,005	35–50	955,900	No	189	*Mutiny On The Bounty*	15,000	2,700	*Great Ziegfeld*	19,300	2
Sherman	New Haven	WB	2,076	35–50	601,100	No	193	*Flirtation Walk*	12,000	2,600	*Follow The Fleet*	15,000	2
Aldine	Philadelphia	WB	1,416	35–55	618,300	No	35	*Dodsworth*	19,000	2,700	*Dodsworth*	44,500	3
Boyd	Philadelphia	WB	2,338	35–55	1,388,700	No	83	*Anthony Adverse*	31,500	6,500	*Anthony Adverse*	71,500	3
Earle	Philadelphia	WB	2,728	40–65	1,650,800	Yes	112	Eddie Cantor/*One Exciting Adventure*	31,000	9,500	Eddie Cantor/*One Exciting Adventure*	31,000	1
Fox	Philadelphia	Independent	3,457	40–65	1,875,200	Occ.	81	stage show/*Thanks A Million*	35,000	8,000	Vincent Lopez Orch/*Private Number*	75,500	4
Roxy	Philadelphia	Independent	4,683	35–75	310,800	Yes	9	Jack Benny/*Woman In Red*	43,800	26,000	Jack Benny/*Woman In Red*	43,800	1
Stanley	Philadelphia	WB	3,009	35–55	1,486,800	No	84	*San Francisco*	31,000	7,500	*San Francisco*	85,000	4
Penn	Pittsburgh	Loew's	3,487	25–50	1,506,400	Occ.	103	Ted Lewis Orch/*Hide out*	31,000	4,000	*Top Hat*	33,000	2

(continued)

Table 11.6 Appendix: the sample cinema set for the period October 1934 to October 1936 *(continued)*

Cinema	*City*	*Affiliation*	*Seats*	*Price range (cents)*	*Box-office over the period ($)*[a]	*Regular live acts? (yes/no)*	*Films screened*	*Best week's film/act*	*Best week's box-office ($)*	*Worst week's box-office ($)*	*Best run film/act*	*Best run box-office ($)*	*Best run weeks*
Stanley	Pittsburgh	WB	4,000	25–50	1,722,500	Yes	103	Jack Benny+Mary Livingstone/ *Behold My Wife*	34,500	3,200	Folies Bergere Unit/*G-Men*	52,500	2
Warner	Pittsburgh	WB	1,800	25–40	496,600	No	192	*San Francisco*	11,000	1,000	*Mutiny On The Bounty*	16,200	2
Broadway	Portland	Parker	1,956	25–40	556,050	Occ.	125	*Libeled Lady*	11,500	2,500	*Libeled Lady*	24,100	3
Paramount	Portland	Hamrick-Evergreen	3,066	25–40	611,650	Occ.	133	Marx Bros./*Ten Dollar Raise*	10,800	3,300	*Curly Top*	17,000	2
UA	Portland	UA-Parker	962	25–40	580,100	No	63	*Mutiny On The Bounty*	10,700	2,400	*Mutiny On The Bounty*	34,400	6
Albee	Providence	RKO	2,394	15–40	762,350	Occ.	131	*Top Hat*	17,000	2,300	*Top Hat*	39,000	2.5
Majestic	Providence	Fay	2,262	15–40	775,550	No	187	*Anthony Adverse*	13,000	3,800	*Curly Top/Silk Hat Kid*	20,000	2
State	Providence	Loew's	2,500	15–40	1,149,900	Occ.	160	*Great Ziegfeld*	23,500	5,000	*Great Ziegfeld*	33,500	2
Strand	Providence	Independent	1,500	15–40	751,899	No	192	*Klondike Annie/Her Master's Voice*	14,300	2,000	*Trail Of The Lonesome Pine*	16,800	1.5
Ambassador	St. Louis	Fanchon and Marco	3,018	25–55	158,500	Occ.	21	*Belle Of The Nineties*	16,000	4,000	*Belle Of The Nineties*	31,000	2
Fox	St. Louis	Fanchon and Marco	5,036	25–55	171,500	Occ.	14	*County Chairman*	18,000	6,000	*One Night Of Love*	40,000	4

Missouri	St. Louis	Fanchon and Marco	3,516	25–40	94,700	Occ.	31	stage show / *Marines Are Coming / Strange Wives*	6,000	3,000	stage show / *Marines Are Coming / Strange Wives*	12,000	2
Shubert	St. Louis	Fanchon and Marco	1,710	25–40	160,000	Occ.	23	*Gay Divorcee / Lives Of A Bengal Lancer*	15,000	6,000	*Lives Of A Bengal Lancer*	26,000	2
State	St. Louis	Loew's	3,050	25–55	233,000	Yes	14	*David Copperfield*	17,000	8,000	*Chained*	32,000	2
Golden Gate	San Francisco	RKO	2,800	30–40	1,460,800	Occ.	80	Eddie Cantor / *Last Outlaw*	34,000	9,000	*Top Hat*	58,400	3
Orpheum	San Francisco	Independent	2,900	30–40	806,050	No	104	*One Night Of Love*	20,000	2,000	*One Night Of Love*	68,000	8
Paramount	San Francisco	Fox	2,735	30–40	1,205,800	No	152	*San Francisco*	28,000	6,000	*San Francisco*	67,000	3
Warfield	San Francisco	Fox	2,657	35–65	1,893,600	Occ.	1Q9	*Forsaking All Others*	29,400	4,750	*Libeled Lady / Sitting On The Moon*	54,000	4
Fifth Ave	Seattle	Hamrick-Evergreen	2,420	25–40	936,525	No	105	*Rose Marie*	17,200	3,800	*San Francisco*	35,600	3
Liberty	Seattle	Jenson & Von Herberg	1,800	15–35	542,300	No	109	*Broadway Bill*	12,200	1,700	*Mr Deeds Goes To Town*	82,700	14
Music Box	Seattle	Hamrick-Evergreen	970	25–40	411,950	No	99	*Roberta*	9,100	1,800	*Roberta*	28,500	6
Paramount	Seattle	Hamrick-Evergreen	3,000	25–35	594,650	Occ.	193	French Folies / *Annie Oakley*	12,800	2,300	*Gorgeous Hussy / Star for The Night*	17,300	2
Music Box	Tacoma	Hamrick-Evergreen	1J00	15–35	444,357	No	206	*San Francisco*	8,000	2,600	*San Francisco*	11,800	2

(continued)

Table 11.6 Appendix: the sample cinema set for the period October 1934 to October 1936 *(continued)*

Cinema	*City*	*Affiliation*	*Seats*	*Price range (cents)*	*Box-office over the period ($)*[a]	*Regular live acts? (yes/no)*	*Films screened*	*Best week's film/act*	*Best week's box-office ($)*	*Worst week's box-office ($)*	*Best run film/act*	*Best run box-office ($)*	*Best run weeks*
Roxy	Tacoma	Jenson & Von Herberg	1,200	25–35	422,837	No	168	*China Seas*	9,000	2,800	*Broadway Bill*	12,400	2
Columbia	Washington	Loew's	1,000	25–40	486,200	No	97	*In Old Kentucky* or *Baboona*[b]	8,000	2,000	*In Old Kentucky*	14,500	2
Earle	Washington	WB	2,240	25–70	1,929,500	Yes	110	Jan Garber Orch/*Mr Deeds Goes To Town*	26,000	6,000	Jan Garber Orch/*Mr Deeds Goes To Town*	26,000	1
Fox/Capitol	Washington	Loew's	3,433	25–60	2,326,500	Yes	106	vaudeville/*Rendezvous*	30,000	15,000	stage show/*Naughty Marietta*	56,500	2
Keith's	Washington	RKO	1,500	25–60	1,014,100	Occ.	77	*Top Hat*	24,500	2,500	*Top Hat*	68,500	5
Palace	Washington	Loew's	2,700	25–60	1,724,500	No	68	*Gorgeous Hussy* or *Mutiny On The Bounty*	28,000	9,000	*San Francisco*	54,000	3
Total			*282,674*		*121,112,628*								

Sources: *Film Daily Yearbooks* for 1936 and 1937; *International Motion Picture Almanac* for 1936–7 and 1937–8; *Variety*, weekly for the period.

Notes:

a. Money values for Montreal are expressed in U.S. dollars.

b. The makers of the documentary animal drama *Baboona*, Martin and Osa Johnson, were present during the week of the film's screening.

Chapter 12

Tino Balio

SELLING STARS: THE ECONOMIC IMPERATIVE

IN THE ERA OF VERTICAL INTEGRATION, the star system affected all three branches of the industry. A star's popularity and drawing power created a ready-made market for his or her pictures, which reduced the risks of production financing. Because a star provided an insurance policy of sorts and a production value, as well as a prestigious trademark for a studio, the star system became the prime means of stabilizing the motion-picture business. At the production level, the screenplay, sets, costumes, lighting, and makeup of a picture were designed to enhance a star's screen persona, which is to say, the image of a star that found favor with the public. At the distribution level, a star's name and image dominated advertising and publicity and determined the rental price for the picture. And at the exhibition level, the costs of a star's salary and promotions were passed on to moviegoers, who validated the system by plunking down a few coins at the box office.

In economic terms, stars created the market value of motion pictures.[1] To understand how this worked, we must remember three things. First, affiliated theater chains were located in different regions of the country, so that to reach a national audience the majors had to exhibit one another's pictures. Second, the majors rented their pictures to exhibitors a season in advance of production. And third, the majors used a differential pricing policy. No set price could be charged for the top-grade product because the market for this type of picture was difficult to ascertain. Charging a percentage was riskier than charging a flat fee, but in so doing, a distributor could reap the rewards of a box-office surge.

How did the majors determine the rental price for a picture, which is to say, the percentage terms for a new picture? They used star power—the ability of screen personalities to attract large and faithful followings. In practice, a distributor simply pointed to the past box-office performance of a star to justify the rental terms for his or her forthcoming pictures. An economist might say the distributor used star differentiation to stabilize the demand curve for class-A product.

Star differentiation did more than stabilize rentals; it also permitted the distributor to raise prices. Demand elasticity explains the phenomenon. "Demand elasticity measures the sensitivity of demand in relationship to quantity and change in price. Theoretically, if demand can be fixed by product differentiation, it then becomes less sensitive to increases in price."[2] Thus if new picture contained a star with a proven box-office record, an exhibitor would likely be willing to pay a higher rental for it, feeling certain that the risk was worth it.

The majors buttressed this method of pricing by instituting elaborate and costly publicity campaigns that revolved almost exclusively around stars. Because these campaigns were designed to peak simultaneously with the release of a new picture, as will be discussed later, they funneled audiences into first-run houses. Owned almost exclusively by the Big Five, these flagship theaters charged the highest ticket prices and generated 50 percent of the domestic rentals. Elaborate publicity campaigns served an added function; a successful launching of a new release helped establish its market value in the subsequent-run playoff.

The contract controls

The economics of the star system is a necessary prelude to understanding how the studios safeguarded their most precious assets. The studios devised an ingenious legal document to control their high-priced talent, the "option contract." This is how the contract worked. In signing an aspiring actor or actress, the studio used a contract that progressed in steps over a term of seven years. Every six months, the studio reviewed the actor's progress and decided whether or not to pick up the option. If the studio dropped the option, the actor was out of work; if the studio picked up the option, the actor continued on the payroll for another six months and received a predetermined raise in salary. Note that the studio, not the star, had the right to drop or pick up the option. The contract did not provide reciprocal rights, meaning that an actor or actress could not quit to join another studio, could not stop work, and could not renegotiate for more money. In short, the contract effectively tied a performer to the studio for seven years.

The option contract did more than that: it had restrictive clauses that gave the studio total control over the star's image and services; it required an actor "to act, sing, pose, speak or perform in such roles as the producer may designate"; it gave the studio the right to change the name of the actor at its own discretion and to control the performer's image and likeness in advertising and publicity; and it required the actor to comply with rules covering interviews and public appearances. Another restrictive clause concerned picture assignments. If the aspiring star refused an assignment, the "studio could sue for damages and extend the contract to make up for the stoppage."[3]

The studios argued that the option contract was not as inequitable as it seemed because developing talent was expensive and risky. If a new player clicked, the studio was justified in wanting to cash in on its investment. If a new player showed little or no promise, it made no sense for the producer to carry him or her for seven years. Be that as it may, stars exercised little control over production. Some stars had story-approval rights and could refuse to appear in an unsympathetic or unflattering role, but in that event, the studio simply assigned the role to another performer. And once into a picture, a star had no say in the interpretation of his or her role, let alone the script, since that was largely the prerogative of the director.[4]

Forming the Screen Actors Guild

Actors had remained relatively docile employees until 1933, when producers responded to the bank moratorium by threatening to close the studios unless talent accepted a 50 percent pay cut for eight weeks. Following the lead of the screenwriters, a group of eighteen actors, among them Ralph Morgan, Alan Mowbray, and Boris Karloff, formed the Screen Actors Guild (SAG) on 30 June 1933. Over the next few months membership grew slowly to around fifty. But when producers drafted the Code of Fair Competition and blamed the financial difficulties of their studios on the star system, actors signed up in droves.

The moguls had written into the Code provisions barring star raiding, curbing the activities of agents, and limiting salaries. To prevent star raiding, a provision permitted a studio to keep a star in tow at the end of a seven-year contract by exercising the right of first refusal, which "amounted to professional slavery," said the guild. To curb agents, producers planned on organizing a general booking office to broker talent. "All agents, in order to deal with the booking office, would have to be licensed by the booking office. This would put the actors' representative completely under the thumb of the producer, make every contract a one-sided bargain, and in the end reduce compensation," said the guild. And to limit salaries, studios wanted to cap earnings at $100,000 a year.[5]

What really turned around SAG membership was a protest meeting held at the El Capitan theater in October 1933. Eddie Cantor, the new president of the guild, told the audience that the Academy was unable to represent the full interests of the actors and that the producers' salary ploy was unconstitutional. One of the highest-paid entertainers in the business, Cantor had endeared himself to actors at an organizational meeting held earlier when he said, "I'm here not because of what I can do for myself, but to see what I can do for the little fellow who has never been protected and who can't do anything for himself. If that's not the spirit of everyone here, then I want to leave." When Cantor called for "a 100% actor organization," five hundred actors out of the more than eight hundred in the audience "flocked to the stage to sign membership blanks in the new Screen Actors' Guild." Among the prominent names signing up were Adolph Menjou, Fredric March, Robert Montgomery, Jimmy Cagney, Miriam Hopkins, Jeanette MacDonald, and Paul Muni.[6]

Responding to the Code, SAG filed a brief with the NRA answering the charge that actors were being overpaid. By way of a preface, the brief said that "history shows that no agreement with producers is worth the paper it is written on"; that Hollywood's code of ethics "is the lowest of all industries"; and that "every dishonest practice known to an industry ... has been resorted to by the producers against the actors." After this indictment, the brief presented statistics on actors' salaries. In 1933 one quarter of the employed actors made less than $1,000, about half made less that $2,000, and approximately three-quarters made less than $5,000. These salaries were gross incomes. Ten percent of an actor's salary went to an agent, and a significant amount went to maintain a proper wardrobe, which at the time was part of an actor's working tools. Concerning performers in the highest income brackets, the brief underscored the fact that earning power lasted a short while: "If one takes a glance at any group of extras of today, he will find many of the stars of yesterday."[7]

Eddie Cantor paid a visit to President Roosevelt, who was his personal friend, to plead the actors' case. To avert a highly publicized labor dispute that might adversely affect public acceptance of the NRA, the president suspended the obnoxious provisions in the motion-picture code by executive order. However, during the days of the NRA, SAG failed to receive recognition as bargaining agent for the actors. Nor did SAG receive recognition when the National Labor Relations Act was signed into law in 1935; Hollywood's response to the act was simply to ignore it.

But two years later, the Screen Actors Guild threatened to call a strike and finally won recognition on 15 May 1937. "The victories have been victories for the rank and file. For themselves the stars have asked and won next to nothing," said the *Nation*.[8] The rank and file won minimum pay rates, guarantees of continuous employment, and twelve-hour rest periods between calls. Although successive contracts won benefits for all classes of performers, the relationship of the actor to the production process remained unaltered; in fact, it was never an issue. The concessions had relatively minor economic impact on the studios, which explains why they were implemented.

The star system in place

The acting profession in Hollywood consisted of four classes of performers. Supporting players performed the least important parts in pictures. Employed for as short a time as a week, a supporting player did not receive screen credit or even the assurance that his or her part would not be cut before the picture was released. Stock players were either promising beginners or experienced old-timers and, as a group, formed a large talent pool from which the studio rounded out the cast of a picture. They received contracts of six months or longer and were paid from $50 to $350 a week. Featured players performed the principal roles and received screen and advertising credit. Their contracts went from year to year and specified a minimum and maximum number of pictures and a salary of so much a picture.[9]

Stars constituted the elite class. Like other classes of performers, they were required to play any part the studio designated, although the biggest names might have the right to refuse a specified number of projects they deemed unsuitable. Paul Muni, for example, used his considerable prestige to accept only those scripts that dealt in some way with social problems. Stars might also have the right to approve the cameraman, but rarely their directors. Greta Garbo, for example, would work with only one cameraman, William Daniels. Stars received option contracts that lasted as long as seven years and were paid on a per-picture basis. As privileged members of the studio, they received a range of perks that could even affect the paint on their dressing room walls. Examples of such perks are fixed working hours of not more than eight out of any twenty-four-hour period with time out for afternoon tea; a dressing room decorated to the star's satisfaction; arrangements for a personal maid or valet; and "greater prominence" in advertising, meaning that the star's name had to be placed above all others in all advertising, in letters larger and of greater prominence than any other name.

"Freelance" actors, as the term implies, hired themselves out to any studio and worked on a picture-by-picture basis. Their ranks numbered about forty stars and featured players, among them Fredric March, Ronald Colman, Jean Arthur, Aline MacMahon, Adolphe Menjou, Edward Everett Horton, and Constance Bennett. These players were almost continually in demand, especially by independent producers.[10]

The total number of contract players in Hollywood during a given year came to around five hundred; the total on a studio's roster varied from around fifty to one hundred. Of the grand total, thirty or so received star billing. A similar number would fade out for a while or forever. As they faded, stars usually remained under contract, but inevitably they were dropped from the lists. Then they had the option of offering their services to independent producers or of becoming freelance supporting players. Sometimes, a waning star received a new lease on life at another studio. "Countless players have made good on the second Hollywood bounce," said *Variety* and identified Warners as the "champ builder-upper." In 1935, Warners had working for it a dozen players dropped by other studios, among them George Brent, Ricardo Cortez, William Gargan, Hugh Herbert, Bette Davis, and Paul Muni.[11]

MGM had the largest and most prestigious stable of stars in Hollywood, which enabled it to produce nearly a third of the top-grossing films every year. Taking maximum advantage of MGM's talent pool, Irving Thalberg instituted a "galactic" system of casting a picture, whereby two or more stars were teamed to increase its box-office power. Among such vehicles are GRAND HOTEL (1932), which starred Greta Garbo, John Barrymore, Lionel Barrymore, Wallace Beery, and Joan Crawford; RASPUTIN AND THE EMPRESS (1932), which teamed the three Barrymores; and DINNER AT EIGHT (1933), which listed eight top names.[12]

After the economic shakeout of the Depression, stars' salaries rarely rebounded to the extravagant levels of the booming twenties. Leo Rosten's survey of actors' salaries showed that in 1939 only 54 class-A actors out of 253 earned over $100,000. Claudette Colbert topped the list, with earnings of $426,944, followed in descending order by Bing Crosby

($410,000), Irene Dunne ($405,222), Charles Boyer ($375,277), and so on down the line. However, in "startling contrast" to these earnings, Rosten's survey revealed that the median average salary for the group was $4,700, which meant that half the actors earned $4,700 or less and that half earned $4,700 or more. Excluded from these calculations were the pittances paid to movie extras. In short, the acting profession in Hollywood remained a poorly paid one.[13]

The fact that most directors, writers, and stars were temperamentally ill equipped to bargain with hardheaded producers explained the existence of the agent. Close to 150 registered agents worked in Hollywood. A dozen or so firms did most of the business, among them the William Morris Agency, Joyce–Selznick, Charles K. Feldman, and Leland Hayward. "In any business as sprawling, loose, and disjointed as show business," said *Fortune*, "there must be an intermediary between the possessors of talent and the users of talent." Agents represented actors, directors, and writers. As one agent described his job, "My occupation is representing clients, placing them advantageously, getting them the highest salaries I can, and maintaining the best possible working conditions for them." Other agents provided "personal representative" and "management" services pertaining to almost every facet of the client's career. Whatever his function, an agent took a 10 percent cut of all wages earned by his client during the term of the contract. This 10 percent fee, fixed by the Screen Actors Guild, was the maximum an agent could charge for his services.[14]

During the days of the NRA, producers tried to curb the power of agents by outlawing star raiding, but an executive order from the White House prevented them. The studios soon devised a way to get around the order. Talent was always scarce. *Variety* reported that "the complexities of casting . . . are so great that no single plant can cast its own productions from its contract list."[15] It gave as a reason talking pictures, which made individualized roles much more important to the acting ensemble than they had been in the silents. Studios developed young talent and recruited personalities from the stage and radio, both at home and abroad, but nothing proved sufficient to meet all their needs. Rather than raiding one another to bolster star rosters, the majors found it easier and just as effective to lend one another talent.

As always, economics played a role. Try as they might, studios found it impossible to keep high-priced talent busy all the time. An idle star was a heavy overhead expense. Why not "loan out" the idle star and recoup the overhead? Studios devised various formulas to determine the fee: the most common one was to charge a minimum fee of four weeks salary plus a surcharge of three weeks; another was to charge the basic salary for however long the star was needed plus a surcharge of 25 percent.[16] Loan-outs kept RKO competitive and enabled Columbia and Universal to maintain their status as members of the Little Three. For example, Columbia borrowed most of the big names it needed for the Capra pictures—Claudette Colbert and Clark Gable for IT HAPPENED ONE NIGHT, Gary Cooper for MR. DEEDS GOES TO TOWN, and Jimmy Stewart for YOU CAN'T TAKE IT WITH YOU and MR. SMITH GOES TO WASHINGTON.

Top-ranking independent producers like Selznick, Goldwyn, and Wanger, who released through UA, also regularly borrowed players. The majors were willing to lend them stars because these producers had longtime connections with the industry and had successful track records. Myth has it that the majors used loan-outs to discipline stars and to keep difficult people in line. But this argument does not make much sense, because it implies that a studio would risk its investment in a star by allowing him or her to appear in a second-rate picture produced by an inferior company. Actually, most stars were on the lookout for challenging parts and wanted the right to play them anywhere. One agent reported, "I have secured for a number of my clients contracts which permit them to play in one outside picture a year, on terms which they negotiate independently. Usually such permission enables the artist to appear in some favorite story or work for some favorite director, and the novelty of an interlude on a different lot breaks the monotony of constant association with too-familiar faces."[17]

Loan-outs frequently revitalized flagging careers. For example, Clark Gable and Claudette Colbert were in lulls when MGM and Paramount, respectively, sent them to Columbia to star in IT HAPPENED ONE NIGHT. Because of the film's success, their careers took off. Bette Davis revealed her capabilities as a mature actress of great gifts when Warners loaned her to RKO to star in OF HUMAN BONDAGE. And on loan-out to Universal to star in MY MAN GODFREY, Carole Lombard found herself in her biggest success yet, which sent her asking price skyrocketing and won her contracts from several studios.[18]

Periodically, courageous stars challenged the system by demanding bigger salaries, better roles, and more respect. The big battles took place at Warners and involved two issues: the right of a studio to treat a star as chattel, as a mere investment that could be milked for all he or she was worth (product maximization); and the right of the studio to tack on suspension time at the end of a contract.

James Cagney started the first battle when he walked out of the studio at the start of a picture in 1932, claiming he was working too hard for too little money. Cagney's original contract with Warners, which was negotiated by William Morris in 1930, paid him $350 a week to start and then rose in increments over the life of the contract. After Cagney made a name for himself in such pictures as THE PUBLIC ENEMY, TAXI!, and THE CROWD ROARS, Warners gave him a bonus that raised his base salary to $1,400 a week. But Cagney was not mollified; he wanted a new contract that started at $3,000 per week. Cagney said he based his stand "on the fact that [his] pictures, for the time being, are big moneymakers—and that there are only so many successful pictures in a personality. And don't forget that when you are washed up in pictures you are really through. You can't get a bit, let alone a decent part."[19]

After the dispute went to arbitration, Warners awarded Cagney a new contract that started at $1,750 a week. Warners also orally agreed that Cagney was required to make a maximum of four pictures a year. In 1935, Cagney again filed suit to break his contract. His stated rationale was that he had made fourteen pictures in three years and that contrary to the billing provision in his contract, he had received second billing on DEVIL DOGS OF THE AIR at certain theaters. Cagney was then earning $4,500 a week. Cagney told the court that "four pictures are enough for any actor whose career has advanced as far as mine . . . When I signed the contract I understood my production schedule was to be limited to that number . . . I feel an actor wears out his welcome with the public if he appears in too many pictures. In other words, the audiences get their fill, and turn in another direction."[20]

At the core of Cagney's discontent was the studio's practice of typecasting him in what he referred to as "dese, dem, and dose" roles and in inferior pictures in which he was teamed up with Pat O'Brien. The court ruled in favor of Cagney, stating that Warners had breached Cagney's contract.[21] Claiming that the so-called advertising breach was inadvertent and casual, Warners appealed the case to the California Supreme Court.

Pending the outcome, Cagney went to work for Grand National, a new Poverty Row studio. Cagney had approached the majors but had gotten nowhere because if the state supreme court were to reverse the decision, any studio that employed him would be subject to damages and might lose the entire amount it had spent on production. Cagney made two pictures for Grand National. Cagney received better assignments afterward, among them BOY MEETS GIRL (1938), ANGELS WITH DIRTY FACES (1938), THE OKLAHOMA KID (1939), and THE ROARING TWENTIES (1939). However, Cagney remained recalcitrant and, when conditions were right, became one of the first stars to go into independent production.

Bette Davis, another Warners star, became dissatisfied with her roles and walked out on her contract in 1936. Bernard Dick described Davis' relationship with the studio as

> one of the stormiest . . . that ever existed between a studio and its star; graphically, it would resemble a fever chart. Davis would no sooner make one good film than she would be assigned to a series of poor ones. As if to punish her for making OF HUMAN BONDAGE (1934) on loanout at RKO because she could not find a decent script at her own studio, Warners released HOUSEWIFE (1934) immediately after her triumph as Maugham's Mildred Rogers. After DANGEROUS [which won her an Oscar for best actress] and THE PETRIFIED FOREST came THE GOLDEN ARROW and SATAN MET A LADY (all 1936), which left her "unhappy, unfulfilled."
>
> (Bernard Dick, ed., *Dark Victory* [Madison: University of Wisconsin Press, 1981], p. 14)

Returning to the studio after OF HUMAN BONDAGE, Davis received a new contract that raised her salary to $1,350 a week from $1,000, but the contract merely guaranteed feature billing in her pictures and did not reduce the number she had to make each year. Through her agent, she sent the following list of demands to the studio: a five-year contract at a salary that escalated from $100,000 to $220,000 per year; the right to make no more than four pictures a year; star or co-star billing; the services of her favorite cameramen; and "three months' consecutive vacation each year with the right to do one outside picture."[22]

Jack Warner suspended her, characterizing her demands "as exorbitant and impossible." He added that "she would remain on the suspended list until she returned to the studio to live up to the terms of her contract."[23] The studio cut off her salary and announced its intention to tack on to her contract the time she spent on suspension. To circumvent the ban, Davis accepted a leading role in a picture to be produced in London by the independent producer Ludovico Toeplitz. Warner took her to court by filing an injunction. After a brief hearing in which the Warners lawyer characterized Davis as a "naughty little girl who wants more money," the court granted the injunction, ruling that Davis must confine her services exclusively to Warners.

Davis accepted the judgment and decided on another approach. Through her attorneys she told Jack Warner that she would return to work without any "modifications" of her existing contract, but she also politely reiterated her case. Said Thomas Schatz, "[Jack] Warner held firm but he got the message; Davis might have lost this skirmish, but the war would go on. And Warner himself, having been without his top actress for nearly a year, was ready to compromise."[24] To demonstrate his good intentions, Jack Warner bought her a property she had wanted, JEZEBEL. And Davis demonstrated her mettle by winning her second Academy Award playing the lead role. But "just when she thought the pattern of one mediocrity for every masterpiece had been altered," she was given COMET OVER BROADWAY as her next picture. Davis refused the assignment and went on suspension in April 1938. After a month, she and the studio resolved their differences.

Now at the pinnacle of her career, Davis received a new contract that started out at $3,500 per week and gave her star billing. However, "by explicitly detailing her duties to the studio," the contract was in some respects more restrictive: "Davis had to 'perform and render her services whenever, wherever and as often as the producer requested.' Significantly, these services included interviews, sittings for photographs, and the rest of the elements the studio could orchestrate in its differentiation strategy." Thereafter, Davis would receive more money and would make fewer pictures, but "she never did earn the right to choose her roles or to have a say in her publicity. On the contrary, as Davis' name grew larger on theater marquees, the studio consolidated more control over her career."[25]

Not until Olivia de Havilland took Warners to court over its suspension practices was a star able to break any of the offensive terms of the seven-year option contract. Ruling in 1943 that Warners had violated the state's antipeonage laws, the court decided in de Havilland's favor, a decision the state supreme court affirmed in 1944. Schatz calls the decision "a watershed event in Hollywood's history, a significant victory for top stars and a huge setback for the studios. No longer could Warners or any other studio tack on suspension time to the end of a contract, thereby preventing an artist from sitting out and becoming a free agent."[26] It is difficult to attach such significance to the case, since every other provision of the option contract designed to keep stars in their place remained in force. It was not until the breakdown of the studio system itself during the fifties that the balance of power tipped in favor of the star.

Star development

To build a roster of stars, a studio relied on what Douglas Gomery calls "the spillover effect" of personalities recruited from professional theater, vaudeville, radio, and other forms of entertainment. Hollywood's raid on Broadway began in earnest in mid 1928 and captured a new generation of actors that included James Cagney, Bette Davis, Claudette Colbert, Irene Dunne, Clark Gable, and Mae West, among others. By 1934, *Variety* reported that "Hollywood is now 70% dependent on the stage for its film acting talent up in those brackets where performers get screen credit."[27]

Although vaudeville was on its last legs, it supplied Hollywood with many early sound stars, among them Al Jolson, Eddie Cantor, the Marx Brothers, Joe E. Brown, George Burns and Gracie Allen, W. C. Fields, Bert Wheeler and Robert Woolsey, Will Rogers, and Mae West.[28] And radio which was becoming increasingly popular throughout the decade, supplied the movies a steadily supply of its stars—Kate Smith, Rudy Vallee, Bing Crosby, and Ed Wynn. Even grand opera provided a few names. As musicals gained in popularity during the decade, studios widened their search and signed opera stars Grace Moore, Lily Pons, Nino Martini, and Gladys Swarthout, among others.

In her anthropological study of Hollywood, Hortense Powdermaker said she was surprised to learn that "while most executives swear by the star system, it is not a part of Hollywood custom to plan coherently even for the stars." She stated a commonly held belief, but it needs revising. During the conversion to sound, Warners and Paramount devised a cost-effective method to test stage and radio talent. After refurbishing their East Coast studios, the companies cast stage and radio personalities in shorts—for example, in "canned vaudeville" or comic skits. If they passed muster, they were offered long-term contracts and sent packing to California for exploitation in feature films. Mae West is the best example of this strategy. After a successful test, Paramount gave her a contract that permitted her to write as well as to star in her own vehicles. In such hits as I'M NO ANGEL (1933) and BELLE OF THE NINETIES (1934), she helped pull Paramount out of the Depression to become the highest salaried woman in the United States in 1934.[29]

Radio became a national pastime during the thirties. Some radio stars had enjoyed renown on the stage, and others became radio originals, created by the new medium. In either case, Hollywood wanted to absorb them. However, producers faced two problems: how to transfer a popular star from a medium that is primarily aural and that has its own set of performance conventions to a medium that is both aural and visual and that has a narrative tradition; and how to adapt a radio star to a motion-picture audience. Did the two audiences have the same demographic composition, age spread, and tastes? If they were different, to what extent? And how could one find out?

Hollywood devised two strategies to adapt radio stars to the movies. The first entailed building a full-length narrative around a radio personality and promoting the vehicle in the conventional manner—that is, by focusing all the attention on the star. Paramount tried this approach on Kate Smith after her appearance in THE BIG BROADCAST (1932). In this picture, she played herself and sang a number much like she did on her radio show. To launch her as a full-fledged movie star, Paramount designed a vehicle, HELLO, EVERYBODY (1933), in which she played a farm girl who breaks into radio as a singer in order buy off land grabbers threatening to take her family's property. It flopped. Using a similar strategy, Universal tried to launch Myrt and Marge, a popular mother–daughter radio team, in a B musical called MYRT AND MARGE (1934). In its review of the picture, *Variety* associated the team with "a whole list of radio folks who went to Hollywood, made one picture, and apart from a piece of change, did themselves little good."[30] With the single exception of Bing Crosby, this strategy of star development failed because radio personalities either lacked the acting skills to sustain a feature-length movie or performed roles that were out of character.

The second strategy to adapt radio stars was devised by Paramount to make THE BIG BROADCAST (1932).[31] A loosely woven musical inspired by the all-star cast and multiple-plot structure of GRAND HOTEL, the film provided the minimum excuse for a series of radio stars to perform their familiar routines. Using a radio station as the motivation for guest appearances, Paramount showcased a series of radio personalities—among them Bing Crosby, George Burns and Gracie Allen, the Boswell Sisters, and Kate Smith—by having each star introduced on camera by the same announcer who handled that function on the star's weekly show. The formula had enough going for it to spawn a sequel, INTERNATIONAL HOUSE (1933), and then a series.

Because the personas of stars in one medium did not always carry over into motion pictures, studios obviously had to build personalities from scratch. Describing the star-making process, W. Robert La Vine said:

> A star was not born, but made. Hair was bleached or dyed, and, if necessary, to "open" the eyes, eyebrows were removed and penciled in above the natural line. Studio-resident dentists, expert at creating million-dollar smiles, capped teeth or fitted them with braces. Cosmetic surgery was often advised to reshape the nose of a new recruit or tighten her sagging chin. A "starlet" was taught how to walk, smile, laugh, and weep. She was instructed in the special techniques of acting before a camera, perfecting pronunciation, and learning how to breathe for more effective voice control. Days were spent in wardrobe, situated in separate buildings within the studio communities.
>
> (*In a Glamorous Fashion: The Fabulous Years of Hollywood Costume Design* [New York: Charles Scribner's Sons, 1980], p. 27)

To devise an appropriate screen image for an aspiring star, a studio would cast the player in a series of roles and test audience response to each by consulting fan mail, sneak previews, reviews, exhibitors' comments, and the box office. In essence, producers attempted to mold their protégés to fit consumer interest. Once the correct formula was found, the ingredients would be inscribed in narratives, publicity, and advertising. Take the case of Bette Davis. Davis appeared in a series of unremarkable pictures from 1932 to 1935 as Warners searched for the correct formula. In one of her first assignments, CABIN IN THE COTTON (1932), Davis played a southern belle who attempts to titillate and cajole a poor sharecropper (Richard Barthelmess) into betraying his friends. Her vivid performance as a coquette elicited a strong audience response and led *Variety* to name her a box-office leader for 1932.

In her first starring role, Davis plays a liberated career woman in EX-LADY (1933) who is forced to choose between remaining a free-spirited single "modern woman" or becoming an "old-fashioned" married woman. The picture was panned by the critics, and Davis was dropped from *Variety*'s charts. In 1934, Warners loaned out Davis to RKO to play the part of

Mildred, a mean, sluttish waitress who seduces and destroys a medical student (Leslie Howard) in OF HUMAN BONDAGE, a character that *Variety* described as a vamp. Audiences loved her in the role. Now that a successful match between narrative and actress had been found, Davis showed signs of becoming an unqualified star.

Returning to Warners, Davis was assigned to play a vamp in a Paul Muni vehicle, BORDERTOWN (1934). She received featured billing. Cathy Klaprat points out that "although *Bordertown* appeared on Warners' production schedule in June 1934, prior to the release of *Bondage* . . . Davis was not cast as Marie until after her triumph." Ads for the picture asked, "Who will be the real star of the film: Bette Davis or Paul Muni?" Campaign books told exhibitors to place signs in their lobbies asking patrons if Davis should have received star billing. Fans must have answered yes because for her next assignment, Warners starred Davis in DANGEROUS (1935) playing another vampish role, an alcoholic actress. Anticipating that DANGEROUS would be a hit, Warners held up its release until the last week in December, the final week to qualify for an Oscar nomination. Warners' strategy was designed to keep the picture fresh in everyone's mind during the evaluation and voting period for the awards. It worked; Davis won her first Academy Award for best actress. Although Davis' Oscar was generally regarded as a consolation prize for not being nominated for OF HUMAN BONDAGE, the process of fitting actor to character as determined by audience demand had been a success.[32]

Having discovered the correct role, the next step was to create a fit between a star's personal life and his or her screen persona. The intent was to convince audiences that the star "acted identically in both her 'real' and 'reel' lives."[33] Fusing actor and character was the function of a studio's publicity department. To begin the process, the department manufactured an authorized biography of the star's personal life based largely on the successful narrative roles of the star's pictures. This material was disseminated to fan magazines, newspapers, and gossip columns. The department then assigned a publicist to the star to handle interviews and to supervise the star's makeup and clothing for public appearances. Finally, the department arranged a sitting for the star with a glamor portrait photographer to create an official studio image.

During Davis' blond coquette period, from 1932 to 1934, pressbooks touted her as a sexy blond and showed photos of her wearing bathing suits, low-cut gowns, and revealing blouses. Little personal information was revealed, indicating that Warners was still searching for the right image. After OF HUMAN BONDAGE, her publicity changed. Davis was no longer displayed as a blond in come-hither poses; her hair became darker and she wore tailored suits. Authorized stories also changed. *Modern Screen*, for example, "avowed that Davis was fiery, independent, and definitely not domesticated (all qualities she displayed in her films)." *Motion Picture Classic* portrayed her as "hard-boiled and ruthless, determined to get what she wants (all traits which motivate many of Davis' actions in her vamp films)."[34] Similarly, stories transposed character relationships from her films to her personal life. For example, one article asked, "Will Bette wed George Brent?" and displayed a publicity still from JEZEBEL showing the two in an embrace. The question would remain on everyone's minds because Brent played Davis' leading man ten times.

The influence of the star system on the narrative structure of classical Hollywood cinema was profound. Classical Hollywood cinema was protagonist-centered, and studio practice dictated casting a star in the principal role. The interrelationship of the two are described by Cathy Klaprat:

> The goals and desires of the protagonist generally motivate the causal logic of the action and, consequently, the structure of the narrative, the components of which included plot, the behavior of the characters in their relationship to the

star, as well as the settings for the action. Thus, we can see that if the protagonist was constructed by the traits and actions of star differentiation, then the narrative was structured by the star.

("The Star as Market Strategy," in Tino Balio, ed., *The American Film Industry*, rev. ed. [Madison: University of Wisconsin Press, 1985], p. 369)

To illustrate this practice, take the case of JEZEBEL (1938). A vehicle designed for Bette Davis, JEZEBEL was based on the play by Owen Davis and produced to capitalize on the interest in David O. Selznick's GONE WITH THE WIND, which was in development. William Wyler directed. Like GONE WITH THE WIND, JEZEBEL takes place in the Civil War-era South and has a headstrong heroine as the protagonist. An examination of the development of the property from a play, through four screenplay drafts, to the finished film reveals how the studio tailor-made the project to fit Bette Davis' screen persona as a vamp.

Owen Davis' play is a historical drama depicting society in transition in the antebellum South; in contrast, the motion picture is essentially a character study. In the development of the shooting script, the structure of the story was changed from a late to an early point of attack that initiates the action at the moment of conflict. In essence, the studio de-emphasized the historical milieu and concentrated on the melodrama. The action has a bipartite structure built exclusively around Bette Davis' role of Julie Marsden. The first half shows us a reckless, daring, but basically sympathetic Bette Davis before her great mistake, appearing in a red dress at the Olympus ball. The second half shows us the results of that mistake; her apology to Pres (Henry Fonda) dressed in white, her revenge, and her redemption.

This structure allows us to witness character development in much the same way that the passage of time from the pre-Civil War era to Reconstruction reveals change in GONE WITH THE WIND. Take, for example, the way Julie is introduced. In the play, Julie simply walks on stage and is greeted by the other characters. In the film, Julie rides up to the plantation on a wild horse, which a young black servant has trouble controlling after she dismounts. Strutting up to the porch, she stops a moment to shout orders at the carriage driver and then whips her riding crop over her shoulder and hikes up her train before entering. We have been introduced to a confident and self-assured Julie Marsden. When she strides through the living room greeting the guests, the contrast between Julie's riding habit and her guests' more formal attire reveals her disregard for social conventions, an attitude that is upsetting to some of the women present.

Another early incident is the Olympus ball sequence. In the play, the dress incident is talked about, not enacted. Julie tells us it happened "one night" rather than at an important society event. The "wrongness" of the dress was simply a matter of personal taste—Pres simply did not like the daring low front of the dress—rather than a flaunting of social convention. And Julie's memory of the incident focuses on a personal hurt rather than a public scandal. In the film, the dress incident reveals not merely a conflict between two lovers but also a struggle between Julie and New Orleans society. Aunt Belle of the film tells us that wearing a red dress to the ball would "insult every woman on the floor." Thus, the Olympus ball sequence has taken on greater significance: it pits Davis against society in a way that is consistent with the "outcast" quality of Davis' roles in OF HUMAN BONDAGE and her other films.

Davis earns the name Jezebel by manipulating the events leading to the duel that takes Buck Cantrell's (George Brent) life. In the play, Julie has a rather passive role in the affair. Pres tells her he still loves her and even kisses her, but when they are discovered by his wife, Amy, he is embarrassed and makes his apologies, at which point Julie slaps him, an understandable reaction. In the film, the incident has been substantially changed to make Pres an innocent victim of Davis' advances. Pres makes no declaration of love to Julie, nor does he reveal any romantic feelings for her; it is Davis who takes the initiative. She dominates the conversation and taking him unawares, kisses him passionately. Pres tears himself away from her, and as he stalks out of the room, a close-up shows the anger and contempt he feels. Humiliated and rebuffed, Julie decides to get her revenge. She manipulates Buck into challenging Pres to a duel by intimating that Pres, under the influence of brandy, improperly made advances. Narrative strategies similar to these were used to construct the Davis persona in pictures such as DANGEROUS, THE LETTER, IN THIS OUR LIFE, and DECEPTION

After developing an effective narrative formula for a star, a studio would naturally want to cash in on its good fortune by repeating the formula in as many vehicles as possible. In this fashion a star could be "milked dry like a vein of gold is pinched out," said *Variety.*[35] The best performers chafed at being stereotyped. Repeating the same role deadened the spirit and prevented a talented performer from reaching his or her full artistic potential. However, studios justified this practice with the explanation that star development was risky and required an enormous amount of money. To tamper with success after having discovered an effective formula would be foolhardy. Yet, if a studio cast a star in the same role again and again, it ran the risk of satiating audience demand. The problem became how to extend the life of a star while simultaneously producing sufficient numbers of vehicles to diffuse the high salary costs.

To conserve resources, producers relied on product variation. As Darryl Zanuck put it, "there is no reason why, with proper care, a star cannot remain popular well beyond the traditional span of five years. To that end, care must be exercised in story selection. Vehicles must be varied." In practice, this meant diversifying character traits of roles "while at the same time invoking the familiar expectations associated with star differentiation"—the same-but-different principle.[36]

Warners had the most success varying Bette Davis' vamp roles by offcasting her as the good woman. The practice started in the forties. For example, THE GREAT LIE (1941) contains a triangle plot formula, but now Davis played a sympathetic woman with maternal and noble instincts. Mary Astor took the unsympathetic part, a brittle, selfish, and spoiled woman. However, publicity continued to refer to Davis as a vamp. One ad reads, "*Contrary* to the former Davis pattern, Bette Davis' new film does not find her killing anyone or acting nasty."[37]

Promotion at the production level

Nothing was too private if it interested the public, nothing was too trite if it got copy, and nothing was too exaggerated if it sold a ticket. It was this kind of material that freely found its way into any newspaper, magazine, or radio station in the country. Good, bad, or ridiculous, someone would be willing to read it, enter it, or buy it. No single audience was ever exposed to all the promotional material created for a motion picture. Publicity announcing a premiere drew an exclusive opening-night crowd. Previews of coming attractions whetted the appetites of loyal fans. Magazine articles, Sunday features, and news items stoked the interest of casual moviegoers. And, of course, movie ads in the local newspaper kept audiences in touch with the current fare. These were the common, expected ways to present the latest Hollywood feature to the audience and together with publicity stunts, trivia contests, merchandising tie-ins, and the like, they constituted motion-picture promotion.[38]

The publicity department of a major studio was organized like the city room of a newspaper. Publicity directors such as Howard Strickling at MGM, Edward Selzer at Warners, Tom Bailey at Paramount, Harry Brand at 20th Century-Fox, and Russell Birdwell at Selznick-International functioned as editors who assigned stories and reviewed finished copy before it was released. They also personally handled front-office news concerning such matters as the hiring and firing of key studio personnel, the acquiring of important properties, and the financial affairs of the company. A suicide, a messy divorce, or a scandal turned their job into public relations with the goal of protecting the image of the studio or of salvaging the reputation of a star.

Working under the publicity director were the unit reporters, who covered the big pictures, and publicists, who were assigned to individual stars. Unit reporters were almost always former newspapermen and rarely earned more than $150 a week for the usual six-day week they put in. A unit reporter prepared a synopsis of the plot and special-interest stories about the production and its stars for use by newspapers. A publicist's job was to handle the public relations of a top star and sometimes even his or her private financial affairs.

To satisfy the huge demand for information about stars, other sections of the publicity department supplied fashion layouts for fan magazines and planted tidbits with gossip columnists Louella Parsons, Sheilah Graham, and Hedda Hopper, among others. Gossip columnists typically commented on affairs of the heart. They seldom dug up information on their own but relied on news from the studios. Studios and stars were happy to cooperate since even unfavorable publicity in a gossip column kept the name of a star in print and the myth of Hollywood alive.

The still-photography department supplied the iconography for promotion. Photographs were the physical artifacts of the motion-picture experience. During each major production, a still photographer took stills of every key scene to be used for lobby displays, advertising, and poster layouts. To service pulp magazines and newspapers, the studio needed high-contrast photos in shallow focus.

For fan magazines and glossy publications, the studio needed higher-quality photos, which were the special province of glamor portrait photographers. These photographers had the task of capturing in a single image the screen persona of a star. As John Kobal put it, "they had nothing to do with making movies, but everything to do with the selling of the dream that movies meant." The most skilled of these portrait photographers—Clarence Bull at MGM, George Hurrell at Warner Bros., Ernest Bachraeh at RKO, and Eugene Robert Richee at Paramount—experimented with "backgrounds, shapes, textures, lighting, and produced a unique genre which not only served a specific function, but—unlike many of the films of the period—survives as an art form."[39] All photos were taken by an eight-by-ten-inch large-format camera. After a negative was processed, artisans retouched it to eliminate excessive weight around the waist, hips, throat, and shoulders by scraping the negative and stippling it (adding new dot patterns to the negative). They also removed lines and skin flaws from the face. After all physical imperfections were removed, the negative was airbrushed to give the face an alabaster appearance on the prints.

Glamour portraits served many purposes, but the main one was satisfying the needs of fans. Margaret Thorp observed the scene and said:

> Fan mail comes into Hollywood's studios daily by the truck load. Special clerks are needed to sort it on each lot. A conservative estimate puts the letters addressed to players at a quarter of a million a month. A top star expects about three thousand a week . . . The bulk of the mail, at least 75 per cent, is made up of requests for photographs or for some personal souvenir.
>
> (*America at the Movies* [New Haven, Conn.: Yale University Press, 1939], p. 98)

Fan magazines were the most voracious consumers of publicity. *Photoplay*, *Modern Screen*, *Silver Screen*, and other such magazines had monthly circulations of nearly a half million each. Each magazine usually contained at least one set of photos that were specially printed on gravure presses to ensure high-quality reproduction. Fans wanted color photos that were clear and suitable for framing on a bedroom wall or pasting in a scrapbook, and were willing to pay extra for such magazines.

Stories in fan magazines dealt with romance, marriage, children, divorce, and death. Adult stars such as Greta Garbo, Carole Lombard, and Errol Flynn made the best copy, along with such child stars as Shirley Temple, Deanna Durbin, Judy Garland, and Mickey Rooney. "Every aspect of life, trivial and important," said Thorp, was "bathed in the purple glow of luxury." The magazines revealed that Bing Crosby bred racehorses, that Shirley Temple had 250 dolls, and that Joan Crawford had a famous collection of sapphires and diamonds. Clothes were "endlessly pictured and described, usually with marble fountains, private swimming pools, or limousines in the background." Regardless of the luxury depicted, one purpose of these articles was to bring the image of the star down to earth: "She takes care today to make it known that she is really a person of simple wholesome tastes, submitting to elegance as part of her job but escaping from it as often as possible. It soothes the fans to believe that luxury is fundamentally a burden."[40]

Hollywood's publicity machinery was thus designed to mesh comfortably with merchandising tie-ins. If advertising attempted to associate consumer products with romance, marriage, and sexual fulfillment, it found a handmaiden in the movies. Ranked second only to food products in the amount spent on advertising, the cosmetics industry signed stars to appear in literally hundreds of thousands of ads—"ads which dutifully mentioned the star's current film"—making cosmetics synonymous with Hollywood.[41] Stars also dutifully provided testimonials for ads, hawking soap, deodorants, toothpastes, lotions, hair preparations, and other toiletries.

Fashion specialists tried to popularize the latest creations to come out of Hollywood's costume design shops. "*Vogue* printed Adrian's sketches for *Camille* (1936) and *Marie Antoinette* (1938). Luise Rainer primped for photos in a room full of spectacularly plumed hats from *The Great Ziegfeld* (1936). Marlene Dietrich appeared in layouts in splendid Russian sables and brocades designed by Travis Banton for *The Scarlet Empress* (1934)." Gimmicks such as these were supposed to generate commercial spin-offs. For example, Adrian's "little velvet hat, trimmed with ostrich feathers," which Garbo wore "tilted becomingly over one eye," created a vogue for the Empress Eugénie hat. "Universally copied in a wide price range, it influenced how women wore their hats for the rest of the decade," said Edward Maeder. Walter Plunkett's wardrobe for GONE WITH THE WIND "produced a merchandising blitz unequaled in the history of period film publicity tie-ins. Brassieres and corsets, dress patterns, hats and veils, snoods, scarves, jewelry, even wrist watches, were marketed as 'inspired' by the film."[42] No dress in the 1930s was as copied as the Scarlett O'Hara's barbecue dress.

The power of the movies to popularize fashion styles was harnessed by Bernard Waldman and his Modern Merchandising Bureau. According to Charles Eckert, he played the role of fashion middleman for most of the major studios:

> By the mid-1930s Waldman's system generally operated as follows: sketches and/or photographs of styles to be worn by specific actresses in specific films were sent from the studios to the bureau (often a year in advance of the film's release). The staff first evaluated these styles and calculated new trends. They then contracted with manufacturers to have the styles produced in time for the film's release. They next secured advertising photos and other materials which would be sent to retail shops. This ad material mentioned the film, stars, and studio as well as the theaters where the film would appear.
>
> ("The Carole Lombard in Macy's Window," p. 8)

Waldman branched out by franchising a business called Cinema Fashions Shops. By 1937, more than four hundred official Cinema Fashion Shops were in operation (only one was permitted in each city) and an additional fourteen hundred stores were handling a portion of the star-endorsed style lines. As a result of Waldman's merchandising efforts and the studios' willingness to publicize costume designs, the names of the leading studio designers, notably Adrian of MGM, Orry-Kelly of Warners, Edith Head of Paramount, and Walter Plunkett of Selznick, became as familiar to shoppers as the stars themselves.

Hollywood added a new dimension to its advertising efforts during the thirties by taking full advantage of radio. If Hollywood wanted to exploit radio stars in the movies, radio also wanted to exploit movie stars for its own purposes. By the mid thirties the two industries had developed a symbiotic relationship. The radio audience had an "insatiable interest in the stars, scripts, and formulas developed by the movies," said Michele Hilmes, but the two networks could not effectively exploit this interest until 1935, when AT&T removed the double transmission rates, which required networks to pay for the transmission of a broadcast both to and from New York. When AT&T dropped the rates, NBC and CBS built new studios in Los Angeles and "produced a veritable deluge of programming."[43]

Hollywood talent and source material was used in four types of programming. Musical variety shows such as Maxwell House Coffee's "Show Boat," "The Rudy Vallee Show," and "Kraft Music Hall" combined big names, lesser stars, and regular performers in a mix of music, comedy, dialogue, and vignettes. Dramatic series such as Campana Balm's "First Nighter" and DuPont's "The Calvacade of America" showcased top dramatic stars. "Cavalcade of America," which presented historical dramatizations, "built up a reputation for thorough and accurate research as well as dramatic appeal," said Hilmes and "attracted stage and screen actors who had formerly remained aloof." Among the stars who portrayed historical figures on the program were Clark Gable, Raymond Massey, Charles Laughton, Lionel Barrymore, Dick Powell, Tyrone Power, and Edward G. Robinson. Hollywood gossip columns hosted by Louella Parsons, Hedda Hopper, Walter Winchell, and others found an eager listening audience for their tales of Hollywood life. From the studio's perspective, radio gossip and talk shows were found to be as effective as promotion in the print media.[44]

The movie adaptation was particularly effective as a publicity device. The "Lux Radio Theater," hosted by Cecil B. DeMille, represented "the culmination of its type" and remained one of the most popular shows on the air. The show was divided into three acts with breaks for commercial messages and interviews conducted by DeMille with that evening's stars. Most adaptations were broadcast after the release of a film and served to boost theater attendance. Among the hit pictures adapted for the program were DARK VICTORY, with Bette Davis and Spencer Tracy; THE THIN MAN, with William Powell and Myrna Loy; and MR. DEEDS GOES TO TOWN, with Gary Cooper.

Promotion at the distribution level

Promotion emanating from the studios was concerned primarily with creating and maintaining star images. Promotion at the distribution level was concerned mainly with advertising and publicizing new releases. The publicity campaign for a picture was formulated in Hollywood as it went into production and was developed and executed in New York at a company's distribution headquarters prior to its release. Hollywood and New York were both concerned with promoting stars and with hyping the new release, but promotion at the distribution level was aimed at motion-picture exhibitors as well as the public.

A promotion campaign began prior to the release of a new picture and was aimed at multiple segments of the audience. Although motion pictures were theoretically designed to reach an undifferentiated mass audience, a uniform campaign would be unsuccessful. Hollywood had already determined that a gala premiere could launch a picture with the biggest bang. And it was a longtime practice to stage such galas in New York and Hollywood.

To distribute its regular class-A product, the majors used tried-and-true practices and procedures. Prior to the release of a picture, the distributor placed ads in trade papers such as *Variety* and *Motion Picture Herald* to kindle exhibitor interest. After an exhibitor booked a picture, he received the official pressbook, containing order forms for posters, lobby cards, and stills and pre-written publicity stories that he could place in the local newspaper.

Advertising reflected the exchange values studios used to differentiate their products. According to Janet Staiger, these values consisted in part of stars, genres, "realism," authenticity, and spectacle.[45] The pressbook for Warners' CAPTAIN BLOOD (1935) illustrates just how these values were incorporated into motion-picture advertising. Directed by Michael Curtiz and featuring two up-and-coming young stars, Errol Flynn and Olivia de Havilland, CAPTAIN BLOOD initiated Warners' swashbuckler series. To call attention to the spectacle in the picture, catchlines on posters read:

> See—A Whole City Built in Splender to Show You How "Blood" Razed it with Cannonfire!
>
> See—The White Slave Markets of the Caribbean Reproduced in All Their Infamy to Show You Why "Blood" Hurled Defiance at an Emperor!
>
> See—Priceless Galleons Launched and Manned to Show You How "Blood" Blew Them to Bits!

To call attention to the novelty, a publicity release related that "the entire studio found itself struck with amazement over the quaintness of the costumes." To call attention to the authenticity, a story for a Sunday feature recounted the history of piracy and discussed legendary pirate heroes. Referring to the appeal of the genre, a catchline read "Exciting as Your Childhood Dreams . . . Thrilling as the Ring of Steel on Steel . . . Romantic as Red Sails in the Sunset!"

The pressbook promoted Errol Flynn as a "carefree adventurer and a rogue to opportunity." A pre-written story entitled "Errol Flynn's Life One of Astounding Adventure" revealed that Flynn was a former Olympic boxer, a wanderer, and a British stage actor. Described as "tall and handsome, lean and brown, with a flair for romance and a craving for excitement," Flynn is quoted as saying, "I would give a leg to play [the part of Blood], but I figure I haven't a chance. I'm an unknown." As Margaret Sullivan noted, "It was the perfect statement to follow two full columns extolling his exploits. He is the young unknown fighting for success against all odds. He expects nothing to be given to him but given a chance he will prove himself."[46]

Notes

1 Cathy Klaprat, "The Star as Market Strategy: Bette Davis in Another Light," in Tino Balio, ed., *The American Film Industry*, rev. ed. (Madison: University of Wisconsin Press, 1985), pp. 351–76.

2 Ibid., p. 354.

3 Ibid., p. 375–76.

4 Leo C. Rosten, *Hollywood: The Movie Colony, the Movie Makers* (New York: Harcourt, Brace, 1941), pp. 331–32.

5 David F. Prindle, *The Politics of Glamour: Ideology and Democracy in the Screen Actor Guild* (Madison: University of Wisconsin Press, 1988), p. 24; "Actor Report to NRA," *Variety*, 18 January 1935, p. 11.

6 "100% Screen Actor Guild," *Variety*, 10 October 1933, p. 7; "Actor's Report to NRA," *Variety*, 18 January 1935, p. 11.
7 "Actor Lash Film Execs," *Variety*, 8 January 1935, p. 3.
8 M. Thompson, "Hollywood Is a Uniontown," *Nation*, 2 April 1938, p. 383.
9 "Hollywood Middleman," *Saturday Evening Post*, 27 June 1936, p. 96.
10 "Only 10 Exclusive Stars," *Variety*, 3 July 1935, p. 23; "Less Than 2,000 Players Work," *Variety*, 7 September 1938, p. 2; "Hollywood Middleman," *Saturday Evening Post*, 29 August 1936, p. 68.
11 "Less Than 2,000 Players Work," *Variety*, 7 September 1938, p. 2; Joe Bigelow, "New, Old and Out Stars," *Variety*, 23 June 1931, p. 1; "H'wood's 2d Chance Stars," *Variety*, 24 April 1935, p. 3.
12 "Thalberg Claims Longer Life," *Variety*, 22 September 1931, p. 5.
13 Rosten, *Hollywood*, p. 345.
14 Alva Johnston, "Hollywood's Ten Percenters," *Saturday Evening Post*, 8 August 1942, pp. 10, 36; "The Morris Agency," *Fortune*, September 1938, pp. 71–73; "Hollywood Middleman," *Saturday Evening Post*, 27 June 1936, p. 94.
15 "Less Than 2,000 Players Work," *Variety*, 7 September 1938, p. 2.
16 Ibid., p. 2.
17 "Hollywood Middleman," *Saturday Evening Post*, 29 August 1936, p. 17.
18 "Lively Trading Keeps Name Players on Move," *Variety*, 7 July 1937, p. 4.
19 "Warners Adamant," *Variety*, 19 April 1932, p. 3; quoted in Shipman, *The Great Movie Stars*, p. 93.
20 "James Cagney Called Bad Boy," *Variety*, 11 March 1936, p. 3.
21 "WB Statement Characterizes Cagney Decision," *Variety*, 18 March 1936, p. 2.
22 Thomas Schatz, *The Genius of the System: Hollywood Filmmaking in the Studio Era* (New York: Pantheon Books, 1988), pp. 219–20.
23 "Bette Davis in Salary Tiff with WB," *Variety*, 8 July 1936, p. 3.
24 Schatz, *The Genius of the System*, p. 220.
25 Klaprat, "The Star as Market Strategy," p. 376.
26 Schatz, *The Genius of the System*, p. 318.
27 "Star Changes up to Now," *Variety*, 18 February 1931, p. 28; "Film Names 70% Stage," *Variety*, 23 January 1934, pp. 1, 23.
28 "H'wood's Debt to Vaude," *Variety*, 24 June 1936, p. 3.
29 Hortense Powdermaker, *Hollywood, the Dream Factory* (Boston: Little, Brown, 1950), p. 246; Douglas Gomery, *The Hollywood Studio System* (New York: St. Martin's Press, 1986), p. 41.
30 "Myrt and Marge," VFR, 13 June 1933.
31 Kevin Heffernan, "Product Differentiation: Paramount's Use of Radio Talent, 1932–34" (Unpublished paper, University of Wisconsin, 1990).
32 Klaprat, "The Star as Market Strategy," p. 360.
33 Ibid., p. 360.
34 Ibid., p. 363.
35 "Funny Guys Last Longest," *Variety*, 30 June 1937, p. 5.
36 Ibid., p. 29; Klaprat, "The Star as Market Strategy," p. 372.
37 Ibid., p. 372.
38 Margaret Sullivan, "A Promotional Study of *Captain Blood*" (Unpublished paper, University of Wisconsin, 1974) and "The Art of Selling the Star System: A Study of Studio Publicity" (Unpublished paper, University of Wisconsin, 1977).
39 John Kobal, ed., *Hollywood Glamour Portraits* (New York: Dover Publications, 1976), pp. v–vi.
40 Margaret Farrand Thorp, *America at the Movies* (New Haven, Conn.: Yale University Press, 1939), p. 74.
41 Charles Eckert, "The Carole Lombard in Macy's Window," *Quarterly Review of Film Studies* 3 (Winter 1978), pp. 10–11.
42 Edward Maeder, *Hollywood and History: Costume Design in Film* (Los Angeles: Los Angeles County Museum of Art, 1987), pp. 81–82, 87, 88.
43 Michele Hilmes, *Hollywood and Broadcasting: From Radio to Cable* (Urbana: University of Illinois Press, 1990), p. 67.
44 Hilmes, *Hollywood, and Broadcasting*, pp. 63–70.
45 David Bordwell, Janet Staiger, and Kristin Thompson, *The Classical Hollywood Cinema: Film Styles and Mode of Production to 1960* (New York: Columbia University Press, 1985), p. 99.
46 Sullivan, "A Promotional Study of *Captain Blood*."

PART III

Representation, technology, production and style, 1926–46

Introduction

Focusing largely on the period between the advent of sound and the end of World War Two, this section of the Reader deals with the adoption and impact of the Production Code, sound and music, cinematography (both monochrome and colour) and innovation in the field of animation. During the course of this introduction, a summary account of the process of production will also be provided. It begins, though, with Richard Maltby's article on 'The Production Code and the Mythologies of "Pre-Code" Hollywood' (Chapter 13), not only because of the Code's industrial and institutional significance, nor only because of the ways in which its adoption and implementation have often been misunderstood or misrepresented, but also because the tenets of the Code were to mark the subject matter and the representational nature of Hollywood's films, with occasional modifications, for over thirty years. To that extent these tenets and their modes of application were as responsible for Hollywood's 'classical style' as were the principles of continuity editing and the norms of narration and narrative construction outlined by Bordwell and augmented by Keating.

As Maltby points out, a 'Code to Govern the Making of Talking, Synchronised and Silent Pictures' was adopted by the MPPDA on 31 March 1930. As he also points out, the Code 'was a more fully formulated version of a policy document regulating motion picture content, known as the "Don'ts and Be Carefuls", which was compiled in 1927 as a synthesis of the restrictions and eliminations applied by state and foreign censors', which itself drew on earlier documents, and which was essentially a means of ensuring (or of trying to ensure, for obvious commercial reasons) the unhindered circulation of its films in the US and elsewhere abroad.[1] When the 'Don'ts and Be Carefuls' were adopted, Will H. Hays, the MPPDA's president, 'established a Studio Relations Committee (SRC) in Hollywood, charged with administering its code. Under the SRC's direction, pictures were modified after production but before release in order to assuage the concerns of civic, religious or manufacturing interests, but until the 1930s its function was purely advisory. The 1930 Code required that pictures be approved by the SRC before they were released'. Maltby goes on to contest the notion that the films made in the early 1930s were laxly censored. He also contests the notion that the Code was enforced only after pressure from the Catholic Church and the appointment of Joseph Breen, a prominent Catholic layman, as head of the Production Code Administration (PCA) in 1934. Pointing out that the films made prior to Breen's appointment were as subject to the strictures of the Code as those made thereafter, he goes on to argue that there 'was no fundamental shift of Code policy in July 1934':

> The apparent changes brought about by the negotiation with the Legion of Decency were in fact mainly cosmetic ... There was a further tightening up of practice, but this had occurred on at least three other occasions since 1931 ... The differences between movies made in the early 1930s and those made later in the decade are undeniable, but the change was gradual rather than cataclysmic, the result of the development of a system that was constructed by experiment and expedient in the first half of the decade and maintained in the second.

This system was designed to facilitate the production of films 'shaped', as Ruth Vasey has put it, 'by the demands of unclassified exhibition and wide-scale distribution' and thus 'for "vertical" consumption across barriers of age, experience, and predilection in a single site of exhibition, and "horizontal" consumption across diverse geographical and cultural territories'.[2] It had two governing principles, as Maltby explains: 'One was stated in the Code itself: that "No picture shall be produced which will lower the moral standards of those who see it." Under this law a strict moral accountancy was imposed on Hollywood's plots, by which a calculus of retribution or coincidence invariably punished the guilty and declared sympathetic characters innocent'. The other 'permitted producers to deny responsibility for a movie's content, through a particular kind of ambiguity, a textual indeterminacy that shifted the responsibility for determining what the movie's content was away from the producer to the individual spectator'. In this way, 'as the Code's first administrator, Jason Joy, explained, studios had to develop a system of representational conventions "from which conclusions might be drawn by the sophisticated mind, but which would mean nothing to the unsophisticated and inexperienced"'. And in this way, the Code provided a means by which, 'as Lea Jacobs has argued ...", offensive ideas could survive at the price of an instability of meaning", allowing producers to deny responsibility for a movie's content'.

As such, the Code could generate unstable combinations of stylistic clarity (as guaranteed by the protocols of continuity editing and classical narration) and narrative opacity (as produced by omissions, ambiguities and dislocations in the depiction of events, in the relationship between events, whether on screen or off, and in the actions, reactions and motives of the characters involved). And as such it 'functioned, however perversely, as an enabling mechanism at the same time that it was a repressive one'. In inviting audience members to interpret the action on screen according to their desires, inclinations and experiences (and their levels of familiarity with Hollywood movies), in trading on opacity, ambiguity and indeterminacy at local levels at the same time as they traded on the certainty of narrative resolutions that would reassert and re-establish 'a deterministic moral order',[3] and in constructing fictional worlds that were typically conservative,[4] typically benign and typically ethnocentric but also typically and overtly unreal,[5] Hollywood's films were as capable of subtlety and sophistication as they were of simple-minded hokum – and indeed often of both at the same time.

There are a number of additional points to make here. The first is that the Code and a number of its applications were modified in the late 1930s and during and after the war.[6] The second is that the political, imperial and ethnocentric conservatism evident in most Hollywood films (particularly in the 1930s) was as much a function of the imperial and ethnocentric nature of its most important overseas markets (and of its domestic market in the Southern states) as it was of Hollywood itself.[7] And the third is that the films made under the aegis of the Code were in many respects not so very different from those made earlier on in the 1910s and the 1920s. It is perfectly true that occasional instances of nudity can be found in 1910s and 1920s films. It is also true that some films made in the late 1910s and 1920s sought to draw on the tenets of naturalism and other forms of contemporary modernism and hence sought to challenge the benign and providential tenets of nineteenth century melodrama and popular narrative comedy. But these were exceptions. Mainstream Hollywood films were designed for unclassified exhibition and widespread consumption in the 1910s and the 1920s too. Although there were concerns about the detailed depiction of criminal acts, extreme forms of physical violence were rare, swearing was rare (though prejudicial racial epithets were probably more common) and characters were never shown engaged in sexual acts other than hugging or kissing – and these were nearly always restricted to heterosexual adults of the same or of similar 'race'. Finally, in combination with styles of performance that often involved a mix of apparent over- or under-reaction, ellipsis and metaphor were probably as prevalent in the 1920s as they were in the 1930s and 1940s, especially in romantic comedies and dramas, and particularly in the 'sophisticated' and 'screwball' guises of the former.[8]

Housed on the fourth floor of the MPPDA office at North 5504 Hollywood Boulevard, the offices of the PCA were located near the studios' principal lots. As noted earlier on, these lots contained offices and canteens, outdoor sets and indoor stages, props and costumes, libraries and editing facilities, cameras, lights and other production equipment, laboratories for processing film stock, and secretarial staff and specialised production and service personnel, all of them organised into departments.[9] During the late 1910s and early 1920s, studio craftspeople became unionised, and the establishment of various labour, trade and professional associations encouraged the exchange of knowledge and information and the discussion and dissemination of practices and standards. Among the former were the International Brotherhood of Electrical Workers (IBEW) and the United Brotherhood of Carpenters and Joiners, who 'achieved control over general construction', and the International Alliance of Theatrical Stage Employees (IATSE), who 'acquired jurisdiction for the actual process of shooting films – constructing property and scenery, handling lighting, and creating miniatures'.[10] Among the latter were 'the three most important agencies that guided technological innovation: the American Society of Cinematographers [ASC] (founded 1918), the Society of Motion Picture Engineers [SMPE] (founded 1916) and the Academy of Motion Picture Arts and Sciences [AMPAS] (founded 1927)'.[11]

The SMPE drew its members from the suppliers and manufacturers of equipment such as the Moviola Company, Eastman Kodak and DuPont (who manufactured film stock), Bausch & Lomb (who manufactured lenses), Bell & Howell (who manufactured cameras, projectors, printers and perforators), the National Carbon Company (who manufactured carbon rods and arc lamps), the Mole-Richardson Company (who manufactured incandescent lights), the Mitchell Camera Corporation and the Technicolor Corporation, as well as from members of the film industry.[12] It was based exclusively in New York until 1929, when it opened an additional branch in Hollywood. Coincident with the gradual winding down of sound production in New York and the establishment of sound departments in Hollywood, the SMPE played a major role alongside AMPAS (and alongside the research departments at Western Electric and RCA) in standardising sound recording and reproduction technologies, in disseminating knowledge and information, and in encouraging innovations and improvements. As Helen Hanson and Steve Neale point out in 'Commanding the Sounds of the Universe' (Chapter 14), this process did not come to a stop after the 'transitional era' that lasted from 1926 to 1931 and that has been the focus of most accounts of sound in Hollywood in the classical era. Indeed, in 1937, Barratt Kiesling estimated that around five hundred improvements in sound recording and reproduction had been made at MGM alone. Paying detailed attention to the use of sound and to sonic motifs in films such as *I Am a Fugitive from a Chain Gang* (1932), *Top Hat* (1935) and a trio of 1940s thrillers – *Cat People* (1942), *The Seventh Victim* (1943) and *Phantom Lady* (1944) – Hanson and Neale seek to underline the numerous ways in which sound was used to provide compositional coherence, innovative local effects, and generically appropriate soundscapes, some of which, they suggest, they either shared with or borrowed from radio.

The coming of sound altered the nature, experience and organisation of musical composition, musical accompaniment, and musical production and reproduction too, and the provision of music for sound films, like the provision of sound itself, also augmented the roster of studio employees and the number and nature of studio departments. During the early years of sound, music was largely used in specific and limited ways. Continuous scores were rare. Music tended to be intermittent and 'sourced' (seen and heard to emanate from the films' fictional worlds, hence the vogue for musicals, musical settings and the appearance of on-screen musicians).[13] When musicals fell out of favour in 1930, the influx of musicians, composers and arrangers that began in the mid- to late 1920s came to a halt and many found themselves out of work.[14] However, songs remained a key component in vaudeville comedies, comic operettas and other kinds of films in 1931 and 1932, and although scores tended to remain sparse or intermittent, the extensive use of music by Mendelssohn, Verdi and Wagner in *A Farewell to Arms* and the extensive scores written by Max Steiner for *Bird of Paradise* and *The Most Dangerous Game* (all 1932) indicated that changes were underway and that the era of what would come to be known as 'the classical Hollywood film score' was relatively imminent.[15] *A Farewell to Arms* received an Oscar for 'sound recording' in 1933. The following year, *One Night of Love* (1933) was the first film to receive an Oscar for its 'music score'. The Oscar was awarded to the 'Columbia Studio Music Department, Louis Silvers head' and is thus among the first pieces of published evidence we have that music departments were established in at least some of the Hollywood studios in or by 1933. Before turning to the classical film score itself, it is worth saying something about them.

According to Roy M. Prendergast, there 'was usually a music building on the studio lot containing executive offices, cubicles for composers to work in, a music library, and a sound stage where the scores were recorded'. The departments themselves 'consisted of a compartmentalized labor force: executives and their secretaries; bookkeepers, librarians; rehearsal pianists; composers, arrangers, and orchestrators; copyists, proofreaders, and orchestral musicians'.[16] According to Mervyn Cooke: 'The organization of a typical department fell into three sub-departments, all of which were answerable to the music director. One dealt with research and preparation, including negotiating clearance for the use of pre-existing music, apportioning the relevant budgets, and maintaining libraries of sheet music and disc recordings'; another 'was responsible for production, i.e. the scheduling of recordings and the employment of all staff and performers participating in them, including not just conductors and orchestral players but also those required for rehearsal purposes, such as voice coaches'; the third 'dealt with postproduction and general scoring matters, including the preparation of cue sheets that listed music required in individual films … Both production and postproduction dealt with the studio's sound department, which was responsible for all aspects of recording, dubbing and soundtrack mixing'.[17] As Cooke goes on to point out, four to six weeks was the normal time allotted for the composition and scoring of a feature film.[18] 'Normal initial procedure was for director, music editor and composer to meet for a "spotting session" in which they agreed the quantity, location and type of musical cues for a film on the basis of a screening of its rough cut.' These cues 'were denoted by their position in each ten-minute reel of film … and composers generally wrote and recorded music on a reel-by-reel basis. Pressure of time compelled most composers to use one or more orchestrators to prepare performing parts on the basis of short-score sketches'.[19] The score itself was synchronised, conducted and recorded, usually on a reel-by-reel basis and usually in two- or three-minute sections, as a copy of the film was projected on screen on a sound stage.[20]

Scores themselves varied of course, but by the mid-1930s they were generally marked by a set of stable conventions. Along with further details about the process of production, these are explored by Kathryn Kalinak in 'The Classical Hollywood Film Score' (Chapter 15). Kalinak focuses in detail on Erich Wolfgang Korngold's score for *Captain Blood* (1935).[21] She draws particular attention to the differences as well as the similarities between scores for silent and sound films, noting the extent to which sound films 'reversed the crucial relationship in silent film between continuity and selectivity. In silent film continuous music took precedence over diegetic fidelity' (fidelity to the nature of the fictional world and its constituents). But in sound films musical accompaniment, however extensive, was nearly always intermittent, and its presence always subordinate to the need for audible dialogue. On the other hand, if the sound score 'was freed from the necessity of seamlessness', it 'inherited from the silent film its function as arbiter of narrative continuity'. Hence, as she demonstrates in detail, it was used to smooth over potentially tenuous links or potentially disruptive gaps in the narrative chain (flashbacks and transitions, dream and montage sequences, and so on and so forth). It also served to underline the tempo of narrative actions, to 'provide characterization, embody abstract ideas, externalize thought, and create mood and emotion'. And it served as well to indicate or underline the historical, geographic and generic nature of settings, to augment spectacle and the expressive possibilities of the human voice, and in all these ways to engage audiences in the films, their stories, and the feelings, fates and actions of their characters. In so doing, its medium was orchestral, its idiom was late romantic and its basic structural feature was the leitmotif.[22]

With the exception of numbers for musicals, which were usually recorded prior to shooting, the recording of scores and the mixing of scores with other elements of the sound track generally occurred toward the end of the production process, as already noted, though further changes could be made to the films, their scores and their soundtracks as a result of responses to previews.[23] The process itself began in New York, where company heads and other senior executives would meet with the head of production to draw up a schedule of A films, B films and specials for the forthcoming season and to allot budgets for each of these categories, as explained by Lewis in Part II (Chapter 9). Once back in Los Angeles, productions would be planned, budgets apportioned, schedules drawn up, and treatments and drafts of scripts would be produced. A copy of each shooting script would be sent to the PCA for initial consideration and to an assistant director, who would use it to itemise the requirements of each of the studio's departments in the form of a 'set list' (broken down into interior or exterior settings, daytime or night-time settings, and studio stage, location or back lot production sites), a 'cast list' (broken down into bit-part players, extras and principal performers), a 'costume or wardrobe plot' (a résumé of all the costumes needed for every performer) and a 'shooting breakdown' (a summarised version of the script incorporating all the essential details of each and every location, each and every back lot, and each and every studio set).[24]

These in turn would form the basis of a shooting schedule, which would be devised in consultation with the producer and which would include any plans for the filming of locations, stunts or action scenes by a second unit. The shooting schedule would be sent for costing to the production department and thence to the departments responsible for research, the casting of extras, bit-part players and stand-ins for the stars, set design, costume design, the building of models and the painting of backdrops, make-up, hair-dressing, sound, music, transportation, the managing of locations, the provision of electricity, the provision of special effects (including the simulation of rain, hail, fog and snow), and the provision of process shots, stock shots and montage sequences. With the addition of a 40 per cent overhead, these costings would be compared with the initial budget, the schedule adjusted accordingly, and a final version of the shooting script produced. Provisional or actual production dates would then be set. By the mid-1930s, most A features would be scheduled to take between six to eight weeks to shoot and a further six to eight weeks to cut and dub.[25]

Once the schedule and script had been finalised, the principal actors cast, the designs for sets and costumes agreed by the producer and director (along at this or a later stage with the assignment of a principal cinematographer and his assistants), the director would usually begin rehearsing the actors and planning the treatment of the film in detail.[26] Meanwhile, in consultation with the director, the assistant director would organise make-up and hairdressing tests for the principle performers, the fitting of costumes, the selection of bit-part players, stand-ins and extras, the provision of on-screen conveyances, livestock and props, the production of 'sides' (passages of the script containing lines of dialogue for the actors), and the assignment of a script clerk, teams of editors, sound recordists, electricians, grips and property men and other members of the production team. Production normally began early on Monday mornings (and each day thereafter), with the application of make-up, the dressing of hair, the dressing and lighting of sets, the checking and placing of cameras, microphones and other equipment, and the rehearsal and preparation of the actors and technicians for the day's first take. According to director John Cromwell and cinematographer John Arnold, three takes for each set-up and eight sets-ups per day were considered the norm (though in practice the number of takes and set-ups would often vary from shot to shot and from day to day, depending on the complexity of individual shots, the amount of 'protection' sought from re-takes and multiple set-ups, and the difficulties of filming on location).[27]

At the end of each day, the editor would dispatch the footage shot and the sound recorded that day for processing by the laboratory, and assemble the footage recorded and shot the previous day – usually known as the 'dailies' or 'rushes' – for viewing by the editor, director, producer and cameraman, and sometimes some of the principal actors. Decisions would be made about the selection of takes, the ways in which they might be edited, and the extent to which additional takes or re-takes were required. When filming was finished, the editor and his or her assistants would edit the film in consultation with the director and the producer (and sometimes the head of the studio), inserting montage sequences and stock or process shots, and coordinating the images with the sound and the music.[28] At this point, an initial print would be struck for preview and any alternations made in response. Once approved and awarded a Seal of Approval by the PCA, release prints would be struck from a final version of the negative and dispatched for distribution.[29]

As is clear from this account and the sources it draws on, the process of production was marked throughout by decisions and choices – and thus by hierarchies of knowledge, power, authority, responsibility, expertise and skill – and it is these aspects of the process that Patrick Keating explores in 'Shooting for Selznick' (Chapter 16).[30] Focusing on the role played by cinematographers in and on the films produced independently by David O. Selznick, highlighting the nature and extent of Selznick's interventions in the production process, and addressing himself to the scope for stylistic variation and choice, Keating conceptualises these issues in terms of a set of distinctions and functions: those between the craft of cinematography (a set of professional ideals and norms), the assigned task (shooting a scene, filming an actor, enhancing a mood) and what he calls 'the work as executed' (the sum of each and every concrete decision: 'where to put the lights, when to modulate their intensity, how to model their resulting shadows, and so on and so forth'). In doing so, he cites numerous examples, paying particular attention not only to Selznick's demands, but to the ways in which different cinematographers responded to the tasks they were set in accordance with their own predilections, skills and habits as well as their own conceptions of their craft.

An additional factor here was the practices, norms and standards established by the ASC, the extent to which they changed over time as new or improved cameras, lights and film stocks were introduced, the extent to which they dovetailed with generic norms (high-key lighting for comedies and musicals, 'criminal lighting' for horror films, gangster film, thrillers and Gothic romances), and the extent to which they gave rise to new stylistic trends (soft and shallow focus and lengthy takes in the late 1920s and early 1930s, deep focus, sharp focus and equally lengthy takes in the late 1930s and 1940s).[31] In the meantime, another source of visual style was colour, another factor the practices, norms and standards pursued by the Technicolor Motion Picture Corporation.

As Scott Higgins points out in 'Order and Plenitude' (Chapter 17), colour had been applied to prints of films (and occasionally used to record them) since the 1890s. Most of the techniques and processes used either involved specialised projection equipment (as well as specialised cameras and film stock) or adding colour to prints after filming. However, in 1922 Technicolor pioneered 'a subtractive color process that layered cyan and magenta images on a single piece of release stock. The result was a two-color image that could roughly approximate flesh tones and rendered other colors as various shades of blue-green and red-orange', and that could be shown using standard projectors. A second version of this process, one which enabled colour to be printed on a single side of film stock, was devised in 1928, giving rise to a flurry of Technicolor features in 1929 and 1931.[32] This was followed in 1932 by a full three-colour process, which the Corporation exploited 'by selling its services as a lab to producers', by renting out its cameras and cinematographers, and by continuing to provide a Color Advisory, giving guidance on colour design and devising a 'color chart' for each and every scene, set and character in each and every Technicolor film made between the early 1920s and the early 1950s. Focusing in detail on a number of examples, Higgins details the interrelationship between this guidance, variations and trends in Technicolor usage, variations in genre, additional technological improvements, and the work of colour-conscious producers such as Selznick and colour-conscious directors such as Vincente Minnelli and Douglas Sirk. Working in Hollywood in the 1950s as well as the 1940s, Sirk and Minnelli both witnessed the advent and adoption of Eastman Color and other colour systems, and both contributed to a trend toward bolder, 'more aggressive' colour designs.

Bold colour designs were by no means restricted to 'live-action' features. They were a hallmark of animated films in the 1930s, the 1940s and the 1950s too. Indeed, as Higgins points out, and, as is alluded to by Mark Langer in 'The Disney–Fleischer Dilemma' (Chapter 18), it was Walt Disney who pioneered a version of Technicolor's three-colour system with *Flowers and Trees* in 1932, and it was Disney who produced *Snow White and the Seven Dwarfs* (1937), which was both the first animated film of feature length and the first animated film of feature length in Technicolor. Langer places these particular innovations within three distinct but overlapping contexts: competition between Walt Disney Productions and the Fleischer Studios over the creation of successful cartoons, marketable cartoon characters and popular cartoon series; competition over the development and introduction of new animation technologies; and the need to differentiate their product. Focusing in particular on the Stereopticon Process devised at the Fleischer Studios in 1933 and the Multiplane camera devised at Disney in 1937, both of which were designed to produce a vivid impression of spatial depth, both of which were expensive and cumbersome, and both of which were either abandoned or used with decreasing frequency in the 1940s, Langer uses these and other innovations not only to exemplify his arguments, but also to question 'great man' theories of technological development and teleological models of technological progress. In doing so, he refers briefly to the introduction of a number of widescreen technologies in the late 1920s and their abandonment in the early 1930s. As we shall in the next and final part of the Reader, some of these technologies (or versions of them) were revived with great success twenty years later.

Notes

1 In *The World According to Hollywood, 1918–1939* (Exeter: University of Exeter Press, 1997), 10, 18–20, 49–99, Ruth Vasey stresses the importance of foreign markets, the extent to which silent films were routinely shortened or censored in overseas countries, and the extent to which sound films were neither as inherently malleable nor as easily altered. The Production Code was thus as much a product of sound as it was of wider factors.

2 Vasey, *The World According to Hollywood*, 225.

3 Vasey, *The World According to Hollywood*, 3.

4 With the exception of Hollywood's wartime resistance films (see note 5), acts of justifiable political rebellion were usually set in the past, though it should be pointed out that Hollywood's films nearly always represented contemporary prison regimes as brutal and unjust and rebellious prisoners and escapees as sympathetic.

5 This is nowhere more apparent than in the indeterminate 'mythical kingdoms' that stand in for specific foreign countries in many Hollywood films in the 1930s and that function as a means of avoiding protest from foreign governments over the (mis)representation of their governments and peoples. See Vasey, *The World According to Hollywood*, esp. 210–24. It is worth noting here that when Hollywood began producing films about its allies and about the resistance to Nazi occupation in a number of European countries in World War Two, traces of mythical kingdom iconography were often combined with explicit national stereotypes in the unreal settings provided by studios and studio backlots. Wherever they were set, such stereotypes pervaded Hollywood's espionage and combat films as well, where they were often accompanied by a positive and carefully chosen mix of domestic ethnic stereotypes. See in particular, Jeanine Basinger, *The World War II Combat Film: Anatomy of a Genre* (New York: Columbia University Press, 1986); Koppes and Black, *Hollywood Goes to War*; and Schatz, *Boom and Bust*, 239–57. It is also worth noting that on occasion, as in *Fury* (1937), knowledge of the Code's strictures about controversy and race could enable audiences to interpret a film about white characters as a film about the lynching of African-Americans. In this context, it is perhaps unsurprising that those films that sought to address racial prejudice in a more overt manner, or that simply sought to be more liberal, tended to do so either by adopting the trope of passing or by including African-American extras and bit-part players in scenes in an otherwise unmotivated way. Examples of the former include *Show Boat* (1936 and 1951), *Imitation of Life* (1934 and 1959), *Gentleman's Agreement* (1947) and *Pinky* (1949). In addition to *Fury*, examples of the latter include *The Best Years of Our Lives* (1946) and *Berlin Express* (1948). Unlike the performances by African-Americans in 1940s musicals, which tended to be clustered in segments that could be cut for distribution in the Southern States, the scenes involving African-American bit-part players and extras were usually not so easily removable.

6 The history of the Code in the postwar era is discussed later on in this Reader (Chapter 13). For further discussion of the Code in the late 1930s and 1940s, see Clayton R. Koppes 'Regulating the Screen: The Office of War Information and the Production Code Administration', in Schatz, *Boom and Bust*, 262–81, and Vasey, *The World According to Hollywood*, 219–24. For specific discussion of the Code and its effects on the representation of physical violence, see Stephen Prince, *Classical Film Violence: Designing and Regulating Brutality in Hollywood Cinema, 1930–1968* (New Brunswick, NJ: Rutgers University Press, 2003).

7 Vasey, *The World According to Hollywood*, 137–40, 148–52.

8 See Jacobs, *The Decline of Sentiment*, 79–126, for a discussion of 'Sophisticated Comedy' in the 1920s. See also her discussion of 'The Seduction Plot' and 'The Romantic Drama', 180–215 and 217–73, respectively.

9 It is worth noting here the MGM lot was 117 acres in size and included 23 sound stages, standing exterior sets including a small lake and harbour, a park, a miniature jungle and streets of houses. Out of a total of 4,000 studio employees in 1934, there were 61 stars and featured players, 17 directors and 51 writers under contract. By comparison, Fox had a 260 acre lot and Columbia a single block with only 3 sound stages.

10 Bordwell, Staiger and Thompson, *The Classical Hollywood Cinema*, 312.

11 Bordwell, Staiger and Thompson, *The Classical Hollywood Cinema*, 252.

12 Bordwell, Staiger and Thompson, *The Classical Hollywood Cinema*, 255–58.

13 William Darby and Jack Du Bois, *American Film Music: Major Composers, Techniques, Trends, 1915–1990* (Jefferson, NC: McFarland, 1990), 7–14. According to Cooke, *A History of Film Music*, 51, early exceptions to the vogue for sourced and intermittent music in sound films in the US included the continuous scores for *Lights of New York* and *The Terror* (both 1928).

14 Prendergast, *Film Music*, 24–25.

15 Darby and Du Bois, *American Film Music*, 13–14, 18–23; Max Steiner, 'Scoring the Film', in Naumburg (ed.), *We Make the Movies*, 219–20.

16 Prendergast, *Film Music*, 36.

17 Cooke, *A History of Film Music*, 71.

18 Cooke, *A History of Film Music*, 73.

19 Cooke, *A History of Film Music*, 74.

20 Steiner, 'Scoring the Film', 226–29.

21 Korngold was one of a number of émigré composers from Europe, many of them Jewish refugees. In addition to Max Steiner, others included Bronislau Kaper, Miklós Rósza, Dmitri Tiomkin and Franz Waxman. For detailed discussion of their work, and of the work of other composers in the classical era, see Darby and Du Bois, *American Film Music*, 15–375.

22 For further discussion of the principal idioms, practices and composers of music for Hollywood films, see George Burt, *The Art of Film Music* (Boston: Northeastern University Press, 1994). For further discussion of individual scores in the 1930s and 1940s, see David Cooper, *Bernard Herrmann's The Ghost and Mrs Muir: A Film Score Handbook* (Metuchen, NJ: Scarecrow Press, 2005); Kate Daubney, *Max Steiner's Now, Voyager: A Film Score Guide* (Westport, CT: Greenwood Press, 2000); 'Adolph Deutsch on *The Three Strangers* (1946)', in Mervyn Cooke (ed.), *The Hollywood Music Reader* (Oxford: Oxford University Press, 2010); Peter Larsen, *Film Music* (London: Reaktion Books, 2005), 98–122; 'Frederick W. Sternfeld on Hugo Friedhofer's *The Best Years of Our Lives* (1947)', in Cooke (ed.), *The Hollywood Music Reader*, 301–16; Ben Winters, *Erich Wolfgang Korngold's The Adventures of Robin Hood: A Film Score Handbook* (Metuchen, NJ: Scarecrow Press, 2007). For discussion of music in animated shorts, see Scott Bradley, 'Personality on the Soundtrack' and 'Conversations with Carl W. Stalling', in Cooke (ed.), *The Hollywood Music Reader*, 101–06 and 107–16. The involvement of directors in the choice and use of music in their films was extremely rare. For what may have been a unique exception, see Kathryn Kalinak, *How the West Was Sung: Music in the Westerns of John Ford* (Berkeley: University of California Press, 2007).

23 The process of pre-recording scores for musicals is discussed in Steiner, 'Scoring the Film', the previewing of films by Jesse L. Lasky, 'The Producer Makes a Plan', in Naumburg (ed.), *We Make the Movies*, 232–35 and 14–15, respectively.

24 Robert Edward Lee, 'On the Spot', in Naumburg (ed.), *We Make the Movies*, 91–95. Most of studios also owned ranches. These ranches housed livestock and more permanent or extensive outdoor sets. They also served as outdoor locations.

25 John Arnold, 'Shooting the Movies', Clem Beauchamp, 'The Production Takes Shape', John Cromwell 'The Voice Behind the Megaphone', and Lee, 'On the Spot', in Naumburg (ed.), *We Make the Movies*, 170–71, 64–79, 62, and 99 respectively. Beauchamp notes, 68, that the 'average studio' allotted 'approximately one hundred thousand feet of negative, sixty thousand feet of positive, and a like amount of sound track for each picture'. Shooting schedules on Poverty Row were much shorter. At Republic they were usually about two weeks, at Monogram about eight days. See Davis, *The Glamour Factory*, 12.

26 Freidman, 'The Players Are Cast', and Hans Dreier, 'Designing the Sets', in Naumberg (ed.), *We Make the Movies*, 106–16 and 80–89, respectively. For more on art department and set design, see Heisner, *Hollywood Art*, 25–303. For an account of screenwriters and screenwriting in the 1930s and 1940s, see Stempel, *Framework*, 70–135. Stempel has a number of interesting observations to make about studio styles and the influence of central producers. For example, he notes that Thalberg at MGM 'eventually worked his way toward ... a style that was in practice uneven and often disjointed. Partially this came out of his determination to find scenes that worked, that played. Sometimes there are connections between scenes and sometimes there are not. Often moments built up to either never appear or else go by very quickly'. By contrast the screenwriting style at Twentieth Century-Fox under Darryl Zanuck was focused on the narrative line: 'Zanuck would cut away the fat to the narrative bone,

leaving out the emotional moments Thalberg made the heart of the film'. See *Framework*, 75 and 78.

27 John Arnold, 'Shooting the Movies', and John Cromwell, 'The Voice Behind the Megaphone', in Naumburg (ed.), *We Make the Movies*, 168–69 and 61–62, respectively.

28 Annie Bauchens, 'Cutting the Film', and Cromwell, 'The Voice Behind the Megaphone', in Naumburg (ed.), *We Make the Movies*, 199–215 and 62, respectively. Most Hollywood directors did not have the contractual right to a final cut in the 1930s and 1940s, and their involvement in editing and post-production clearly varied. The role played by editors and the nature of decisions they made have rarely been discussed in any detail. Most discussions of editing in classical Hollywood films assume it to be either merely a matter of ensuring adherence to the tenets of classical continuity or else a manifestation of directorial style. For at least some exceptions, see the discussions of Clarence Kolster's editing style at Tiffany and Universal in the early years of sound, Dorothy Spencer's editing of the scene in which Ringo (John Wayne) proposes to Dallas (Claire Trevor) in *Stagecoach* (1939) and Margaret Booth's editing of the wedding scene in *Camille* (1936), in Donald Fairservice, *Film Editing: History, Theory and Practice* (Manchester: Manchester University Press, 2001), 261, 269–73 and 274–76, respectively. See also Edward Dmytryk, *On Film Editing* (London: Focal Press, 1984).

29 For detailed accounts of specific productions, see Lutz Bacher, *Max Ophuls in the Hollywood Studios* (New Brunswick, NJ: Rutgers University Press, 1996); Peter Baxter, *Just Watch! Sternberg, Paramount and America* (London: BFI, 1993), 15–84; Rudy Behlmer, *America's Favorite Movies: Behind the Scenes* (New York: Ungar, 1982); Robert Carringer, *The Making of Citizen Kane* (Berkeley: University of California Press, 1996 edition); Aljean Harmetz, *The Wizard of Oz* (New York: Limelight, 1977) and *Roundup the Usual Suspects, The Making of Casablanca: Bogart, Bergman, and World War II* (New York: Hyperion, 1992); and Vertrees, *Selznick's Vision*. See also the introductions to each of the volumes in the Wisconsin/Warner Bros. Screenplay Series edited by Tino Balio. For lists of the principal stars and departmental members in each of the principal Hollywood companies, see Joel W. Finler, *The Hollywood Story* (London: Octopus, 1988), 256–75.

30 Selznick's productions were based and filmed at a studio lot on Washington Boulevard in Culver City. Samuel Goldwyn's productions (and those of a number of other independents) were filmed at the Pickford–Fairbanks Studios (otherwise known as 'The Lot'), which was located at the corner of Formosa Avenue and Santa Monica Boulevard. The production process itself was more or less identical to that of the larger production companies, though the number of films being planned or shot at any one time would have been far fewer and, as is evident in Keating's account, Selznick, along with Goldwyn and most of the other independent producers, would have overseen and supervised the production of the films themselves. For more on Goldwyn, see A. Scott Berg, *Goldwyn: A Biography* (New York: Knopf, 1989).

31 Bordwell, Staiger and Thompson, *The Classical Hollywood Cinema*, 267–93, 306–8, 341–52; Keating, *Hollywood Lighting*, 107–97, 222–66; Neale, *Genre and Hollywood*, 170–73 (London: Routledge, 2000); Salt, *Film Style and Technology*, 184–86, 196–97, 201–08, 214–15, 225–27, 229–40.

32 For further details on colour and colour films in the 1920s and early 1930s, see Hall and Neale, *Epics, Spectacles and Blockbusters*, 62–68, and Koszarski, *An Evening's Entertainment*, 127–30.

Chapter 13

Richard Maltby

THE PRODUCTION CODE AND THE MYTHOLOGIES OF 'PRE-CODE' HOLLYWOOD

MOST PEOPLE KNOW TWO THINGS ABOUT the Hays Code. One is that the bedrooms of all married couples could contain only twin beds, which had to be at least 27 inches apart. The other is that although the Code was written in 1930, it was not enforced until 1934, and that as a result, the 'pre-Code cinema' of the early 1930s violated its rules with impunity in a series of 'wildly unconventional films' that were 'more unbridled, salacious, subversive, and just plain bizarre' than in any other period of Hollywood's history.[1]

Neither of these things is true.

The 'Code to Govern the Making of Talking, Synchronized and Silent Motion Pictures' adopted by the Board of Directors of the Motion Picture Producers and Distributors of America, Inc. (MPPDA), in March 1930 contained a set of 'General Principles' and a list of 'Particular Applications'. In the section on 'Locations', it observed that: 'The treatment of bedrooms must be governed by good taste and delicacy'. Such statements obviously themselves required interpretation, and much of the work of the Code's administrators, before and after 1934, was a matter of negotiating the application of the Code's generalised statements to particular instances. In developing the detailed operation of the Code, its administrators relied heavily on their knowledge of the practices of the various national, state and municipal censor boards that regulated two-thirds of Hollywood's American market and almost every major foreign country. Hollywood's married movie stars slept in single beds to meet a requirement of the British Board of Film Censors.

The Code is remembered with nostalgic contempt for the trivia of its requirements, and it has often been blamed for Hollywood's lack of realism and political timidity. These charges both overestimate and underestimate its influence. Hollywood was, is, and always will be a cinema censored by its markets and by the corporate powers that control those markets. The Code contributed significantly to Hollywood's avoidance of contentious subject matter (which included miscegenation as well as other forms of sexual behaviour; revolutionary political beliefs as well as defamatory representations of domestic and foreign governments), but it did so as the instrument of an agreed industry-wide policy, not as the originating source of that policy. Within its own sphere of influence, however, the Code was a determining force on the construction of narrative and the delineation of character in every studio-produced film after 1931. Public arguments about the Code's application – over Clark Gable's last line in *Gone with the Wind* (1939), for example – have themselves tended to be over trivia, and have supported claims that the Code was a trivialising document. The agreements that underlay the Code have received much less attention. But they amounted to a consensus between the industry's corporations and legislative and civic authorities over what constituted appropriate entertainment for mass audiences in America and, by default, the rest of the world.

Hollywood's 'self-regulation' was not primarily about controlling the content of movies at the level of forbidden words or actions, or inhibiting the freedom of expression of individual producers. The cultural anxieties that brought the Code into being addressed more fundamental social issues than a few bawdy Mae West jokes, the length of a hemline, or the condoning of sin in an 'unmoral' ending. Rather, they concerned the cultural function of entertainment, and possession of cultural power and the policing of the ideological apparatus of representation that was Hollywood. The Production Code was a sign of Classical Hollywood's cultural centrality, when its familiar Utopias were such common currency that the nation's guardians insisted on judging movies by the content of their characters and according to a standardised morality. Its history is a history of the attempts by cultural élites to exercise a controlling surveillance over the mass culture of industrial capitalism.

From 1907, when the first motion picture censorship ordinance was introduced in Chicago, public debate revolved around concerns about the effects of entertainment on children and young adults, and the question of whether the motion picture industry was morally fit to control the manufacture of its own products or whether this was more appropriately undertaken by state or federal agencies. The middle-class possessors of cultural power have consistently demonstrated a concern that forms of entertainment aimed primarily at working class audiences should be regulated so that they support, through their representations, the existing hegemony. This concern has been expressed primarily as an anxiety about the effects of such entertainment on children, and specifically on the learned behaviour of adolescents. The anxiety has been that adolescent males will learn criminal behaviour and adolescent females sexually permissive behaviour from various forms of commercial entertainment. Behind this bourgeois concern about their offspring has been a deeper class-based anxiety about the extent to which the sites of entertainment provided opportunities for the heterogeneous mixing of classes. These anxieties have appeared throughout the century, regularly looking, as it were, for a new form of representation to fix upon. Their terms of expression, however, remain remarkably fixed, and largely impervious to evidence. The debate over censorship, then, is a debate over social control, and as such it is seldom to be explained by the particular content or structure of the entertainment form that provokes any given instance of it. At the root of this debate is the persistent conflict inherent in popular culture, which is a hegemonic institution deriving its regenerative energies from the cultural expressions of the under classes, and seeking to sanitise and regulate those expressions through the imposition of convention.

The Supreme Court declared the censorship of motion pictures to be constitutional in 1915, and by 1921 seven of the 48 states had implemented systems of state censorship. In addition, municipal authorities from Vermont villages to Chicago operated local censorship boards, usually administered by the police with varying degrees of rigor, predictability and frequency. More than 60% of domestic sales, together with virtually the entire foreign market, were made in territories under 'political censorship'. State censorship boards exerted an essentially negative power to cut or ban films; as both the industry and its critics acknowledged, they could not positively influence the quality of Hollywood production. By its nature, 'political censorship' was concerned with what happened at the site of exhibition, not the site of production. Exhibitors often readily acquiesced in the practice of censorship as a means of ensuring that movies offended as few of their community's cultural and legislative leadership as possible, in order that their commercial operations might continue unhindered.

Exhibitor organisations commonly supported the establishment of censor boards as the best means of protecting their commercial interests.[2]

With the prominent Republican politician Will H. Hays as its first President, the MPPDA was established in 1922 to safeguard the political interests of the industry's emerging production-distribution oligopoly. While censorship and the regulation of content was an important aspect of the MPPDA's work, its central concern was with managing the industry's public relations and with containing the threat of legislation or court action that would impose a strict application of the anti-trust laws on the industry. The Association pursued policies of industrial self-regulation not only in regard to film content but also in matters of arbitration and intra-industry relations and in its negotiations with government, attaching the movies to the 'associative state' fostered by Herbert Hoover's Department of Commerce.

Hays presented the MPPDA as an innovative trade association at the forefront of corporate organisational development, largely responsible for the industry's maturation into respectability, standardising trade practices and stabilising the relationship between distributors and small exhibitors through Film Boards of Trade, arbitration, and the Standard Exhibition Contract. The establishment of 'the highest possible moral and artistic standards of motion picture production' was in one sense simply an extension of this practice, but it also implicitly accepted that 'pure' entertainment – amusement that was not harmful to its consumer – was a commodity comparable to the pure meat guaranteed by the Food and Drug Administration. For all industry parties, issues of oligopoly control and trade practice were much more important than censorship. But questions of censorship were of greater public interest, and could also be resolved at less economic risk to the majors. These factors encouraged the MPPDA to displace disputes over the industry's distribution of profits onto another arena quite literally – from economic base to the ideological superstructure of movie content.[3]

Although the Association succeeded in preventing the spread of state censorship after 1922, its attempts to abolish existing boards failed, so that its mechanisms for self-regulation comprised an additional, rather than a replacement, structure. Hays sought to persuade his employers that they could not 'ignore the classes that write, talk, and legislate': their movies had not merely to provide a satisfactory level of entertainment for their diverse audiences, but also to offend as small a proportion of the country's cultural and legislative leadership as possible. His public relations policy affiliated the MPPDA with nationally federated civic and religious organisations, women's clubs and parent–teacher associations, aiming to make 'this important portion of public opinion a friendly rather than a hostile critic of pictures', and contain the legislative threat posed by their political lobbying power. To establish self-regulation as a form of industrial self-determination, the industry had to demonstrate that, as Hays put it, 'the quality of our pictures is such that no reasonable person can claim any need of censorship'. In part he achieved this by conceding that there was no dispute over the need to regulate entertainment or over the standards by which it should be regulated, only over who possessed the appropriate authority to police the ideological apparatus of representation.

The 'Code to Govern the Making of Talking, Synchronized and Silent Motion Pictures' adopted by the MPPDA in March 1930 was a more fully formulated version of a policy document regulating motion picture content, known as the 'Don'ts and Be Carefuls', which was compiled in 1927 as a synthesis of the restrictions and eliminations commonly applied by state and foreign censors. The 'Don'ts and Be Carefuls' was itself an elaboration of previous documents, including the Standards of the National Board of Review, which dated from 1909, and the 'Thirteen Points' of the MPPDA's precursor trade association the National Association of the Motion Picture Industry, written in 1921. When the MPPDA adopted the 'Don'ts and Be Carefuls', Hays established a Studio Relations Committee (SRC) in Hollywood, charged with administering its code. Under the SRC's direction, pictures were modified after production but before release in order to assuage the concerns of civic, religious, or manufacturing interests, but until 1930, its function was only advisory. The 1930 Code required that pictures be approved by the SRC before they were released.

Recent suggestions that the Code existed on paper but not in practice between 1930 and 1934 are as mythological as most stories about the Code, but they do tell us a peculiar truth about Hollywood history. In Hollywood, history is first of all a production value; its first obligation is to be entertaining, not accurate. Will Hays was fond of declaring that 'No story ever written for the screen is as dramatic as the story of the screen itself', and most of what passes for film history continues to be written under the curious expectation that the history of entertainment must itself be entertaining. Hollywood's history is too often written as a melodrama of rags to riches to heartbreak, or of creative virtue triumphing over the moustachioed villains of corporate capitalism and the moral repressions of conspiratorial reactionaries.

There are two versions of the 'pre-Code cinema' myth. They tell the same story but interpret it very differently. One held sway for about as long as the Production Code itself did, from the mid-1930s until the mid-to-late 1960s. Like the Code this 'official' history served the industry's interests. According to this version, Hollywood was established by immigrants untutored in the finer manners of corporate capitalism and occasionally in need of reminding about their civic responsibilities. During the 1920s, Hays worked with civic and religious groups to improve their opinion of the movies, a policy that culminated in the writing of the Production Code in 1930. But as every Hollywood melodrama requires, a misfortune – the Depression – intervened. Needing to maintain income in the face of declining audiences, producers returned to their old sinful ways, exploiting their audiences' baser instincts with a flood of sexually suggestive and violent films. Without adequate powers to enforce the Code, the MPPDA was unable to prevent this, and the crisis was only averted by the Catholic Church, which established the Legion of Decency in April 1934 and threatened to boycott Hollywood. Almost immediately, the producers surrendered, agreeing to a strict enforcement of the Code under the administration of prominent Catholic layman Joseph Breen.

After the Code itself had been abandoned in 1968, a second version of this history came to predominate. The events in this second version were the same as in the official history, but their values were inverted, most evocatively in historian Robert Sklar's description of the early 1930s as Hollywood's 'Golden Age of Turbulence'. Instead of Hollywood the fallen woman being rescued from sin and federal censorship by virtuous hero Joe Breen riding at the head of the Legion of Decency, Sklar argued that:

> In the first half decade of the Great Depression, Hollywood's movie-makers perpetrated one of the most remarkable challenges to traditional values in the history of mass commercial entertainment. The movies called into question sexual propriety, social decorum and the institutions of law and order.[4]

This was an extraordinary claim. Why would an industry that claimed to be the fourth largest capitalist enterprise in the United States, intricately linked to Wall St. finance capital, produce 'one of the most remarkable challenges to traditional values in the history of mass commercial entertainment' at the very moment of perhaps the greatest social and political instability the US had experienced? Such an improbable account of the industry's activities can gain credence only because it provides a version of history that many of Hollywood's critics are eager to accept. The idea of a 'Pre-Code cinema' conforms to the need to situate Hollywood within a critical melodrama of daring creative heroes and reactionary villains, because the only version of Hollywood its critics can truly love is an 'un-American' anti-Hollywood, populated by rebel creators challenging and subverting the industrial system. Ironically, much contemporary criticism has been as concerned to investigate the subversive potential of Hollywood cinema as were the anti-Communists of the 1950s, and although they come to praise the authors' subversion, not to incarcerate them, their critical methodologies are strikingly similar.

Robert Sklar was writing in the early 1970s, when conventional wisdom suggested that few written records had escaped the studio shredder. Within a decade, however, film scholars gained access to several major archives containing a surfeit of documents detailing the bureaucratic operations of the Dream Factory. The Production Code Administration (PCA) Archive is one of the richest of these sources, describing the negotiations between PCA officials and the studios, movie by movie, script draft by script draft. In complete contradiction to the mythology of the Code's not functioning during the early 1930s, its records reveal that this period actually saw by far the most interesting negotiations between the studios and the Code administrators over the nature of movie content, as the Code was implemented with increasing efficiency and strictness after 1930. Throughout the period, movie content was changed – sometimes fundamentally – to conform to the Code's evolving case law.

Amongst what one Code Administrator called the 'picayunish' details of censorship, the PCA files also contain evidence of the rationale for the industry's policy of self-regulation, and they make it clear that the institution of censorship in Hollywood was not primarily about controlling the content of movies at the level of forbidden words or actions, or inhibiting the freedom of expression of individual producers. What the Code was, what it did and why it came into existence can only be partially understood by thinking of it in terms of its effects on production, or indeed by assuming that the social crisis over cinematic representation in the early 1930s was caused by the content of motion pictures. The Production Code was a corporate statement of policy about the appropriate content of entertainment cinema, that acknowledged the possible influence of movies on the morals and conduct of those who saw them, and accepted, however reluctantly, the industry's responsibility for the moral well-being of its audiences. It was primarily concerned with the relationship between Hollywood and its audience rather than with content, and in defining that relationship as one of acknowledged responsibility, the Production Code was the industry's most important documentary statement that the motion picture 'art industry' was inherently different from the book, magazine or theatre business, in that producers were constrained by their obligations to the moral protection of their audiences.

But if the Production Code itself was a statement of principle, the implementation of that principle was a complex practical matter, involving extensive negotiation over procedure and detail. In any form of censorship there is an inevitably large gap between general statement and practical application; the language of general statements is invariably vague, and for good reason. If censorship is to be at all a sensitive instrument of social or hegemonic control, it must be constantly adjusted to the nuances of the immediate situation. The general principles that govern its operation must, therefore, be sufficiently vague in their expression to permit variation in their application. In a fashion industry, such as the movies, the application of censorship is, like all other considerations, a matter of fashion: hemlines go up and down, but the movement of a hemline does not require the rewriting of a code of conduct. The operation of such procedures has an unfortunate effect on the public reputation of those involved in administering them: the negotiations are, indeed, nearly always over relative trivia. Censorship therefore appears always to be concerned with the trivial details of hemlines, and thus is constantly subject to ridicule: a hemline six inches above the knee tends to deprave and corrupt, but one five inches above does not. The censor's lot is not a happy one.

In the 1990s, a number of authoritative books – Lea Jacobs' *The Wages of Sin*, Tino Balio's *Grand Design: Hollywood as a Modern Business Enterprise*, Ruth Vasey's *The World According to Hollywood* – established quite unequivocally that the old account of the Production Code's history had to be discarded, since it is demonstrably incorrect to suggest that movies made between 1930 and 1934 were 'uncensored'.[5] Individual recommendations might be disputed, often in hyperbolic language, but the Code's role in the production process was not a matter of contention, and studio personnel did not resist its implementation. Instead, this period saw the more gradual, more complex and less melodramatic evolution of systems of convention in representation. But the Frankenstein monster of Pre-Code cinema continues to rise from its grave.

There are two explanations for the persistence of this myth. One has to do with the commercial value in repackaging material from Hollywood's past to suit present entertainment needs. In its new guise, 'Pre-Code cinema' has been re-invented as a critics' genre, much like '*film noir*' or 'melodrama', with no roots in industry practice. The growth of satellite and cable television stations such as Turner Classic Movies has provided new outlets for the circulation of movies previously almost unseen since their initial release. The promiscuous scheduling policies of these stations encourage the construction of vague generic systems of classification, around which seasons can be thematically strung. Because the industry suppressed the circulation of some early 1930s movies after 1934 in conformity with the 'official' history of the Code's implementation, many movies from this period have only recently surfaced from obscurity. Sklar's account of the Golden Age of Turbulence relied on an analysis of about 25 movies, or approximately one percent of Hollywood's total output of feature pictures between 1930 and 1934. The critical canon of 'pre-Code cinema' to be found in the schedules of American Movie Classics, the virtual pages of Reel.com and the plot synopses of several recent books is now perhaps ten times that size. Regardless of these movies' cultural status at the time of their initial release, they continue to be critically configured as a 'Forbidden Hollywood', a subversive body of work that represents, as one book puts it, a 'road not taken' by later Classical Hollywood.[6]

By comparison to the commercial investment in conceptions of genre by the audio-visual, broadcasting and publishing industries, there is no investment in the veracities of a more complex historical narrative. Instead, just as some stars and some movies have acquired a cult critical status unrelated to their box-office earnings, some periods of Hollywood's history – the late 1940s, 1968–74, as well as the early 1930s – have come to be understood as more reflective of their cultural moment than others. Curiously, these tend to be periods of economic uncertainty, declining audiences and 'turbulence' in Hollywood's conventions of representation. On the face of it, it is difficult to see why movies produced during such periods should be regarded as more *zeitgeistig* than those produced in periods of larger, more stable audiences and under more secure representational regimes, but these periods attract critics because they are often seen as giving rise to new forms, like *film noir*, that have an obvious appeal as objects of critical study. In what almost inevitably become self-fulfilling analyses, these periods of turbulence come to be perceived as doubly rewarding for study, since their innovative movies were also *zeitgeistig* – one historian describes *I Am a Fugitive from a Chain Gang* as 'the perfect expression of the national mood in 1932: despair, suffering, hopelessness . . . The film *was* 1932: hopelessness'.[7]

These accounts propose an odd, hybrid history, suggesting that movie genres have their own internal histories of textual relations and stylistic influence, but that at the same time they also *reflect* the social and political history of their moment. Early 1930s gangster movies, for example, are deemed to tell us something about early Depression America although critics disagree about whether audiences were meant to see gangsters as rebellious social bandits or as monstrous emblems of the rapacious capitalism that had caused the Depression. Such interpretations serve as a means to keep these old movies relevant for new audiences, but their bowdlerisation of history and their romantic refusal to recognise that the motion picture industry operates like any other major capitalist enterprise ensures that most

Hollywood history remains no more than a form of entertainment. Even as they are incorporated into a larger version of social history, the movies remain under the obligation to entertain: in history textbooks describing the Depression, accounts of *I Am a Fugitive from a Chain Gang* or *Scarface* appear as diverting boxed features on 'social realism in the movies', alleviating the tedium of unemployment statistics and banking reform.

The early 1930s is, indeed, one of Hollywood's Golden Ages of Turbulence, like the early 1970s and the early 1990s, when a combination of economic conditions and technological developments destabilised the established patterns of audience preference, creating opportunities for greater experimentation and variation from Hollywood's established norms. This variation, however, occurred within strict limits and existed, in large part, to test, negotiate and reconfigure the boundaries of Hollywood's conventions. The two principal factors that brought this situation into being were the revolution in content, source material and mode of production brought about by the adoption of sound technology after 1928, and the economic collapse of the leisure market after the 1929 Wall Street Crash.

In the late 1920s, as sound cinema spread out of the cities into neighbourhood and rural cinemas, sections of the American middle class grew increasingly vocal in their reaction against what they saw as the moral excesses of the post-war decade. The spread of sound seemed to confirm the widespread conviction that movies were a major source of influence on the behaviour, attitudes and morals of their audience, particularly the young and uneducated. The movies' relatively permissive representations of sex and violence became one of the sites at which an increasingly insecure Protestant provincial middle class felt its cultural hegemony, its command of public life was threatened by the incursions of a modernist, metropolitan culture – a largely Jewish and Catholic culture – which the provincials regarded as alien. Throughout the 1920s Broadway had been castigated for its 'realism' and 'sophistication', particularly in its representation of sexual mores and improprieties. With the coming of sound, and Hollywood's increasing adaptation of Broadway plays, provincial morality perceived that the threat had moved much closer to home. Broadway's dubious dialogue and 'sophisticated' plot material was now playing on Main Street for the children to see and hear.

The industry's financial crisis drove it to concentrate on making product for its most profitable market, the young urban audiences in the first-run theatres owned by the major companies. Complaints about the shortage of movies suitable for children or the overproduction of 'sophisticated' material unacceptable to small-town audiences were a form of market response to the shortage of appropriate content for other sectors of the audience, but they were most often couched in moralistic terms, and attached to demands for federal censorship.

There is little evidence that there was any widespread concern *among moviegoers* about the moral quality of the entertainment they consumed in the early 1930s. There is, however, a good deal of evidence of concern about *moviegoing* in the period, although the groups and people most vociferously complaining about the moral viciousness of Hollywood were not themselves part of the audience. Contrary to the mythology of 'Pre-Code cinema', the early 1930s was in fact a period of increasing moral conservatism in American culture, in which the movie industry, along with other institutions of representation, failed to keep pace with a growing demand for a 'return to decency' in American life. The protests about movies by women's organisations and Parent–Teacher Associations was a moral panic expressing class and cultural anxieties at a time of social, economic and political uncertainty; movie content was the site of this moral panic, rather than the cause of it.

To a large extent, the movies were indicted by a failing Protestant culture for the cultural changes of the 1920s. As the movies had been a prominent success in the 20s, they were a prominent target of the general questioning of business morality that followed the Crash. The

Protestant attacks on the industry combined a critique of the movies' moral content with an even more savage critique of the immorality of the industry's business methods – the distribution contracts that underpinned the majors' control of both production and exhibition and forced immoral product on independent neighbourhood exhibitors obliged to show 'sex-smut' regardless of his or his community's preferences.

Faced with an alliance of small exhibitors, small-town Protestant conservatives and Progressive reformers wanting to extend Federal Regulation, and unable to recruit a sufficiently authoritative Protestant voice to endorse its program of self-regulation, the MPPDA turned to the Catholic Church as one of its oldest and most faithful friends. The Legion of Decency was by no means the first large-scale activity relating to movies undertaken by the Catholic Church. Throughout the 1920s, Catholic groups had co-operated enthusiastically with Hays, and they remained aligned with the MPPDA in the late 1920s when Protestant and other civic groups began to demand federal regulation of the industry. In the absence of reliable support from Protestant bodies, the MPPDA began to offer the Catholic Church an opportunity to act as a moral and cultural broker between the city and provincial Protestant morality. Prominent Catholics were involved in writing and promoting the Code; Joe Breen began working for the MPPDA in 1930, and was effectively in charge of the Code's administration for at least a year before it was allegedly implemented in July 1934.

The Catholic Church seized the opportunity to 'clean up' the movies as part of a wider project of cultural assertiveness, connected to their emergence into greater political prominence. Under the banner of 'Catholic Action', Catholic intellectuals put themselves forward as the saviours of American ideals, maintaining an instrumentalist view of culture – that it should demonstrate how people ought to behave, rather than what they did – far more effectively than any other group. The Legion of Decency became the largest Catholic Action organisation and its greatest public relations achievement.

The real danger the industry faced in the early 1930s was from the passage of legislation outlawing block booking and imposing federal regulation of the industry's business practices. For all industry parties, the issues of monopoly control and trading practices were economically much more important than questions of censorship. But questions of censorship were of greater public interest and concern, and could, if necessary, be resolved at less risk to the majors' monopoly interests. The MPPDA's awareness of this encouraged it to displace the public debate from the economic base of distribution practices to the ideological superstructure of movie content. But since movie content itself was not the fundamental cause of the crisis, the crisis could not be resolved by changing content alone. Rather, the crisis in the public perception of the industry had to be resolved through the industry's manipulations of its public relations.

By late 1933, when The Payne Fund Studies into 'Motion Picture and Youth' added further fuel to the arguments in favour of government regulation of the industry, it had become evident that the industry could redeem itself only by a public act of atonement before an identifiable moral authority, and through the Legion of Decency the Catholic Church provided it with an opportunity. The Legion of Decency campaign was neither a spontaneous expression of public opinion nor a conspiracy to establish Catholic control over the movies. The 'organised industry' acquiesced in the limited Catholic attack on its morals in order to protect its more fundamental economic interests, while the Legion claimed the glory of reforming the movies for Catholic Action. In going through a public act of contrition, the industry succeeded in separating the issues of trade practices and profitability from the question of moral content, and silencing the voices calling for more fundamental reforms.

There was no fundamental shift of Code policy in July 1934. The apparent changes brought about by the negotiations with the Legion of Decency were in fact mainly cosmetic, and had much more to do with the movies appearing to make a public act of recantation than with substantive changes in the practice of self-regulation. There was a further tightening up of practice, but this had occurred on at least three other occasions since 1931, and Breen was not given any new or arbitrary powers to cut or ban movies. The differences between movies made in the early 1930s and those made later in the decade are undeniable, but the change was gradual rather than cataclysmic, the result of the development of a system of conventional representation that was constructed by experiment and expedient in the first half of the decade and maintained in the second. From this perspective, Mae West, the great exponent of the double entendre, can be seen as a half-way point between the naive sophistication of, say, 'the unspeakable Constance Bennett' in *The Easiest Way* (1931) and the highly sophisticated innocence of the discourse on sexuality in late 1930s screwball comedies, Astaire–Rogers musicals or, for that matter, the equally sophisticated innocence, what Graham Greene called the 'dimpled depravity', of Shirley Temple in *Wee Willie Winkie* (1937). The Code forced Hollywood to be ambiguous, and gave it a set of mechanisms for creating ambiguity, while viewers learned to imagine the acts of misconduct that the Code had made unmentionable.

This system of representation had two governing principles. One was stated in the Code itself: that 'No picture shall be produced which will lower the moral standards of those who see it'. Under this law a strict moral accountancy was imposed on Hollywood's plots, by which a calculus of retribution or coincidence invariably punished the guilty and declared sympathetic characters innocent. The Code's other principle permitted producers to deny responsibility for a movie's content, through a particular kind of ambiguity, a textual indeterminacy that shifted the responsibility for determining what the movie's content was away from the producer to the individual spectator. The early 1930s saw the gradual development of a set of representational conventions for the treatment of 'objectionable' material, from which, as the Code's first administrator, Jason Joy, explained, studios had to develop a system of representational conventions 'from which conclusions might be drawn by the sophisticated mind, but which would mean nothing to the unsophisticated and inexperienced'. As Lea Jacobs has argued in *The Wages of Sin*, the Code provided a means by which 'offensive ideas could survive at the price of an instability of meaning', allowing producers to deny responsibility for a movie's content. Much of the work of self-regulation lay in the maintenance of this system of conventions, and as such, it operated, however perversely, as an enabling mechanism at the same time that it was a repressive one.

The events of 1934 ensured that everyone in the industry recognised that the Code must be accepted as a convention of representation. Having conceded the limitations of its boundaries, in the second half of the decade producers and audiences alike could explore the possibilities as well as the constraints of the convention, much as they had done before, but with much less public hindrance. Textual indeterminacy became a feature of Hollywood's representation of sexuality, emerging in complex and oblique codes that resembled neurotic symptoms or fetishes. Even under Breen, the censorship of sexuality remained an imperfect procedure of repression, with the repressed always returning, as Freud had promised, in distorted form, given a sufficiently imaginative audience. The rules of both conduct and representation under these conditions were perhaps most cogently articulated by F. Scott Fitzgerald's Monroe Stahr in *The Last Tycoon*, explaining to his scriptwriters how the audience is to understand their heroine's motivation:

> At all times, at all moments when she is on the screen in our sight, she wants to sleep with Ken Willard . . . Whatever she does, it is in place of sleeping with Ken Willard. If she walks down the street she is walking to sleep with Ken Willard, if she eats her food it is to give her enough strength to sleep with Ken Willard. *But* at no time do you give the impression that she would even consider sleeping with Ken Willard unless they were properly sanctified.[8]

Individual PCA recommendations might become causes of dispute, frequently expressed in a highly conventionalised hyperbolic language of absolutes, but the role that the PCA played in the production process was not a matter of contention, nor was the implementation of the Code 'bitterly fought' by the studios. Writers and producers commonly left material that they knew would be cut in scripts sent to the PCA, often in hopes of negotiating something else through, and frequently shots or sequences that the PCA initially objected to survived in the final film. By the same token, when Breen informed Jack Warner that the story of *Each Dawn I Die* (1939) 'in its present form, is thoroughly and completely unacceptable from the standpoint of the Production Code, and ... enormously dangerous from the standpoint of political censorship, both in this country and abroad', he was merely opening negotiations over modifications.[9] Although they often took place at a high decibel level and in retrospect often seem bizarre (in order to dilute the 'flavor suggestive of propaganda for radicalism' in *Winterset* [1936] Breen suggested that RKO 'substitute the word "lunatic," or some other word, for the word, "capitalist"'), what took place between the PCA and the studios were genuine negotiations in which concessions were made by both parties in pursuit of a common objective. Breen's insistence that the PCA was 'regarded by producers, directors, and their staffs, as participants in the processes of production' was perfectly accurate: in a typical minor instance, Hal Wallis' assistant Walter MacEwen recognised that Breen's main aim in having Warners alter one word in Joan Blondell's final song in *Gold Diggers of 1933* was to prevent the routine being 'mutilated' by censor boards.[10]

The establishment of the PCA did not abate the process of negotiation over what constituted satisfactory material for films. The Code's provisions on crime were modified on a number of occasions during the later 1930s to eliminate themes of kidnapping or suicide, to prohibit scenes showing law-enforcing officers dying at the hands of criminals or the display of machine guns. On the whole, however, these questions of Code enforcement were relatively minor: the studios had acquiesced in the PCA machinery and, with occasional displays of resistance, acquiesced in its decisions. More importantly, public opinion had, with few and inconsequential exceptions, recovered from its moral panic and accepted the Association's or the Legion's account of the industry's rescue from the abyss in 1934.[11]

The difficulties the Association faced over film content in the late 1930s were largely the result of its successes in imposing a definition of entertainment as recreation. Protestant criticism about Catholic domination of Code machinery was concerned less with theology than with the Code's preoccupation with sex at the expense of broader issues of social morality. Hays, however, confidently asserted in 1938 that the industry could afford 'the soft impeachment' that it provided nothing more than 'escapist' entertainment, since 'entertainment is the commodity for which the public pays at the box-office. Propaganda disguised as entertainment would be neither honest salesmanship nor honest showmanship'.[12] As Breen's dealings with the studios became more assertive after 1934, his correspondence made fewer distinctions between a decision under the Code, advice regarding the likely actions of state or foreign censors, and the implementation of 'industry policy' in response to pressure groups, foreign governments and corporate interests. Industry policy was, like self-regulation, designed to prevent the movies becoming a subject of controversy or giving offence to powerful interests, but events such as MGM's decision in 1936 to not to produce a film version of Sinclair Lewis' *It Can't Happen Here*, and Catholic protest over *Blockade* (1938), which was set in the Spanish Civil War, led to accusations from liberals within the industry and outside that 'self-regulation ... has degenerated into political censorship'. Breen defended his practice of linking this strategy with Code enforcement: since he saw the PCA as representing a national consensus on political issues as well as moral ones, he denied that there was anything 'sinister' in his rejecting material that characterised 'a member of the United States Senate as a "heavy"; or ... in which police officials are shown to be dishonest; or ... in which lawyers, or doctors, or bankers, *are indicted as a class*'.[13] Such actions, however, succeeded only in making the PCA controversial precisely because of its success in keeping controversy from the screen.

In July 1938 the Department of Justice filed the anti-trust suit against the majors that would result, ten years later, in the dismantling of their distribution-exhibition oligopoly, and with it, the studio system of production. The suit implicated the PCA in the majors' restrictive practices, alleging that the majors exercised a practical censorship over the entire industry through the Code, restricting the production of pictures treating controversial subjects and hindering the development of innovative approaches to drama by companies seeking to use innovation as a way of challenging the majors' monopoly power. In the face of the anti-trust suit, MPPDA Washington Bureau chief Ray Norr insisted, the object now was 'to *limit* the jurisdiction of the Motion Picture Production Code in various respects'. In 1939 the PCA's jurisdiction was restricted so that there was a clear distinction between its administration of the Code and its other advisory functions. Although PCA officials continued to voice concern over whether such subjects as *Confessions of a Nazi Spy* constituted appropriate screen entertainment, the MPPDA's response to the anti-trust suit was to encourage, or at least acquiesce in, the use of politically more controversial content as a way of demonstrating that the 'freedom of the screen' was not hampered by the operations of the PCA. When Hays declared the industry's support for 'pictures which dramatized present-day social conditions', in his Annual Report in March 1939, Margaret Thorp suggested that this change of tone marked 'the day the motion picture industry extended an official welcome to ideas'. More realistically, it reflected an accommodation to political necessity not unlike those that had occasioned the MPPDA's actions in 1930 and 1934. Where the dominant voices to which the Association was attempting to adjust film content earlier in the decade came from moral conservatives, most clearly orchestrated by the Catholic Church, the events of the late 1930s suggested that mechanisms for the control of content had become too extensive, so that it could not so effectively fulfil its function as the currency of negotiation among parties who felt that the movie business was their business.

Industry trade practices, which were to a large extent the hidden agenda behind much of the activity around censorship in the decade, were frequently alleged to have an inhibiting effect on the industry's preparedness to experiment 'with less popular themes aimed at smaller, more specific audiences'. Undoubtedly, the industry's oligopoly structure inhibited experimentation, and the Production Code contributed to that effect. But the industry had a more sophisticated understanding of the preferences of its several audiences than it was given credit for. It did not produce experimental films not because it refused to differentiate among its audiences, but because there was an insufficiently large audience for such productions.[14] The Production Code did not cause the lack of aesthetic experimentation or political radicalism in Hollywood product. Rather, it was itself a symptom of the underlying cause. The Code was a consequence of commercialism, and of the particular understanding of audiences and their desires that the industry's commercialism promoted. For Hollywood to produce movies different from those it actually produced would have needed changes far more substantial than the alteration or even abolition of the Code; it would have needed a re-definition of the cultural function of entertainment, and that was a task beyond the limits of responsibility the industry set itself.

The cultural function of cinema and the permissible boundaries of representation were under constant negotiation inside the movie industry and in a wider public arena in the early 1930s, and the history of those debates is quite exactly inscribed in the movies themselves. The movies, of course, now exist independently of that history, and can be enjoyed and examined without reference to it. To reduce the history of their circulation to a simplistic melodrama of subversion and repression, however, is to perpetuate the misunderstanding of American cinema history as nothing more than a form of entertainment.

Notes

1 Thomas Doherty, *Pre-Code Hollywood: Sex, Immorality and Insurrection in American Cinema, 1930–1934* (New York: Columbia University Press, 1999), back cover.

2 *Motion Picture Almanac* (Chicago: Quigley Publications, 1931). See also Garth Jowett, *Film, The Democratic Art: A Social History of American Film* (Boston: Little, Brown, 1976), pp. 113–19.

3 'Article 1, By-Laws of the MPPDA', reprinted in Raymond Moley, *The Hays Office* (Indianapolis: Bobbs-Merrill, 1945), 227.

4 Robert Sklar, *Movie-Made America: A Cultural History of American Movies* (New York: Random House, 1975), 175.

5 Lea Jacobs, *The Wages of Sin: Censorship and the Fallen Woman Film, 1928–1942* (Madison: University of Wisconsin Press, 1991); Richard Maltby, 'The Production Code and the Hays Office', in Tino Balio, *Grand Design: Hollywood as a Modern Business Enterprise, 1930–1939* (New York: Scribner's, 1993); Ruth Vasey, *The World According to Hollywood, 1918–1939* (Exeter: University of Exeter Press, 1997).

6 *Forbidden Hollywood* was the title given to a series of Laserdisks featuring early 1930s Warner Bros. movies released by the Turner corporation in the late 1990s, and subsequently released on DVD; Doherty, *Pre-Code Hollywood*, 2.

7 Robert McElvane, *The Great Depression: America, 1929–1941* (New York: Times Books, 1984), 208, 213.

8 F. Scott Fitzgerald, *The Last Tycoon* (Harmondsworth, 1960), 51–52.

9 Jack Vizzard, *See No Evil: Life Inside a Hollywood Censor* (New York: Simon and Schuster, 1970), 63–64. On October 5, 1934, Breen advised Vincent Hart, in charge of the East Coast office of the PCA, to 'sneer back' at producers, 'raise hell with them – threaten to punch them in the nose, etc. If you do this two or three times, I think you will have little trouble of this kind thereafter'. PCA *Crime Without Passion* file. For a colourful contemporary account of Breen's behaviour, see J.P. McEvoy, 'The Back of Me Hand to You', *The Saturday Evening Post*, December 24, 1938, 8–9, 46–48. Breen to Jack Warner, November 22, 1938, WB *Each Dawn I Die* production file.

10 Breen to B.B. Kahane, June 11, 1936. RKO *Winterset* production file; Breen to Hays, June 22, 1938, MPA 1939 Production Code file; Walter MacEwen to Wallis, October 12, 1936, WB *Gold Diggers of 1933* production file.

11 Gregory D. Black, 'The Production Code and the Hollywood Film Industry, 1930–40', *Film History* 3;2 (1989), 180–82; 'Summary of emendations and interpretations of the Production Code on crime', December 20, 1938; MPA 1938 Production Code file. The high volume of crime films resulted from the heavy use of crime formula stories in B-features. Breen to Hays, September 6, 1939, MPA 1939 Production Code file.

12 Quoted in Margaret Farrand Thorp, *America at the Movies* (London: Faber, 1946), 161.

13 Breen, June 22, 1938, MPA 1939 Production Code file.

14 Jowett, 202; David O. Selznick to Jock Whitney, September 6, 1939, *Gone with the Wind* Censorship File, Selznick Papers, quoted in Leonard J. Leff and Jerold L. Simmons, *The Dame in the Kimono: Hollywood, Censorship, and the Production Code, from the 1920s to the 1960s* (New York, Grove Press, 1990), 100–101; Walter Wanger, in *Time*, quoted in Kenneth Clark to Hays, May 2, 1939, MPA 1939 Production Code file.

Chapter 14

Helen Hanson and Steve Neale

COMMANDING THE SOUNDS OF THE UNIVERSE: CLASSICAL HOLLYWOOD SOUND IN THE 1930s AND EARLY 1940s

IT HAS LONG BEEN A CONTENTION IN film sound scholarship that the study of film is visually biased. Nowhere is this more evident than in existing accounts of classical Hollywood cinema, in which sound styles have largely been conceived in terms of broad technological standards, unchanging general principles and overarching norms, and where variations in practice have either remained 'unheard' or have only been identified in a handful of special cases. This can partly be attributed to the periodisations that have governed film historiography. As Donald Crafton has observed, there are few divisions more obvious than the one between sound and silent film, few periodisations more evident than the 'transitional era': the years between 1926 and 1931 in which a number of new sound technologies were adopted, used for a variety of different purposes, then refined, rendered more flexible, and finally honed to comply with the norms of the classical Hollywood feature film.[1] However, while scholars have now revealed the extensive roles played by music and sound in the 'silent era', little attention has been paid to the twenty-year period that followed the transition, a period conceived as one of stability and incremental improvement prior to the adoption of magnetic recording and stereophonic sound in the late 1940s and 1950s (itself an even more neglected topic).[2] Here we focus on the period between the early 1930s and the mid 1940s, seeking to detail some of the general principles and practices that governed sound films and the construction of their soundtracks and to pinpoint the uses made of sound in a number of different films. Prior to doing so, though, it is important to summarise some of the key developments and characteristics of the era that preceded it.

The transitional era

The transition to sound led to an influx of sound engineers and sound technicians into the US film industry. Many of these men (and it seems they were indeed all men) came from the radio and telephone industries. They were experts in the science of sound recording but were largely unacquainted with the norms and procedures of film production. As a result the studios and their employees found themselves seeking to reconcile two distinct aims: recording and reproducing high quality sound (science) and facilitating the production of effective films (art) within the limitations of existing sound technologies. As has probably best been summarised by Kristin Thompson and David Bordwell, sound technology in the mid-to-late 1920s was marked by a number of limitations and thus posed a number of problems:

> The first microphones were omnidirectional and thus picked up any noise on the set. Cameras, which whirred as they filmed, had to be placed in soundproof booths. Moreover, initially all the sounds for a single scene had to be recorded at once; there was no 'mixing' of sound tracks recorded separately. If music was to be heard, the instruments had to play near the set as the scene was filmed. The microphone's placement limited the action. It was often suspended over the set on heavy rigging; a technician moved it about within a limited range, trying to point it at the actors.[3]

In attempting to overcome some of these difficulties, and in order to restore the stylistic flexibility hitherto provided in silent films, scenes were sometimes shot using long and complex takes or a number of cameras filming the action simultaneously from a variety of different positions. However, complex takes were time consuming and cameras were harder to move when muffled to reduce the sounds they made, and 'multiple-camera shooting did not exactly duplicate silent film style. The presence of the bulky booths made it difficult to get precise framings . . . Also, since the cameras tended to be lined up in front of the scene [in order to ensure that some cameras were not accidentally filmed by others], portions of the space often overlapped between shots'.[4]

In order to tackle problems such as these, the Academy of Motion Picture Arts and Sciences (AMPAS) created the Academy School in Sound Fundamentals. Here studio employees were trained by sound experts and technological problems were discussed and solved. The Academy disseminated information and guidance in its Academy Technical Digest series, then published the papers it regarded as key in *Recording Sound for Motion Pictures: AMPAS* in 1931.[5] Along with articles in the *Journal of the Society of Motion Picture Engineers* (*JSMPE*) and *American Cinematographer* (*AC*), and along with the guidance provided by Standards Committees, information was shared, new developments were reported on, and problems were identified and solved. By the 1930–31 season, filmmakers were able, according to Crafton, to augment or diminish 'the sound at their disposal. It could be brought in or out, up or down, made "expressive" or "inaudible" as desired. Voices and effects could be "synthetic", the term used for dubbing and adding sounds in post-production. Music could well up and fade back to underscore action and mood'.[6] Omnidirectional microphones, which tended to pick up unwanted noise, had been replaced by unidirectional ones, lighter microphone booms had made the process of recording dialogue and sound more flexible, and 'by late 1932, music, voices and sound effects could be registered separately and later mixed onto one track'.[7] Edge numbers could be printed on sound and image negatives, which enabled more precise synchronisation, and among other things actors 'no longer had to move gingerly or to enunciate their dialogue slowly, and thus the lugubrious pace of many early talkies gave way to a livelier rhythm'.[8]

This did not mean, though, that stylistic or technological innovation stopped or that developments in sound came to an end. Research on sound and sound technology was conducted by commercial sound organisations as well as by AMPAS and the SMPE throughout the 1930s and 1940s, and this research yielded steady and continuous improvements in all aspects of sound recording, editing, playback and reproduction, among them refinements to microphone technology and sensitometric improvements in the quality of films stocks and photographic emulsions.[9] Writing in 1937, Barratt Kiesling estimated that around five hundred improvements in sound recording and reproduction had been made at MGM alone since the introduction of sound and that 'for the entire industry such variations would run into thousands'.[10] Thus while the post-transitional era is best characterised as one of evolution rather than revolution, it was one in which the 'soundscapes' of Hollywood's films were constantly changing as sound personnel sought to provide solutions to problems, to find effective ways of telling stories, and to augment the presentation of narrative settings and events; musical numbers; moments of comedy, drama or romance; and scenes of spectacle, tension, danger and suspense.

The post-transition era

By 1931, the range of sound technologies used in Hollywood had stabilised as the major studios opted for sound-on-film 'double' systems that permitted greater control over sound quality. This led to divisions of labour and specialisation, with post-production practices and the editing of dialogue emerging as distinct and different roles.[11] It also augmented the importance of dubbing, which allowed poor sound recordings to be re-recorded or re-mastered and thus afforded greater precision and greater control. In 1931, J.J. Kuhn of Bell Telephone Laboratories suggested that the possibilities of re-recording were now the main focus of research and development, and Kenneth Morgan noted that re-recording was 'already a fundamental part of the production of sound pictures', permitting as it did the recreation of 'auditory sensations', 'the localization of a definite viewpoint', and 'in some instances', the employment of 'unnatural sound emphasis for dramatic reasons'.[12] Morgan went on state that 'most of the pictures released are now 50 to 100 per cent re-recorded' and that

> the dubbing process must be recognised as a most expedient technical, economic, and dramatic asset in sound production. It permits production to proceed without delays that otherwise might be necessitated if all sounds had to be recorded at once. It facilitates the editing of pictures, especially the attainment of uniform level and proper balance of sound. It is an acoustical art in itself, and represents a new form of composition in sound where the artist's keyboard commands the sounds of the universe.[13]

Carl Dreher, RKO's Director of Sound, made a number of similar points in an article entitled 'Recording, Re-Recording and Editing of Sound' in *JSMPE* that same year. Dreher was a key figure in contemporary debates about sound. He wrote the foreword to the AMPAS collection as well as a number of articles in *JSMPE*. In 'Recording, Re-Recording and Editing of Sound' he compares the construction of soundtracks to 'special process photography': 'The blue cloth against which the foreground action is photographed corresponds to the silent background of dialog in sound. In re-recording, any desired sound background may be supplied, just as in special-process photography any desired visual background can be secured *only* by re-recording'.[14] 'Assume, for example', he wrote,

> that it is desired to introduce the call of a meadow lark behind open-air dialog between two lovers. It is readily possible to engage lovers for the purposes of the screen, but it would be difficult to secure the song of the lark at the proper time and place. It is easy, however, to obtain a meadow lark sound track and re-record it back of the dialog at the precise intervals desired and the exact loudness which would be most appropriate.[15]

As well as emphasising the role played by dubbing, Dreher here also promotes the importance of well-stocked effects libraries with a wide variety of recorded sounds appropriate to a diverse array of narrative settings and situations. As re-recording and dubbing became standard practice, so too did the collection of sounds. Sometimes specific sounds were specially recorded or a mixture of 'synthesised' sounds (now known as Foley sounds) were specially produced. In either case, wrote Dreher, 'music, traffic noises, door slams, shots, glass crashes, screams, animal noises, automobile noises, and almost every other conceivable sound' should be available, 'properly catalogued and filed for use when the demand arises'.[16] Dreher's advice was clearly heeded, not just at RKO, but at MGM, where sound effects editor Milo Lory recalled that the sound effects library contained everything 'from a garter snap to an atomic bomb', and at Twentieth Century-Fox, where Joe Delfino estimated that he had in excess of 3,000 'noise makers' in his 'den of din'.[17]

The emergence of practices such as these fostered new forms of thinking about sound work. Dreher captured the dual roles played by sound personnel when he recommended that while the first concern of the sound man should be 'the transmission units and dynes per square centimetre' that guarantee good recording, it was also desirable for sound men to understand the 'story values' of production (provided that they did not become 'what is known in the art as a "script meddler"').[18] In other words, as James Lastra points out, sound personnel moved away from a conception of sound reproduction as governed by 'fidelity values' to a model of sound 'shot through with the hierarchies of dramatic relationships'.[19] Facilitated by the development of edge numbering and the introduction of a three-track Moviola in 1933,[20] dialogue, music and effects could now be recorded separately in order to facilitate flexible reassembly and the matching of sound and image at the editing stage, resulting in what Altman, Jones and Tratoe have called the 'multiplane' soundtrack.[21]

For Altman, Jones and Tratoe, the multiplane soundtrack is one in which intelligible dialogue is paramount and music and effects remain the background unless brought to the fore for specific reasons. However, when watching and listening to classical Hollywood films, it is clear that audible dialogue is indeed paramount, but also that there are more reasons, more occasions and more opportunities for music and effects than their model implies. Even humdrum sounds are needed to characterise environment or time of day or to accompany routine activities such as walking in and out of rooms and buildings, and as Kathryn Kalinak points out in her contribution to this Reader, music in classical Hollywood films is modulated for a number of reasons, sometimes dying away or fading in the mix not just to facilitate the audibility of dialogue, but also to facilitate the perception of silence or ordinary or extraordinary sounds such as traffic noise or the firing of canon – or perhaps the singing of a meadowlark. Moreover dialogue itself can be 'orchestrated' so as to build sonic patterns, rhythms and crescendos or mixed with other sounds to do the same thing. Notable examples of both occur in *His Girl Friday* (1940), in which the dialogue sequences between Walter Burns (Cary Grant) and Hildy Johnson (Rosalind Russell) turn at various points into 'duets' of overlapping or simultaneous speech,[22] and in which the sequence in which Earl Williams (John Qualen) escapes from prison begins with silence and ends with the sounds of speeding cars, wailing sirens and running feet. Points like this are borne out by contemporary practitioners. Screenwriter Tamar Lane instructed would-be writers of scripts to consider 'the *audible* effect of the film' and to 'appeal to the ear as well as the eye',[23] going on to detail a number of ways in which they could use sounds to evoke 'the time, place, mood and atmosphere of the story', imply off-screen activity, and add pace and shape to visual action.[24]

These perspectives evidence a repertoire of sound techniques and an understanding of how they could be deployed in the service of classical Hollywood films. What follows is a discussion of the ways in which examples from a number of films in the post-transitional era use sound to evoke 'time, place, mood and atmosphere', imply off-screen activity, add pace and shape to visual action, and form motivated links, patterns and motifs. In pursuing this discussion, we will draw from time to time on Murray Schafer's work on 'soundscapes'. For Schafer a soundscape is 'any acoustic field of study': 'We may speak of a musical composition as a soundscape, or a radio program as a soundscape. We can isolate an acoustic environment as field of study just as we can study the characteristics of a given landscape'.[25] Schafer's work is highly relevant to film. He conceives of soundscapes as being composed of different sonic strata: keynote sounds, signal sounds, and soundmarks. The keynote of a soundscape 'identifies the key or tonality of a particular composition'. It is 'in reference to this point that [other sounds] take on ... meaning'.[26] Keynote sounds thus form a 'ground' against which signal sounds stand out. 'Signals are foreground sounds', some of which constitute 'acoustic warning devices: bells, whistles, horns and sirens'. And finally a 'soundmark is a community sound which is unique or possesses qualities which make it specially regarded or noticed by people in that community'.[27] (An obvious example of a soundmark in a film would be the sound of the tolling church bell in Tombstone in John Ford's 1946 Western, *My Darling Clementine*.)[28] The pertinence of some of these terms will, we hope, be evident some of the following examples.

I am a Fugitive from a Chain Gang

I am a Fugitive from a Chain Gang was produced by Warner Bros. and initially released in 1932. It was therefore made at a time when a number of innovations in sound technology and sound practice had been introduced. *Fugitive* has been widely discussed as a film that typifies the Warners' 'house style' in the early 1930s, its story 'ripped from the headlines' and adapted from the real-life account of fugitive Robert Burns. A World War I veteran from the South, Burns fell on hard times on his return to civilian life, served time in a Georgia chain-gang, escaped, rehabilitated himself, then was betrayed to the authorities. He returned to Georgia to complete his sentence but managed to escape a second time (and was still at large when the film was made). The adaptation of his memoir (in which Burns is renamed 'Jim Allen' and played by Paul Muni) is vividly evocative of the settings he describes and visceral in its inflection of the action – and sound plays a major part in both.

The film opens with a sound bridge, using the sound of a ship's whistle to effect a transition from the ship bearing Allen and his comrades back from the war in Europe. Whistles form a sound motif throughout the film. As here, they are associated initially with journeys and transport, then later on with instruction, routine and discipline, and warnings and danger, serving overall to dramatise tensions between movement and stasis, freedom and confinement. Allen's later pronouncement that 'I don't want to be spending the rest of my life answering a factory whistle instead of a bugle call ... I want to get out, away from routine', serves verbally to underscore this motif and its significance as well as to foreshadow the 'routine' that later marks his life on the chain-gang. Having been involved later on in robbing a diner, Allen is caught with stolen money and sentenced to hard labour in a prison camp. The camp is characterised throughout by keynote sounds (the clanking of chains that accompany the movements of the prisoners, the monotonous sounds made by picks and hammers) and an interplay between signal and keynote sounds regulates the actions of prisoners, who are awakened by the striking of a large metal triangle and whose work is marked by the repeated sound of pick axes striking rock. In several sequences, the sounds of labour are looped, creating a sonic background devoid of development and direction and inferring unending duration and toil.

Given his hatred of routine and the brutal treatment meted out by the prison guards, Allen resolves to escape. He persuades Sebastian (Everett Brown), one of his fellow inmates, to strike his shackles, then bends them so that he can later slip them off. He bides his time but eventually makes his move when the gang are at work on a section of railroad. Keynote sounds of hammers and picks form a sonic background as Allen and Bomber (Edward Ellis), another inmate, conspire in whispers and as Allen explains his plan to get permission to relieve himself as a means of making his escape. At this point, while cutting back and forth between Allen slipping off his shackles and a guard watching the gang, a different organisational principle begins to govern the film. The volume of the hammers is manipulated to correspond to respective positions of guard (who is higher up and nearer to the gang) and Allen (who is lower down and further away), thus using levels of sound to mark the proximal relations and positions of the characters. Bomber manages to distract the guard's attention and Allen starts to run. The guards' steam whistle breaks through the keynote sounds of labour, sounding the alarm. The guards and their bloodhounds give chase, and the chase itself is dynamically structured by cross-cutting between Allen and his pursuers. Again volume is used to infer proximity and distance as the baying of the hounds is heard differently from the respective positions of Allen and the guards. The sequence climaxes with the alignment of sound with Allen's point of audition. As the bloodhounds gain on him, he splashes waist-deep in the river. Desperate to hide, he breaks a reed which he plans to use as a means of breathing and plunges below the surface. As he does so, the soundtrack is silenced. The dramatic and sonic contrast between shots and sounds of baying hounds and the silent immersion of Allen below the water line is repeated several times before the guards give up the chase and Allen is able to escape.

This is just one of a number of sequences in *Fugitive* that draws on established sonic motifs while also providing striking and inventive uses of sound in specific scenes and sequences. These were not brought about by accident; nor were they the product of creative spontaneity. Early drafts of the script show quite clearly that scriptwriters Brown Holmes and Sheridan Gibney gave considerable thought to the ways in which sound could be used.[29] And information from production records reveal that the filming and recording of the chase sequence was carefully planned. The sounds of the chase (dogs, movements through the forest) were recorded as 'wild' sound, allowing its dynamics to be captured for subsequent editing, while the underwater shots were recorded without sound at a later date.[30] Unfortunately, evidence regarding the use and design of sound in Hollywood's films is by no means always available, either because papers, scripts and memoranda do not survive or are currently unavailable, or because those that survive do not specify sonic effects, patterns or motifs. This does not mean, though, that such patterns or motifs do not exist, as a film like *Top Hat* (1935) serves to testify.

Top Hat

Produced at RKO, *Top Hat* is an Astaire–Rogers musical. As such it has been discussed many times: as the second in a highly successful series, as a showcase for the dancing of Fred Astaire and Ginger Rogers, and as a showcase for the numbers written for film by Irving Berlin. It has also been discussed as an example of an 'integrated' musical, as a musical in which the song and dance numbers further the narratives, articulate the situations to which they give rise, and develop or further articulate the relationships between the characters involved. However, it has rarely if ever been discussed as a film that uses sound – and the distinctions and similarities between sound, noise, music and silence – as a governing motif.[31]

This motif is established in the film's first scene. As we discover as the scene unfolds, Jerry Travers (Astaire), a professional singer and dancer from America, has been hired by impresario Horace Hardwick (Edward Everett Horton) to perform in Hardwick's new show in London. Jerry is staying at Horace's club, whose ambience is underlined by a sign announcing that 'Silence Must Be Observed In the Club Rooms' and by the silence that marks the club rooms themselves as we dissolve to a shot showing its members reading newspapers. A waiter traverses the room with two glasses of wine on a silver salver. He clinks the glasses as he arrives at one of the tables in order to signal his presence. Its occupants react with silent disapproval. 'I beg your pardon', says the waiter as he places the glasses on a table then turns to walk away. The camera follows him, then dollies in to show the pages of a newspaper whose reader is at this point hidden behind them. The reader clears his throat. The members react with annoyance. The reader is Jerry, who is revealed when he peers out timidly from behind the pages of his paper, looks around, raises his paper, peaks out from behind the pages again, looks increasingly bored, then inadvertently rustles the paper's pages as he tries to turn them over. Further disapproval from the members ensues. Jerry attempts to turn the pages again without much success. At this point Horace arrives, greeting the cloakroom staff with a cheery 'Good Evening'. However, his greeting is too loud. He has forgotten where he is. He lowers his voice as he repeats his greeting and explains that he needs Jerry to come to his hotel to work on his show. He goes in to the club room to tell Jerry. Jerry is delighted; Jerry has been rescued. As he leaves he beats out a loud volley of tap steps by the door.

In Horace's hotel room Jerry explains his plans for one of the numbers in the show, singing, slapping and tapping his way through 'No Strings (I'm Fancy Free)'. However, the sounds he is making are keeping Dale Tremont (Ginger Rogers) awake in the room below. After a series of phone calls to Horace and the hotel management, Dale decides to express her anger in person when plaster from her ceiling begins to fall into her bed. She marches into Horace's room. Horace has left to find the hotel manager. But Jerry is there. Dale is angry; Jerry is smitten. However, Dale has softened somewhat by the time she leaves, and when she returns to her bed, Jerry sprinkles sand on the floor in Horace's apartment and dances a gentle soft-shoe shuffle in order to help her sleep.

Although she smiles as she drifts off to sleep, Dale is still sufficiently annoyed to ward off Jerry's advances when next they meet. However, Jerry is persistent. Taking the place of a hansom cab driver, he drives her to a park. She is not amused. She gets out of the cab to go horse riding. Jerry follows in his cab. After a while, the sky turns dark, lightning flashes and rain begins to fall. Sounds of thunder and rain fill the soundtrack and Dale takes shelter in a bandstand. Jerry follows her: 'May I rescue you?' he asks. 'No thank you', she replies. But a clap of thunder sends her into his arms. She backs away. 'Are you afraid of thunder?' he asks. 'Oh no, just noise', she replies. Then another clap of thunder inaugurates the beginning of 'Isn't This a Lovely Day (To Be Caught in the Rain)', which is sung by Jerry to Dale. Both are seated in the bandstand. Dale softens. She begins to whistle in accompaniment. They get up to dance. Their initial hesitations give way to sweeping spins and turns, accelerated rhythms, and perfectly coordinated taps, prompted near the end by a final almost joyous clap of thunder.[32]

This, though, is not the end of the story. Complications follow when Dale thinks Jerry is Horace, the man who is married to one of her female friends but whom she has not ever seen. Her reaction is capped by the sound as well as the sight of the slap she gives him when she sees him. After the ensuing show sequence and the aural as well as the visual dazzle of the 'Top Hat' number itself, the sound motif is more or less abandoned. Indeed, one of the key sequences in this part of the film is the sequence of emotion-laden silence that follows 'Cheek to Cheek' (itself a number in which, in contrast to 'Isn't it a Lovely Day', the sounds of dancing are minimal). However this sequence can itself be seen as an integral counterpoint to the sounds and noises – and indeed the stuffy rather than profound silence in the club scene – in the film's first half. And the sound-and-noise motif is reprised near the end when Jerry dances taps in the room above Dale's in order to recall their initial meeting prior to sorting out their misunderstandings, declaring their love, and dancing together again in the film's final sequence.

By their very nature, musicals offered specific opportunities for particular kinds and uses of sound. However, as Harold Lewis pointed out in 1934, other genres offered opportunities too, opportunities afforded, indeed mandated, by the nature of their stories and by the stylistic traits with which they were associated. Thus comedies require a 'high key' tonal quality to aid the intelligibility of fast-paced dialogue and action, dramas should be marked by 'restraint' and 'subtlety' and thus mixed in a lower key, and 'melodrama' (by which he means mysteries, crime films, horror films and thrillers) 'requires strongly contrasted sound-treatment, even as it requires strongly contrasted photographic treatment'.[33] Lewis was President of the Society of Sound Engineers, which was formed in 1934. His recommendations both coincided with and helped determine the numerous technological amendments, adjustments and improvements that marked the mid-to-late 1930s and the early 1940s. These included a new 'dubbing rehearsal channel' that could accommodate six sound tracks at once and that was equipped with a larger viewing screen than those on standard Moviolas.[34] They also included a potentiometer, a resistor used to control the volume of up to eight different tracks in the mixing stage. This device worked with a linear motion rather than a rotary one, which meant that several potentiometers could be used simultaneously, enabling easier control of a greater number of channels.[35] Both played their part in the fabrication of generic worlds and in the production of generic and cyclic tropes, among them a trope that might best be called 'footsteps in the street' – footsteps rather different in kind from those of a dancing Astaire or a dancing Rogers, but footsteps equally conceivable as signal sounds and equally subject to sonic manipulation in the service of generically appropriate effects.

Footsteps in the street

During the course of the 1940s, 'footsteps in the street' became a powerful sonic metaphor in a number of urban crime, horror and mystery melodramas, among them *Cat People* (1942), *The Seventh Victim* (1943) and *Phantom Lady* (1944). In each case what might ordinarily be thought of as keynote sounds become signal sonic motifs, motifs central not only to the composition of the following sequences, but to the generically mandated effects to which their use gives rise. In *Cat People*, a woman, Alice (Jane Randolph) walks down a dimly-lit street at night. She passes alternately through deep pools of darkness and bright splashes of light thrown by a series of street lamps. She is alone. Her high heels rap out a regular tempo on the pavement. She and the cinema audience are aware that another woman, Irena (Simone Simon), is following behind her. The tempo of this second set of footsteps is faster, lighter, more urgent. The distance between these two women closes then the second set of footsteps suddenly stops. Alice registers the silence behind her. Feeling vulnerable and threatened she looks behind, around and above, trying to identify the whereabouts of the now-silent figure. Her footsteps speed up, becoming irregular, chaotic and distressed. Uncannily the street now appears to be empty. Suddenly a threatening 'growl' is heard, its source suddenly revealed as a bus enters the frame and grinds to a halt with a hiss of its airbrakes. Shaken and scared, Alice gets on the bus. 'You look as if you seen a ghost', says the driver. 'Did you see it?' she replies. But there is nothing to be seen.

In *The Seventh Victim*, a woman, Jacqueline (Jean Brooks) walks down a dimly-lit street at night. She passes alternately through deep pools of darkness and bright splashes of light thrown by a series of street lamps. She is alone. Her high heels rap out a regular tempo on the pavement, mixed on the sound track with musical markers of tension, danger and suspense. Jacqueline and the cinema audience are aware that she is being followed. Her passage through the street is marked by rising musical chords on the soundtrack as she looks into the dark spaces around her. A crash is heard behind her. It turns out to be caused by a dog kicking over a dustbin. Then there is ambient silence. Jacqueline looks around for her pursuer. But there is nothing to be seen. As she begins to hurry, her panic is underlined by the flurries of woodwind in the musical score and by close-ups of her feet. As she passes a darkened doorway, a man's face looms into the light, marked by a sudden blast of brass in the score. The tempo of her footsteps increases. She careers across the street, stopped by a screech of brakes as a taxi cab stops to avoid knocking her over. Then there is ambient silence. This pattern is repeated for a third and final time, building gradually as she passes the stage door of a theatre. The sounds of music and laughter emanating from the theatre contrast with Jacqueline's isolation outside. She attempts to hide in the shadows. Sliding her hand along a wall, the tips of her fingers encounter the arms of her pursuer. Her face freezes in fear. Another brass 'stinger' on the soundtrack marks her fear as her assailant grabs her arm, and the score dies away to highlight the sharp menacing 'click' of his flick-knife. The ambient silence is suddenly shattered by a shrill peal of laughter as a group of theatre performers spills out from the stage door and into the street and Jacqueline seizes the opportunity to elude her assailant by joining them. The troupe moves into a bar whose ambience is marked by the sounds of laughter and the playing of a honky-tonk piano. These sounds signify communality. But Jacqueline remains outside in the street, alone and afraid.

In *Phantom Lady*, a man, Mack (Andrew Toombs, Jr.), comes out of a bar and walks down a wet, dimly-lit city street at night. He passes alternately through deep pools of darkness and bright splashes of light thrown by a series of street lamps. He is alone. His shoes rap out a heavy and regular tempo on the pavement. He is aware that he is being followed by a woman, Kansas (Ella Raines). Her high heels rap out a different and lighter tempo. She follows him up a set of iron stairs and onto the platform of an elevated railway. The timbre of their respective footsteps is such that the audience is aware of the wetness of the street and the other surfaces they traverse as they walk. Mack waits on the platform. Kansas waits on the platform. The audience waits in suspense. The sound of an approaching train can be heard. Mack moves along the platform to stand behind Kansas. The sound of the train increases in volume as the train nears the platform. Mack steps forward in order to push Kansas under the train. As he does so though, the tension is broken and Kansas is saved when the click and the whirr of a turnstile indicate that another woman is now on the station platform too. However, the danger and suspense are by no means over. The sequence of suspense is resumed and tension builds once more as Mack leaves the station and Kansas follows him again through the pools of light and shadow in the wet city streets. The tempo of her footsteps becomes more insistent, faster, more urgent. Mack finally stops, turns and confronts her, speaking the first lines of dialogue in over five minutes: 'What are you after?' Thinking that Mack is accosting her, a group of men offer to intervene. 'This is between us', she tells them. Then she turns to Mack. 'You've got something to tell me', she says. But Mack won't talk. He turns to flee, stepping off the sidewalk and into the street. The sickening sounds of a crash and a scream end the sequence as Mack is run down by taxi cab.

With barely a word being spoken, these sequences all provide intense renderings of places, shifting character situations, and fear, suspense and death. Along with the careful synchronisation of sound and image, the precision with which sounds and silence are mixed and balanced creates powerful patterns of sensation. The settings in all three sequences are night time city streets. In addition to creating sonic rhythms, timbres, tempos and pitches, the audibility of the footsteps helps establish the relative emptiness of the streets and the presence (or the sudden absence) of the characters involved. In addition, and just like the members of the audience, these characters are engaged in intense listening experiences. In all three sequences one character is following and another is being followed and their proximity to one another is invested with danger and fear and the threat of death. The interplay of image, sound and music focuses in particular on footsteps, with repeated shots of the characters' feet. But this is not cross-cutting as conventionally understood, for it does not involve showing the relative location of the characters. The repeated shots of pairs of feet is an instruction not so much to look but to listen. It is also notable that these sequences unfold with little or no dialogue. It is sounds, not words or music, that dominate the sound mix.

Meanwhile, the varying tempos and rhythms of each set of footsteps render the relative proximity of danger and convey essential information about the each of the steppers. Their bodily dimensions, their gender, even their mental states are rendered through pitch and through irregularities of rhythm – stumblings, shufflings, trips, and transitions in pace from walking to running. The sounds of the footsteps also help suggest the nature of the surfaces of pavements and roads. All three sequences were filmed on set rather than on location. But the footsteps have been carefully rendered by what we would now call Foley artists in order to convey the nature of these surfaces: pavements wet and dry, iron staircases, railway platforms, and so on and so forth. Finally, it should be noted that all three sequences build, dissipate and rebuild tension and suspense by alternately showing and concealing the locations of the characters relative one another. As the sound of footsteps is the key means by which those pursued locate their pursuers, these sequences all use silence to create suspense as characters and audiences alike strain to hear and judge the relative proximity of danger and death. It is thus through silences as well as sounds, sounds as well as images, noises as well as shadows, that sequences like this work.

These examples are drawn from films produced at RKO (*Cat People* and *The Seventh Victim*) and Universal Studios (*Phantom Lady*). Both of these studios were committed to the advancement of sound practices and to investment in new sound technologies. In 1937, Homer G. Tasker led a symposium on 'How Motion Pictures Are Made' which was hosted by Universal, and Bernard B. Brown, who was Universal's Director of Sound and who oversaw sound on *Phantom Lady*, was a regular contributor to *JSMPE*. RKO was set up in order to use RCA's Photophone sound system, and Carl Dreher, RKO's director of sound, was, as we have seen, an authority on sound. Meanwhile at Universal, sound mixer Edwin Wetzel, who also wrote for *JSMPE*, worked with Tasker and others on the re-recording and mixing of soundtracks and effects for *Bride of Frankenstein* (1935) and other Gothic horror films. These too were a major influence, even on *Cat People*, *The Seventh Victim* and the other B films produced at RKO by Val Lewton, which are generally and rightly regarded as eschewing visually spectacular Gothic monsters, but which are nevertheless as dependent for some of their effects on the silences and sounds of earlier Gothic films (silences and sounds that include the howl of the Wolfman, Dracula's noiseless movements and Middle European accent, the monster's grunts and shufflings in the Frankenstein films – and, of course, menacing passages of quiet and piercing screams of terror).

Lewton's films are as famous for their use of sound as they are for their use of light and shadow, their powers of suggestion, and their preoccupation with loneliness and death. But where do the footsteps in these and other melodramas come from? Edmund G. Bansak suggests that Lewton was influenced by 'Victorian ghost stories and Depression-era radio plays'.[36] He also points out that the shock effect of the 'hissing bus' in *Cat People* was devised by sound editor Mark Robson, who had worked with Orson Welles, 'the former boy wonder of radio', on the soundtrack of *Citizen Kane* (1941).[37] But the footsteps, which were equally the product of Robson and his team, may have had another source in radio too.

Radio sounds and radio footsteps

Like the film industry, the radio industry in the US had been fabricating sounds for years. According to Lucille Fletcher, the 'first sound effect in a radio drama went over the air . . . on

August 3, 1922. That evening, Station WGY, Schenectady, broadcast 'The Wolf', a drama by Eugene Walter. At one point in the action the director of the play, Mr. Edward H. Smith, slapped a couple of pieces of two-by-four together, to simulate the slamming of a door'.[38] By the 1940s, the Columbia Broadcasting System (CBS) was spending 'a hundred thousand dollars a year exploring the nuances of everyday clicks, rumbles, echoes, squeaks and plops. At NBC headquarters a staff of twenty-five engineers and their assistants toils day and night in soundproof studios, seeking to add to the studio's repertoire of ticks and crashes'.[39] Just as in the film industry, sound engineers and 'sound-effects directors' such as Walter Pierson at CBS and Ray Kelly at the National Broadcasting Company (NBC) assembled well-stocked libraries of sounds and worked to improve the quality of sound in general.

During the1930s, the airways were peppered by comedies, music shows, soap operas, and hundreds of adaptations, among them 'The 39 Steps', 'A Tale of Two Cities' and 'The War of the Worlds', and 'Dracula', numerous Edgar Allen Poe stories, and a number of other Gothic tales. They were joined in the early 1940s by *Inner Sanctum Mysteries*, *The Weird Circle*, *The Shadow*, *Suspense* and many other examples of what *New York Times* radio critic John Hutchens called 'the shockers'.[40] As Neil Verma points out, 'Hutchens argued that there are advantages to the grim atmosphere of which shock radio liberally partook',[41] not the least of which were the occasions they provided for sonic effects: 'Doors squeak ... triggers click, doleful music wafts through the night' – and 'footsteps pace off the measured step of doom'.[42] One of the most famous examples of these shockers was 'The Hitch-Hiker', which was initially aired as part of an episode of *The Orson Welles Show: Almanac* (aka *Orson Welles's Almanac*) in 1941 then reworked for *Suspense* in 1942. 'The Hitch-Hiker' was written by Lucille Fletcher, who is cited above. Fletcher was married to composer Bernard Herrmann. Herrmann worked closely with Orson Welles, who not only starred in initial version of 'The Hitch-Hiker' but, as already noted, who also worked with Mark Robson that same year.[43] We have no evidence that 'The Hitch-Hiker' or any other shocker was directly responsible for the 'footsteps in the street' motif in *Cat People* or in any other film. But we would suggest that radio was a possible source and we would argue that it contributed to the motif's circulation. Either way, the sonic crossovers between radio and film in the 1930s and 1940s have, like many other aspects of sound, yet to be fully explored.

Notes

1 Donald Crafton, *The Talkies: American Cinema's Transition to Sound, 1926–1931* (New York: Scribners, 1997), 1.

2 In 1940, Disney used a stereophonic system called Fantasound, which was developed by RCA and used for roadshow presentations of *Fantasia*. As far as we are aware, this was the first use of stereo sound in a Hollywood film.

3 Kristin Thompson and David Bordwell, *Film History: An Introduction* (New York: McGraw-Hill, 2003 edition), 216.

4 Thompson and Bordwell, *Film History*, 216.

5 Lester Cowan (ed.), *Recording Sound for Motion Pictures: AMPAS* (New York: McGraw-Hill, 1931).

6 Crafton, *The Talkies*, 355. Crafton provides a detailed account of technical and aesthetic developments in the transitional era in 'Three Seasons: The Films of 1928–31', the second part of his book from which this quote is drawn.

7 Thompson and Bordwell, *Film History*, 241.

8 Thompson and Bordwell, *Film History*, 241.

9 Rick Altman, 'The Evolution of Sound Technologies', in Elisabeth Weis and John Belton (eds), *Film Sound: Theory and Practice* (New York: Columbia University Press, 1985); Edward W. Kellogg, 'History of Sound Motion Pictures: First Installment', *Journal of the Society of Motion Picture Engineers (JSMPE)*, June 1955; 'History of Sound Motion Pictures: Second Installment', *JSMPE*, July 1955; 'History of Sound Motion Pictures: Final Installment', *JSMPE*, August 1955, all reprinted in Raymond Fielding

(ed.), *A Technological History of Motion Pictures and Television* (Berkeley: University of California Press, 1967).

10 Barratt Kiesling, *Talking Pictures: How They are Made and How to Appreciate Them* (Richmond: Johnson Publishing Company, 1937), 205.

11 James Buhler, David Neumeyer and Rob Deemer, *Hearing the Movies: Music and Sound in Film History* (New York: Oxford University Press, 2010), 306; W.C. Harcus, 'Making a Motion Picture', *JSMPE*, vol. 17 no. 5 (1931), 802–10; Barry Salt, *Film Style and Technology: History and Analysis* (London: Starword, 1992 edition), 179–218.

12 J.J. Kuhn, 'A Sound Film Re-Recording Machine', *JSMPE*, vol. 17 no. 3 (1931), 326–42; Kenneth Morgan, 'Dubbing', in Cowan (ed.), *Recording Sound for Motion Pictures*, 145–46, 150. Morgan was a recording manager at Electrical Research Products Incorporated (ERPI).

13 Morgan, 'Dubbing', 154.

14 Carl Dreher, 'Recording, Re-Recording and Editing of Sound', *JSMPE*, vol. 16 no. 6 (1931), 759. Emphasis in original.

15 Dreher, 'Recording, Re-Recording and Editing of Sound', 749.

16 Dreher, 'Recording, Re-Recording and Editing of Sound', 759.

17 Milo Lory, 'Early Sound and Music Editors', Oral History interview with Irene Lory Atkins, American Film Institute, Louis B. Mayer Library Collection, OH 27, 1975, 51. Anon, 'Noise for Sale', *Cinema Progress*, vol. 12 no. 1 (1937–38), Margaret Herrick Library, Core Collections File, no. 28.

18 Carl Dreher, 'Sound Personnel and Organisation', in Cowan (ed.), *Recording Sound for Motion Pictures*, 345–46.

19 James Lastra, *Sound Technology and the American Cinema: Perception, Representation and Modernity* (New York: Columbia University Press, 2000), 207.

20 Harcus, 'Making a Motion Picture', 310; David Bordwell and Kristin Thompson, 'Technological Change and Classical Film Style', in Tino Balio, *Grand Design: Hollywood as a Modern Business Enterprise, 1930–1939* (New York: Scribners, 1993), 124–25.

21 Rick Altman, McGraw Jones and Sonia Tratoe, 'Inventing the Cinema Soundtrack: Hollywood's Multiplane Sound System', in James Buhler, Caryl Flynn and David Neumeyer (eds), *Music and Cinema* (Hanover, NH: Wesleyan University Press, 2000), 339–59.

22 *His Girl Friday* was directed by Howard Hawks, in whose films speech rhythms are nearly always used to indicate the compatibility or incompatibility of characters. For more on rhythms and tempos in general in *His Girl Friday*, see Lea Jacobs, 'Keeping Up with Hawks', *Iris*, no. 32 (1998), 402–26.

23 Tamar Lane, *The New Technique of Screen Writing: A Practical Guide to the Writing and Marketing of Photoplays* (New York: McGraw-Hill, 1937), 4. (Emphasis in original.)

24 Lane, *The New Technique of Screen Writing*, 4.

25 Murray R. Schafer, *The Soundscape: Our Sonic Environment and the Tuning of the World* (Rochester: Destiny Books, 1977), 7.

26 Schafer, *The Soundscape*, 9.

27 Schafer, *The Soundscape*, 10.

28 A variant on this particular soundmark occurs in *The Man Who Shot Liberty Valance* (1962), Ford's bitterly ironic take on the history of the West and its community in whose opening sequence the church bell in the town of Shinbone is silent and what we hear instead is the tolling of the bell on a railroad train, a symbol of all that has changed – and all that has been lost in the process.

29 John O'Connor (ed.), *I am a Fugitive from a Chain Gang* (Madison: University of Wisconsin Press, 1981), 197.

30 *I am a Fugitive from a Chain Gang*, Daily Production and Progress Reports, 11 August and 3 September 1932, *I am a Fugitive from a Chain Gang* Production Folder, document FY10040, Warner Bros. Archives, School of Cinematic Arts, University of Southern California.

31 According to the Internet Movie Database (IMDb), the sound editor on *Top Hat* was George Marsh, the sound effects editor was Robert Wise, the principal recordists were Hugh McDowell Jr., Eddie Harman, Clem Portman and John E. Tribby, and the music recordist was P.J. Faulkner Jr. Of these, only Marsh, Faulkner Jr. and McDowell Jr. were credited.

32 Like the other dance sequences in this and other Astaire films, the nature and structure of this sequence is discussed in detail by John Meuller in *Astaire Dancing: The Musical Films* (London: Hamish Hamilton, 1986), 80–81.

33 Harold Lewis, 'Getting Good Sound is an Art', *AC*, vol. 15 no. 2 (1934), 65, 73–74.

34 Homer G. Tasker, 'A Dubbing Rehearsal Channel', *JSMPE*, vol. 29 no. 3 (1937), 286–92.

35 K.B. Lambert, 'An Improved Mixer Potentiometer', *JSMPE*, vol. 37 no. 9 (1941), 283–91.

36 Edmund G. Bansak, *Fearing the Dark: The Val Lewton Career* (Jefferson, NC: McFarland, 1995), 137.

37 Bansak, *Fearing the Dark*, 137.

38 Lucille Fletcher, 'Squeaks, Slams, Echoes, and Shots', in John D. Kern and Irwin Griggs (eds), *This America* (New York: Macmillan, 1942), 200. (Originally published in *The New Yorker*, 13 April, 1940.)

39 Fletcher, 'Squeaks, Slams, Echoes, and Shots', 200.

40 John K. Hutchens, 'The Shockers', *New York Times*, 8 November 1942, X12.

41 Neil Varma, 'Honeymoon Shocker: Lucille Fletcher's "Psychological" Sound Effects and Wartime Radio Drama', *Journal of American Studies*, vol. 44 no. 1 (2010), 145.

42 Hutchens, 'The Shockers', X12.

43 Fletcher went on to write the radio and film scripts for 'Sorry, Wrong Number' (in 1944 and 1948 respectively). By this time the number of crime shows, radio adaptations of crime films, and crime films adapted from radio shows were legion. We would like to thank Frank Krutnik for providing some of the sources we have drawn on in this section and for providing a copy of his forthcoming article, 'Theatre of Thrills: The Culture of Suspense'.

Chapter 15

Kathryn Kalinak

THE CLASSICAL HOLLYWOOD FILM SCORE

THE ACADEMY OF MOTION PICTURE ARTS and Sciences added the originally composed film score as an award category in 1934. (Steiner won his first Oscar the next year for *The Informer.*) By the mid-thirties musical accompaniment which depended on original composition and incorporated the selective nondiegetic use of music became the dominant practice, replacing older traditions (Chaplin's use of continuous music in his early sound films, for instance) and alternative practices (Warner Brothers' cycle of social realism films which omitted music entirely). What would become known as the classical model was almost immediately absorbed into Hollywood practice, its production both reflecting and reinforcing the structure of the studio system.

Studio production was not new in the sound period. Silent filmmakers depended upon the collective resources of a studio to produce, promote, and distribute their films. What the sound period did was institutionalize a mode of production which contractually bound the individual agents who create a film to the studio. The classical score developed during a period which saw the solidification of the producer's power. A composer, like other craftspeople employed by the studio, experienced a relationship to any given film that was specific, transitory, and subject to the authority of the studio.

A determining characteristic of the studio model was its efficiency, keeping the assembly line of production moving with a highly diversified division of labor. Personnel were assigned tasks specific to their talents and capabilities and were expected to execute them with skill and speed. The process by which music became part of a film was broken down into its various parts and divided up among personnel trained to complete these assignments quickly. Composers, usually with a music editor, sometimes with other production personnel, spotted the film, that is, identified likely places for musical accompaniment. Composers were then expected to compose, sketching out their ideas on anywhere between two and eight staves of music. These sketches were passed on to an orchestrator or orchestrators, who chose particular instruments to voice specific lines. The terms "orchestrator" and "arranger" were used fairly interchangeably in Hollywood, although technically an arranger orchestrated a preexistent piece of music (like "Dixie" in *Gone With the Wind*). On "B" films, with abbreviated production schedules, this process was often subdivided even further with teams of composers and/or orchestrators dividing up the work.[1] Copyists transcribed the finished score into parts for the orchestra. The conductor conducted the recording session which synchronized score to film. This division of labor was built for speed and efficiency. In practice, however, the distinctions between the various components of the process were not always clear.

The division between composition and orchestration was the most frequently blurred. Although most composers did not orchestrate their own scores, it is a misconception to assume that they did not exert a determining influence in this regard. In fact orchestration was a sore point among those composers in Hollywood (and there were many) who were clearly capable of orchestrating their work, but were prevented from doing so by a system which used them for their compositional talents. Other composers established a relationship with an orchestrator who would become so familiar with the composer's orchestral style that he could reproduce it instinctively. Hugo Friedhofer, a composer who began his career in Hollywood as an orchestrator for Steiner and Erich Wolfgang Korngold, worked on seventeen of Korngold's eighteen original film scores. Friedhofer claims that he knew what Korngold wanted, even if it was not notated in the score.

Many composers, however, insisted on conducting. While it was the common practice for a studio's musical director to conduct a score, especially on "B" pictures, it remained the prerogative of the composer on the big-budget, prestige productions. Korngold, Steiner, Alfred Newman, and Bernard Herrmann usually wielded the baton for the scores they composed.

The Hollywood production model also dictated the time frame in which composers were expected to work. Scoring started after a film was in rough cut, and had to be completed in the time it took to finish a print for distribution. Typically this period would extend from four to six weeks, but a more accurate indicator of the actual time a composer was allotted is a continuum. One end represents the shortest amount of time humanly possible for the composition of a score. Max Steiner claimed he lived without sleep for the eight days it took him to compose and conduct *The Lost Patrol* (1934). The other end represents the maximum amount of time over the roughly six-week limit that a composer could negotiate. Erich Wolfgang Korngold's contract for Warners, for instance, allowed him longer than the usual six weeks to complete a score. In addition, it was not uncommon for one composer to work on several productions simultaneously or for several composers to work on one production simultaneously. These conditions did not promote allegiance to a single film, but rather fostered primary allegiance to the studio.

The classical Hollywood film score proliferated in the thirties but in an interdependence with the growing power of the studio. The hierarchy which positioned management over labor permeated every facet of a film's production from its visual style to its music. In the case of the score, the music department often became the intermediary between studio and composer. A few composers, like Max Steiner at RKO and Alfred Newman first at United Artists and then at Twentieth Century-Fox, were also heads of music departments themselves and had easier access to and more clout with studio chiefs as a result. Ultimately, however, this structure of accountability affected the composer's responsibility for the score and gave the studio more direct control over how a score would be fashioned. David O. Selznick's famous memos often contained intricate instructions on the music. Hal B. Wallis' music notes for *Captain Blood* ran five single-spaced pages. Some studios even instituted the policy that all title music be written in a major key. (Irving Thalberg once wrote a memo on main titles requesting MGM composers to "kindly refrain in the future from using minor chords.")[2]

This intervention marks a major difference between the classical film score and its silent predecessor. The mediating devices which regulated the conventions of the silent film score (manuals, columns in trade papers, musical encyclopedias, and cue sheets) were, for the most part, produced independently of the studio whose authority had only an indirect and limited impact on the music. It was over the silent film's orchestral scores that studios wielded the kind of power that they exercised in the sound film. But even here the studio was not invulnerable. Theater organist Gaylord Carter recalls a battle with Paramount over the score for *The Student Prince* (1927) in which the studio capitulated to pressure by accompanists to use Sigmund Romberg's music instead of the studio's score.[3]

The classical Hollywood film score

The classical Hollywood film score can best be understood not as a rigid structural or stylistic manifesto but rather as a set of conventions formulated to sustain and heighten the active reality of the classical narrative film. A score can be termed classical because of its high degree of adherence to these practices. This is not to say that all scores composed in Hollywood fit this model or that the model didn't change in response to innovation and experimentation. A conventional practice for scoring films, however, did develop during the first decade of sound production. By the mid-thirties its principles found acceptance and proliferated. Composers who negotiated successful careers in Hollywood relied upon the principles which garnered Academy Awards and prestigious assignments for its practitioners.

One of the most influential of those practitioners was Erich Wolfgang Korngold whose first original score for Warner Brothers' *Captain Blood* (1935) was regarded, along with the work of Max Steiner and Alfred Newman, as the apex of musical achievement in Hollywood during the crucial first decade of sound production. Scores like *Captain Blood* set the trend for originally composed, full-length, symphonic scores modeled on a nineteenth-century musical idiom. The remainder of this chapter will be devoted to this score, to an analysis of selected cues which demonstrate key techniques and practices that characterize the Hollywood model.

The classical Hollywood score differed from its silent predecessor in two important ways. First, it reproduced every diegetic reference to music in the film, unlike the silent score which reproduced such references only selectively. Second, the classical score provided intermittent musical accompaniment rather than a continuous stream of music. A contemporary textbook on composing for film and television puts it this way: "A constant flow of music gets in the way."[4] In fact, the classical Hollywood film score actually reversed the crucial relationship in silent film between continuity and selectivity. In silent film continuous music took precedence over diegetic fidelity. The reproduction of diegetic reference to music was intermittent, left to the prerogative of individual accompanists. In sound film it was diegetic fidelity that took precedence over continuous music. Musical accompaniment was intermittent, its placement determined by the individual composer.

Music and structural unity

One of the most important ways in which music became conventionalized in the classical model was through shared perceptions about when music should be heard. Formal concerns provided immediate and identifiable points of entry for musical accompaniment. The silent film score both compensated for the silence of the theatrical auditorium and sustained narrative continuity. The sound film score was freed from the necessity of seamlessness, but inherited from the silent film its function as arbiter of narrative continuity. It is not surprising, then, that composers of the classical score gravitated toward moments when continuity is most tenuous, to points of structural linkages on which the narrative chain depends: transitions between sequences, flash-forwards and flashbacks, parallel editing, dream sequences, and montage. Virgil Thomson articulates the problem from a composer's point of view: "The cinema is naturally a discontinuous medium. . . . [Music] should envelop and sustain a narrative, the cinematographic recounting of which is after all only a series of very short incidents seen from different angles."[5] In the classical system of narration the reconstruction of time and the redefinition of space were controlled through lighting, camera placement and movement, and techniques of continuity editing. An overlooked element in this process is music which helps render the presence of these other elements in the classical narrative system invisible.

Martin Skiles's text on composing for film and television cautions the would-be composer that "he will be faced with having to literally fill the gap with some music ... without its application, scenes such as these could have a disastrous effect on the overall story."[6] This structural reciprocity between narrative continuity and the musical score can be seen in the practice of the bridge, a musical transition which elided gaps occasioned by spatial and temporal disjunctions. Music compensates for potentially disruptive shifts in the form of continuous playing and frequently through reliance upon extended melody or established musical forms. This reciprocity governed the use of musical accompaniment in situations ranging from those as simple as the straight cut to those as complex as the montage. An example which falls between these extremes is a transition in *Captain Blood* precipitated by the shift from Jacobean England, where Dr. Peter Blood (Errol Flynn) has been convicted of treason, to the British colonies in the Caribbean, where he is sent as a slave. It is precisely this kind of narrative moment which engendered musical accompaniment.

The transition is effected in a sequence of twelve shots facilitated by editing techniques which establish the passage of time, such as the lap dissolve, and a superimposed text which establishes the shift in geographic location. Music negotiates this spatiotemporal ellipsis in a number of ways. In this case Korngold uses a standard ABA form built around two motifs associated with Peter Blood: the exposition of theme A (heard here in the minor), followed by a contrasting theme B, and a return to the original theme A.[7] To provide maximum continuity, music begins *before* the shift in time and place, immediately following the delivery of King James's speech decreeing that rebels awaiting execution be sent to the colonies as slave labor. The first theme or A section, played initially in a standard symphonic setting with the violins carrying the melody, enters at a low volume which is noticeably increased during the lap dissolve to the next shot, a ship at sea with a superimposed text explaining its destination. The musical cue continues uninterrupted through the succeeding lap dissolve to a close-up of one of the prisoners aboard ship. Here the Blood motif is repeated in an instrumental variation (horns with accompanying woodwinds and strings). A sequence in the ship's hold follows with a contrasting melody, the B section, played in the low registers of the strings and woodwinds. At the lap dissolve which covers the ellipsis in time and shift in geography, the A theme returns in its initial orchestral setting. A dialogue cue confirms the progress of the ship ("Ahoy, on deck. I've sighted Port Royal"). As the establishing shot of the port replaces the ship on the screen, the score offers the exotic motif representing the city. Rather than break the melodic integrity of the cue, Korngold inserts this new material into a pause in the Blood motif. The disjunction between the establishing shot of Port Royal and the following shot of Blood on board ship is negotiated by several techniques: a dissolve to a reverse eyeline match from Blood's point of view, a repeated word from the previous dialogue cue ("It's a truly royal clemency we're granted, my friends"), and the repetition of the Blood motif in a developmental variation. The sequence concludes with close-ups of Blood's impassioned speech about injustice and a return to the closing phrase of his motif. The reliance upon an established form, the use of continuous music, and the dependence on extended melody rather than short phrases creates the effect of unity reinforcing the narrative at a potentially disruptive moment.

The montage made even greater demands on the music. Quick cuts seldom matched, superimpositions of sometimes three or four images including overlays of text and graphics, and the absence of dialogue, diegetic sound, or both, taxed the conventions of narrative construction, particularly the representation of time. Music lent its rhythmic temporality to such sequences, regulating the flow of images. While it was not necessary or even desirable in the classical narrative system to accompany every transition with music (there were a number of factors which went into this determination including length, narrative content, and proximity of other musical cues), it was almost impossible to allow a montage to exist without accompaniment.[8]

In *Captain Blood* one such montage compresses the exploits of Peter Blood's pirate career. The structural function of the bridge, to provide continuity in moments of spatial-temporal disjunction, is extended here to cover a fifty-eight-second sequence with multiple superimpositions of image and text. The montage is preceded by a long shot of Blood as he commits himself to a life of piracy on the high seas. Music, which sneaks in under the men's rousing cry of approval, again precedes the actual beginning of the montage, eliding the boundary between the unobtrusive, invisible style of narrative exposition and the technical artifice of the montage. One element fundamental to that artifice is the manipulation of the soundtrack. This often takes the form, as it does here, of a suspension of diegetic sound. Shipboard battles, for instance, are seen but not heard. Even a close-up of Blood clearly speaking has no diegetic sound. Music functions here not unlike its counterpart in the silent film, compensating for and covering this void.

In the same way that music compensates for the absence of diegetic sound in the montage, it also compensates for its spasmodic rhythms. Korngold's musical cue, marked *allegro furioso* (furiously fast), embodies a kind of metronomic function in a quick, continuous pulse which provides a rhythmic framework for a series of edited images lacking temporal and spatial coherence. Finally, music's continuous presence, structurally organized around repetition of the Blood motif, offers its own structural unity as a frame for that of the images. Music thus provides a variety of functions in the montage: it elides the boundary between narrative exposition and virtuosic technical display; it compensates for the absence of diegetic sound; it marks out a regular rhythm often lacking in the complex structure of the montage; and it helps to provide unity through conventions of musical form.

Music and narrative action

If formal concerns provided the film composer with a definable set of access points for musical intervention, the starting point for the creation of a musical cue was the image. Its content was gauged in two ways: that which was explicit in the image, such as action, and that which was implicit, such as emotion or mood. What Hollywood film scoring conventionalized was a set of practices for guiding composers in responding to the image musically.

From the nuance of facial expression and bodily gesture to the sweep of movement across the screen, music punctuated narrative. At the heart of this practice was the relationship between music and action. Structural properties of music, such as tempo or rhythm, were harnessed to the visual representation of movement in order to create a particular speed or rhythm. Explains Hugo Friedhofer: "You know the old idea—the horse runs: the music runs."[9] Sequences most likely to be scored in this manner were those which embodied a consistent tempo and/or rhythm which was visually discernable and could easily be matched to a musical accompaniment. In *Captain Blood* there are numerous examples of musical tempo matched, though not necessarily synchronized, to narrative action. Three examples of such cues occur within the first five minutes of the film. The first, a horseback ride at breakneck speed, is accompanied by a passage Korngold marked *ancora piu mosso* (even more agitated), the tempo of the music reproducing and reinforcing the frenzy of the ride. Within a few moments the opposite effect is achieved when Korngold uses the deliberate tempo of the *largo* (slowly) to accompany the lugubrious procession of accused traitors toward prison. Between these two cues is yet a third in which Korngold scores the entrance of loyalist troops with a cue he marked *quasi marcia* (like a march), matching the tempo of the music to the pace of the soldiers' step. Throughout the film similar musical cues contribute to the action, from the extended *allegro* (fast) and *presto* (very fast) in the duel between Blood and his rival Levasseur (Basil Rathbone), to the *sforzando* chord, known as a stinger, which marks the sword's fatal thrust. In fact, swashbucklers like *Captain Blood* contained so many action sequences that Korngold initially refused the studio's follow-up, *The Adventures of Robin Hood* (1938), because of its imposing action format and the sheer amount of music he would need to compose. As he wrote Hal B. Wallis, "*Robin Hood* is no picture for me. . . . I am a musician of the heart, of passions and psychology; I am not a musical illustrator for a 90% action picture."[10]

The relationship which existed between music and narrative action did not preclude the possibility of deliberate disjunction for thematic resonance. (There is a striking sequence in *Force of Evil* (1948), for instance, where the protagonist races down a staircase looking for his dead brother to a musical accompaniment of very slow tempo.) But the use of music for this purpose in the classical film score was limited. A more common practice involved using tempi or rhythm in disjunction with narrative action in order to increase a pace perceived lacking in the images. Explains Max Steiner: "There may be a scene that is played a shade too slowly which I might be able to quicken with a little animated music; or, to a scene that is too fast, I may be able to slow it down a little and give it a little more feeling by using slower music."[11] Skiles's text offers this practical advice for those scenes which "played a shade too slowly." "Intelligent chord progressions, changes of color . . ., the use of such melodic percussion instruments . . . can give the feeling of movement an [*sic*] sustain the audience interest through periods such as these."[12]

In *Captain Blood* Korngold uses the standard musical device of the *ostinato*, a repeated melodic or rhythmic figure, to propel scenes which lack dynamic and compelling visual action. One such sequence involves a clandestine meeting between Blood and his fellow slaves to plan their escape. By necessity dialogue is whispered and character movement is minimal. Korngold begins the musical cue before the cut to the meeting. As guards patrol the grounds, a simple *basso ostinato*, a repeated figure in the bass line, creates a rhythm for the sequence. Following the lap dissolve to the interior of the slaves' quarters, this *ostinato* is joined by two distinctive musical figures, one for cello and tuba, the other for trombones, which add rhythmic intensity to a pan over the motionless men. As the meeting begins, the volume of the music gradually decreases and eventually drops out so that information crucial to the plot—relayed through dialogue—can be heard unobstructed. Music returns during the emotional scene between Blood and a fellow slave and continues to the end of the sequence.

At its extreme the practice of matching music to narrative action resulted in the direct synchronization between visual action and music known as Mickey Mousing. Although it is often associated with silent film accompaniment, Mickey Mousing is a product of sound technology, named for the animated cartoons in which it developed. Mickey Mousing was a well-established practice by the mid-thirties, and it is particularly prominent in the scores of Max Steiner. In *The Informer*, for instance, the protagonist, Gypo Nolan, lumbers down the street in direct synchronization to the leitmotif Steiner wrote for him, every step accompanied by a note of music. To the modern listener, Mickey Mousing may seem excessively obvious and even distracting, but ironically it was a practice founded in the very principle of inaudibility. The vocal track in classical cinema anchors diegetic sound to the image by synchronizing it, masking the actual source of sonic production, the speakers, and fostering the illusion that the diegesis itself produces the sound. Mickey Mousing duplicates these conventions in terms of nondiegetic sound. Precisely synchronizing diegetic action to its musical accompaniment masks the actual source of sonic production, the offscreen orchestra, and render the emanation of music natural and consequently inaudible. Musical accompaniment was thus positioned to effect perception, especially on the semiconscious level, without disrupting narrative credibility.

Mickey Mousing, like other elements of the classic score, was harnessed to the narrative, catching an action to solidify its importance. Korngold did not depend on Mickey Mousing in his scores (his lengthy cues made its use problematic), but there is a textbook example in *Captain Blood.* As Colonel Bishop is thrown overboard, a rising musical figure imitates each swing of his body as the men prepare to fling him aloft. His entry into the water is caught with a stinger.

Music and emotion

Music, however, not only responded to explicit content, but fleshed out what was not visually discernible in the image, its implicit content. In this capacity, music was expected to perform a variety of functions: provide characterization, embody abstract ideas, externalize thought, and create mood and emotion. For the fitter of film music in the silent era visual reference was the surest indicator of content. In the sound era the dialogue also helped mark intent. But ultimately the interpretation of implicit content was left to the individual prerogative of the composer. Frank Skinner's admonition to students of Hollywood scoring echoes the familiar advice dispensed to silent film accompanists: "train yourself to grasp situations fast."[13]

Scenes that most typically elicited the accompaniment of music were those that contained emotion. The classical narrative model developed certain conventions to assist expressive acting in portraying the presence of emotion, primarily selective use of the close-up, diffuse lighting and focus, symmetrical *mise en scéne*, and heightened vocal intonation. The focal point of this process became the music which externalized these codes through the collective resonance of musical associations. Music is, arguably, the most efficient of these codes, providing an audible definition of the emotion which the visual apparatus offers. Samuel Chell compares the musical score in classical Hollywood narrative film to the laugh track of television situation comedy.[14] Music's dual function as both articulator of screen expression and initiator of spectator response binds the spectator to the screen by resonating affect between them. The lush, stringed passages accompanying a love scene are representations not only of the emotions of the diegetic characters but also of the spectator's own response which music prompts and reflects.

Composers of the Hollywood film score were drawn almost compulsively to moments of heightened emotional expression which afforded them what they perceived to be the most direct access to the spectator. Steiner said of his score for *The Informer*: "I put the music in the harp when McLaglen sold this guy down the river and in the very end, I had him sing when his mother forgave him in church. It brought a few tears."[15] That power of music to elicit emotional response is often ascribed to music's ability to transcend the limitations of the visual image. Elmer Bernstein typifies this position when he argues: "music can tell the story in purely emotional terms and the film by itself cannot." For Bernstein film is "a visual language and basically intellectual. You look at an image and you then have to interpret what it means, whereas if you listen to something or someone and you understand what you hear—that's an emotional process."[16] This position, frequently articulated by film composers, posits music's ability to transcend cognitive mediation as the source of its emotional appeal. But it may be more accurate to attribute music's power not to its presumed innocence but to the fact that its forms of mediation are less immediately discernible.

Korngold's score for *Captain Blood* contains some interesting examples of how music is used in the classical film score to create emotion. Predictably, many of these scenes are related to the romantic plot concerning Blood and the niece of the governor of Jamaica, Arabella Bishop (Olivia de Havilland). At cross-purposes for most of the film, these two lovers acknowledge their feelings for each other only at the last possible moment. Thus Korngold's music responds not to what is explicitly stated in the dialogue (they are either coy or insulting to each other), but to what is implicit in their demeanor and reinforced by conventional expectations of classical narrative (that two attractive stars of the opposite sex belong together). When Blood, for instance, responds to the inquiries of Miss Bishop with a cavalier attitude and high-handed remarks, the music reassures us that he is hardly as indifferent as his demeanor suggests. Similarly, when Arabella Bishop tells Blood that she hates and despises him, the soaring violins of the love theme soften, even negate, her rejection, pointing to her true feelings, thinly disguised beneath the surface. Music draws out the emotional content of the scene, hidden from the characters but not from the spectators.

The leitmotif Korngold used to represent the love between Peter Blood and Arabella Bishop facilitates such a reading. It depends upon several standard devices for emotional expression: dramatic upward leaps in the melodic line; sustained melodic expression in the form of long phrases; lush harmonies; and reliance upon the expressivity of the strings to carry the melody. Because of the nature of the romantic entanglement in this film, Korngold gives the love theme extended expression not when the lovers are together on the screen, but when they are apart, during "private" moments when they are free to rhapsodize about each other. One such example occurs when Blood sends Arabella Bishop ashore moments before the final sea battle. The love theme enters beneath Bishop's dialogue and is turned up to full volume as the dinghy bearing her toward safety moves farther and farther away from the ship. Her gaze, as well as Blood's in the reverse shot, is encoded as melancholic by the extended reprise of the love theme.

Music and subjectivity

The preceding example is particularly interesting for in it music encodes not only emotion but point of view. The classical narrative model developed a number of conventions for internal thought including voice-over, specific editing patterns (the eyeline match especially combined with the dissolve), and musical accompaniment. Together or separately, these techniques offered an analogue for a character's consciousness. That Bishop is explicitly thinking of Peter Blood, whom she has left behind, is denoted through editing and camera movement. A dolly-out on Blood aboard ship is positioned as the object of her eyeline match. What she is thinking about him, however, is only implicit in the scene, made explicit by a combination of generic expectation (given the conventions of Hollywood romance, what else would she be thinking of?) and the presence of the love theme.

Earlier in the film there is a sequence which this one mirrors. Blood, freed from his imprisonment, is poised on the brink of a pirate career. He stands on the deck of a captured Spanish galleon as his crew prepares to depart, and for a moment he gazes at the faraway shore. A dissolve to a close-up of Bishop at the dock reveals the contents of his mind and reinforces what we have already presumed—that his gaze indicates regret about leaving her. A soaring string melody which modulates into the love theme authorizes this interpretation.

Music and mood

The creation of mood and atmosphere also relied on the ability of the composer to discern implicit content and respond with appropriate music. "To understand moods in music and to be able to grasp a mood in a pictorial situation,"[17] a skill crucial to the silent film accompaniment, remained central to the sound composer's task. Establishing shots were the most typical point of access for atmospheric music, especially those which made direct reference to geographic location or historical period. Mood music tapped the power of collective associations to create the time and place represented in the image. Korngold had ample opportunity for atmospheric music in *Captain Blood*, its Caribbean setting providing exotic locales and its historical context offering a wide spectrum of national types. Korngold's Port Royal cue, for instance, uses unusual instrumentation to define the exotic nature of the port (triangle, celesta, harp, and muted trumpets), and assigns percussive instruments the job of carrying the melody. (A celesta is a percussive instrument resembling a small upright piano.) Other examples of music's function in establishing geographic context or national identity include the motif heard during the Spanish pirates' conquest of Port Royal and the motif composed for Tortuga. Both are built upon the pronounced syncopated rhythms associated with Latin cultures. Korngold takes advantage of even the smallest opportunity in *Captain Blood.* A variation on the opening motif of "God Save the Queen," labeled "The Good King William," can be heard when either England or King William is mentioned. The lowering of the French flag is accompanied by a variation on the bugle call from Bizet's *Carmen.*

The classical score, however, relied not so much on actual imitation of music indigenous to other cultures as on a more generic concept of exoticism. Fred Karlin and Rayburn Wright in their 1990 manual for film and television composing, *On the Track: A Guide to Contemporary Film Scoring*, describe it as "quasi authenticity."[18] Instruments not typical of the symphonic complement and rhythmic patterns associated with foreign cultures represented, often interchangeably, anything from Tortuga, Peter Blood's seventeenth-century pirate haven, to Skull Island, the twentieth-century home of King Kong. A prominent example is the cue Korngold composed for the Spanish "Street Scene." Here Korngold exploits the horns playing in parallel thirds characteristic of the Mexican mariachi bands which developed during the nineteenth century to represent Spanish pirates roaming the Caribbean during the seventeenth century. Similarly, "Tortuga" suggests the exoticism of the pirate haven through an exceptionally unusual complement of instruments. Hand cymbals, tambourines, triangles, harp, celesta, saxophones, vibraphone, and guitar are joined by a cimbalom, a Hungarian instrument in the dulcimer family frequently associated with gypsies! (Watch for the cimbalom solo diegetically produced by a musical pirate.)

Atmospheric or mood music was not limited to that generated by the spatial and temporal necessities of the plot. Music was called upon to create a mood sometimes only dimly suggested by the images and thus ready the spectator for a heightened response to narrative development. One of the most important of these moods is suspense, and the Hollywood film score depended upon a core of musical conventions, a shorthand if you will, for encoding tension. A sequence which takes place during Blood's enslavement on the Bishop plantation exemplifies many of these practices. A prisoner who has attempted escape is about to be branded as a fugitive traitor. Korngold's music intensifies the suspenseful potential of the scene through the use of *tremolo* strings, cymbal rolls, horn flutters, an exceptionally high flute part, and *crescendo.* At the moment when the iron sears the man's flesh, the *crescendo* reaches a *fortissimo*, and a *sforzando* attack marks the action transpiring out of the camera's range.

Another technique Korngold uses to create tension is the *ostinato.* Repetition of a specific musical figure can provide a rhythmic intensity to fill an otherwise unremarkable scene. An *ostinato* can also create tension through sheer accumulation, a kind of musical Chinese water torture. Korngold heightens this effect during the men's escape when he introduces a descending *basso ostinato* to infuse their stealthy and circumspect movements with urgency.

Music and dialogue

An important constraint on the presence of music was dialogue. Privileged above all other elements of the soundtrack, dialogue had priority in the classical system's hierarchy of audibility. Music, even that diegetically produced, must not detract from the power of dialogue in the exposition of narrative. In the crucial early period of sound film, underscoring, or musical accompaniment to dialogue, was avoided. The artificiality of nondiegetic music was particularly at odds with the "naturalism" of the spoken word, and the earliest recording techniques prohibited mixing dialogue and music. By the early thirties technology permitted limited underscoring, but the process proved difficult and often produced distortion. Leonid Sabaneev, in his text for film composition translated in 1935, argued that "music should cease or retire into the background when dialogues and noises are taking place. Except in rare instances, it blends but poorly with them."[19] Yet as early as 1931, a film like *Cimarron* used musical accompaniment under dialogue (Sabra Cravat's monologue about her husband, Yancey, is accompanied by music).

Conventions for underscoring developed to bring the expressive possibilities of music to the human voice. These included relying on the strings; avoiding the woodwinds, whose timbre tends to obscure dialogue; avoiding extremes in register; using melody and avoiding counterpoint; and relying on simple rhythms, slow tempi, and low volume. All of these conventions, however, could be violated for a specific thematic purpose. Nonetheless, conventions for underscoring were tacitly assumed throughout the industry. Even contemporary scoring practices reflect these conventions. Karlin and Wright's manual, for instance, advises would-be composers to avoid textures that are "intrusive." Above all, "Don't overwrite."[20]

A recurrent refrain in Wallis' music notes for *Captain Blood* concerns the audibility of dialogue. Wallis indicated where music should be lowered in volume to accommodate the spoken word or eliminated entirely. Typical of his instruction is the following note. "The music when BLOOD's ship sails, we keep all the music in, but every time BLOOD gives a command or there is any dialogue from WOLVERSTONE or PITT, or any of them, drop the music way down and bring the voices up and then raise the music when the dialogue is over."[21] As Wallis puts it: "Under the dialogue, let's hear the dialogue."[22]

Korngold was particularly noted for underscoring, and in *Captain Blood* he experimented with a variety of approaches from the sustained melodic underscoring of the love scenes to the uniquely constructed cues which punctuated key words or phrases in a line of dialogue. One of the longest sustained sequences between Peter Blood and Arabella Bishop, a horseback ride, is entirely underscored. Violins at low volume carry the melody with simple harmonic accompaniment in the strings. There is not even a pause for the dramatic slap which greets Blood's kiss, the modulation to the Blood motif and the change in instrumentation from strings to a horn solo shifting the mood. Similarly, the sequence on Tortuga which results in the partnership between Blood and Levasseur is almost continuously underscored. Here the only concession to dialogue is volume. The unusual instrumental texture, however, remains, unmodified to accommodate the spoken word. Not surprisingly, Korngold eliminates music entirely from Blood's final close-up and crucial interior monologue.

Perhaps the most interesting examples of Korngold's underscoring are the unique cues he customized for specific lines of dialogue, punctuating vocal rhythms with musical rhythms to heighten a line's delivery. Blood's final order as the escaped slaves make ready to get under way is an interesting example: "Break out those sails and watch them fill with the wind that's carrying us all to freedom!" The first words are spoken against a rising sequence in the violins, which create a sense of anticipation. At the pause after the phrase "Break out those sails," an orchestral chord punctuates Blood's speech, almost as if it were the musical equivalent of the comma. The next phrase is "punctuated" similarly, with chords after "and" and "watch" speeding up the delivery of the line and lending an urgency to Flynn's intonation. The final two cords punctuate the line after "the wind." A drum roll culminates on the last word in the line, "freedom," obviously drawing attention to its importance. The tension set up by the drum roll is resolved when the fully orchestrated version of the Blood motif returns at the conclusion of Flynn's dialogue.

For his next film, *Anthony Adverse* (1936), Korngold would develop this technique, composing music just under the pitch of the voices and "rushing" it into pauses left open in the dialogue.[23] As Korngold explained the process: "I wrote the music in advance, conducted—without orchestra—the actor on the stage in order to make him speak his lines in the required rhythm, and then, sometimes weeks later, guided by earphones, I recorded the orchestral part."[24]

The precision of such underscoring is amazing considering that in recording sessions Korngold shunned the standard devices for synchronization. The common technique for insuring coordination between music and image (and the one favored by Max Steiner) was the click track, basically a soundtrack with audible clicks. The precise tempo necessary to fit music to image in a particular sequence was prerecorded on a magnetic track in the form of a series of holes. When run over a playback head these holes produced pops or clicks, creating an audible metronome fed to the conductor and musicians via earphones. Since the click track was variable, any tempo could be produced. Two other devices used to facilitate the synchronization of musical effects were the punch and the streamer. Both were visual cues on the image track itself. A punch was a perforation in the film which caused a flash of light during projection. It could be seen without looking up from the score. A streamer was a diagonal line scratched into the film's emulsion, creating a vertical line which moved across the screen. Punches and streamers readied the conductor for especially difficult or tricky points of synchronization by signaling their approach. Korngold did not use the click track, relying instead on his eye, an innate sense of timing, and an occasional punch or streamer as "insurance."[25] He avoided even a simple stopwatch and was renowned for his sense of timing. "If a sequence called for forty-two and two-thirds seconds of music, he would write a piece of music and conduct it so that it would fill forty-two and two-thirds seconds."[26]

One final example warrants analysis. In the cue which accompanies King James's speech about the fate of the rebels and which precedes the transition to the colonies, Korngold violates the conventions for unobtrusive underscoring. Here the incongruity of sonorities between the bass voice of actor Vernon Steele and a flute high in its register is heightened by unusual instrumentation such as the celesta and the vibraphone, a nonmelodic and arresting musical line, and dissonant harmonies. Korngold has obviously exploited these devices to provide a distracting and discomforting musical background for the callous cruelty of James and his decree.

Music and spectacle

Although dialogue received the highest priority on the classical soundtrack there were moments when music was privileged in the same way, that is, when dialogue and sound effects were mixed to accede to its priority. The creation of spectacle in the classical narrative model afforded music this position, where virtuosic technical display was heightened by the substitution of music for sound. The closer narrative moved toward pure spectacle and away from the naturalistic reproduction of sound, the more music moved toward the forefront of conscious perception in compensation. Wallis' music notes reflect this relationship: "Lift it way up on that long shot of the boat and right up to where the man says 'Ahoy! Port Royal!' The music should lift up there on those silent shots."[27]

Both the montage and transition analyzed earlier provide examples of music's reciprocity with spectacle, music responding to the presence of spectacle with continuous playing and increased volume. Another example occurs during the sequence in which Blood's ship makes ready to sail from Port Royal. It begins with the narrative justification necessary in the classical model for the eruption of spectacle. In a series of quickly edited shots, Blood barks orders to his crew to ready the ship for departure, and the spectacle of sails unfurling and the ship under way immediately follows. The leap from narrative to spectacle, evidenced in the increased length of shots, the virtuosic camera angles and movements (a stunning vertical pan of the ship under full sail), and close-ups of the film's stars, is marked by the music. Naturalistic sound drops out, and the music, turned up to full volume, provides the only sonority on the soundtrack.

The greatest opportunities for sustained musical expression, however, were extra-diegetic, the main title and end title cues. Freed from the restraints of the diegesis, title music was structured more by the conventions of musical form than by the dictates of narrative development. Typically a main title presents the principal leitmotifs later developed in the score. The end title, the musical accompaniment to the final credits, was more freely structured and appreciably shorter. It often comprised little more than a memorable restatement of an important leitmotif.

The lengthiest, uninterrupted musical cue in the classical narrative film was typically the main title, which because it preceded the actual diegesis had increased power to set atmosphere and mood for the entire production. The main title for *Captain Blood* was arranged by Ray Heindorf of Warners Brothers' music department. It was quite common within the industry to turn the main title over to an arranger who created an overture from the composer's own material, saving the composer for more important work. Although it is somewhat atypical of Korngold (he did arrange the main titles for most of the other scores he wrote), such was the case with *Captain Blood.* Rhythm, instrumentation, harmony, and melody work together to create a heroic opening which announces as clearly as the pirate ship and crossed swords emblazoned on the title cards that this is a swashbuckler. The main title begins with a musical flourish suggesting the heraldry to come: a drum roll, cymbal crash, and string *glissandi.* Heard immediately after this introduction is the leitmotif associated with Captain Blood, arranged as a horn fanfare. Brass instruments with their connotations of the military are a classic convention for the heroic. Here harmonic structure adds stability and weight (the use of triads, the most fundamental chord in tonal harmony, and strong cadences), as does the use of a major key. Dotted rhythms add vitality to a melodic line which incorporates several dramatic upward leaps. The contrasting B section introduces a second motif quite different in character. The strings carry this new melody which incorporates even rhythms and lyrical phrasings.

Typically, the main title was conceived in terms of the structure of concert music, here a variation of the *sonata-allegro* which encapsulates exposition, development, and recapitulation of one or more themes. Following the exposition is a developmental section in which the musical material is developed through instrumental variation: woodwinds with *pizzicato* (or plucked) string accompaniment and later strings with a horn countermelody. A bridge passage which modulates upward in gradual *crescendo* builds to the climactic moment when the main theme returns for a final reprise in its original instrumentation. A brief musical coda concludes the main title.

By 1935 it had become commonplace to connect the main title to a film's first diegetic cue in order to create a continuous musical background. In *King Kong*, the main title cue extends twenty seconds into the diegesis. *Captain Blood*'s first cue extends from the initial orchestral flourish which opens the main title to the final chord which concludes the first sequence, over a minute into the film. Such continuity covered the entrance of nondiegetic music, naturalizing its presence in the diegesis. End title music functioned in reverse, the last diegetic cue connected to the end title cue. This musical practice helped to negotiate the rift between the hypnotic darkness, which facilitates an absorption into the cinematic image, and the startling bright light of the theater, which disrupts it.

The placement of music

As important to the classical film score as conventions for where to place music were conventions for how to place it, that is, techniques for introducing music into the narrative. Steiner admitted that "the toughest thing for a film composer to know is where to start, where to end; that is, how to place your music."[28] By the mid-thirties, conventions for placing nondiegetic music into the diegesis coalesced, guided by the principle of inaudibility and the technological realities of sound editing and mixing. The key was introducing music into the narrative without calling conscious attention to it. As Herbert Stothart, a contemporary of Korngold, put it, "If an audience is conscious of music where it should be conscious only of drama, then the musician has gone wrong."[29] Karlin and Wright advise that "music starts most effectively at a moment of shifting emphasis."[30] Thus music might be introduced at a scene change or a reaction shot or on a movement of the camera. A sound effect could also distract the spectator from the music's entrance. In *Captain Blood*, for instance, the sound of Colonel Bishop whipping Blood's horse or the sound of cannons aboard the captured Spanish galleon masks the entrance of musical cues. Another technique involved sneaking, an industry term for beginning a musical cue at low volume usually under dialogue so that the spectator would be unaware of its presence. Many of *Captain Blood*'s musical cues at transitions and montages depend upon this kind of entrance. The accompaniment for the pirate montage in *Captain Blood*, for example, begins not at the montage itself but under the masking effect of the pirates' noisy cheer preceding it. Even a distracting visual action could cover the entrance of music. The dramatic drawing of Blood's sword in the duel or the unexpected arrival of James's troops at the rebels' hiding place allows for music to enter with some volume. Ending a musical cue proved easier. In general, it was timed to conclude with the end of the sequence it accompanied, the shift in time and place covering the sudden absence of music. Cues ending within a sequence were masked by dialogue or sound effects or typically depended on tailing out, a term which designates a gradual fade-out of sound.

The idiom of the classical score

The classical Hollywood film score developed an idiom for its expression based on musical practices of the nineteenth century, particularly those of romanticism and late romanticism. The silent film score's reliance upon these practices offered a clear precedent for their use in the sound era, and composers in the crucial decade of the thirties, themselves trained in the late romantic style, reinforced the connection. Max Steiner's grandfather owned the Theater an der Wien, one of Vienna's operatic showplaces, and by the age of sixteen Steiner himself had written an operetta which ran for a year. Korngold was a child prodigy (Mahler pronounced him a genius at age ten) and by the age of seventeen had written two operas widely performed throughout Europe. Dimitri Tiomkin, Franz Waxman, Bronislau Kaper, and Miklós Rózsa were all émigrés who brought with them a musical predilection for the nineteenth century. In fact with the exception of American-born Alfred Newman, the development of the classical Hollywood film score in the crucial decade of the thirties was dominated by a group of composers displaced from the musical idiom in which they had been trained. It was in Hollywood that they were able to reconstitute what John Williams has called "the Vienna Opera House [in] the American West."[31]

Certainly historical factors contributed to Hollywood's adoption of the romantic idiom. It may be more than tradition, however, that coupled them. Film is a discontinuous medium, made up of a veritable kaleidoscope of shots from different angles, distances, and focal depths, and of varying duration. Romanticism, on the other hand, depends upon the subordination of all elements in the musical texture to melody, giving auditors a clear point of focus in the dense sound. Given the high value placed on the spectator's focus on, indeed absorption into, the narrative by the classical Hollywood narrative film, the romantic musical idiom may be its most logical complement.

The musical idiom of the nineteenth century influenced the conventions of orchestration and harmony basic to the classical film score and determined its medium as symphonic. Romanticism and late romanticism relied on lyrical melody as a means of expression. Hollywood's adaptation of this model was characterized by the transcendence of melody, doubling of individual parts, and a reliance on strings to carry melodic material. The romantic attraction to the vast possibilities of orchestral color provided a natural model for film composing which combined orchestral color and the power of collective musical association for thematic resonance. In *Captain Blood*, for instance, Korngold scores the principle leitmotif for Peter Blood as a brass fanfare, relying on the power of the horns to suggest heroism; uses the pathos of two violins for the love theme; and exploits the gypsy associations of the cimbalom to evoke the exoticism of the pirates on Tortuga.

Finally, the nineteenth century bequeathed to the classical film score a symphonic medium. As it had done with other conventions of its nineteenth-century prototype, Hollywood adapted the late-romantic orchestra of ninety-plus players for the recording studio, recreating the deployment of instruments typical of a late-romantic symphony, but reducing the number of players, sometimes by half. The orchestra for *King Kong*, for instance, reported to be as high as eighty musicians, actually numbered about forty-six.[32] In 1933 Warner Brothers had eighteen players under contract; by 1938, the year Korngold conducted *The Adventures of Robin Hood*, its orchestra had expanded to only about sixty players.[33] Orchestra size was influenced by genre, from the costume epics which warranted a big sound to the contemporary dramas which evoked a leaner one. A swashbuckler like *Captain Blood*, of course, warranted the former, a full, rich, symphonic sound. In fact, Korngold's neoromantic score for *Captain Blood* so precisely reconstitutes the musical idiom of late romanticism that most listeners cannot distinguish between Korngold's original composition and the two extended selections he borrowed from Franz Liszt.[34] Other musical idioms, such as jazz and pop, found their way into the classical film score during the forties and fifties. Despite the divergence of the musical idioms on which they are based, jazz and pop scores preserved the structure of the classical model. More important, their failure to adopt the pervasive romantic idiom was to some extent narratively justified. Alfred Newman and Alex North both used jazz for films set in contemporary New Orleans, *Panic in the Streets* (1950) and A *Streetcar Named Desire* (1951), respectively. The famous theme song in *High Noon* (1952) is accompanied by a guitar, the quintessential Western instrument. Contemporary musical elements eventually took on a life of their own in response to their growing marketability, but at least initially, pop and jazz grew out of the demands of the narrative.

The leitmotif

A final characteristic of the classical film score is the importance it places on its own structural unity. Marlin Skiles's text on film scoring cautions that "scoring involves more than merely mood-matching, for the music should be capable of standing up alone as an integrated whole."[35] The most typical though not the only method to accomplish an overall unity was the use of the leitmotif. The precedent of silent film accompaniment with its structural imperative that "every character should have a theme," and the operatic backgrounds of Steiner, Korngold, and others, facilitated a practice that linked musical themes to character, place, object, and abstract idea. To the sound film score, with its piece meal construction and gaps in musical continuity, the leitmotif offered coherence. According to Karlin and Wright in *On the Track*: "The development of motifs is a powerful compositional device for the film composer, allowing him to bring an overall sense of unity to his score and still leave room for variety."[36] Through repetition and variation, leitmotifs bound a series of temporally disconnected musical cues into an integrated whole. Further, leitmotifs functioned in an interdependence with the visual text. Music responded to the dramatic needs of the narrative and in turn clarified them, sealing music and visual text into mutual dependence. Finally, leitmotifs heightened spectator response through sheer accumulation, each repetition of the leitmotif bringing with it the associations established in earlier occurrences.

One of the primary functions of the leitmotif was its contribution to the explication of the narrative. As Steiner explains: "Music aids audiences in keeping characters straight in their minds."[37] To this end composers created musical identifications for characters, places, and even abstract ideas in a film. In Korngold's score for *Captain Blood* there are leitmotifs for Peter Blood as well as for King James and King William; for all the important locations, Port Royal (also used to accompany the governor of Jamaica), Tortuga, Virgen Magra, England, and France; for the love between Peter Blood and Arabella Bishop; and for the torturous slavery on Colonel Bishop's plantation (which doubles as a motif for Colonel Bishop). An analysis of the leitmotifs associated with Peter Blood demonstrates how the leitmotivic score depended upon the principle of repetition as a means to structure diverse and discontinuous musical cues and the principle of variation as a means to clarify the visual text.

The leitmotif Korngold composed for Peter Blood is presented initially in the main title. It is an extended theme consisting of two shorter motifs: the heroic brass fanfare analyzed earlier and a second, contrasting, lyrical string melody. Throughout the film Korngold separates and recombines these motifs, the A motif underscoring moments of Blood's heroism and the B motif emphasizing his more human and vulnerable side. The initial appearance of the heroic motif coincides with Blood's first appearance in the film. Summoned at his home in the middle of the night to attend a wounded rebel, Blood agrees to treat a traitor even though he disagrees with his politics. The initial statement of the motif is reminiscent of the martial opening, but it is cast here at low volume in the lower register of the horns and succeeded by reprises first in the woodwinds and then in the strings. Instrumentation here reflects and defines narrative content. The brasses are used under soldier Jeremy Pitt's desperate pleas for aid, but strings underscore Blood's declaration of pacifist principles. These variations set up the groundwork for a climactic return of the original brass orchestration of the motif later in the film. Each repeated variation, and there are several in the first third of the film, accompanies some aspect of Blood's heroic persona. Thus the leitmotif operates on two levels simultaneously: in terms of the narrative it helps to create the heroism of the protagonist; in terms of the music it creates both a thread of repetition and variation which binds the score, and also, through anticipation of a return to the original version, a climactic center for the score's overall design.

Blood's dedication and self-sacrifice are embodied in the contrasting B motif. This initially appears in the sequence where Blood ministers to the wounded rebel. (He will be tried and convicted as a traitor as a consequence.) The volume of the music is low, the tempo is slow, and the instrumentation is primarily string. In fact, much of the pathos of the drama emanates from instrumentation: a solo violin can be heard when Blood refuses to abandon his charge.

For the transition which takes Blood and his fellow convicts from their home in England to their enslavement in Jamaica, Korngold re-combines the two motifs. The heroic A motif returns to the score here in an orchestral setting, but significantly it is transposed to the minor and its tempo is slowed. These changes infuse the cue, marked *lento con dolore* (slowly with sadness), with a sense of dislocation contributing to the mood of uneasiness that attends the unfortunate change in Blood's circumstances. The contrasting B motif forms the B section, accompanying Blood's ministrations to his fellow slaves. Like the A motif, it is played in an unfamiliar setting: woodwinds (flutes and clarinets) and strings played in a low register. When the A motif returns it does so with a horn solo. It is then fractured into developmental variations, coalescing into a familiar phrase only as the sequence ends.

During Blood's enslavement on the Bishop plantation short variations of both motifs accompany moments of Blood's daring (outwitting the governor to return to the stockade in time for the escape, for instance) and compassion (attending to his beaten and battered friend, Jeremy Pitt, and risking his life to do so). The climactic musical center of this film occurs when Blood is free at last. As captain of a captured Spanish galleon he becomes what the film's title foretells. Visual and musical text mark the scene as momentous. Inspiring camera work (a series of low-angle close-ups) combines with the dynamic return of the A motif in the original brass fanfare of the main title (here "plumped" by the addition of some strings). It is followed, as it is in the main title, by the contrasting B motif. This aural reminder of Blood's emotional vulnerability is heard during the shot-reverse shot dissolve where he imagines Arabella Bishop's melancholy at his departure.

At this point in the score a new leitmotif is introduced for Blood's career as a pirate. The leitmotivic score achieved an integral structure through repetition and variation. This structure was tightened through relationships constructed among the various motifs. Steven Wescott has called this quality "interconnectedness," achieved through "the common use of prominent melodic contours and intervals" and "shared melodic and rhythmic motives."[38] I would add shared instrumentation and harmonic construction. The relationship between the Captain Blood A motif and the pirate motif provides a case in point. The most apparent connection between them is instrumentation. Both motifs depend on the brasses. (The initial appearance of the pirate motif is played on a French horn with brass accompaniment.) In addition, each motif is built on a melodic contour derived from arpeggiated triadic chords. Finally, each motif incorporates a rising third at the end of the first major phrase.

From the point at which Blood embarks on his pirate career until the end of the film when he accepts a commission in King William's navy, it is this motif which most often accompanies his appearances. Korngold preserves the martial definition of the pirate motif with most of the variations highlighting some type of brass instrument and occurring at moments which define Blood's success as a privateer. Developmental variations of the Captain Blood A motif occur only briefly in the second half of the film and can be heard during the montage of Blood's piracy as well as during Blood's commands to ready the ship for its voyage to Port Royal. Transposed to the minor the Captain Blood motif lends a solemnity to Levasseur's death. But a full reiteration of the Captain Blood motif in its original setting is withheld from the second half of the film and reserved for the climactic naval battle where Blood eschews the life of a pirate for the patriotism of king and country. Significantly, the contrasting B motif appears only minimally in Blood's pirate life, suggesting perhaps that the qualities it represents in Blood have diminished. This motif can be heard in connection with his constant affection for Arabella Bishop, first when Blood unknowingly allows the ship on which she is traveling to pass unmolested, and second during a shipboard meeting between them.

When Blood accepts the pardon and naval commission from King William, the pirate motif relinquishes its place to the Captain Blood motif which dominates the busy musical texture of the ensuing sea battle. With the decisive conclusion of the struggle, the pirate motif disappears from the score entirely and the Blood motif returns to its initial preeminence accompanying the final two sequences in the film. There are two telling variations here: the first in the minor for the pirates' disarmament and the second in a slow, string setting appropriate for the embrace of newly appointed governor Peter Blood and Arabella Bishop. This final variation modulates into the brass fanfare of the opening credits for the end title.

Notes

1 For an interesting personal recollection of team scoring see David Raksin, "Holding a Nineteenth Century Pedal at Twentieth Century-Fox," in *Film Music I,* ed. Clifford McCarty (New York: Garland, 1989), 167–81. See also David Raksin, interview with Irene K. Atkins, 6 December 1977–15 February 1978, transcript, Yale Oral History Project, Yale University Library, New Haven, 120–28.

2 Ernst Klapholz, quoted in "Interview with Arthur Lange and Ernst Klapholz," Cue *Sheet* 7, 4 (December 1990), 159.

3 Rudy Behlmer, "Tumult, Battle, and Blaze: Looking Back on the 1920s—and Since—with Gaylord Carter, the Dean of Theater Organists," in *Film Music I* 27–28.

4 Robert Nathan, quoted in Marlin Skiles, *Music Scoring for TV and Motion Pictures* (Blue Ridge Summit, PA: Tab Books, 1976), 38.

5 Virgil Thomson, "A Little about Movie Music," Modern *Music* 10, 4 (1933): 188.

6 Skiles, *Music Scoring for TV and Motion Pictures* 94.

7 All of the musical examples in this chapter originated as my transcriptions. I was able to compare these against the full score in the Warner Brothers Archive/University of Southern California Cinema-Television Library (hereafter referred to as Warners). All future references to Korngold's score will be quoted from this copy. My analysis is based on the restored 119-minute original version of the film, not the 99-minute edited version that is in general distribution.

8 Director William Friedkin, in describing the unusual effects he wanted in the score for *The Exorcist* (1973), said, "I wanted the music to come and go at strange places and dissolve in and out. No music behind big scenes. No music ever behind dialogue.... Only music in the montage sequences...." Quoted in Randall Larson, *Musique Fantastique: A Survey of Film Music in the Fantastic Cinema* (Metuchen, NJ: Scarecrow, 1985), 312.

9 Friedhofer, quoted in Skiles, *Music Scoring for TV and Motion Pictures* 234.

10 Erich Wolfgang Korngold to Hal B. Wallis, 11 February 1938, Warners Rudy Behlmer suggests Korngold changed his mind when he heard of the meeting between Chancellor Schuschnigg of Austria and Hitler, effectively preventing him from returning to Austria to pursue his career. See Behlmer, *America's Favorite Movies* (New York: Ungar, 1982), 84–85.

11 Steiner, "Scoring the Film," 223.

12 Skiles, *Music Scoring for TV and Motion Pictures* 94.

13 Frank Skinner, *Underscore* (New York: Criterion Music, 1960), 191.

14 Samuel Chell, "Music and Emotion in the Classic Hollywood Film: The Case of *The* Best Years *of Our Lives*," *Film Criticism* 8, 2 (Winter 1984): 38.

15 Steiner, "Music Director," 396.

16 Elmer Bernstein, quoted in Thomas, *Music for the Movies* 193, quoted in Simon Frith, "Mood Music: An Inquiry into Narrative Film Music," Screen 25, 3 (1984): 84. For a discussion of the concept of emotional realism in narrative film scoring, see Frith 83–84.

17 Skinner, *Underscore* 191.

18 Karlin and Wright, *On the Track* 497.

19 Leonid Sabaneev, *Music for the Films*, trans. S. W. Pring (London: Pitman and Sons, 1935; repr. Arno, 1978), 20.

20 Karlin and Wright, *On the Track* 132.

21 Cutting notes, Hal B. Wallis, 10 December 1935, Warners.

22 Ibid.

23 Verney Arvey, "Composing for the Pictures by Erich Korngold: An Interview," *Etude* 55, I (January 1937), 15.

24 Erich Wolfgang Korngold, "Some Experiences in Film Music," in *Music and Dance in California*, ed. Jose Rodriguez (Hollywood: Bureau of Musical Research, 1940), 137.
25 Komgold's son, Ernst W. Korngold, told me his father didn't use a click track, but he did remember punches and streamers from his father's recording sessions. Their use was rare, however, and they were employed not to establish the rhythm of a scene but as a kind of insurance for specific musical cues that Korngold was largely able to synch by ear and eye.
26 Thomas, *Music for the Movies* 131.
27 Cutting notes, Hal B. Wallis, 10 December 1935, Warners.
28 Steiner, "Music Director" 393. See also *Film Music Notes* 2, 4 (January 1943), where he claims that "the great problem of composing for the films [is]—to give the score continuity, to keep the audience unconscious of any break. . . ."
29 Herbert Stothart, "Film Music," in *Behind the Screen*, ed. Stephen Watts (London: Barker, 1939), 143–44.
30 Karlin and Wright, *On the Track* 49.
31 John Williams, quoted in Tony Thomas, "A Conversation with John Williams," Cue Sheet: *The Journal of the Society for the Preservation of Film Music* 8, I (March 1991): 12.
32 Evans (*Soundtrack*), Thomas (*Music for the Movies*), and Goldner and Turner (*Making of King Kong*) erroneously report the size of the orchestra as eighty players. Fred Steiner, who conducted the rerecording in 1976, established the original size through research on the score. See Fred Steiner, liner notes, *King Kong*, sound recording, Entracte ERS 6504, 1976.
33 Archive, American Federation of Musicians, Local 47, Los Angeles.
34 Korngold borrowed selections from two symphonic poems by Franz Liszt: *Prometheus* and *Mazeppa*. *Prometheus* is used briefly during the Spanish conquest of Port Royal and is quoted at length during the duel between Blood and Levasseur. *Mazeppa* is used during the final sea battle and briefly during the Spanish conquest of Port Royal. Because Korngold did hot compose all of the cues for the film, he insisted on the credit "Musical Arrangements" in the main title.
35 Skiles, *Music Scoring for TV and Motion Pictures* 123.
36 Karlin and Wright, *On the Track* 176.
37 "Music in the Cinema," *New York Times*, 29 September 1935, 4.
38 Steven D. Wescott, "Miklós Rózsa's *Ben-Hur:* The Musical-Dramatic Function of the Hollywood *Leitmotiv*," in *Film Music I* 191.

Chapter 16

Patrick Keating

SHOOTING FOR SELZNICK: CRAFT AND COLLABORATION IN HOLLYWOOD CINEMATOGRAPHY

ANY SUCCESSFUL WORK OF HOLLYWOOD cinematography is the sum total of countless choices and decisions—choices about where to put the lights, how many to use, their relative brightness, the use of diffusion, the placement of flags and other shadow-casting devices; choices about camera position and camera movement; decisions about film stock; decisions about development and printing, and so on and so forth. In the studio system, the responsibility for making most of these decisions rested with the cinematographer. With responsibility came a modest amount of freedom: since no producer or director could monitor every decision, cinematographers could, in theory, explore original approaches to their art. Yet the opportunity to experiment was always limited. Indeed, the whole concept of the studio system is built on a certain amount of regulation—whether through implicit norms or explicit rules, each member of the filmmaking team was expected to have a carefully controlled understanding of the contribution he or she could make. This dialectic between the responsibility of individual cinematographers to shape film style and the power of the studio system to control their work is central to any attempt to understand Hollywood cinematography.

Elsewhere, I have approached this problem from the point-of-view of cinematographers as a group, with a particular emphasis on the American Society of Cinematographers (ASC) as an institution that came to define a set of shared ideals and norms.[1] Here, I want to approach the problem from a different angle, from the perspective of a producer. David O. Selznick consistently hired the top cinematographers in the business, among them George Barnes, James Wong Howe, Gregg Toland, Harold Rosson, Lee Garmes, Stanley Cortez, and Joseph August. If any cinematographers had individual styles, they did. Yet Selznick was reputed to be the most micro-managing producer in the business. If any producer had the power to control the look of his films, he did. Assuming that Selznick wanted to shape his films' style, what tools allowed him to do so? In what ways did Selznick's own aesthetic agenda converge or conflict with the agendas of his cinematographers? And what can a case study of Selznick tell us about the cinematographer's place in the larger context of classical Hollywood cinema?

In order to best position the role of individual style in Hollywood studio cinematography, we need an interactive model—a model that situates the individual cinematographer within the context of larger institutions, such as the studio and the ASC, while acknowledging that these institutions could never fully regulate the way a single cinematographer might solve a particular stylistic problem, even when managed by someone as detail-oriented as Selznick. Such a model should draw on, and contribute to, two existing lines of inquiry in film scholarship. The first is the ongoing discussion surrounding questions of authorship and collaboration in film. One of the most common responses to the auteur theory's idolization of the director has been a conscientious effort to examine the contributions of other members of the filmmaking team, such as producers, writers, and cinematographers.[2] In an extreme form, this approach simply reproduces the flaws of the auteur theory, finding hidden geniuses struggling against the system in otherwise ignored corners of the industry. A more judicious line of argument shifts the object of study from authorship to collaboration. Taking the collaborative nature of film authorship as a given, we can assume that any one film is the result of the intersection of many competing ideas, agendas, and styles. Thus Robert Carringer's in-depth study of the making of *Citizen Kane* (1941) demonstrates how Orson Welles drew on the talents of a wide range of collaborators, from the cinematographer to the art director to the special-effects team.[3] Thomas Schatz takes a similar approach, though he prefers the term "negotiation" to "collaboration" in order to suggest the power struggles involved whenever a team of workers with different ideals and agendas come together to make a film. Schatz places particular emphasis on the role of producers like Irving Thalberg, Darryl F. Zanuck and Selznick in establishing standards at their respective companies. But both he and Carringer provide a model for understanding the studio system as an organizational framework bringing together a range of complementary and occasionally competing contributions.[4]

A second, and related, line of scholarly inquiry concerns the relationship between norm and variation within the classical Hollywood style. In *The Classical Hollywood Cinema*, Bordwell, Staiger and Thompson emphasize the coherence of the Hollywood style. While the devices of filmmaking may change (before and after the transition to sound, for instance), classical Hollywood cinema retains its commitment to certain functions—most notably, the function of clear narration. The authors allow for a certain amount of variation within the system, even as they affirm the dominance and stability of the classical style.[5] Other scholars, such as Richard Maltby, express reservations about the idea that classical Hollywood filmmaking ever achieved an advanced level of coherence, proposing instead that individual films were often quite opportunistic in construction.[6] The study of Hollywood's stylistic norms is related to the previously discussed studies of collaboration because the process of bringing different professionals together to create a film depends on a context of shared norms. If all the practitioners adhere to a certain set of norms, then we should expect the process of collaboration to proceed fairly smoothly; however, if individuals and groups cannot agree on certain normative ideals, then the process of confrontation could easily become a struggle.

This essay will draw a three-part distinction that can help explain what role a producer might have in shaping and controlling cinematographic style while still acknowledging the importance of the individual cinematographer and institutions like the ASC. My proposal develops certain ideas borrowed from the art historian Michael Baxandall. Discussing the building of the Forth Bridge, Baxandall points out that the architect was given a "charge": build a bridge. The architect's work will necessarily be shaped by more general conceptions of what the charge "building a bridge" should mean. But with any particular bridge, there are inevitably more specific problems to consider. How long should this particular bridge be? How deep is the river? What materials are available? And so on and so forth. Baxandall calls this cluster of task-specific problems the "brief." Together with the bridge itself, the charge and the brief form a triangle of issues for the art historian to consider.[7] In a similar way, we can examine cinematography at three different levels of generality. At the most abstract level, we can think of cinematography as a craft: a set of norms and ideals defining what the job should be when practiced with artistry and skill. These ideals usually are drawn from the culture at large, as mediated by Hollywood institutions. To understand the craft is to understand certain standards that all films should uphold. Alternatively, we can examine specific tasks that were given to the cinematographer and his crew to perform: photograph this star in this particular film, and get it done in this amount of time. Although notions of craft will inform the assignment of particular tasks, to think about cinematography at the level of the task is to operate at a more specific level than the level of craft ideals. Still, no matter how detailed, the task at hand will always underdetermine the look of the resulting film, requiring us to consider the work as it was actually executed. Keeping the ideals of the craft in mind, the cinematographer and his crew would need to translate the requirements of the task into concrete decisions about style: where to put the lights, when to modulate their intensity, how to model their resulting shadows, and so on and so forth.

These three distinctions—the craft of cinematography, the assigned task, and the work as executed—can help us understand the different functions of the ASC, the studios, and the individual cinematographers. The ASC's primary goal was defining the craft of cinematography. Articles in *American Cinematographer* do not tell the cinematographer how to shoot any one particular film, but they do work to establish a widely shared sense of what it is to shoot a film in the first place. To achieve its mission, the ASC reminds the cinematographer that his craft is an art form, capable of fulfilling culturally respectable artistic functions like expressivity, realism, and compositional clarity. By contrast, the primary function of the studio (defined broadly to include majors like Twentieth Century-Fox as well as more specialized production companies like Selznick's) was to assign cinematographers specific tasks, articulated in the form of scripts, budgets, schedules, memos, and other forms of bureaucratic instruction. Meanwhile, the individual cinematographer was responsible for negotiating the potentially conflicting mandates of the craft and the task at hand in order to execute the work. Of course, it is too simple to say that the craft-task-work distinction can be mapped onto the ASC-studio-cinematographer grouping so neatly; we should expect to find considerable areas of overlap. This is why a case study of Selznick's relationship with the cinematographers he employed can be so useful—the examination of a particular case can help clarify where the distinctions apply, and where they need to be nuanced and rendered more complex.

Selznick is a good candidate for study, not only because his papers are readily available, but also because his status as one of Hollywood's most powerful producers ultimately reveals the limits of a producer's power. As we will see, Selznick did indeed have several strategies for controlling the look of his films, including: hiring and firing, working with the writer and the production designer during the pre-production stage, controlling the budget, dictating memos with specific instructions about what to do (and what not to do), making recommendations about printing, and ordering retakes of particularly important shots until they were lit and composed to his satisfaction. This seems like a powerful set of tools, and indeed it was. But Selznick's power over film style was never absolute. He relied on his cinematographers to interpret his instructions creatively, and cinematographers took the opportunity to put the ASC's ideals into practice.

Selznick's approach to hiring cinematographers was similar to his approach to casting—just as casting is a matter of finding the right fit between performer and role, hiring a cinematographer was a matter of finding the right fit between cinematographer and story. Perhaps the task was simpler for producers at the major studios. Over at Fox, Zanuck could always choose from Fox's regular pool of staff cinematographers, people like Leon Shamroy or Arthur C. Miller. At Warners, a producer like Hal Wallis could pick among Arthur Edeson, Sol Polito and Tony Gaudio. Eventually, Selznick did hire cinematographers to be part of his team, but in general he looked to hire the best cameraman available, whether the cinematographer was working as a free agent or available on loan from a major studio. So he might borrow Toland from Goldwyn for one project, then work with Warners' Ernest Haller on another. In making these decisions, Selznick operated from the assumption that most cinematographers were specialists in a particular kind of work. One cinematographer might be good at exteriors, and another one good at interiors. One cinematographer might be good at wide shots, and another one good at close-ups. For Selznick, the secret to hiring cinematographers was to get the one who was best suited to the story at hand; if the story had a lot of variety, then the ideal solution was to hire multiple cinematographers to cover all the possibilities. On *The Garden of Allah* (1936), Selznick tried to get Technicolor to send him two cinematographers, since he thought that one, Howard Greene, was better with exteriors, while another, William Skall, was better with interiors. Meanwhile, Selznick wanted the Technicolor man to focus on technical issues while hiring Hal Rosson, a black-and-white specialist, to make sure that the photography was sufficiently glamorous for Marlene Dietrich.[8]

Unsurprisingly, most cinematographers did not share Selznick's enthusiasm for this mix-and-match approach. Working cinematographers usually took exception to the idea that cinematographers were technical one-trick ponies—each with his own narrow, constricting set of abilities. On *The Garden of Allah*, all the memos suggest that Greene and Rosson ended up hating each other (though they did share a special Academy Award for their work on this film). On a deeper level, this animosity can be attributed to the fact that Selznick's approach to hiring contravened some of the most basic assumptions that cinematographers held about their craft. Their professional organization, the ASC, was endeavoring to develop a public identity of the cinematographer as a highly skilled professional artist, and one of the recurring themes in their publication *American Cinematographer* was the idea that a cinematographer's primary skill was the efficient command of a variety of visual styles. A good cinematographer should be able to shift from comedy to drama, from interior to exterior, from wide shot to close-up, without any loss of quality. Here is John Arnold, head of MGM's camera department and president of the ASC, on the subject:

> No picture is going to sustain exactly the same mood throughout all of its many hundred scenes and set-ups. Even in the heaviest drama we have moments of romance, or of robust humor. [. . .] Properly balancing between the individual requirement of the scene itself and the sustained mood of the production as a whole gives the Cinematographer a greater problem than any faced by Director or Actor. When you see a story of many moods photographed so smoothly that you are not conscious of photography, rest assured you have seen the work of a master of the camera.[9]

For Arnold, a cinematographer has to combine versatility with subtlety. Hiring multiple cinematographers might be an easy way to produce visual variety, but it risked incoherence. The disparity between Selznick's and Arnold's views extends from fundamental differences in the way each conceives of the craft of cinematography. For Selznick, the cinematographer's craft is a set of skills which the smart producer can use to control the look of a film. For Arnold, the cinematographer's craft is an art, with all the expressive resources of theater or painting.

Instead of taking chances with free agents, most studios preferred to keep cinematographers on long-term contracts. One advantage of this approach was that cinematographers grew skilled at meeting the photographic needs of the studio's contract stars. Selznick eventually came around to this way of thinking, hiring George Barnes and Stanley Cortez to ensure glamorous photography for marquee stars like Joan Fontaine, Ingrid Bergman, and Jennifer Jones. These stars were assets, and Selznick needed to have cinematographers available to protect them. His memos are quite explicit on this point—in one of them he complains that he and his production team are underutilizing Cortez and Barnes, "despite the fact that we are carrying them mostly for the protection of our stars."[10] This sensibility, the cinematographer as visual caretaker, applies not only to films produced at Selznick's own companies, but also to films that Selznick packaged together to sell to others. For instance, he assembled *Jane Eyre* as a vehicle for Joan Fontaine, then sold it to Fox, whom he encouraged to hire Cortez or Barnes.[11] The point is not simply that Selznick wanted to profit by loaning out his contract cinematographers; it was also that Selznick wanted to ensure that Joan Fontaine would be protected by a cinematographer who was capable of flattering her, thus looking out for her interests—and Selznick's. Eventually, Fontaine was photographed by Barnes, who had already demonstrated his ability to light her in *Rebecca* (1940). In this way, Selznick used hiring decisions to shape the look of films that he did not even produce.

Yet hiring was ultimately an indirect way of controlling the look of a film. Once hired, a cinematographer would have a certain amount of freedom, since Selznick could not be on set to approve or veto every decision. As a way of making his intentions known explicitly, one solution was to put camera and lighting instructions in the script: most studio-era scripts contain surprisingly detailed camera instructions, and even some recommendations about lighting. For instance, the final shooting script for *The Prisoner of Zenda* (1937) contains the following instructions for shooting the duel scene:

> INT. GREAT HALL
> shooting towards opening of passage.
> The two men come out fighting from the top step. Rudolf has his back to CAMERA. He is allowing Rupert to force him back in the fight.
> CLOSE SHOT – THE DUELLERS' SHADOWS
> thrown on the wall by the firelight (shooting from the fireplace)
> CLOSE SHOT – THE DUELLERS
> Their figures in the f.g., lit only by firelight – in profile; and behind them on the wall, their great shadows.[12]

The filmmaking team (including director John Cromwell and cinematographer James Wong Howe, replacing Bert Glennon) did follow some of these instructions, though not always perfectly. The more important point is that instructions about lighting in the script are the exception, rather than the rule. In the entire script for *Zenda*, there are only a handful. Even the detailed instructions about camera placement were not absolute mandates; departures from the shooting script's camera notes were quite common, both at Selznick's company and at others.[13]

Perhaps because the shooting script proved to be a weak tool for the control of visual style, Selznick exerted even more control by rethinking the very process of determining the visual style of a film in pre-production. This involved the creation of a new position: the production designer. The most celebrated example here is William Cameron Menzies, whom Selznick hired to work on premiere projects like *The Adventures of Tom Sawyer* (1938) and *Gone with the Wind* (1939), endowing him special responsibility for the visual style of the films.[14] Because Menzies would create sketches of his ideas before the scenes were filmed, Selznick had the opportunity to approve or revise the visual design in advance, thereby increasing his ability to control the production itself.

Cinematographers had been pleading for more attention to pre-production for years. Here is Paramount cinematographer Victor Milner on the subject:

> The real art of lighting for mood and tempo must depend primarily upon the individual Cinematographer's artistic sense, and upon his ability to visualize in terms of lighting. Since the majority of Cinematographers have developed this

> faculty to a marked extent, it would be of incalculable benefit to the Industry if more Producers and Directors made it a rule to consult the Cinematographer earlier in the preparatory stages of production, and to allow him more ample time to familiarize himself with the script before actual filming commences.[15]

Milner argued that decisions about lighting and camerawork were too important, and in some cases too complex, to be made amidst the hectic atmosphere of a shooting day. Instead, he advocated the proposal that they be made ahead of time after careful pre-production discussions. Milner's agenda was to increase the cinematographer's power and prestige by giving him a more intellectual role—not just the leader of a crew executing a series of technical tasks, but an artistic figure who could act as an equal in the design of a film. Again, it is important to see this statement as something more than just an effort to give the cinematographer more time to work; it is also an attempt to give the cinematographer a new identity: an identity as a visual storyteller, on par with the director and writer. In this context, Selznick's creation of the position of production designer must have seemed rather cruel to working directors of photography. Selznick was adopting the cinematographers' view that the visual design should be mapped out ahead of time, but denying them a leading role in the decision-making process. In effect, cinematographers were reduced to an even more subordinate creative role: that of executing the drawings created by the production designer and approved by the producer. Instead of treating cinematographers as visual architects, Selznick was treating them as visual carpenters—albeit highly skilled ones.

Cinematographers were able to resist this creative marginalization, as all the pre-production in the world cannot determine what will happen once production begins. For *Since You Went Away* (1943), Selznick prepared a script dotted with various lighting suggestions and instructed production designer Bill Pereira to sketch them. But memos indicate that cinematographer Stanley Cortez often ignored them and went about lighting the scenes his own way. This approach could produce some beautiful images, but it also entailed risks. Cortez received more than one strongly worded memo from Selznick himself, who complained that Cortez was trying to push the silhouette effects much too far—in one romantic scene (Figure 16.1), for example, it is hard to see Robert Walker's face.

One of Selznick's memos was particularly blunt: "[Cortez] has made a lot of suggestions which were splendid from the standpoint of camera effect but idiotic from the standpoint of the scene. I understand that this is because he does not read the script."[16] In other memos, Selznick worried that Pereira's lighting sketches were not detailed enough, and he claimed that Cortez went about lighting in a trial-and-error fashion—trying out different options rather than planning a strategy in advance. Eventually, Cortez was drafted into military service and replaced by Lee Garmes. After comparing the work of the two cinematographers, Selznick estimated that Cortez had cost the production an extra $300,000.[17]

This figure may be exaggerated, but it brings us to a third means by which a producer might exercise control over lighting and camerawork—by controlling the budget. The cinematographer and the crew required a lot of tools to get their jobs done, and those tools were by no means always cheap. So far, I have stressed the ways that Selznick sought to control the cinematography, but here is an area where Selznick seemed to offer a great deal of freedom, in the form of generous resources for spectacular effects. Selznick was willing to pay substantial costs of shooting in Technicolor and, where necessary, to tolerate lengthy shooting schedules. The memos indicate that Selznick preferred to work with fast and efficient cinematographers like Gregg Toland, but he would also hire relatively slow DPs like Cortez and Charles Rosher when the story called for their signature aesthetic talents. Indeed, it is possible that otherwise fast filmmakers would slacken their pace when working with Selznick. Director William Dieterle and cinematographer Joseph August worked together at RKO,

Figure 16.1 *Since You Went Away* (1944), directed by John Cromwell

where they shot 16 set-ups per day on *The Hunchback of Notre Dame* (1939) and 18 set-ups per day on the relatively low-budget *All that Money Can Buy* (1941). Several years later, they worked together on Selznick's *Portrait of Jennie* (1948). On paper, the production did not present the same challenges as *Hunchback*, with its huge cast and its enormous cathedral set, but Dieterle and August ended up shooting a more leisurely eight set-ups per day.[18] This could certainly be taken as evidence of Selznick's eagerness for control: perhaps his micro-managing took its toll on efficiency. But it could also be seen as evidence of that fact that filmmakers could take advantage of Selznick's perfectionism in order to avoid the blistering pace that a large but sometimes desperate studio like RKO required.

In exchange for Selznick's generosity with time and resources, filmmakers had to endure Selznick's most notorious method of control: his memos. For all their notoriety, the power of the memos can be overstated, especially when it comes to cinematography. Although Selznick could dictate volumes of ideas when it came to screenplay construction, he would comment on the cinematography only when it was very bad or very good. The cinematographer did not even receive the memos directly; often, they were addressed to the director or the production designer. In any case, as the example of Stanley Cortez illustrates, the cinematographer with a particularly thick skin could always try ignoring Selznick's instructions, almost daring the producer to fire him. As with Selznick's other tools, the memo was an indirect means of control; it was just another way of trying to nudge the cinematographer in a certain direction.

A good example of Selznick's approach can be found in a memo on *The Prisoner of Zenda*. Bert Glennon worked for several weeks on the film, but Selznick replaced him with James Wong Howe. Howe was required to reshoot several of Glennon's scenes. Before they were reshot, Selznick wanted to make sure that Howe would avoid repeating Glennon's mistakes, so he dictated a very detailed memo that was sent directly to Howe:

> On the Terrace Sequence, I am really hopeful that the set we discussed will permit you to get some really beautiful photography that will give us an idyllic quality in the love scene down the path and near the pool. I am counting on you to avoid the stagey appearance of most exterior sets that are built on the stage; to give us real night photography that will not hesitate to lose figures in deep shadows at the same time that it gives enough light for expressions on faces.[19]

Selznick went on to spell out instructions for several other scenes, providing evidence of the level of precision that Selznick could provide. He does not tell Howe what lights to use and where to put them; he says nothing about f-stops, wattage, contrast ratio or lamp diffusion. The technical details are left entirely to Howe—but Selznick is very meticulous when it comes to atmosphere. He articulates a clear idea of the mood that the scene is supposed to create, and he offers some moderately specific instructions regarding the type of image it demands: an "idyllic quality" to set the mood for romance.

Although the primary function of this memo is to clarify the nature of Howe's task, we cannot understand the task in isolation from issues of craft. Selznick's instructions are shaped by his own assumptions about the craft of cinematography, about what it is to shoot a high-quality film. For Selznick, good cinematography will set the right mood for the story, offer a plausible representation of place and time, and focus the spectator's attention on the actors' faces. There is nothing unusual about these ideas; indeed, they overlap rather neatly with the ASC's own account of the cinematographer's craft.[20] We might take this as evidence that the ASC had succeeded in its unstated mission of redefining the identity of the cinematographer; or we might take it as evidence that the ASC had always established its definition with producers in mind, emphasizing the traits that producers might find worthy of praise (and employment).

James Wong Howe was undoubtedly familiar with the ASC's ideas concerning the craft of cinematography. He himself had contributed an article on lighting to the second volume of the *Cinematographic Annual* in 1931.[21] The article contains some original ideas, but it is best seen as an example of institutional discourse, articulating concepts that were commonly accepted among members of the ASC: light should create a sense of roundness, draw the spectator's eye to the most important story point, and so on. Howe's goal is to not to tell anyone how to shoot one particular film; it is to establish certain standards for the craft as a whole, thereby elevating the cinematographer's professional reputation. These generalizations could only take a cinematographer so far. When Howe was assigned the job of finishing *The Prisoner of Zenda*, the task at hand involved several other specific requirements, some carefully articulated by Selznick in his memos, some more or less implicit in the assignment. Howe would need to study the faces of co-stars Ronald Colman and Madeleine Carroll and select the most attractive strategies for lighting them; he would need to figure out which was the dominant mood for each scene, shifting the mood from action to romance at the appropriate times; and, since Howe was replacing the fired Bert Glennon, he would need to shoot footage that matched reasonably well with whatever Glennon footage was to remain in the final film.

Looking at the scene itself, we can see how Howe used his understanding of the craft to translate the task at hand into a fully executed work.

To fulfill Selznick's demand for idyllic imagery to produce a mood of romance, Howe follows a very common cinematographic convention, photographing a romantic scene with a soft style.

The softness takes several forms. There seems to be some diffusion on the lens, giving these highlights an extra glow. The focus is very shallow, especially in the close-ups, rendering the background soft. There is diffusion on some of the lights, especially the key-light for Madeleine Carroll—notice that the shadow on her forehead has a fairly soft edge. And the

Figure 16.2 *The Prisoner of Zenda* (1937), directed by John Cromwell

scene as a whole has very few areas of dark black; instead, these images are composed almost entirely in shades of gray. This is particularly clear in the wide shot, where Howe has added some fog to the background, turning it into a pattern with remarkably gentle gradations of gray.

On the one hand, Howe has followed Selznick's instructions very closely. Selznick wanted the figures to be in shadow, but he also wanted to see their faces. Howe accomplished this contradictory goal by putting the figures in the shadow at the beginning of the scene, then gradually using more light on the closer framings. Selznick also wanted our attention to be taken away from the background, and Howe has fulfilled that mandate by obscuring the set in shadow then casting the background into soft focus. On the other hand, we might say that Howe has not so much followed instructions as he has interpreted them in creative ways. Selznick did not ask for lens diffusion, lamp diffusion, and shallow focus on the close-ups. Selznick asked for an idyllic quality, and Howe figured out that lens diffusion, lamp diffusion, and shallow focus on the close-ups could provide Selznick with what he wanted.

We can find similar examples of these creative interpretations of instructions at other studios. At Warners, head of production Hal Wallis did not monitor the lighting as closely as Selznick, but he, too, could fire off an angry memo or two when a certain image displeased him. One recurring theme is the need for "sketchy" lighting on crime films, along with "character" lighting for tough-guy stars like James Cagney. Cinematographers like Tony Gaudio, Sol Polito, and (eventually) James Wong Howe would need to translate these more atmospheric ideals into practical terms, perhaps reducing the fill light to create more shadows in the crime film, or moving the key-light to the side to bring out the lines of character in Cagney's face.[22]

When they disapproved of the photography, producers had a few options to consider: they might order a retake, they might accept the film as shot and ask the cinematographer to be more careful in the future, or they could attempt to fix the shot in post-production. It is well known that Selznick had a high tolerance for retakes and pickups, sometimes requesting them the following day if he did not like the dailies, sometimes scheduling days of additional shooting at the rough cut stage. By contrast, high-volume studios like Warners and RKO did not always have the luxury of extensive re-shooting. Whereas Selznick was only producing a few films a year, the majors had to provide a steady supply of films to their theaters, and perfectionism was by no means always an option. Primary shooting might finish later than planned, but once the film was in the can, only the most serious problems would be addressed with re-shooting. Compare the following three films with original shooting schedules of around 30 days. On Warners' *Torrid Zone* (1940), shot by James Wong Howe after he moved to the studio, the 30-day shooting schedule expanded to 43, but then the crew had only one day for re-shooting. On RKO's *The Mad Miss Manton* (1938), shot by Nick Musuraca, the crew finished ahead of schedule, trimming the original 31-day plan to 28. On Selznick's *Intermezzo* (1939), shooting was supposed to proceed quickly, since the crew could simply follow the model of the original Swedish film on which it was based. However, the 28-day shooting schedule expanded to 35 and Selznick ordered 15 days of additional footage. Unsurprisingly the budget ballooned to over a million dollars.[23]

Most of the extra footage for *Intermezzo* was shot for story reasons, not for reasons of style. However, at least one important shot was taken a number of times to ensure that the photography was perfect: the first close shot of Ingrid Bergman. Because Bergman was unfamiliar to American audiences, Selznick knew that the most important close shot of her would be her first. Unhappy with Gregg Toland's photography in this scene, Selznick ordered a retake. The retake was unsuccessful, so Selznick ordered Toland to try a third time. Just to make sure that there would be no mistakes, Selznick wrote an unusually detailed memo:

> We apparently did not learn fully our lessons about her head being down whenever possible; about her one very good side and her one very bad side; about the three-quarter view in which she is so enchanting looking when photographed from her good side; about throwing her into effect lighting whenever possible, even if it is necessary to reach for it.[24]

Selznick goes to this level of specificity because he believes that Bergman needs to be lit differently than most female stars. We can see those differences by comparing Figure 16.3 and Figure 16.4.

In Figure 16.3, the key-light is placed in a frontal position, just to the right of the camera, smoothing out all the features of Carroll's face. There is a flag (a piece of grip equipment with black cloth stretched over a metal frame) over the key-light, producing a shadow on Carroll's forehead, but the effect is quite soft, just strong enough to hint at the nighttime atmosphere that is more obvious in the wide shot. By contrast, the shadows are more aggressive in Bergman's shot, in spite of the fact that it is a daytime interior. The shadow on her forehead almost covers her eyes, and the placement of the key-light creates a visible nose shadow (whereas Carroll's nose shadow is hidden on the far side of her face). The lack of a backlight adds to the effect, as Toland creates separation by contrasting her dark hair and coat with the moderate light on the background. The lighting on Carroll is more typical for a female Hollywood star, but Selznick argued that Bergman's unusual face demanded special treatment—her face does not glow so much as it emerges from the shadows, glamorous and mysterious.[25] Here we see Selznick at his most controlling. He cannot tell Toland exactly which lamps to use, but he can use the combined power of the memo and the retake to push the lighting in a very specific direction. Yet it is significant that this example is the exception,

Figure 16.3 *The Prisoner of Zenda* (1937), directed by John Cromwell

Figure 16.4 *Intermezzo* (1939), directed by Gregory Ratoff

not the rule. Even for Selznick, it was not practical to adopt this policy of unlimited retakes for every shot; only the most important shots would require this level of control.

Short of requesting a retake, Selznick could control the image in post-production by making recommendations about how the lab should print the film. For the early Technicolor film *The Garden of Allah*, Selznick had comments on almost every scene: less red for the wedding, darker for the sandstorm, more yellow for a scene set in sunlight, more shadows for an interior church scene.[26] One of his recurring worries was the fear that a beautiful print might be ruined when screened in a theater with an inadequate projector bulb; more often than not, he encouraged the lab to brighten the print so that film would be easy to see in a dim or ill equipped theater. Here again, Selznick took a position that could put him in conflict with his cinematographers, many of whom (such as Howe and Garmes) were known for their dark, low-key photography. We have already seen that Selznick used his memos to request lighting that was dark, but not too dark. A cinematographer might interpret this ambiguous request as an opportunity to experiment with low-key lighting, only to have Selznick push it back toward the center by ordering a brighter print.

The role of the lab was even more important at traditional studios like MGM or Fox. It is tempting to think of "studio style" as a form imposed by powerful heads of production, but often a studio's style might be found at a different level: the level of routine work procedures developed and standardized by studio personnel over the years. As Barry Salt has argued, light levels differed from studio to studio, in part because the house laboratories had different processing norms. For instance, cinematographers at MGM in the late 1930s used a great deal of light, overexposing the negative, which would then be under-developed, resulting in prints with glamorous pearly grays. Cinematographers at Warners used less light, underexposing the negative; the lab would compensate by over-developing, producing a more high-contrast print. Perhaps the most rigid studio of all was Twentieth Century-Fox. Toward the end of the decade, Daniel Clark, the Head of the Camera Department and a former President of the ASC, led a move to standardize the light meters employed at the studio. Using these meters, cinematographers were then required to set their key-lights at a certain level (150 foot-candles), and to set the camera's f-stop at 3.5. The result would be an image with consistent exposures on the faces and (even before *Citizen Kane*) a moderate amount of depth of field.[27] Since Selznick did not have his own laboratory, he relied on lab services offered by studios such as MGM or RKO. But it is doubtful that Darryl F. Zanuck or Hal Wallis could dictate all of their labs' policies, either. The pressure for production efficiency encouraged standardization, not everyday micro-managing. Still, the lab's control of the cinematographer's work could never be absolute. Even the restrictive Fox system afforded the cinematographer a great deal of freedom in lighting. He could still make individual decisions about the placement and motivation of key-lights; about the relative intensity of fill lights, backlights, and background lights; and about the use of diffusion, flags, and other grip equipment to model the light. Meanwhile, most cinematographers were allowed to consult with the lab when they needed special processing.

As this discussion of Selznick and his cinematographers suggests, the cinematographer always had some room for individual decision-making. Indeed, the system required it. In spite of Hollywood's reputation as a "factory," every film was ultimately unique, with slight or significant variations in story, cast, crew, and budget. Even an obsessive micro-manager like Selznick could not control every detail of the filmmaking process; he had to rely on cinematographers to make the appropriate adjustments in lighting and camerawork for each particular film. Some systems, like Fox's standardized lighting plan, could make the process more predictable, but cinematographers were allowed, and encouraged, to modulate a film's style to ensure that the mood was the right fit for the story.

Drawing these distinctions among the craft, the task, and the executed work can meter out the specifics of how the process of collaboration worked. In some cases, Selznick's instructions seemed more like restrictions. Clearly, *Intermezzo* is not Toland's most innovative work—even if we set aside his extraordinary work from the 1940s, the film is not as daring as *Dead End* (1937) or *Wuthering Heights* (1939), both shot for director William Wyler at Goldwyn. James Wong Howe's work on *The Prisoner of Zenda* is a superb example of lighting being modulated to set the right atmosphere for a story with multiple moods, but it does not seem to contribute to the cinematographer's career-long concern with realism—a concern that he would pursue more thoroughly in the *Life* magazine-inspired photography of the Warner Bros. film *Air Force* (1943), and much later in the harsh imagery of *Hud* (1963).[28] By contrast, other cinematographers used their Selznick projects to produce some of their best work. As we have seen, Selznick's request for silhouette effects in *Since You Went Away* gave cinematographer Stanley Cortez the opportunity to experiment with remarkably low-key lighting for a romantic drama. Selznick later tried to rein in the experimentation, but the film still contains some beautiful effects. Similarly, on *Gone with the Wind*, Selznick encouraged his cinematographers to ignore Technicolor's demand for high-key lighting and produce more shadowy images evocative of the film's gas-lit time period. Selznick ended up having to write a scalding memo when his cinematographers took the suggestion too far, but the end result was a new standard in Technicolor cinematography.[29] Here, it is too simple to say that the cinematographers were doing innovative work in spite of Selznick; after all, they were, in a sense, following the producer's instructions. It is just that the producer's instructions were always, inevitably, in need of interpretation. The task assigned will always underdetermine the look of the film. The work as executed is a result of the cinematographer's attempt to fulfill his craft ideals in the context of an always ambiguous assignment.

As a final example, consider Joseph August's remarkable photography for *Portrait of Jennie*. In the late 1930s, cinematographers began using tiny spotlights, which could produce just enough light to register an exposure on the new fast film stocks. Articles in *American Cinematographer* explained how these lights could be used to produce lighting effects, such as the effect of candlelight on a wall.[30] While Toland's use of fast film stocks to produce deep-focus ended up getting more attention, this move toward smaller lighting units was an equally important step in the development of cinematographic style in the 1940s. This general trend, endorsed by the ASC, intersected with August's career-long interest in experimenting with figure-lighting. Whereas other cinematographers, like Arthur Miller, used the small spotlights for effect-lighting, August began to use them to pick out the details of the performers' faces. We can find evidence of August's experiments at RKO, on films like *Gunga Din* (1939) and *All that Money Can Buy* (1941), but *Portrait of Jennie*, his last film, is his most daring foray into precision lighting. Selznick allowed August to participate in pre-production meetings, and some of his suggestions made their way into the final shooting script. August had the further advantage of working with the director William Dieterle; they had collaborated several times before, and the film's expressive style shows the influence of Dieterle's onetime colleague in Germany, F.W. Murnau. With such favorable conditions for collaboration, August had an unusual opportunity to employ different effects. In one scene, Eben (Joseph Cotten) enters his art studio and finds Jennie (Jennifer Jones) looking at his paintings. When she looks up at him, her face is initially in shadow, but then a small spotlight lights up her face, and her eyes begin to glow with eye-light. It is a remarkably precise effect—so precise that the exposure on the nearby background does not change at all. In another scene, shown in Figure 16.5, Eben is standing in his studio during the daytime.

The primary light on his body has been flagged off, creating a shadow over his face, but August has compensated for the shadow by targeting a small spotlight at Cotten's face, giving us just enough light to see his eyes. Both examples are quite unusual, since August makes

Figure 16.5 *Portrait of Jennie* (1948), directed by William Dieterle

almost no attempt to motivate them as plausible lighting effects, and yet the examples do not demonstrate rebellion so much as they demonstrate August's desire to accomplish certain goals valued by cinematographers and Selznick alike. With his little spotlights, August manages to preserve the atmosphere of mystery that is the dominant mood of the film, while following the industry guideline that the actor's eyes should always be visible. Indeed, we might say that August has successfully solved the puzzle that Selznick had posed to almost all of his cinematographers—make the images dark, but not too dark. On *Since You Went Away*, Cortez had attempted to solve this puzzle by pushing the limits of the silhouette effect. On *The Prisoner of Zenda*, Howe's solution involved a shift toward brighter tonalities as the scale moved closer to the subject. For *Intermezzo*, Toland eliminated the backlight and cast a shadow over the subject's forehead, just high enough to provide illumination for the eyes. Drawing on his years of experience with small spotlights, August had offered a fourth solution to the problem.[31] It is at this nearly atomic level that we are most likely to find individual cinematographic style in the classical Hollywood system—not in opposition to the task assigned, but in the different ways of fulfilling those tasks, while staying true to a larger conception of the craft.

In conclusion, we do not need to turn cinematographers into traditional "auteurs," rebelling against the system, to recognize that they had the ability to explore certain stylistic problems on an individual level. Instead, we can see studio system filmmaking as an interactive process, with the cinematographer's individual stylistic concerns contributing to the studio's specific needs, and to the industry's larger sense of what the craft of cinematography could be. While the ASC's primary function was defining the nature of the craft, and the studio's function was assigning specific tasks, this analysis of Selznick's role shows that these distinctions were never absolute. Selznick himself had a well-defined notion of the craft, calling for cinematography that was pictorially striking, expressive, and glamorous all at once, but he could not execute that vision himself. What he could do was to articulate specific tasks—tasks that often took the form of impossible problems like "make it dark but not too dark"—and hire the best cinematographers he could get to solve them.[32]

Notes

1 I develop this argument in Patrick Keating, *Hollywood Lighting from the Silent Era to Film Noir* (New York: Columbia University Press, 2010).

2 For some of the classic controversies around authorship, see Andrew Sarris, *The American Cinema: Directors and Directions, 1929–1968* (New York: Dutton, 1968), and Richard Corliss, *Talking Pictures: Screenwriters in the American Cinema, 1927–1973* (Woodstock, NY: The Overlook Press, 1974). For a more recent discussion of the different ways authorship has been theorized in film studies, see Janet Staiger, "Authorship Approaches," in David A. Gerstner and Janet Staiger (eds), *Authorship and Film* (New York: Routledge, 2003), 27–57. For a discussion of the producer's status as a possible auteur, see Matthew Bernstein, "The Producer as Auteur," in *Auteurs and Authorship: A Film Reader*, ed. Barry Keith Grant (Malden, MA: Blackwell, 2008), 180–89.

3 Robert L. Carringer, *The Making of Citizen Kane*, rev. ed. (Berkeley: University of California Press, 1996). Carringer has also discussed these issues in his essay "Collaboration and Concepts of Authorship," *PMLA* vol. 116, no. 2 (2001): 370–79.

4 Thomas Schatz, *The Genius of the System: Hollywood Filmmaking in the Studio Era* (New York: Pantheon, 1988), 12. Another important study of the relationship between studios and auteur directors is Rick Jewell, "How Howard Hawks Brought Baby Up: An Apologia for the Studio System," in *Howard Hawks: American Artist*, ed. Jim Hillier and Peter Wollen (London: British Film Institute, 1996), 175–84.

5 See, especially, David Bordwell, "The Bounds of Difference," in *The Classical Hollywood Cinema*, by David Bordwell, Janet Staiger, and Kristin Thompson (New York: Columbia University Press, 1985), 70–84.

6 Richard Maltby, *Hollywood Cinema: An Introduction* (Cambridge, MA: Blackwell, 1996), 30–35.

7 See Michael Baxandall, *Patterns of Intention: On the Historical Explanation of Pictures* (New Haven: Yale University Press, 1985), 29–32.

8 I discuss the case of *The Garden of Allah* in more detail in Keating, *Hollywood Lighting*, 205–6. Another example of this approach occurred on *Indiscretion of an American Wife* (1953). De Sica's regular cinematographer Aldo Graziati photographed the wide shots, while the young British cinematographer Oswald Morris was brought in to handle Jennifer Jones's close-ups. According to Morris, this situation produced a very tense and angry set. See Oswald Morris with Geoffrey Bull, *Huston, We Have a Problem: A Kaleidoscope of Filmmaking Memories* (Lanham, MD: The Scarecrow Press, 2006), 182–86.

9 John Arnold, "Why Is a Cinematographer?" *American Cinematographer* 17 (November 1936): 462. I also discuss this passage in Keating, 161–62.

10 David O. Selznick to Ray Klune, November 25, 1942, Selznick Files, Studio Files 1938–43—Stanley Cortez. David O. Selznick Collection, Harry Ransom Humanities Research Center, University of Texas at Austin. This collection of documents will be referred to as the Selznick Files.

11 See David Thomson, *Showman: The Life of David O. Selznick* (London: Abacus, 1993), 373.

12 The final shooting script (screen play credited to John L. Balderson, adaptation by Wells Root, additional dialogue by Donald Ogden Stewart) can be found in the Selznick Files, Script Development 1935–46: *The Prisoner of Zenda*. Of course, the term "final" was never absolute for Selznick. The Selznick files contain at least two different scripts marked "final shooting script," and the script was constantly revised during filming.

13 A glance at the daily production reports for Selznick's films reveals that, on a typical production, more than half of the set-ups were classified as "added scenes"—that is, set-ups departing from the literal instructions in the shooting script. Selznick's films were not unusual in this regard; many production reports at Warners and RKO show a similar pattern, with the majority of set-ups being considered added scenes.

14 For a discussion of Menzies's distinctive visual style, see David Bordwell, "William Cameron Menzies: One Forceful, Expressive Idea," *David Bordwell's Website on Cinema*, March 2010, http://www.davidbordwell.net/essays/menzies.php.

15 Victor Milner, "Creating Moods with Light," *American Cinematographer* 16 (January 1935): 14.

16 David O. Selznick to Ray Klune, October 23, 1943, Selznick Files, *Since You Went Away*—Cameramen.

17 On Pereira's sketches, see David O. Selznick to Ray Klune, October 23, 1943, Selznick Files, *Since You Went Away*—Cameramen. For Selznick's complaints about Cortez's slow pace, see David O. Selznick to Richard Johnston, December 24, 1943, Selznick Files, *Since You Went Away*—Cameramen.

18 For *All That Money Can Buy*, I consulted the daily production reports in the RKO Radio Pictures Studio Records, 1922–53, *All That Money Can Buy*—Production Files, University of California—Los Angeles, Performing Arts Special Collections. For *The Hunchback of Notre Dame*, see the daily production reports in the RKO Radio Pictures Studio Records, 1922–53, *The Hunchback of Notre Dame*—Production Files. For *Portrait of Jennie*, see Selznick Files, Production Reports—*Portrait of Jennie*. Calculations are based on the reports for principal photography by the first unit.

19 David O. Selznick to James Wong Howe, May 5, 1937, Selznick Files, *Prisoner of Zenda*—Cameramen.

20 For an example of an ASC publication offering a boldly aesthetic definition of cinematography, see Victor Milner, "Painting with Light," in *The Cinematographic Annual*, vol. 1, ed. Hal Hall (Los Angeles: American Society of Cinematographers, 1930), 91–108.

21 See James Wong Howe, "Lighting," in *The Cinematographic Annual*, vol. 2, ed. Hal Hall (Los Angeles: American Society of Cinematographers, 1931), 47–59.

22 I discuss two of Wallis's memos in more detail in Keating, *Hollywood Lighting*, 129, 138–39.

23 Selznick complains about the escalating costs of *Intermezzo* in a memo to Henry Ginsberg, July 11, 1939, Selznick Files, Production Files 1938–43: *Intermezzo* Costs. For *Intermezzo*'s escalating budget, see the reports in the Production Files 1938–43: *Intermezzo* Daily Budget Reconciliation. The information for *Torrid Zone* comes from *Torrid Zone*—Daily Production and Progress Report, Warner Bros. Archive, University of Southern California. The daily production reports and the budget for *The Mad Miss Manton* can be found in RKO Radio Pictures Studio Records, 1922–53, *The Mad Miss Manton*—Production Files, University of California—Los Angeles, Performing Arts Special Collections. The film appears to have been a "tweener"—a film with a relatively modest budget, but enough star power to be booked as an A-film if the possibility arose.

24 David O. Selznick to Gregory Ratoff and Gregg Toland, July 28, 1939, Selznick Files, *Intermezzo*—Cameramen.

25 For more on Hollywood's different strategies for lighting men and women, see Keating, *Hollywood Lighting*, 30–55, 127–33.

26 David O. Selznick to Hal Kern, October 30, 1936, Selznick Files, *Garden of Allah*—Technicolor.

27 See Barry Salt, *Film Style and Technology: History and Analysis*, 2nd ed. (London: Starword, 1992), 196. Salt argues that Toland's experiences shooting *The Grapes of Wrath* at Fox encouraged him to extend his experiments with deep-focus photography. For a discussion of light levels at various studios, see William Stull, "Surveying Major Studio Light Levels," *American Cinematographer* 21 (July 1940): 294–96, 334.

28 For a discussion of Howe's lighting strategies over the course of his career, see Todd Rainsberger, *James Wong Howe: Cinematographer* (La Jolla, CA: A.S. Barnes, 1981).

29 David O. Selznick to Henry Ginsberg, March 3, 1939, Selznick Files, *Gone with the Wind*—Cameramen. I discuss this memo in Keating, 208. For an in-depth discussion of the color cinematography in the film, see Scott Higgins, *Harnessing the Technicolor Rainbow: Color Design in the 1930s* (Austin: University of Texas Press, 2007), 172–207.

30 See Arthur Miller, "Putting Naturalness into Modern Interior Lighting," *American Cinematographer* 22 (March 1941): 104–5, 136. See also Tony Gaudio, "Precision Lighting," *American Cinematographer* 18 (July 1937): 278, 288. The latter article appeared after Eastman introduced a new stock in 1935, but before the new stocks of 1938.

31 This discussion of style in terms of problems and alternative solutions is influenced by David Bordwell, *On the History of Film Style* (Cambridge, MA: Harvard University Press, 1997), 149–57.

32 Thanks to Steve Wilson for giving me the opportunity to develop these ideas for a talk at the Harry Ransom Center at the University of Texas at Austin, and to the audience at the On, Archives! Conference in Madison, Wisconsin, where I developed the ideas further. Thanks also to Steve Neale, Lisa Jasinski, the Warner Bros. Archives at the University of Southern California, the archives at the Harry Ransom Center at the University of Texas at Austin, and the Department of Special Collections at the University of California, Los Angeles.

Chapter 17

Scott Higgins

ORDER AND PLENITUDE: TECHNICOLOR AESTHETICS IN THE CLASSICAL ERA

COLOR IN CLASSICAL HOLLYWOOD CINEMA was a supple means for directing attention, punctuating action, amplifying emotion, signaling subtext, and augmenting visual spectacle. It was also one of the most carefully monitored and painstakingly crafted aspects of the moving image. This was possible largely because Technicolor had shaped color aesthetics through cautious negotiation and experimentation. Technicolor's engineers, cinematographers, and consultants sought to weave color into the fabric of classical style, a goal that was more or less accomplished during the 1930s. Filmmakers of the 1940s and 1950s refined, developed, and strategically departed from the Technicolor aesthetic, deepening and broadening color's contributions. When Eastman Color supplanted Technicolor in the 1950s, the aesthetic grip was eased but not erased. Technicolor design proved a resilient and flexible system for harnessing chroma to the demands of storytelling, and its influence can be felt through the end of the 1950s.

Though color has been part of moving pictures since their inception, it was not integrated into Hollywood's formal repertoire until the 1930s. Hand coloring and tinting and toning, which had their origins in pre-cinematic magic lantern practices, provided spectacular flourish and rudimentary narrative coding during the silent era, but neither proved fully compliant with unobtrusive narration. Color produced by these techniques remained external to diegesis; it originated from outside the film's worlds. Pathé's introduction of stencil tinting between 1905 and 1908 helped rationalize the labor-intensive practice of coloring films frame-by-frame, and it was used until the early 1930s to create images that, at their most detailed, could resemble hand-tinted postcards. Still, hand coloring was an extravagant embellishment; expense and stylization kept it out of regular studio use.

Tinting and toning release prints by running them through dyes and chemical agents was an equally artificial but far more efficient means of achieving color. By 1914 and the rise of the feature film, standard conventions were in place: blue signaled night, red indicated fire and passion, magenta designated romance, green was used for nature and gruesome scenes, amber indicated lamplight, and so forth.[1] The uniform washes of color added variety to the viewing experience, denoted basic shifts in time, space, or dramatic tone, and could even swell into an expressive register. Cecil B. DeMille, for example, combined red tinting with experimental lighting effects to sweep the frame in a hot sinister glow for the introduction of

Sessue Hayakawa's villainous character in the *The Cheat* (1915). With between 80 and 90 percent of all prints tinted or toned by the early 1920s, color was a familiar facet of cinema.[2] Despite its ubiquity, tinting and toning remained more of an accompaniment to monochrome imagery than truly integral. Such color could have dramatic and emotional resonance, but it remained a quality laid over the image, never quite emanating from within it.[3]

Photographic color processes, on the other hand, could boast accurate reproduction of the hues before the camera, a quality more amenable to classical absorption. Before the 1920s, the major photographic color processes were additive; they required specialized projection equipment that could mix the colors of two or more distinct images on the screen. Along with technical deficiencies such as color fringing and low illumination, this requirement doomed photographic color to the status of a "special attraction."[4] In 1922, Technicolor solved this problem by pioneering a subtractive color process that layered cyan and magenta images on a single piece of release stock. The result was a two-color image that could roughly approximate flesh tones and rendered other colors as various shades of blue-green and red-orange. Color was stylized, but it was more accurate than additive techniques. Since these films could be shown on any standard projector, mainstream distribution channels more easily accommodated them, and this facilitated the production of full-length color features such as *The Toll of the Sea* (1922) and *The Black Pirate* (1926) as well as black-and-white films with color sequences such as *The Phantom of the Opera* (1925). Technicolor offered two two-color processes: the first between 1922 and 1927 involved printing images on either side of the release stock, and the second between 1928 and 1932, printed cyan and magenta on top of one another on the same side of the stock.

Its two-color systems put Technicolor on Hollywood's map and helped lay the groundwork for the company's studio-era dominance. Herbert Kalmus, Daniel Comstock and W. Burton Wescott formed Technicolor in 1915 with the aim of bringing full color reproduction to mainstream features. According to engineer James Arthur Ball:

> In the earliest days of Technicolor development we recognized that the ultimate goal . . . must be a process that would add a full scale of color reproduction to existing black-and-white product without subtracting from any of its desirable qualities, without imposing complications upon theater projection conditions, and with a minimum of added burden in the cost of photography and in the cost of prints.[5]

Technicolor's two-color systems necessarily compromised between the available magenta and cyan. Saturated and distinct reds, blues, greens, and yellows were impossible. However, careful design could produce extraordinary subtle color effects. *The Toll of the Sea*, in particular, balances pastels with soft browns for subtle and gently graded compositions. When Anna May Wong's character awaits her lover in a rose garden she is dressed in aqua green and framed by reddish accents. Composition and soft-focus lend her close-ups a sense of balance and completeness that masks the narrow chromatic range. The overwhelming bulk of two-color production, though, exploited color as a temporary novelty. This was especially the case during the brief boom in two-color production that coincided with the transition to sound. Musicals like *King of Jazz* (1930) managed a series of brilliant color effects, but they never developed schemes for integrating color into the basic vocabulary of narrative filmmaking. As its novelty faded, the color craze quickly bottomed out. Twenty-eight Technicolor features were produced between 1929 and 1931, but only four were produced in 1932 and 1933, the last being Warner Bros.' *Mystery of the Wax Museum*.[6]

Innovation of three-color Technicolor

By the end of the early sound boom, Technicolor was already moving ahead with full-color cinematography. Ball solved the problem of adding a third color component in May of 1932 with the completion of the three-color camera and the introduction of Technicolor Process number 4, commonly known as three-strip Technicolor.[7] The three-color camera exposed three negatives, each registering a different portion of the spectrum. Each negative was then optically printed onto a special matrix stock. The matrices performed like rubber stamps in transferring dye to the release stock, known as the blank.

Technicolor exploited this process by selling its services as a lab to producers; the technology itself was strictly guarded and controlled. As such, studios were spared research and development costs, and color motion pictures would remain specialized fare. To help meet studios' demands for practical, efficient, and artistically pleasing color, Technicolor wielded command over film aesthetics. The Technicolor package included rental of the cameras, employment of Technicolor's cinematographers (until the early 1940s), and oversight by the Color Advisory Service, which was headed by Natalie Kalmus. When it was set up in the 1920s, the service was meant to aid production personnel in designing for the limited range of the two-color system. With the advent of three-color, Technicolor promoted its service as a means of avoiding the purported excess of the late 1920s and early 1930s. Natalie Kalmus emphasized the importance of her department by pointing to "the early two-color pictures" in which "producers sometimes thought that because a process could reproduce color, they should flaunt vivid color continually before the eyes of the audience."[8] Technicolor publicized the department as vital because three-color had "greatly increased the demands of precision in color control in order that the fine gradations of color now available on the screen may comprise a pleasing harmony."[9] Since three-color had so substantially increased the filmmaker's palette, regulation became more essential.

From the 1930s into the 1950s, Kalmus and her crew of consultants oversaw the color design of every Technicolor production and Kalmus' contract stipulated that she receive screen credit as color consultant on each of Technicolor's features.[10] Beginning in 1937, work was parceled out between Kalmus and her associates, chief among them Henri Jaffa, Morgan Padelford, William Fritzche, and Robert Brower.[11] According to Kalmus, the department would review scripts and generate a "color chart for the entire production" that accounted for "each scene, sequence, set and character." The goal was to produce a color score, like a musical score, that "amplifies the picture" by matching color to the "dominant mood or emotion" of a sequence, thus "augmenting its dramatic value."[12] According to Leonard Doss, a Technicolor consultant during the late 1940s and early 1950s, and, in 1989, the department's lone survivor, a color consultant's responsibilities involved five major steps:

> First, they read the script, then researched and planned out the appropriate color schemes (meeting occasionally with Kalmus). Second, they met with the producers to set up a budget and schedule. (A larger production budget would warrant several consultants being assigned to a film.) Third, they met with the costume department, since the interior set was generally designed around the colors worn by the protagonists. Fourth and fifth, they met with the studio's Art and then Props Departments to guarantee that the props and sets would reinforce the color schemes planned for each shot and scene.[13]

These practices continued until the early 1950s, when Eastman Kodak introduced Eastman Color, which was sold directly to producers on the open market.

Silly symphonies: Disney and early Technicolor

In 1932, Herbert Kalmus faced the problem of enticing a doubting industry to reconsider color. Because it would take some time to build enough three-color cameras and convert the plant to the new process, and because producers were unwilling to invest in the untested system, Technicolor turned to the field of animation.[14] Animation cells were filmed using the successive exposure of frames through colored filters and did not require the three-color camera.[15] Kalmus found a customer in Walt Disney who entered into a two-year exclusive contract for three-color animation and premiered the process in the Oscar-winning short *Flowers and Trees* late in 1932.

Disney had launched the 'Silly Symphonies' series in 1929 to exploit sound. After *Flowers and Trees*, it became a platform for Technicolor too. In important respects, Disney's color shorts anticipated the direction of color aesthetics in 1930s cinema. Disney artists favored pastels and saved strongly saturated hues for accents. Backgrounds were kept desaturated and cool to assure that the main characters would have chromatic prominence. The balance between decoration and clarity, so central to all Technicolor design, was first encountered at Disney. Characters were kept centered, and the artists would introduce decorative and coordinating accents into the frame's margins, as with the flowers that consistently appear in the lower corners in *Flowers and Trees*. Disney artists were adept at coordinating color and drama. After the red and black scourge of forest fire in *Flowers and Trees*, soft pastels reappear and a rainbow frames the film's romantic couple. Color temperature and lighting effects also supported mood. *The Flying Mouse* (1934) offers perhaps the first instance of what would become a Technicolor staple, the expressive balancing of color temperature. Steely blue moonlight contrasts with orange-gold lamplight when the warm hearth of mother mouse's pumpkin home beckons her wayward son to abandon his dreams of being a bat.

These seven-to-eight-minute shorts were a training ground for Disney, whose animators could indulge in a free play of color more adventurously than could live-action filmmakers. Shorts like *Funny Little Bunnies* (1935) push the limits of comprehension by packing the frame with an unsettling amount of movement and detail. In one shot, rabbits in red and lavender wheel giant baskets on carts laterally across the foreground while two worker-bunnies in red and blue operate a sluice gate that releases red, green, orange, and blue eggs into the basket in the rear. Meanwhile two more bunnies with red and blue cummerbunds juggle purple, blue, yellow, and red jelly beans into the passing basket. Movement is lateral, vertical, and axial; action and color are so complicated that they risk outrunning the capacities of human perception.

Three-color for live action

Live-action Technicolor faced more severe technical obstacles, especially when it came to lighting. In 1936, when the average black-and-white set illumination ranged from 250 to 400 foot-candles, Technicolor levels ranged from 800 to 1000.[16] Before 1950 Technicolor's appetite for light remained huge, averaging at 400 to 500 foot-candles of illumination while black-and-white had reduced levels to around 75 to 100 foot-candles by 1940.[17] Joe Valentine, writing about his cinematography for *Joan of Arc* in 1948, highlighted the difference between color and monochrome: "the highest key I ever used in black-and-white photography turned out to be the lowest key of lighting in the history of Technicolor. . . . ten lighting units must be used in color for two in black and white."[18]

In addition to brighter light, Technicolor required a color temperature equivalent to daylight. To achieve this, producers turned to arc lights, which had not been in general use since the 1920s.[19] Cinematographers working in Technicolor also faced stricter oversight than their black-and-white colleagues. The Technicolor lab demanded well-exposed negatives that properly registered color values. While black-and-white lighting was routinely balanced by eye, Technicolor insisted on foot-candle meters to determine levels.[20] Less flexible lighting robbed cinematographers of the control over highlight and shading that they enjoyed when working in black-and-white, as Ronald Neame, among others, pointed out.[21]

With monochrome as the measure against which Technicolor would be held, cinematographers and designers engaged in a dialogue about how best to control color so as to fulfill classical ideals of motivated spectacle, narrative centering, and relative unobtrusiveness. Herbert Aller summarized the situation plainly in *International Photographer* when he applauded the Technicolor cinematographers of the 1930s who "carried with them the inalienable thought that the audience when leaving the theater must not say the story lagged for the sake of color."[22]

Conventions for handling color developed swiftly during the 1930s as filmmakers worked through three consecutive modes of design. Three-color style began with a demonstration mode (1934–35) that kept color on display and ostentatiously dramatic. After the demonstration films, the restrained mode (1936–37) closed down stylistic play and channeled color into less pronounced roles. As three-color gained acceptance, and as a body of conventions for binding it to the classical style developed, the palettes and the range of effects broadened resulting in the assertive mode of color design (1938–1940). Through all three modes, the most lasting uses of color followed the paths carved for other stylistic devices. These included the spectacular embellishment of transitions, the gentle directing of attention, the momentary highlighting of actions, the development of motifs, and the general correlation of color with the mood or tone of the drama.[23]

During the 1940s and early 1950s, the restrained and assertive modes remained in play as options, generally associated with genre; drama and prestige pictures might cling to restraint while adventures and musicals boldly embraced assertion. Directors like Vincente Minnelli, Nicholas Ray and Douglas Sirk drew on Technicolor practice, mixing modes and pushing conventions to produce virtuoso designs. The Technicolor look was varied and multifaceted, but on the whole remarkably consistent.

Demonstration in the 1930s

The demonstration mode is exemplified by the first two three-color live-action productions produced by Pioneer Pictures: the short film *La Cucaracha* (1934) and the first Technicolor feature *Becky Sharp* (1935). Pioneer Pictures was founded by Merian C. Cooper and John Hay Whitney in 1933 with the aim of bringing live-action three-color to the theatrical market. Whitney had invested heavily in Technicolor stock and hoped to boost the company's value. *La Cucaracha* and *Becky Sharp* were prototypes for color filmmaking: first and foremost they were designed to test and promote the Technicolor process. Given Whitney's financial interest in Technicolor, their profitability was probably a secondary concern. *La Cucuracha* managed a modest profit and an Oscar for best Comedy Short Subject, but *Becky Sharp* was a critical and box office failure.[24] Together, however, they demonstrated the possibilities of color and garnered attention from mainstream producers. Both granted color a ruling position rather than negotiating its place among established techniques. The demonstration mode was highly noticeable but not well suited to the long-term integration of three-color into classical style.

The aesthetics of demonstration were crafted by Robert Edmond Jones and Rouben Mamoulian. Jones, a famous Broadway designer hired by Pioneer, claimed, "the difference between a black-and-white film and a Technicolor film is very like the difference between a play and an opera."[25] Mamoulian, who directed *Becky Sharp* and shared design credit with Jones, said he "tried to make the dramatic and emotional use of color play a vital part."[26] Jones and Mamoulian touted color's expressive power, but their designs were clumsy. The films develop rather elaborate pretexts for emotional effects, as when the lovers in *La Cucaracha* are staged progressively nearer to a red lamp as they argue: the scarlet light intensifying their passion. *Becky Sharp*'s color centerpiece occurs midway through the film at the Duchess of Richmond's ball during which Napoleon attacks and British soldiers rush out to meet him. The scene was oft cited by Mamoulian to exemplify his dramatic use of color. As the British soldiers heed the call to battle, Mamoulian shifts between bold blues and reds, culminating in a swell of brilliant warm color. First, a line of soldiers in sky blue capes sweep down a corridor in two shots that present a spectacular color mass streaking diagonally across the frame. These compositions are then repeated with soldiers in fiery red capes. Moments later the entire frame is washed in red light as soldiers run down a drive beneath a red streetlamp. While the organization of color heightens dramatic urgency, the patterning is boldly forced onto the situation; staging and editing are bent toward serving chromatic display.

La Cucaracha and *Becky Sharp* keep color active and noticeable. *La Cucaracha* packs the frame with yellow, cyan, and magenta contrasts while *Becky Sharp* provides strictly neutral backgrounds upon which strong accents vie for attention. The result in both cases is a constant parade of color that sometimes resembles a test pattern. If this approach erred, it was in giving color too high a priority in the hierarchy of film style. In making stylistic choices, the filmmakers consistently favored alternatives that relied on color. Instead of making color functional *within* the classical system, the demonstration mode showcases color, makes it extrusive. As Jones' comparison of color to opera suggests, it was a case of trying to make Technicolor sing before it could talk. This was not necessarily a failing because the demonstration mode helped to make a clear case that Technicolor *could* contribute meaningfully to narrative film. Andre Senwald, in his review of *Becky Sharp* for the *New York Times*, aptly described Technicolor's challenge in the wake of the demonstration films: "the problem is to reduce this new and spectacular element to a position, in relation to the film as a whole, where color will impinge no more violently upon the basic photographic image than sound does today."[27]

Technicolor's general principles

Just as *Becky Sharp* was hitting theaters, Natalie Kalmus issued her major statement on color aesthetics. Entitled "Color Consciousness," her essay was originally presented as a lecture in May 1935 and read in her absence by Kenneth MacGowan as part of an Academy of Motion Picture Arts and Sciences panel about *Becky Sharp*. It was reprinted in the Academy's *Technical Bulletin*, in the August issue of the *Journal of the Society of Motion Picture Engineers*, and again in 1938 as a chapter entitled "Colour" in Stephen Watts' anthology *Behind the Screen: How Films are Made*.[28] Kalmus' timing was telling. Since she was largely absent from discussions of the demonstration films, "Color Consciousness" reclaims territory by consolidating aesthetics under her authority. Her comments define the Color Advisory Service's long-term approach to color, but they are also highly rhetorical, seeking to prove her department's command of the new medium. Thus she indulges in a literal color vocabulary, claiming, for instance, "by introducing the colors of licentiousness, deceit, selfish ambition, or passion" to the color red, "it will be possible to classify the type of love portrayed with considerable accuracy."[29] It is unlikely that such overt semantic coordination ever informed Technicolor design, but Kalmus' explanation is vivid and could be easily grasped and visualized by potential customers.

"Color Consciousness" is compelling at a broad level, as a set of four general principles for helping color serve narrative. Kalmus' widest ranging directive was that color should support the mood or tone of the story. The main goal of creating color charts, according to Kalmus, was to ensure that color "subtly convey dramatic moods and impressions to the audience, making them more receptive to whatever emotional effect the scenes, action, and dialog may convey."[30] Her second general rule held that excessive use of bright, saturated color should be avoided in favor of more "natural," harmonious and less intense schemes. In Kalmus' words: "A super-abundance of color is unnatural, and has a most unpleasant effect not only upon the eye itself, but upon the mind as well."[31] Successful design would balance assertive hues with an array of neutrals. The third principle links this kind of harmony to the problem of directing attention. Warmer and brighter shades should emphasize only narratively important information, otherwise Kalmus advised cool, neutral colors: "the law of emphasis states in part that nothing of relative unimportance in a picture shall be emphasized. If, for example, a bright red ornament were shown behind an actor's head, the bright color would detract from the character and action."[32] The maximum point of contrast should be associated with the principal players in a scene while those who play "relatively unimportant roles" blend with the background.[33]

Finally, Kalmus advised filmmakers to "consider the movement in the scene in determining its color composition because the juxtaposition of color is constantly changing due to this movement."[34] This directive was aimed at the problem of complementarity. She explained, for example, that orange would "appear more red than it really is" next to a blue green because "each color tends 'to throw' the other toward its complement."[35] Restraint and attention to harmony were the keys to avoiding unwanted juxtapositions. Kalmus' definition of harmony limited complements and played contrasting hues against neutrals so as to curb distraction. This rule is especially important to the restrained mode, but even films with more assertive palettes, like *The Adventures of Robin Hood* (1938), exhibit attention to juxtaposition, sometimes revealing complements to underscore a moment of action.

Kalmus aimed to align color with the classical functions of film style. Harmonizing color to avoid striking juxtapositions, concentrating assertive hues on the protagonists, and associating color with mood would have the net effect of preventing Technicolor from becoming intrusive. The rules were drafted to guarantee that color, like lighting, sound, camera-movement, and editing, would hold the viewer's attention to the narratively important elements of the moving image and thus suit the expressive demands of feature production. At the same time, the principles were broad enough to encompass a range of color styles. Color could more or less explicitly underscore narrative development, and color harmonies could run from the sparing use of accents against a background of neutrals to the rich play of stronger hues, carefully handled to keep the key contrast centered on the action. The strength of Kalmus' formulation is that it did not posit an inflexible, single, mode of design.

Restrained design

The feature films that followed *Becky Sharp* exhibit a conservative application of these principles. Color designers subdued variations of hue in favor of an emphasis on tone and value, qualities shared by black-and-white. The insistent foregrounding of *La Cucaracha* and *Becky Sharp* gave way to a less obtrusive look that was judged more easily assimilated with narrative. This is the restrained mode of color design. Films in this mode employ a constricted palette, based heavily on neutrals and browns, from which filmmakers could strategically diverge. Limiting color contrast avoided potentially distracting juxtapositions, and managed quieter color effects that did not so boldly command attention. More generally, in relying on low profile designs, the restrained mode helped color blend with unobtrusive classical form. The mode helped prove that three-color had wider applications than those of a simple novelty. The restrained mode was initiated by Paramount's 1936 release *Trail of the Lonesome Pine*, the first truly successful three-color feature, and it was the dominant style through 1938. Restrained films include the Selznick-International productions *The Garden of Allah* (1936), *Nothing Sacred* (1937) and *A Star is Born* (1937), Warner Bros.' *God's Country and the Woman* and *Valley of the Giants* (1938), Walter Wanger's *Vogues of 1938* (1937), Disney's *Snow White and the Seven Dwarfs* (1937) and Samuel Goldwyn's musical *The Goldwyn Follies* (1938).

Restraint had several important advantages. Where *Becky Sharp* splashed highlights against neutral backgrounds, the restrained designs either eschewed or softened the accents into pastels and varied the neutrals to create more complex texture. Harmony was defined as the avoidance of strong contrast, and this gave restrained films a tightly coordinated look. In films like *The Goldwyn Follies* and *Vogues of 38*, where color was not called into dramatic service, it lent a polished gloss. Pampered star Olga Samara's (Vera Zorina) dressing room in *Goldwyn Follies*, for instance, is rendered in soft pastel tints of aqua and rose that elegantly match costume to set without pronounced contrast. The restrained mode's greatest asset was that against a tight palette small variations become significant. A finely tuned design could draw attention to, and get use from, minute and precise changes in color. This gave filmmakers a finer brush. They could draw attention to texture and detail to create small but spectacular moments, as when Esther Blodgett (Janet Gaynor) steps out of shadow and reveals her cypress green dress when she visits the central casting office in *A Star is Born*. Likewise, relatively limited color accents could punctuate action. When June (Sylvia Sidney) arrives to confront Jack (Fred MacMurray) about her little brother's death in *Lonesome Pine*, green details planted in the brown set converge with her costume and so underline her entrance. Character development could also be signaled through shifts in palette. In *God's Country and the Woman*, when the tough woman lumber camp owner Jo Barton (Beverly Roberts) falls in love with her competitor Steve Russett (George Brent), her formerly brown and tan wardrobe shifts to light blue, and pastel accents appear in her surroundings. Finally repetitions of color could stand out within the finely modulated palette and build motifs. Splashes of autumnal orange in *Lonesome Pine* are first associated with the blossoming romance between June and Jack, but then return toward the end of the film during the funeral of June's young brother, highlighting the emotional distance traveled since the film's start. Red-orange in *A Star is Born* is associated with the glittering Hollywood skyline and the fan magazine Esther keeps in her farmhouse, then connects to the Pacific sunset that frames Norman Maine's death, and finally returns in the searchlights at Esther's triumphal premiere which closes the film. The meaning of this color motif is nebulous, but it connotes the lure and dangers of Hollywood fame.

That said, restrained films did not turn away entirely from color foregrounding. Rather, they tended to concentrate color display in transitions and in other scenes that seemed open to stylization. The effect could be rather forced, as in *Lonesome Pine* when June and Jack share a romantic phone call after she has left the backwoods to study in Louisville. Here the designers exploit the love scene, a conventional occasion for glamour, to display Technicolor with a punch reminiscent of the demonstration mode.

June's hyper-feminine bedroom is appointed in pastel pinks and greens, and ornamented with frilly curtains and soft flowers, all of which startlingly depart from the film's rugged earth-tone mise-en-scene. The overt color unexpectedly spills over into Jack's masculine field office, as he fidgets with a set of colored pencils that sit before a black-and-white glamour photo of June on his desk. Foregrounding was more successfully introduced at the margins of narrative, during establishing shots and montage sequences. Transitional montages, like the tour of Hollywood landmarks that accompanies Esther's arrival at the West Coast in *A Star is Born*, or the elliptical sequence that depicts her studio makeover, nudge color forward by presenting sharp contrast and coordination from shot to shot. In its own way, the restrained mode continued the tension in Technicolor aesthetics first seen at Disney: the pull between unobtrusive support of the story, and the need to display color as an added attraction.

Assertive design in the late 1930s

The restrained mode is the lynchpin to understanding classical Hollywood color, but *The Adventures Robin Hood* was a turning point. For the first time in a three-color feature, the palette was opened wide and intricately organized. Far from a return to demonstration, *Robin Hood* modulates color to effectively direct attention and underscore drama. The film draws on the methods forged in the restrained mode and brings them to a system in which hue rather than tone is the dominant variable. Assertive designs necessarily limited the power of subtle variations, but *Robin Hood* addresses this problem and clears a path for films that followed by varying the functions of color across its running time. Spectacular flash dominates the opening and climax of the film, while the palette becomes more variable and tied to the characters' fortunes during the middle stretch. At different stages of the story the color design emphasizes different kinds of tasks, all the while favoring a wide mix of hues.

With its emphasis on red, blue, yellow, and green, the film unavoidably deals in more complex harmonies. Where restrained films were content to rest with low-contrast coordination between the women's costumes and the men's neutrals, *Robin Hood* thrives on strong complementaries. Assertive colors play off one another to create balanced but varied compositions, suggesting a redefinition of harmony as the equilibrium of hues. When Robin (Errol Flynn) and Marian (Olivia de Havilland) flirt during the forest banquet, for instance, she wears a brownish chili red velvet cape to contrast with his Lincoln green hood and tights. Other accents push the costumes in both directions of coordination and contrast. The clasps on her cape feature deep emerald stones that pick up Robin's greens, but her faint lavender veil contrasts in both hue and value. Costumes harmonize through the intricate play of accents while large areas of each outfit (Marian's shimmering silver dress, Robin's leather brown vest) are more or less neutral. In the assertive mode, compositions could become rich medleys of accents; red details answering patches of green and so forth.

Robin Hood's most prominent innovation was the marriage of Technicolor to the historical adventure genre. The adventure's basic structure strings together a series of large production scenes, offering clear opportunities to keep hue vital and perceptible. *Robin Hood* adapts techniques from both the demonstration films and the restrained mode, including contracting and expanding the palette from scene to scene, extending sequences of establishing shots to showcase colorful sets and costumed extras, and selectively revealing and removing hues within scenes. Yet, in distinction to the demonstration mode, *Robin Hood* does not incessantly prioritize color. Assertive designs tend to be redundant with other cues like lighting, musical score, camera movement, and editing; color has been incorporated rather than clumsily laid on top. Many of the most obvious color effects, like the fuchsia red wine that Prince John's thugs spill as they brutalize a Saxon inn keeper, are localized in transitional and montage sequences, bracketed from the main line of action. The design's variety keeps it from generating subtle motifs, so *Robin Hood* favors punctual contrasts (the momentary convergence of color and narrative). Color thus serves story without relying on the labored patterns and literalism of *La Cucaracha* or *Becky Sharp*. In *Robin Hood*, one senses that the filmmakers had found new confidence; color offers its own arsenal of techniques that could freely blend with the ongoing action. If the restrained mode taught methods for unobtrusively manipulating color, *Robin Hood* illustrates how they could be successfully brought to an extravagant palette.

1939 was a banner year for the Technicolor Corporation. Confidence in three-color was such that Technicolor embarked on a $1,000,000 expansion program and the company's profits finally overtook its losses.[36] Twentieth Century-Fox's *Jesse James* was the highest grossing film of the year and a *Hollywood Reporter* exhibitor poll named it and two other Technicolor features (*The Wizard of Oz* and *Dodge City*) as the most lucrative releases of the year.[37] The year also saw the introduction of an improved negative stock and camera optics, which enhanced sensitivity to light. But 1939's crowning achievement came in December with the premiere of classical Hollywood's most expensive and popular production, *Gone with the Wind*. Although the *New York Times* critic maintained, "color is hard on the eyes for so long a picture," the Hollywood establishment greeted *GWTW* as evidence that color had arrived.[38] At their annual ceremony, the Academy of Motion Picture Arts and Sciences presented two special awards: one to William Cameron Menzies for "outstanding achievement in the use of color for the enhancement of dramatic mood in the production of *Gone with the Wind*," and another to Herbert Kalmus for "contributions in successfully bringing three-color feature production to the screen."[39] Placing Technicolor on a firm commercial footing had taken nearly a decade, but *GWTW* emerged at the end of the 1930s as an emblem of three-color's success.

GWTW embellishes techniques from earlier films, and because the filmmakers could draw on a body of established effects and motivations, color appears more flexible. Generally, the palette expands and contracts around Scarlett O'Hara (Vivien Leigh), modulating with her success or failure as the film develops. The design is dynamic, moving from assertion during the Twelve Oaks barbecue and armory bazaar sequences to severe restraint as Tara recovers from the war. *GWTW* combines the restrained mode's power to develop chromatic motifs with the variety and ornamentation of *Robin Hood*. Most importantly, in *GWTW* color design has been reconceived in relation to cinematographic opportunities opened up by Technicolor's new film stock. The film's real contribution to the history of Technicolor lies in an innovative synthesis of light and color, and this is best demonstrated by its experiments with colored illumination.

Colored lighting brings color firmly within the cinematographer's realm. *GWTW* usually tempers the device by providing plausible motivation and by reducing the range of hue to an opposition of cool and warm light, much as Disney had done in *The Flying Mouse*. For instance, contrasting color temperatures punctuate the appearance of villain Jonas Wilkerson (Victor Jory) as Scarlett's mother Ellen (Barbara O'Neil) returns home at the beginning of the film. When Wilkerson halts Ellen in her doorway he receives very dim bluish highlights and much of his figure is left in darkness, while she, apparently lit by an oil lamp, is bathed in a warm amber key light. When she leaves the frame, the camera dollies toward Wilkerson leaving him in cold white highlights. The shift in color conveys Wilkerson's intrusion into Tara's peaceful candlelit warmth. Colored illumination has been rendered pliant; warm and cool components intermingle, amplifying the kind of tonal contrast valued in monochrome.

GWTW also has a strong experimental streak. The filmmakers test limits, find the maximal level of stylization that can still accommodate functions of classical style. Rhett (Clark Gable) and Scarlett's iconic kiss as they flee the burning of Atlanta is both emblem and apogee of *GWTW*'s experiments in color cinematography. A vigorous mixture of low-key and colored lighting heightens the moment, which is rendered in three successively closer shots. Each cut inward places greater stress on facial detail and on the contrast of red-orange light on Rhett and white highlights on Scarlett. Shifting colors track the power struggle between the lovers. When Scarlett resists Rhett's advances and turns away from him, she dodges out of her key and is enveloped in shadow. Then, when Rhett declares his love, he grasps her face, moving her so that the bright cool hues return. The third shot, a tight close-up, immerses the entire frame in red and black, except for Scarlett's flesh-tones. With his kiss, Rhett blocks out these highlights and tilts Scarlett's head back so that she is suffused with red orange. His movement forward creates a brief surge of chroma at the center of the frame, punctuating the kiss.

This effect is a descendent of Robert Edmund Jones' early experiments in expressive color. Red stands for "passion" in this scene in the same way that it connoted "rage" in *La Cucaracha*, or "danger" in the ball sequence in *Becky Sharp*. In those films, the technique of flooding the frame with red light was used only briefly and uniformity of color resembled tinting and toning. But in *GWTW*, the color effect is integrated with more conventional expressive techniques like low-key modeling, editing, and performance. By incorporating white and bluish highlights, and manipulating deep shadows, the cinematographers lodge conventional methods of shaping the image within the red environment. *GWTW* transforms colored illumination into an expressive and lithe element of style. This is exemplary of Technicolor's aesthetic development during the 1930s. Early experiments presented a range of devices that were then reclaimed for feature production when they could be suitably motivated and made to work with established techniques.

Technicolor in the 1940s and beyond

The basic path of stylistic change during the 1930s was toward the integration, flexibility and confidence in color design. Once Technicolor was seen as an embellishment of existing practices rather than a fundamentally different art, color style progressed fairly rapidly. Technicolor began to modify its protective posture by allowing some studio cinematographers to work on their own and by 1943 most productions were shot by unsupervised studio cameramen. Technicolor found a comfortable niche in Hollywood, beginning with about fourteen features a year from 1940 through 1942 and increasing thereafter, and the range of design options expanded.[40] Filmmakers reworked and combined restrained and assertive modes and their attendant strategies. Warners' military aviation film *Dive Bomber* (1941), for example, adheres very closely to the restrained mode in nearly as strict a manner as *Lonesome Pine*. MGM's biopic *Blossoms in the Dust* (1941) and the musical *Smilin' Through* (1941) are both fairly restrained, but they take up different options within that mode. The former draws more heavily on color scoring and motif building, while the latter uses color as a glamorous setting for Jeanette MacDonald. The successful series of Twentieth Century-Fox musicals starring Alice Faye, among them *That Night in Rio* (1941) and *The Gang's All Here* (1943), also draw on tight harmonies to create a sense of high polish in exposition scenes. The big production numbers, however, showcase assertive and varied palettes. Adventure films such as *The Black Swan* (1942) are reminiscent of *Robin Hood*, if not as insistent in their emphasis on variety.

Technicolor became more varied, but the overarching principles of "Color Consciousness" held sway. The reviewers in *American Cinematographer*'s "Photography of the Month" column in the early 1940s denigrated color designs that dispersed attention. A close-up of Alice Faye in *That Night in Rio* was criticized for "the presence in the extreme background of an extra woman in a too strongly blue gown which, even though extremely out of focus, is still a sufficiently strong tonal intrusion to distract the eye from the star's face."[41] The musical *Louisiana Purchase* (1941) was similarly berated for settings that "like an irresistible magnet drew attention from the players to the background."[42] As late as 1957, the manual *Elements of Color in Professional Motion Pictures* published by the Society of Motion Picture and Television Engineers as a production aid in the early Eastman Color era, testifies to Technicolor's aesthetic longevity. The manual's suggestions for color design follow closely those propagated by Natalie Kalmus decades earlier. In fact, the book suggests that productions employ "color coordinators" to take the place of the now absent Technicolor consultants.[43]

During the 1940s and early 1950s, Technicolor's aesthetics provided a standard of craft practice and an arena for ambitious stylists. Vincente Minnelli's first Technicolor feature *Meet Me in St. Louis* (1944), for instance, largely obeys conventions of narrative centering through coordination and contrast, but with unprecedented complexity.[44] In "The Trolley Song" number, Minnelli inverts Technicolor norms by dressing Esther (Judy Garland) in the least colorful outfit on a bright yellow trolley packed with extras in pink, green, and blue. This allows Minnelli to cram the frame with aggressive color, while ensuring that contrast always underlines Esther's position. Even at its most audacious, Minnelli's design manages to achieve harmony. Esther's light blue gloves, and her green and blue taffeta skirt with pink details balance and echo the accents surrounding her. The "Skip to my Lou" dance similarly marshals varied costumes that generate fluid patterns of connection and divergence with the lead players. Throughout his career, when Minnelli staged scenes of elaborate action he could array strong saturated accents across the frame and keep them in continual motion. The shoeshine number in *The Band Wagon* (1953) and the carnival sequence in *Some Came Running* (1958) constantly evolve color relationships that draw the eye through space. Remarkably, Minnelli's mosaic designs reinforce the central narrative action, but they do so in a perceptually demanding way.

Color designs generally became more aggressive in the 1950s, using dynamic triads and complementaries to punctuate action and steer attention. Douglas Sirk and his production team at Universal, among them cinematographer Russell Metty, made the most of this latitude. Sometimes regarded as radical experiments in expressive color, Sirk's films may be profitably understood as virtuoso inflections of Technicolor aesthetics. Bridging the final years of Technicolor's reign and the beginning of Eastman Color's, Sirk extrapolated from such 1940s Technicolor dramas as *Blossoms in the Dust* and *Leave Her to Heaven* (1946) to forge a hyperbolic and semantically complex style.[45]

Imitation of Life (1959), an Eastman Color film made under Technicolor's aesthetic shadow, exemplifies Sirk's elaboration. The famous hard cut from Sarah Jane (Susan Kohner) laying in the gutter after being beaten by Frankie (Troy Donahue), her yellow dress made dingy by the grime and dim lighting of the street, to Lora Meredith (Lana Turner) reclining in a pink robe as Annie Johnson (Juanita Moore playing Sarah Jane's mother) massages her feet, owes much to Technicolor scoring. Of the primaries, yellow is the hue that loses saturation most quickly when darkened, making it a particularly apt conductor for the emotional charge of Sarah Jane's downfall: nothing degrades like yellow. This color motif has been referenced earlier when Susie (Sandra Dee) promised to buy the yellow sweater for Sarah Jane, who desires to emulate Susie by "passing" as white. When Frankie rages against Sarah Jane's mixed parentage, it is fitting that he should defile the yellow sweater and dress. This kind of color motif had served Technicolor well since the 1930s. Sirk's punctual cut to Lora Meredith's well-lit, pink bedroom throws Sarah Jane's disgrace into sharp contrast, emphasizing the emotional discontinuity between the scenes.[46] The viewer is wrenched from a scene of abasement to one of empty luxury; beautiful surface collides with the ugliness that lies beneath. In *Imitation of Life* color motif and punctuation serve direct emotionally expressive functions at the same time that they seem to embody social critique. Sirk and Metty torque Technicolor conventions to develop a disarming commentary on the action.

In pressing color forward as an expressive register and narrative device, Sirk's work, like Minnelli's, both sums up and transcends classical color aesthetics. During the 1930s filmmakers and studios tempered color's novelty and developed practical methods for managing it. Tinting and toning, and two-color Technicolor had served storytelling, but their stylization limited integration and filmmakers' reservoir of conventions remained shallow. By the 1940s color could be fully orchestrated according to a flexible set of design principles in some variation of the restrained and assertive modes. Ultimately, Technicolor aesthetics proved generative rather than limiting. They assured even average films a sense of order and plentitude that, like music, could operate quietly, beneath the threshold of the viewer's direct attention. Colorists like Sirk and Minnelli used those same methods to intensify color's emotional valence, perceptual demands, and semantic complexity. But even color virtuosi built on the foundation of the classical Technicolor style.

Notes

1. Barry Salt, *Film Style and Technology: History and Analysis* (London: Starword, 1992 edition), 78, 124, 150–51.
2. Salt, *Film Style and Technology*, 150–51.
3. Tinting and toning decreased with the coming of sound, probably because the dyes interfered with the optical soundtrack. In the late 1920s Eastman Kodak introduced Sonochrome release stock that allowed the practice to continue. As late as 1937 MGM released *The Good Earth* in 500 prints toned red-brown. See Roderick Ryan, *A History of Motion Picture Color Technology* (New York: The Focal Press, 1977), 16–17, 19.
4. Salt, *Film Style and Technology*, 79–80; Ryan, *A History of Motion Picture Color Technology*, 26–30.
5. Joseph Arthur Ball, "The Technicolor Process of Three-Color Cinematography," *Journal of the Society of Motion Picture Engineers* (August 1935), 127.
6. Herbert Kalmus, with Eleanore King Kalmus, *Mr. Technicolor* (Abescon, New Jersey: MagicImage Filmbooks, 1993), 82.
7. Kalmus, *Mr. Technicolor*, 91.
8. Natalie Kalmus, "Color Consciousness," *Journal of the Society of Motion Picture Engineers* (August 1935), 147.
9. "Importance of Color Control Service Growing," *Technicolor News and Views* (February 1941), 1.
10. Natalie had been married to Herbert Kalmus between 1903 and 1921, and after their divorce she headed the Technicolor color control department, also referred to as the color advisory service. Accounts of how she gained this position vary, but it is commonly asserted that her job was stipulated as part of the Kalmus' divorce settlement.
11. "Color Director's Work Important," *Technicolor News and Views* (May 1939), 3. "Importance of Color Control," 3.
12. Kalmus, "Color Consciousness," 145.
13. Richard Neupert, "Technicolor and Hollywood: Exercising Restraint," *Post Script* (1990), 23.
14. Kalmus, *Mr. Technicolor*, 94.
15. Richard Haines, *Technicolor Movies: The History of Dye Transfer Printing* (Jefferson, North Carolina: McFarland & Company Inc., 1993), 18–19.
16. "Report of the Studio Lighting Committee," *Journal of the Society of Motion Picture Engineers* (January 1937), 39.
17. Leigh Allen, "New Technicolor System Tested by Directors of Photography," *American Cinematographer* (December 1950), 414.
18. Joe Valentine, "Lighting for Technicolor as Compared with Black and White," *International Photographer* (January 1948), 7.
19. Winton Hoch, "Cinematography in 1942," *Journal of the Society of Motion Picture Engineers* (October 1942), 99.
20. For a detailed discussion of the use of exposure readings in Technicolor cinematography, see Ralph A. Woolsey, "Lighting and Exposure Control in Color Cinematography," *Journal of the Society of Motion Picture Engineers* (June 1947), 548–53.
21. John Huntley, *British Technicolor* (London: Skelton-Robinson, 1948), 55.
22. Herbert Aller, "Color Marches On," *International Photographer* (May 1936), 17.
23. For an extended discussion of these modes of design, and of Technicolor aesthetics generally, see Scott Higgins, *Harnessing the Technicolor Rainbow: Color Design in the 1930s* (Austin: University of Texas Press, 2007).
24. "What? Color in the Movies Again?," *Fortune* (October 1934), 164; "Whitney Colors," *Time* (27 May 1935), 28; Richard Jewell, "RKO Film Grosses, 1929–51: the C.F. Tevlin Ledger," *Historical Journal of Film, Radio and Television*, vol. 14 no. 2 (1994), microfiche supplement.
25. Jones, "A Revolution in the Movies," *Vanity Fair* (June 1935), 13.
26. William Stull, "Will Color Help or Hinder?," *AC* (March 1935), 107.

27 Andre Sennwald, Review of *Becky Sharp, New York Times* (14 June 1935), 27:2.
28 I refer to the *Journal of the Society of Motion Picture Engineers* version of "Color Consciousness."
29 Kalmus, "Color Consciousness," 143.
30 Kalmus, "Color Consciousness," 142.
31 Kalmus, "Color Consciousness," 142.
32 Kalmus, "Color Consciousness," 146.
33 Kalmus, "Color Consciousness," 146.
34 Kalmus, "Color Consciousness," 147.
35 Kalmus, "Color Consciousness," 147.
36 "Technicolor Expansion Program in Operation," *Technicolor News and Views* (April 1939), 1, 3; Fred Basten, *Glorious Technicolor: The Movie's Magic Rainbow* (Cranbury, NJ: A.S. Barnes, 1980), 90; Herbert Kalmus, *Mr. Technicolor*, 118.
37 "3 Technicolor Pictures Lead Exhibitor Poll," *Technicolor News and Views* (January 1940), 1.
38 Frank Nugent, review of *Gone With the Wind, New York Times* (20 December 1939, Amusements), 31.
39 "Special Award Bestowed for Color Process," *Technicolor News and Views* (March 1940), 1, 3.
40 Fourteen full color, live-action Technicolor features were released in 1940, fifteen in 1941, and thirteen in 1942. From 1943 on there was a consistent rise in production each year. Twenty-two features were released in 1943, twenty-four in 1944, thirty in 1946, thirty-two in 1947, and forty-four in 1948, and so on and so forth.
41 "Photography of the Month," Review of *That Night in Rio, American Cinematographer* (April 1941), 169.
42 "Photography of the Month," Review of *Louisiana Purchase, American Cinematographer* (January 1942), 31.
43 The Society of Motion Picture and Television Engineers, *Elements of Color* (Los Angeles: SMPTE, 1957), 44.
44 For detailed analysis of color design in *Meet Me in St. Louis*, see Scott Higgins "Color at the Center: Minnelli's Technicolor Style in *Meet Me in St. Louis*," *Style*, vol. 32 no. 3 (1998), 451–73; and "Deft Trajectories for the Eye: Bringing Arnheim to Vincente Minnelli's Color Design," in Scott Higgins (ed.), *Arnheim for Film and Media Studies* (New York: Routledge, 2010), 107–26.
45 For a discussion of Sirk's color aesthetics in relation to Technicolor norms, see Scott Higgins, "Blue and Orange, Desire and Loss: The Color Score in Far from Heaven" in James Morrison (ed.), *The Cinema of Todd Haynes* (London: Wallflower Press; New York: Columbia University Press, 2006), 101–13; and "Color Accents and Spatial Itineraries" Dossier on Cinematic Space, *The Velvet Light Trap*, no. 62 (2008), 58–70.
46 Dramatic discontinuity is one of the key characteristics of the Hollywood family melodrama isolated by Thomas Elsaesser in his seminal essay "Tales of Sound and Fury: Observations on the Family Melodrama," in Christine Gledhill (ed.), *Home Is Where the Heart Is* (London: BFI Publishing, 1987), 60.

Chapter 18

Mark Langer

THE DISNEY–FLEISCHER DILEMMA: PRODUCT DIFFERENTIATION AND TECHNOLOGICAL INNOVATION

AN EXAMINATION OF COMPETING three-dimensional animation technologies at the Disney and Fleischer studios during the 1930s reveals problems in previous historical accounts of their genesis and use. The first of these technologies was the Stereoptical Process, invented by Max Fleischer and John Burks of the Fleischer Studios, Inc. in 1933. The Stereoptical Process was a three-dimensional setback system arranged horizontally, with the camera in front of the cels and background. Cels containing the animated characters were photographed in front of a three-dimensional set mounted on a turntable. The turntable could be rotated in order to get the effect of a pan or tracking shot. The background set was constructed with a vanishing point at the centre of the turntable so that the further an object was from the lens, the more slowly it appeared to move. When photographed, it appeared as if the two-dimensional cartoon characters were moving within a three-dimensional environment.[1]

The Multiplane camera, developed by a Walt Disney Productions team headed by William Garity, was a vertical arrangement with the camera above the elements to be photographed. Unlike the earlier Stereoptical Process the Multiplane camera used two-dimensional elements for each plane within the background. Artwork of different planes was held in individual light boxes, separated from other artwork by some distance. This made the various foreground and background planes spatially distinct. A greater illusion of depth was achieved by moving the camera down toward the background elements.[2]

Neither technology was particularly efficient. Both were extremely expensive in their use of labour and materials. Light reflections from the surfaces of cels were a major problem with both processes, and the Multiplane camera had particular trouble with dust accumulation on the image surface. The employment of the Stereoptical Process was very time-consuming due to the remarkably long camera exposures required.[3] Why then were these technologies developed and used? Recounting a progression of three-dimensional animation processes from Carl Lederer's apparatus used in the 1910s, Lotte Reiniger and Carl Koch's device employed in the 1920s, through Ub Iwerks's development of a horizontal Multiplane camera before 1934, and Hans Fischerkoesen's appropriation of the Multiplane and Stereoptical technology in the 1940s is a task for other historians.[4] This study concentrates on the problems posed by past considerations of the Disney and Fleischer processes.

Historical representations of the development of the Stereoptical Process and of the Multiplane camera explain the genesis of these new technologies in two major ways. Either the technologies were introduced through the imagination of an inventive genius, or they were developed as a movement towards greater realism in animation. In other words, accounts of the development of the Stereoptical Process and Multiplane camera involve either a belief in the 'great man' theory of history, or a belief that animation history followed an evolutionary or teleological progression towards mimesis.

Historians representing these two groups overlap to a certain extent. Among the 'great man' proponents are such people as Leonard Maltin, Richard Schickel, Ralph Stephenson, Frank Thomas, Ollie Johnston, G. Michael Dobbs, and Leslie Carbaga.[5] Thomas and Johnston state:

> Bill Garity, an expert on camera lenses, was nominal head of the department, but Walt worked with each man on an individual basis... They were called into sweatboxes and story meetings and often just sat around listening, getting the feel of what Walt was after.... Once they were asked to build an arrangement that could hold separate layers of artwork at varying distances from a still camera ... it worked and Walt liked the result and suddenly was talking about building another one, larger and more complicated, that might be used for shooting animation.... And so the first multiplane camera was born.[6]

Similar thinking informs Leslie Carbaga's *The Fleischer Story*, in which the author dubs the Stereoptical device 'The most wondrous of Max's highly acclaimed innovations'. G. Michael Dobbs sees the Stereoptical Process as a physical simulacrum of Max Fleischer's thought: 'Max Fleischer ...', avers Dobbs, '... was very literal-minded, and his method of adding three dimensions to cartoons reflects this belief. You want three dimensions, you use a three-dimensional model as your background!'[7] Historians who promote evolutionary or teleological arguments include Leonard Maltin (again), David R. Smith and Richard Hollis and Brian Sibley. Hollis and Sibley maintain

> ... there was the perennial difficulty of conveying 'depth', something that hadn't mattered in the early comic cartoons but that was of vital importance in creating the realistic mood Walt wanted for *Snow White*. An illusion of depth was eventually achieved by ... the huge but extremely versatile 'multiplane camera'....[8]

In such considerations, the 'great man' theory and teleology coalesce. Great inventors provide the technology to aid in the inevitable march of progress. While one should not wish completely to deny concepts of individual endeavour or progress as historical factors, one must question the assumptions that underlie these concepts. The 'great man' theory views individual creativity as the motor of history. Is innovation an individual act, or does it have an institutional dynamic? Do personal or institutional interests best provide a motive for the creation of these new, expensive and unwieldy technologies?

Secondly, teleological or evolutionary approaches join with the 'great man' theory in projecting history as a line of continuous development in one direction. These approaches reflect a belief in technological innovation as part of an ideology of progress. Something is invented by someone and technological, industrial or social change follows. Raymond Williams has criticized this type of technological determinism for assuming that

> new technologies are discovered by an essentially internal process of research and development, which then sets the conditions for social change and progress.

> The effects of the technologies, whether direct or indirect, foreseen or unforeseen, are, as it were the rest of history.[9]

Absent from such considerations is the possibility that research and development is a symptom, rather than an agent, of social change and progress. While Williams discusses the external dynamic of a system that incorporates technological innovation, this article examines the particular cultural subsets of the American animation industry, with focus on the actions of Walt Disney Productions and the Fleischer Studios, Inc. Broader parallels between this internal dynamic and the film industry at large or the social/economic system as it existed in the 1920 to 1942 period may be implied. However, the external dynamic will not be examined specifically in this study.

In historical representations of the Multiplane camera and Stereoptical Process, the consequence of invention is portrayed as primarily aesthetic rather than social or economic. Oddly enough (in light of the fact that animation does not use a three-dimensional, live performance as its starting point), this aesthetic/historical system is strikingly comparable to that voiced by André Bazin, who stated that 'Cinema attains its fullness in the art of the real'. To Bazin, the basic need completely to represent reality was the driving force behind technological advancement. Thus, technological innovations such as sound, colour and wide screen created a closer relationship between cinema and its surrounding world.[10] Similarly, animation historians view the development of style and technology within American studio animation to 1942 as an unbroken march towards mimesis. Richard Schickel, for example, sees the Multiplane camera as a logical aesthetic extension of Disney's adoption of sound and colour: 'The multiplane camera thus becomes a symbolic act of completion for Disney. With it, he broke the last major barrier between his art and realism of the photographic kind'.[11]

It is the end point of this march that poses a challenge to the assumptions behind these historical methodologies. Both the Stereoptical and Multiplane technologies led to dead ends. Use of the Stereoptical Process was discontinued by 1941, and the Multiplane camera was employed with decreasing frequency through the 1940s. No significant amount of further technical refinement or development was made of either apparatus. Historical methodologies previously applied to three-dimensional animation technology all assume that history unfolds in a rational, continuous manner. The 'great man' theory accounts for the invention of these technologies as a logical extension of a visionary personality. The evolutionary or teleological approaches presume some sort of continuity, whether through the permanent institutionalization of these technologies, or through their organic relationship to some further development. The aesthetic assumptions see these technologies as a logical development towards mimesis. Mimesis was later abandoned in favour of increasingly 'flat' and stylized graphics used from 1942 throughout the US animation industry. As will later be demonstrated, this stylistic change, which is usually credited to such studios as Warner Bros. or UPA, can be observed within the films of those very studios that were allegedly heading toward mimesis. These films include *Bone Trouble* (1940), *Fantasia* (1940), *The Reluctant Dragon* (1941) or *Dumbo* (1941) at Disney; and *Goonland* (1938) and *Mr Bug Goes to Town* (1941) at Fleischer.

Through an examination of both the institutional history and the style of films produced by the Disney and Fleischer studios during the 1930s and early 1940s, I wish to propose the need for alternative historical models of innovation and competition to those used by previous animation historians: models not so dependent on the presumption of a rational order governing the behaviour of individuals, institutions, technologies or aesthetic movements. Such models take into account not only the strategies of institutional interaction between Fleischer Studios, Inc. and Walt Disney Productions, but also the coexistence of different styles within the discourse of each studio.

Evidence suggests that the Disney and Fleischer studios were keenly aware of each other's actions. While by the early 1930s Walt Disney Productions was considered to be the leader in terms of artistic innovation, Fleischer Studios, Inc., through cost control and the popularity of their 'Popeye' character, created animated films that were far more profitable to the Fleischers and their distributor Paramount than were Disney's when distributed by United Artists or RKO. Fleischer Studios, Inc. was the leading animation studio on the east coast of the United States, dominating in prestige and profitability its local competitors Van Beuren and Terry. Max Fleischer resented Disney's having lured away many of his top employees with high salaries, and was conscious of Disney having utilized many Fleischer-developed processes, such as the Rotoscope. Walt Disney Productions took little notice of such west coast competitors as Celebrity Productions, Harman & Ising, or Leon Schlesinger. For example, while requesting screenings for the training of studio staff in 1935, Disney stated: 'I think it is all right to show Fleischer's stuff, but I would keep away from the local product'.[12] While many early animation companies were preoccupied with the development of new technology, by 1934 Walt Disney Productions and Fleischer Studios, Inc. were the only animation studios conducting research and development on a sustained basis. Both companies were highly competitive and noncooperative.[13]

Much scholarship, such as Donald Grafton's and Kristin Thompson's work on the adoption of the cel method, Harvey Deneroff's writings on labour organization, or my own work on studio hierarchies, has emphasized the movement towards commercial standardization in the American animation industry.[14] The need to provide a standard product resulted in other forms of regulation. Most animated films were one reel in length. Standardization encouraged the production of animation series constructed around a central 'star' character, such as Felix the Cat, Ko-Ko the Clown or Mickey Mouse. Even as the dictates of commerce standardized animation production, commerce also motivated a countervailing tendency towards product differentiation. For example, the Warner Bros Negro boy character Bosko has been seen as imitative of Disney's Mickey Mouse in terms of his physical proportions, simple black on white 'inkblot' design, squeaky voice and musical routines.[15] But Bosko was also clearly differentiated from Mickey Mouse by the use of long trousers and the lack of tail and mouse ears. Imitation of the successful product of one company was counterbalanced by the need to distinguish the product of one company from that of another firm.[16] Product differentiation was common in the American animation industry during the 1920s and 1930s but was accomplished primarily through graphic style or character design. Product differentiation by means of technology was far more expensive. During the period, innovation of animation technology was motivated chiefly by competition between the two major animation companies – Fleischer Studios, Inc. and Walt Disney Productions.

Max and Dave Fleischer's first animated cartoon was manufactured in 1915. Although their films used the then common convention of depicting both the artist and the animated character in the same world, they were differentiated from contemporary animated films through the use of the Rotoscope. The Rotoscope allowed an animator to copy live action movement by means of the rear projection of live action film frame by frame onto a piece of translucent glass set into a drawing board. The improved smoothness of movement obtained by this process was a central part of the marketing publicity surrounding the Fleischer brothers' 'Out of the Inkwell' cartoons. As the *New York Times* noted in 1920, the Fleischer protagonist, Ko-Ko the Clown, had

> a number of distinguishing characteristics. His motions, for one thing, are smooth and graceful. . . . He does not jerk himself from one position to another, nor does he move an arm or leg while the remainder of his body remains as unnaturally still as if it were fixed in ink lines on paper.

In 1925, Max Fleischer invented the Rotograph as a further advance in technological product differentiation. The Rotograph was a rear-projection system which allowed superior image quality and ease of construction of scenes combining live action and animated characters.[17]

Disney's entry into national distribution of product also depended on product differentiation. Disney's first series protagonist in the 'Alice Comedies' was a live-action girl played by Margaret Davis. In the 'Alice Comedies' this character was matted into a cartoon world. Such practice differed from that of competing animation series which featured animated protagonists. Although the Fleischer 'Out of the Inkwell' cartoons mixed live action and animation, their protagonists were drawn figures (albeit rotoscoped in the case of Ko-Ko the Clown) in a live-action world. Disney's distributor, Margaret J. Winkler, had been Max Fleischer's distributor. When Fleischer and Winkler parted company upon Max Fleischer's venture into distributing his own product, Winkler arranged to distribute the Disney product. She announced the 'Alice Comedies' in trade papers as 'Kid comedies with cartoons co-ordinated into the action. A distinct novelty'. Contemporary trade reviews made much of the distinctiveness and novelty of a live-action girl in a cartoon world. By emphasizing his live character Alice, rather than an animated rotoscoped character like Ko-Ko, Disney consciously used some conventions of classical Hollywood cinema as a means to differentiate his product from the products of others.[18]

A chronology of technological innovations at the Disney and Fleischer studios demonstrates a competitive pattern of innovation and product differentiation. From 1924 to 1926, Max and Dave Fleischer released their animated sound 'Song Car-Tunes' using the DeForest Phonofilm process. In 1928, Disney utilized the Powers Cinephone process to synchronize *Steamboat Willie*. Fleischer Studios, Inc. returned to sound production with *Noah's Lark* in 1929. Disney then signed a contract with Technicolor, giving him exclusive rights for animation to the three-colour Technicolor process. This was used in Disney's *Flowers and Trees* (1932). The Fleischer Studios, unable to gain access to three-colour technology, made do with the bichromatic Cinecolor and two-colour Technicolor processes. In order positively to differentiate his colour films from those of Disney, Max Fleischer introduced the three-dimensional Stereoptical Process in his first colour film, *Poor Cinderella*, in 1934. After three-colour Technicolor became available to Fleischer Studios, Inc., the company moved to longer animated films with *Popeye the Sailor Meets Sindbad the Sailor* (1936). Walt Disney Productions countered with the development of the Multiplane camera, beginning in 1935 and culminating with its use in *The Old Mill* (1937). Several months later, the Disney studio released the feature *Snow White and the Seven Dwarfs* (1938). By coopting the dominant feature-length format of classical Hollywood cinema, Disney differentiated his product, as he had similarly done with the earlier 'Alice Comedies'. His competitors could follow, or be left with what would be perceived as a less innovative, inferior, product. This is when the house of cards began to collapse. Following on the success of *Snow White and the Seven Dwarfs*, Max and Dave Fleischer drove themselves into insolvency by combining all of their technologies in the feature-length *Gulliver's Travels* (1939) and *Mr Bug Goes to Town* (1941). Disney narrowly escaped doing the same by applying his expensive technologies to the money-losing *Pinocchio* (1940), *Fantasia* and *Bambi* (1942).[19]

With the exception of the anomalous success of *Snow White and the Seven Dwarfs*, technological differentiation of product did not provide benefit to either company. Walt Disney Productions remained viable in the 1930s and early 1940s largely because of income obtained from a stock offer and from ancillary business interests, such as product licensing, books, music and revenue from comic strip and art sales, rather than because of income resulting directly or indirectly from the development of the Multiplane camera.[20]

Much less of the Fleischer Studios' income came from ancillary interests. Although some attempt had been made to market products based on Fleischer animated characters, the merchandising rights for the company's popular series of 'Popeye the Sailor' cartoons were held by King Features Syndicate. Cost controls imposed upon Fleischer Studios, Inc. that limited the company's use of new technology helped maintain that company's profitability. For example, Paramount contracts with the Fleischers specified that 'Popeye the Sailor' cartoons be black and white as late as 1942. This was seven years after Disney had completely converted to Technicolor production. Paramount's constraints kept production costs of 'Popeye' down to $16,500 per film, and effectively limited the use of the Stereoptical Process in these films. Fleischer one-reel colour films were to cost no more than $30,000, with only their two-reel specials and the first 'Superman' cartoon exceeding this figure. Budgets were considerably less than the $40,000 to $50,000 typically spent on a one-reel film by Walt Disney Productions. When these financial controls were relaxed for the Fleischer feature-length response to *Snow White and the Seven Dwarfs*, the Fleischers fell into a trap. *Gulliver's Travels* went over budget, and Fleischer Studios, Inc. entered a crisis from which it never emerged.[21]

Other factors contributed to the fiscal woes of the Fleischer Studios, Inc. and Walt Disney Productions. After suffering a financial blow from a lengthy 1937 strike, the Fleischers were confronted with the expense of moving their company from New York to a new facility in the non-union labour environment of Miami, Florida during 1938. Walt Disney spent much of the profit of *Snow White* on his new studio in Burbank. Shortly after this, Disney also had to cope with a bitter strike in 1941. Profits of both companies were adversely affected by the loss of continental European and Asian markets during World War II, as well as by currency restrictions in the United Kingdom. Some attempt was made to adjust to these conditions by an expansion of activity in Central and South America. Despite these factors, technological innovation was a decisive determinant in the sinking fortunes of the Disney and Fleischer studios. The continuation of technological innovation by these two animation companies contrasted with the practice of the dominant feature-length, live-action film industry. In the era of mature oligopoly, the feature film industry sharply reduced its degree of reliance on technological innovation for purposes of product differentiation following a period of expensive innovation in competing sound, colour, and widescreen technologies from 1926 to 1935. With the standardization of sound and colour technologies, and with the abandonment of widescreen the feature film industry entered into a period of financial stability marked by relatively little technological innovation until after World War II. In comparison with the experience of the Fleischer and Disney companies, the stability of this dominant sector was relatively unaffected by the loss of markets during wartime.

Nevertheless, the importance of technological differentiation to the financial wellbeing of the institution was an article of faith held by Walt Disney and Max Fleischer. Both men clearly believed in an ideology of progress, wherein technological development played a key role. In a 1941 article in *American Cinematographer*, Disney stated: '. . . the public will pay for quality. . . . Our business has grown with and by technical achievements. Should this technical progress ever come to a full stop, prepare the funeral oration for our medium'.[22] Shortly after the collapse of Fleischer Studios, Inc., Max Fleischer recalled his career as a series of successful technical innovations, culminating with his introduction of 'the very first attempt to incorporate a third dimensional effect in cartoons . . . by the "Setback" method of photography'.[23]

Was the Multiplane camera a competitive response to the Fleischer system? Preliminary research strongly suggests that early Multiplane camera development did imitate the Stereoptical Process. Although few documents on Multiplane development exist, Disney employee Ken Anderson recalled that in 1935 he created a three-dimensional model of a door for an early horizontal version of the Disney 3-D process. When photographed with foreground cel images, the result showed Snow White interacting with the three-dimensional set. A search at the Disney Archives has failed to document why this earlier system was abandoned, although the need to avoid patent infringement seems a likely reason. It should be remembered that both Ub Iwerks's three-dimensional apparatus and a later Walt Disney Productions model of the Multiplane camera developed for the last shot of *Fantasia* used the same horizontal format as the earlier Fleischer process. The Disney Horizontal Multiplane camera also had the potential for the use of three-dimensional materials or mockups.[24]

At the time of their introduction, these innovations were perceived as technologically differentiating one studio's product from those of others. 'News stories' planted by Paramount pointed out that Max Fleischer's 'camera wizardry' brought about an advance over the earlier two-dimensional animation system. *Popeye the Sailor Meets Sindbad the Sailor* was publicized as the 'first two-reel, full-color, three-dimensional film'. *The Old Mill* received an Academy Award for best animated short subject, and the Disney studio was awarded another Oscar for 'technical achievement' in the development of the Multiplane camera. As had been the case with the Stereoptical Process, the Multiplane camera was extensively publicized as an important technological innovation.[25]

The publicity and awards given to these technologies might seem to support the evolutionary or teleological argument that their development was an advance on the road to mimesis. Yet, as mentioned above, the Multiplane camera and the Stereoptical Process were used infrequently after a few years, which suggests that they were more of a dead end than the road of progress. The economic consequences of using such expensive technologies simply did not justify their continued use.

Did economics stand in the way of aesthetic evolution? Historical contentions that the Stereoptical Process was developed for reasons of mimesis are arguable.[26] Max Fleischer maintained that he was opposed to mimesis in animation during the time in which he developed the Stereoptical Process.

> During the span from 1914 to 1936, I made efforts to retain the 'cartoony' effect. . . . Let us assume we desire to create the last word in a true to life portrait. We examine the subject very carefully and religiously follow every shape, form and expression.
>
> We faithfully reproduce every light, shade and highlight. Upon completion of this grand effort, we compare our result with a photograph. . . What have we now? Nothing at all. We have simply gone the long way around to create something which the camera can produce in seconds. In my opinion, the industry must pull back. Pull away from tendencies toward realism. It must stay in its own back yard of 'The Cartoonist's Cartoon'. The cartoon must be a portrayal of the expression of the true cartoonist, in simple, unhampered cartoon style. The true cartoon is a great art in its own right. It does not require the assistance or support of 'Artiness'. In fact, it is actually hampered by it.[27]

Examination of the use of the Stereoptical Process in Fleischer films does tend to corroborate Max Fleischer's remarks. In no film is the Stereoptical Process consistently used as a background element. While there may be economic reasons for this, the effect on the screen is a rupture of the films' visual continuity. Through an alternation of three-dimensional and two-dimensional backgrounds, attention is drawn to the Stereoptical Process as a technological gimmick. Rather than reinforcing the realist codes of classical Hollywood cinema, the Fleischer use of the three-dimensional setback system appears to have been employed chiefly as a form of spectacle that contrasted with the traditional appearance of cel animation used in the rest of the film.

The employment of the Stereoptical Process in animated cartoons was consistent with the use of other Fleischer processes with a mimetic potential, such as the Rotoscope. In both

Gulliver's Travels and *Mr Bug Goes to Town*, the human world and the worlds of Lilliput or the insects are defined by their degree of determination by photographic images. Both Gulliver and the 'Human Ones' are rotoscoped. Most of the Lilliputians and the insects are not. The juxtaposition of the two styles emphasizes the artificiality or 'cartooniness' of the film. A similar juxtaposition informs the visual discourse involving the Stereoptical Process.

The Multiplane camera was first developed for use in Disney's 'Silly Symphonies' series, of which William Garity – head of the studio team that developed Multiplane – once stated: 'It is the present intent to maintain this series in the realm of the unreal'.[28] The Multiplane camera maintained a uniform use of flat surfaces with some space separating each level of artwork. Lacking 'real' three-dimensional surfaces (such as angles or curves) that would contrast with the flat plane of conventional animation cels, the Multiplane camera's two-dimensional surfaces may not seem to offer as great a potential for visual discontinuity with conventional cel animation as did the three-dimensional backgrounds employed by the Stereoptical Process. Nevertheless, the most often cited examples of Multiplane use – the long tracking shots in *The Old Mill*, or the camera descent to Pinocchio's doorway as he prepares to leave for school – emphasize the camera's potential for spatial realism.

These three-dimensional effects were not the only reason for the development of the Multiplane camera. According to William Garity, a major reason for the camera's development was that 'almost any scene can be broken down in such a way that lighting, colour and optical control is achieved over any part or all of the scene. This control would not be at all practical if the technique was confined to a single plane'.[29] Special effects were a primary consideration in the design of the apparatus. The Multiplane camera was used for sequences that were often anything but mimetic, or stylistically continuous with the rest of the picture. Disney layout artist Kendall O'Connor recalls that sections of the cartoony 'Dance of the Hours' sequence in *Fantasia* and the surreal 'Pink Elephants' fantasy sequence in *Dumbo* were shot with the Multiplane camera. This was done to take advantage of the superior control that the Multiplane camera gave in achieving a higher degree of artificial stylization. For example, the device's detailed control of light on each plane permitted the use of a better flat black background for the stylized shenanigans of the 'Pink Elephants' sequence.[30] The episodic quality of *Fantasia* emphasized the ruptures and contrasts among disparate styles – such as the abstraction of 'Toccata and Fugue in D Minor', the 'cartoony' burlesque of 'Dance of the Hours', and the realistic drama of natural evolution in 'The Rite of Spring'. As Disney films became more discontinuous in terms of narrative and graphic style during the 1940s, those Multiplane camera sequences that emphasized three-dimensionality increasingly tended to create discontinuity when juxtaposed with more 'cartoony' elements, or with live action, as seen in *Saludos Amigos* (1943), *The Three Caballeros* (1945), or *Song of the South* (1946).[31]

Conventional wisdom has it that American animation was rescued from the aesthetic dead end of realism by Warner Bros and UPA animators. Steve Schneider claims that the stylized backgrounds and movement in Chuck Jones's *The Dover Boys* (1942) heralded a new beginning in American animation as 'the first cartoon since the rise of Disney in which the demands of realism were almost entirely banished. . . . Later in the 1940s, some of the founders of the UPA studio cited the film as an inspiration for their innovations'.[32] Ralph Stephenson identified this new trend in post World War II animation as 'moving away from realism. . . . UP A started this'.[33]

Such considerations look for continuity within the output of a studio. For example, while Schneider hails the change in Jones's style in 1942, elsewhere in his book he observes somewhat contradictorily that Jones's 'Sniffles' cartoons 'stayed heavily under the Disney influence . . . [with] . . . slower and atmospheric pacing . . . realistic Backgrounds, and a striving for "cuteness" throughout'. Jones continued to make 'Sniffles' cartoons in this style through 1946.[34]

Similar problems arise in the categorization of Disney and late Fleischer films as 'realistic'. Histories portray the Disney and Fleischer studios as late-comers to the tendency towards abstraction and stylization pioneered elsewhere. Animation historians confirm the existence of unified styles within the products of these companies, uninfluenced by the production of other animation studios until the artistic mantle passed to Warner Bros. and UPA in the 1942 to 1949 period. Earlier Disney and Fleischer abstraction, or stylizations, such as the apparent breaking and splicing of the film image in Fleischer's *Goonland*, the distorted images in the hall of mirrors in Disney's *Bone Trouble*, the story presented as still sketches in the 'Baby Weems' sequence in *The Reluctant Dragon*, the self-reflexive antics and lyrics in the 'Pink Elephants' sequence of *Dumbo*, or the simple outlines, flat background and electronic, percussive score of the 'Jitterbug' sequence in *Mr Bug Goes to Town*, all appear to exist outside of history. The Stereoptical Process and the Multiplane camera do not obliterate these tendencies – they coexist with them, clash with them, complement them, and even support them.

No single historical methodology can account for all aspects of any item from the past. This study is not intended as a theory of history, even in the restricted context of animation technology. My observations merely suggest how previous treatments of the subject fail to account fully for the development and use of three-dimensional processes. Evolutionary or teleological theories do not account for the discontinuities in the use of the Stereoptical Process and the Multiplane camera, nor does the 'great man' theory account for the institutional structure of the American animation industry. Aesthetic changes do not necessarily occur in a coherent, linear manner. All of these methodologies assume a kind of rational, continuous unfolding of technological history. As this study demonstrates, innovation takes place in a context far more complex and fragmented than that envisaged in previous considerations of the Stereoptical Process and Multiplane camera.

Notes

1 Max Fleischer. US Patent 2,054,414, 15 September 1936; Seymour Kneitel and Izzy Sparber, *Standard Production Reference* (Miami: Fleischer Studios, Inc., 1940), p. 29.

2 C. W. Batchelder, *Multiplane Manual* (Burbank: Walt Disney Studio, c1939), pp. 1–6.

3 'Invents movie 3rd dimension', *NY Morning Telegraph*, 11 March 1934, n.p.; William Stull, 'Three hundred men and Walt Disney. That's the analysis of one reporter', *American Cinematographer*, vol. 19, no. 2 (1938), pp. 50, 58; C. W. Batchelder, 'Multiplane camera lecture for ass't directors', Walt Disney Archive, *Standard Production Reference*, 13 January 1939, pp. 29–31.

4 This study will not attempt to separate the individual contributions of Max Fleischer or John Burks to the development of the Stereoptical Process, nor that of Walt Disney, Ub Iwerks or William Garity to the Multiplane camera. A mention of a single name may depict an exclusive individual, or an entity representing the technological work of several people. Joe Adamson, 'A talk with Dick Huemer', in Danny Peary and Gerald Peary (eds), *The American Animated Cartoon* (New York: E. P. Dutton, 1980), p. 33; Lotte Reinigerr, 'The adventures of Prince Achmed or what may happen to someone trying to make a full length cartoon in 1926', *The Silent Picture*, no. 6 (1970), pp. 2–4; Peter Adamakos, 'Ub Iwerks', *Mindrot*, no. 7 (1977), p. 24; William Moritz, 'Resistance and subversion in animated films of the Nazi era: the case of Hans Fischerkoesen', *Animation Journal*, vol. 1, no. 1 (1992), pp. 5–33.

5 Leslie Carbaga, *The Fleischer-Story* (New York: Nostalgia Press, 1976), p. 71; G. Michael Dobbs, 'Koko Komments', *Animato*, no. 18 (1989), p. 35; Leonard Maltin, *Of Mice and Magic* (New York: Plume, 1980), pp. 109–10; Richard Schickel, *The Disney Version* (New York: Avon, 1968), pp. 164–65; Ralph Stephenson, *The Animated Film* (New York: A. S. Barnes, 1973), p. 37.

6 Frank Thomas and Ollie Johnston, *Disney Animation: The Illusion of Life* (New York: Abbeville, 1981), pp. 262–64.

7 Carbaga, *The Fleischer Story*, p. 71; Dobbs, 'Koko Komments', p. 35.

8 Leonard Maltin, *The Disney-Films* (New York: Crown, 1973), pp. 12–13; David R. Smith, 'Beginnings of the Disney Multiplane camera', in Charles Solomon (ed.), *The Art of the Animated Image: An Anthology* (Los Angeles: The American Film Institute, 1987), p. 41; Richard Holliss and Brian Sibley, *The Disney Studio Story* (New York: Crown, 1981), p. 30.

9 Raymond Williams, *Television: Technology and Cultural Form* (New York: Schocken, 1975), p. 13.

10 André Bazin, *What Is Cinema? Vol. 1* (Berkeley: University of California Press, 1971), p. 15; Robert C. Alien and Douglas Gomery, *Film History: Theory and Practice* (New York: Knopf, 1985), pp. 70–71.

11 Schickel, *The Disney Version*, p. 169. A categorical but ahistorical description of the Multiplane camera and Stereoptical Process can be found in Russell George, 'Some spatial characteristics of the Hollywood cartoon', *Screen*, vol. 31, no. 3 (1990), pp. 290–321.

12 Walt Disney, Memo, 18 September 1935, Walt Disney Archives.

13 While Ub Iwerks did some development of a horizontal three-dimensional process at Celebrity Productions, this effort did not proceed independently of the one at Walt Disney Productions beyond 1934. Max Fleischer, US Patent no. 1.242.674, 9 October 1917; Mickey Mouse as actor a dud at making money', *NY Herald-Tribune*, 12 March 1934; 'Disney's "Pips" eat up profits', *New York Telegraph*, 18 November 1933; A. M. Botsford to Russell Holman, Memo, 1 March 1938, *Gulliver's Travels –* Production File (private collection); Austin Keough Max Fleischer and Dave Fleischer, Agreement, 24 May 1941, pp. 11, 12; Richard Fleischer, interviews with the author, 8 November 1990 and 18 May 1991; Grim Natwick, interview with the author, 28 January 1990.

14 Donald Grafton. The Henry Ford of animation: John Randolph Bray', and 'The animation "shops"', in *Before Mickey: The Animated Film 1898–1928* (London: MIT Press, 1982), pp. 137–216; Kristin Thompson, 'Implications of the cel animation technique', in Teresa De Lauretis and Stephen Heath (eds), *The Cinematic Apparatus* (New York: St Martin's Press, 1980). pp. 106–20; Harvey Raphael Deneroff, 'Popeye the union man: a historical study of the Fleischer strike', unpublished dissertation. University of Southern California, 1985; Mark Langer, 'Institutional power and the Fleischer Studios: the *Standard Production Reference*', *Cinema Journal*, vol. 30, no. 2 (1991), pp. 3–21.

15 Greg Ford, 'Warner Brothers [*sic*]', *Film Comment*, vol. 11, no. 1 (1975), p. 16.

16 Live-action counterparts to this kind of combined imitation/distinction include the Billy West and Harold Lloyd 'Lonesome Luke' spinoffs of the Chaplin tramp. Legal considerations played a part in product differentiation, as evinced by the lawsuit brought against Van Beuren by Walt Disney for copying the Mickey Mouse character. 'Van Buran [*sic*] scoffs at Disney suit', *Motion Picture Daily*, 15 April 1931.

17 'Fleischer advances technical art', *Moving Picture World*, 7 June 1919; 'The inkwell man', *New York Times*, 22 February 1920, p. 9; 'Offers new series of "Out of Inkwell"', unidentified clipping, *c*. 1921, Margaret J. Winkler Papers, Film Study Centre, The Museum of Modern Art, NY; Milton Wright, 'Inventors who have achieved commercial success', *Scientific American*, vol. 136, no. 4 (1927), p. 249; Max Fleischer, US Patent no. 1,819,883, August 1931.

18 Certificate of Incorporation of the Red Seal Pictures Corporation, 7 September 1923, New York City County Clerk's Office; M.J. Winkler advertisement, *The Film Daily*, 11 May 1924; 'Alice's Wild West Show', *The Film Daily*, 16 March 1924; 'Alice's Wild West Show' and 'Alice's Day at Sea', *Motion Picture World*, 10 May 1924; Walt Disney, 'Letter to Margaret Winkler', in David R. Smith, 'Up to date in Kansas City', *Funnyworld*, no. 19 (1978), p. 33; George Winkler, interviews with the author, 25 May 1891 and 26 May 1991; Ron Magliozzi, 'Notes for a history of Winkler pictures', unpublished ms., 1991.

19 The $2,595,379 cost of *Pinocchio* was estimated to be in excess of one million dollars over projected income in the year of its release, while *Dumbo*, which made less use of the expensive Multiplane camera, cost only $600,000 and generated profit. Much of the studio's income was dependent on short subjects, which did not make extensive use of Multiplane, and income from ancillary sources, such as comic strips, licensing fees, etc. Walt Disney Productions lost $1,259,798 in 1940, $789,297 in 1941 and $191,069 in 1942. *Annual Report Fiscal Year Ended September 28, 1940* (Burbank: Walt Disney Productions, 1940, p. 3); *Annual Report Fiscal Year Ended September 27, 1941* (Burbank: Walt Disney Productions, 1941), p. 2; *Annual Report Fiscal Year Ended October 3, 1942* (Burbank: Walt Disney Productions, 1942), pp. 2, 10: 'Disney loss cut', *Variety*, vol. 145, no. 7 (1942), p. 20; 'Walt Disney: great teacher', *Fortune*, vol. 26, no. 12 (1942), p. 154.

20 The big bad wolf, *Fortune*, vol. 10, no. 5 (1934), p. 94; *Annual Report Fiscal Year Ended September 28, 1940*, p. 5.

21 The Stereoptical Process was most generally used in the two-reel 'Specials' and in films of the more expensive 'Color Classics' series, such as *Little Dutch Mill* (1934) or *Hawaiian Birds* (1936). Isolated examples of the Stereoptical Process can be found in lower-budget black and white Fleischer cartoons. Examples include the 'Betty Boop' film *Housecleaning Blues* (1937) or the 'Popeye' cartoons *King of the Mardi Gras* (1935) and *Little Swee' Pea* (1936). More generally, depth effects were provided by exaggerated perspective drawing, as seen in the Talkartoon *Sky Scraping* (1930) or the 'Popeye' *A Dream Walking* (1934). In an attempt to control the cost of production for the Fleischer features, three-dimensional effects were limited to an opening sequence of a ship in *Gulliver's Travels* and a

descending camera track through the model of a city in *Mr Bug Goes to Town*. Most depth effects in the features were achieved through a return to exaggerated perspective drawing. After the returns from *Gulliver's Travels*, the Fleischer Studios, Inc. owed Paramount $100,000 in payment of loans. Following the disastrous release of *Mr Bug Goes to Town*, Paramount was owed $473,000. King Features Syndicate, Inc. and Fleischer Studios, Inc., *Agreement*, 17 February 1937, pp. 15, 19; Botsford to Holman, Keough, Fleischer and Fleischer, *Agreement*, pp. 2, 3, 11, 12; 'The big bad wolf', pp. 91, 94; 'Instead of getting $80,000, Disney says he'll lose 56G on US tax short', *Variety*, vol. 145, no. 10 (1942), p. 19; Richard Murray, *Deposition*, Dave Fleischer v AAP Inc., et al, US District Court Southern District of New York, 6 December 1957.

22 Waft Disney, 'Growing pains', *American Cinematographer*, vol. 22, no. 3 (1941), p. 106.

23 Max Fleischer to Jimmy (Shamus) Culhane, *c*. December 1945, Collection of Shamus Culhane, p. 2.

24 Use of this horizontal version of the Multiplane camera was limited to the last shot of the 'Night on a Bald Mountain/Ave Maria' sequence of *Fantasia*. Ken Anderson, interview by David Smith, 5 September 1975, Walt Disney Archives; Thomas and Johnston, *Disney Animation*, p. 264.

25 'Popeye slams foes into third dimension through Fleischer's camera-wizardry', and 'Popeye knocks Bluto into third dimension', in *Popeye the Sailor Meets Sindbad the Sailor Pressbook* (New York: Paramount 1936), pp. 13, 15; Schickel, *The Disney Version*, pp. 164–65; 'Three-ply Mickey coming: Disney announces new process for tri-dimensional films', *The New York Journal-American*, 28 May 1937; Frank S. Nugent, 'This Disney Whirl', *New York Times*, 29 January 1939, p. 5.

26 See Russell George, 'Some spatial characteristics of the Hollywood cartoon' for considerations of the differences between live action and cartoon realism.

27 Fleischer to Culhane, *c*. December 1945; Richard Fleischer, interview with author, 6 October 1990.

28 William Garity, 'The production of animated cartoons', *The Journal of the Society of Motion Picture Engineers*, vol. 20, no. 4 (1933), p. 309.

29 William Garity, *The Disney Multiplane Crane*, n.d., Walt Disney Archives, *c*. 1938, p. 3.

30 Kendall O'Connor, interview with the author, 24 May 1991.

31 For a closer analysis of the separate discourses in *Dumbo*, see Mark Langer, 'Regionalism in Disney animation: Pink Elephants and Dumbo', *Film History*, vol. 4, no. 4 (1990), pp. 305–21.

32 Steve Schneider, *That's All Folks! The Art of Warner Bros Animation* (New York: Henry Holt, 1990), p. 73.

33 Ralph Stephenson., *The Animated Film*, pp. 48–49. Similar evaluations are found in George, 'Some spatial characteristics ...', p. 306; Maltin, *The Disney Films*, p. 274.

34 Schneider, *That's All Folks!* p. 60.

PART IV

Postwar Hollywood and the end of the studio system, 1946–66

Introduction

The final part of this Reader deals with the period between the mid-1940s and the mid-1960s, a period that witnessed the demise of the studio system and classical Hollywood. A number of factors were involved, some of them with their roots in the late 1930s and the early to mid-1940s, others with their roots in the postwar era. Among the former were an increase in profit-sharing, independent production and freelance employment.[1] Among the latter was a marked decline in cinema-going in the US, which was prompted in part by the postwar baby boom, in part by the resumption of suburbanisation (already evident by 1940 but put on hold during the war), and in part by the spread of television and the growing popularity of other leisure pursuits.[2]

The decline in cinema-going and hence in ticket sales and income in the late 1940s and early 1950s was exacerbated by the restrictions placed by many foreign governments either on the number of films they could import or on the portion of their earnings that could be returned or 'remitted' to their country of origin in any one year, or both. Faced with these developments, the Hollywood studios cut back on production costs and laid off substantial numbers of in-house staff. They also began making more films (or shooting a larger proportion of their footage) on location abroad in order to spend the unremittable portion of their earnings and in order to take increasing advantage of foreign subsidies, locations and labour while complying with foreign quota laws. Meanwhile a small but bourgeoning cohort of independent distributors began importing foreign films for showing in specialised art-house cinemas.

In the meantime in 1948, in the midst of Hollywood's initial postwar crises, the US Supreme Court ruled in favour of the Department of Justice in the Paramount case. While it allowed the Big Five companies to keep their premiere 'showcase' venues, it ordered them 'to divorce themselves of their theatre circuits. It also ordered the divorced circuits to divest themselves of approximately one-half of the 3,137 theaters they owned in 1945'.[3] The last of these divestitures were not completed until 1959. While this was evidence that some of these companies wished to hold on to their theatres, it was also evidence that falling box-office revenues and the costs involved in running or remodelling theatres 'impeded their saleability'.[4] By then, amidst a continuing decline in the numbers of films they themselves produced, the

Big Five, like the new independents, the newly resurgent Little Three, and Republic and Allied Artists (a successor to Monogram Pictures),[5] tended to lease some of their equipment and to draw on the services of specialised providers.[6] And by then Universal had been acquired by Decca Records and RKO by Teleradio, thus setting precedents for a series of corporate takeovers that were to reach a peak in the 1960s, when Decca/Universal was in turn acquired by the Music Corporation of America (in 1962) and when Paramount was acquired by Gulf & Western (in 1966), United Artists by Transamerica (in 1967), and Warner Bros. by Kinney National Services and MGM by Kirk Kerkorian (in 1969).[7]

RKO ceased production in 1957, but it had earlier been involved, along with a newly diversified Disney and most of the other majors, in selling or leasing films to the television networks, in making television programmes, or in both. In the meantime, the remaining majors began to follow the example of United Artists under Robert Benjamin and Arthur Krim. Benjamin and Krim had acquired United Artists in 1950 and effective control in 1951. Unconstrained by the ownership of theatres, studio facilities or studio lots, they began to pursue a policy of funding, co-funding and distributing independent productions and of building long-term relationships with those who produced them.[8] A number of the remaining Hollywood companies followed suit, mixing a diminishing proportion of films produced in-house by staff on long-term contracts with a growing proportion of films produced in-house by a mix of studio employees and freelance personnel and by full-blown teams of independents.[9] Some used the companies' lots in Los Angeles, others affiliated or unaffiliated lots and locations abroad or elsewhere in the US.[10]

All the while, an increasing number of films were released either on a saturation or roadshow basis. Most of the former consisted of action films and Westerns and an increasing number of horror and science-fiction films, pop and rock 'n' roll musicals, and other films adjudged to appeal to an increasingly important teenage market.[11] Most of the latter were large-scale, high-cost adaptations of musicals or books set in the recent or distant past. These films were designed to appeal to adult or family audiences.[12] By increasing the price of admission and insisting on a higher-than-average percentage of the box-office take, those who produced them hoped not only to recoup their substantial production costs but to make substantial profits at a time when income was declining and profits were scarce. In addition to these trends many producers sought further to enhance the appeal of their films to adults by tackling controversial themes, adapting controversial novels and plays, complicating the motives of characters, depicting violence and pain in more vivid ways, dressing actors in more revealing costumes, and evoking a wider range of sexually, socially and psychologically driven forms of behaviour. In the words of one of *Variety*'s by-lines, the credo had become 'Make 'em Epic or Provocative'.[13] In pursuing this policy, films of this sort often showcased the Method and other new strains of acting and helped prompt the incorporation of modernist musical idioms (including those of contemporary jazz) into Hollywood's scores.[14] Along with the other trends noted above, they were also involved in showcasing a new crop of national and international film stars and providing opportunities for a new generation of directors, producers and writers.[15] Reinforced by the Supreme Court's 1952 declaration that 'expression by means of motion pictures is included within the free speech and free press guaranty of the First and Fourteenth Amendments' to the US Constitution, the trend toward adult, art and independent films in particular resulted in evasions of the Production Code and in subsequent modifications to its tenets. It also repeatedly risked the ire of the Catholic Legion of Decency and a number of other pressure groups.

In the meantime, while the decline in attendances either slowed or reversed in 1954, 1955 and 1956, it resumed again in 1957 and reached a then all-time low in the early 1960s.[16] By then the number of new features, shorts and newsreels had drastically declined and the production of serials had ceased.[17] While the number of industry employees was falling to another all-time low,[18] the proportion of independent productions rose from 25 per cent in early 1950 to 70 per cent in early 1959, at which point *Variety* reported that the newest 'change in the changing face of Hollywood is removal of the final barriers of exclusivity of the major studios. With the decision last week of Columbia and the week before of Paramount to open their doors to outside producers – television and theatrical – as paying tenants, the end of stubborn wastefulness has arrived'.[19]

Many of these developments are detailed in 'Individualism versus Collectivism' by Janet Staiger (Chapter 19) and in subsequent contributions to this Reader. Drawing on trade and government sources, Staiger is principally concerned to identify 'a general transition away from a regular output and mass production of films to fewer releases and higher-priced product' in the period between the late 1930s and the late 1950s. However, she is also concerned to stress that 'the mode of production for independent firms differed little from studio production'.[20] While company-owned lots were no longer the dominant sites of production, while in-house departments were no longer the dominant means of organising labour, and while labour was more often freelance, 'workers were still strictly specialised ... Only certain top management positions (producers, directors, writers, and actors) had more flexibility than in the earlier Hollywood mode of production' and financing was still 'based on conceptions of what would sell – now more carefully researched but still an effect of a decision to retain profit making as the controlling goal for the suppliers of capital'.

The contributions by Sheldon Hall (Chapter 20) and John Belton (Chapter 21) deal respectively with the new modes and policies of distribution and exhibition and the new technologies that characterised the postwar era. In 'Ozoners, Roadshows and Blitz Exhibitionism', Hall begins by noting the cultural and demographic changes that marked the late 1940s and 1950s, and the extent to which suburbanisation and an increase in the ownership of automobiles cemented the growth in drive-in cinemas and the decline in routine attendances in traditional city-centre movie houses: 'By 1953', he writes, 'there were nearly 4,000 drive-ins either already open or under construction, making up one-quarter of all exhibition sites and providing one-fifth of domestic rental revenues'. By 1960, 'some 5,000 drive-ins were in operation, compared to around 13,000 hardtops [conventional cinema buildings], contributing 23 per cent of the total box-office take.... Rather than being regarded as a fringe business, drive-ins had by this time established themselves as a mainstay of the industry and enjoyed their share of major-studio first runs'.[21]

Drive-ins participated centrally in the increasing adoption of saturation releasing: the opening of films in multiple venues on a regional or nationwide basis. This practice was not new, but along with the increasingly intensive advertising campaigns with which it was accompanied, the scale of its adoption was. Marking the increasing diversity of Hollywood's policies, products and markets, the practice of roadshowing, the very obverse of saturation releasing, increased too. Roadshowing took various forms, some of them involving the hiring of movie theatres on a 'four-wall' basis, others the use of revamped showcase venues. However, they all involved separate daily performances, higher-than-average ticket prices, reserved seats, advanced bookings, programme brochures 'and often an intermission during the movie itself', as Hall explains. These practices were initially revived 'with the limited release of "art" or prestige pictures aimed at specialised audiences', such as *Henry V* (1946) and *The River* (1951), then 'with what became known as "blockbusters", big-budget super-productions'. These super-productions included *This Is Cinerama* (1952), *Oklahoma!* (1955), *The Ten Commandments* (1956), *The Bridge on the River Kwai* (1957), *South Pacific* (1958), *Ben-Hur* (1959), *Exodus* and *Spartacus* (both 1960), *El Cid* and *King of Kings* (both 1961), *Lawrence of Arabia* and *The Longest Day* (both 1962), *Cleopatra* and *How the West Was Won* (both 1963), *The Fall of the Roman Empire* and *My Fair Lady* (both 1964), and *The Greatest Story Ever Told* and *The Sound of Music* (both 1965).[22] Nearly all these films were extremely long.[23] Longer films became more common in the late 1950s and early 1960s, particularly at the classier end of the adult trend, and the problems this caused ordinary exhibitors was exacerbated by their reluctance to abandon double bills.[24]

Roadshowing was a means of providing several hours of luxurious screen entertainment with a single film. As John Belton points out in 'Glorious Technicolor, Breathtaking CinemaScope and Stereophonic Sound' (Chapter 21), it was also a means of showcasing new technologies. As early as 1948, William Goetz at Universal-International had suggested that 'our business will either have to come up with some new development as startling as sound, or refine its processes, both artistically and commercially'.[25] One of these processes was colour, which was estimated that same year to add approximately 25 per cent to the earnings of a feature film.[26] As a result, the proportion of colour films produced by the majors rose from 4 per cent in 1939 to 21 per cent in 1951. Fed by the advent of a number of new colour processes, this proportion continued to rise, reaching 61 per cent in 1955.[27] Partly because of television, which was still largely in black and white and which by this time was becoming a major market for feature films, the proportion of colour films dropped between 1956 and 1960 before increasing again and nearing 100 per cent by 1968.[28] In the meantime, outside the industry's mainstream, a number of other technologies were developed by independent filmmakers, inventors and entrepreneurs. These included Cinerama, which was pioneered with *This Is Cinerama* on 30 September 1952, and 3-D, which had been tried briefly in novelty short form in the 1920s and 1930s, but which was successfully pioneered in feature-length form in the postwar era with *Bwana Devil* two months later. 3-D required the use of polarised spectacles and four projectors rather than the standard two for continuous screening, but was otherwise showable in ordinary cinemas.[29] Cinerama, on the other hand, necessitated the installation of a special screen and a special system of interlocking projectors in order to exhibit its images in the width, scale and depth required. It also necessitated a specially arranged set of speakers capable of reproducing its seven-channel magnetic sound track.

All three sets of processes were designed to enhance the cinema experience. While colour was familiar, its designs were often bolder, as Scott Higgins has pointed out in 'Order and Plenitude' (Chapter 17). 3-D and Cinerama, though, were much more unusual and were specifically designed to revivify the perception and experience of cinematic space (either visually or visually and aurally).[30] By 1955, 3-D had been abandoned (in part because the use of spectacles proved cumbersome, in part because of the successful introduction of CinemaScope).[31] However, with the backing of the Stanley Warner Theatre Corporation, Cinerama continued to co-produce films in the travelogue format pioneered with *This Is Cinerama* until the late 1950s, when it was used by MGM to film *The Wonderful World of the Brothers Grimm* (1960) and *How the West Was Won*. Despite its particularities, Cinerama's influence was manifest in Hollywood in the 1950s in at least three ways: in reviving the traditional large-scale roadshow; in reviving stereophonic sound (which had been initially pioneered by Disney with *Fantasia* in 1940);[32] and in reviving large-screen and wide-screen formats (which along with large-gauge formats, which were also revived in the 1950s, had been initially pioneered in the mid- to late 1920s and early 1930s).[33]

As Belton goes on to point out, the success of *This Is Cinerama* led both to a number of rival wide-screen formats and to the practice of exhibiting films in 'ersatz' wide-screen 'by cropping films shot in the standard 1.37:1 format to an aspect ratio of from 1.66:1 to 1.85:1 and enlarging the image on the screen by using a wide-angle projection lens'. Amid the growing plethora of formats and systems that characterised the decade, the most successful or influential were CinemaScope (which was introduced by Twentieth Century-Fox in 1953), Vista Vision (a variable ratio format introduced by Paramount in 1954) and Todd-AO (which was developed independently by Michael Todd and others in 1955 and which, like Cinerama, was restricted to the roadshow market). Like Cinerama and CinemaScope (at least in its earliest incarnations),

Todd-AO was marked by its use of stereophonic sound. Being filmed on 65mm film stock and printed on 70mm, it also revived the use large-gauge formats. By the end of the decade, Panavision had more or less displaced CinemaScope, and wide-screen and large-gauge systems and trade names such as CinemaScope 55, MGM Camera 55, Ultra-Panavision 70 and Vistascope had all been pioneered.[34]

Most of the films produced in most of these formats proved impossible to show on television without substantial visual modification, even though TV screenings of films like these were a major source of income for their producers and distributors in the 1960s. This was just one of a number of ironies that marked the changing relationship between Hollywood and television in the postwar era, as Janet Wasko points out in 'Hollywood and Television in the 1950s' (Chapter 22). Noting that the 'film and broadcasting industries have shared a "symbiotic relationship" since the 1920s', Wasko outlines the extent to which the MPPDA and a number of individual Hollywood companies conducted and monitored research, purchased laboratories and regional TV stations, and involved themselves in theatre and subscription television during the 1920s, 1930s and 1940s. Effectively barred from owning television networks by a freeze on the licensing of new television stations between 1948 and 1952, most of the majors were initially unable or unwilling to provide TV programmes or to license the showing of their films. Nevertheless, a number of independents supplied films and programmes, Columbia and Universal produced commercials, series and news, and once the freeze was lifted the other major companies involved themselves more fully both by making programmes and by selling or licensing the showing of their pre-1948 feature films. 'By the end of 1959, 78 per cent of network evening programming originated in Hollywood and over 88 percent of that was filmed', as Wasko points out. A year earlier, the contribution made by television to the revenues of Columbia, Disney, Loew's and Warner Bros. had been estimated at 25, 20, 11 and 30 per cent, respectively. Three years later, in 1961, *How to Marry a Millionaire* (1954), one of the earliest CinemaScope productions, was used to launch 'Saturday Night at the Movies' on NBC. To adapt the phrase used to advertise Cinerama, another new era had begun. Meanwhile, the release of *Spartacus* and *Exodus* marked the beginnings of yet another, not because they were roadshow productions, nor because they were scheduled for television screenings, but because their scripts were credited on screen to Dalton Trumbo.

Dalton Trumbo was one of the 'Hollywood Ten'. After hearings conducted by the House Committee on Un-American Activities (HUAC) in 1949, he had been imprisoned in 1950 and blacklisted as a Communist by the major Hollywood studios. By 1960 he had been writing scripts under pseudonym for nearly a decade. Placing these events within their social and political contexts, Brian Neve, in 'Hollywood and Politics in the 1940s and 1950s' (Chapter 23), traces the history of Hollywood radicalism and of investigations into the film industry by HUAC and others from the 1930s to the 1960s. In doing so, he details their development and draws attention to the individuals and organisations involved. He also discusses the contributions made by blacklistees, leftists and liberals to the cycles of socially conscious adult films that marked the postwar era, among them a number of crime melodramas that are now often labelled as 'films noirs' or 'films gris'. Toward the end of his article, Neve draws attention to the fact that although the crediting of Trumbo marked 'the beginning of the end' of the blacklist era, it 'took thirty years or more for Michael Wilson and Carl Foreman to receive a credit for their contribution to the script for *The Bridge on the River Kwai*'. In this context, it is worth underlining the fact that some blacklistees remained unable to work for a number of years, that many of those who had gone abroad remained there, and that others continued to write under pseudonym for years.[35]

Finally, 'Arties and Imports, Exports and Runaways, Adult Films and Exploitation' (Chapter 24) details the growing internationalisation of Hollywood, the US film industry and, indeed, film production, distribution and exhibition as a whole in the postwar era. As well as drawing attention to the economic issues at stake, it highlights the array of policies, trends and practices involved. Thus it deals not only with art theatres and art films, but with imports of all kinds. In addition to the trend toward shooting on location, it also deals with 'runaway' production and the trend toward co-production with and within Latin America and a number of European countries and their spheres of influence. By the end of the decade, foreign markets were providing a higher percentage of the majors' earnings than domestic ones,[36] adult, art-house and teen-oriented trends continued to place the Production Code under strain, and the number of imports aimed at mainstream audiences was on the rise. The result was a burgeoning foreign market, a shrinking domestic market, and a wider range, though a smaller number of films, in circulation in the US than had ever been in circulation before.

Notes

1 *Variety*, 7 January 1948, 43, reported that there were over 100 independent production companies in 1947, though many of them soon disappeared in the downturn.

2 For facts, figures and details on suburbanisation, demographic changes, consumer spending and television viewing in the postwar era, see John Sedgwick, 'Product Differentiation at the Movies: Hollywood 1946 to 1965', in Sedgwick and Pokorny, *An Economic History of Film*, 188–92. According to Garth Jowett, *Film: The Democratic Art* (Boston: Little, Brown, 1976), 376 and 351, respectively, there was a net increase in movie theatres and attendances in the South between 1948 and 1954 and a marked increase in attendances 'in the summer quarter, when television viewing was its weakest' between 1952 and 1957. Augmented by the growth in drive-ins, summer was by now as important a quarter as winter, which had previously been the key peak period for cinema attendances in the US.

3 Michael Conant, *Antitrust in the Motion Picture Industry: Economic and Legal Analysis* (Berkeley: University of California Press, 1960), 106.

4 Conant, *Antitrust in the Motion Picture Industry*, 111. The vertically integrated majors all signed consent decrees, RKO in 1948, Paramount in 1949, Twentieth Century-Fox and Warner Bros. in 1951 and Loew's/MGM in 1952. They were each given approximately two years to complete divorcement of their theatre circuits and for the circuits to divest themselves of a number of their theatres. But most of them took longer and Loew's/MGM, which was last of the vertically integrated majors to set up a separate exhibition subsidiary, did not complete divorcement until 1959. It should also be noted that two unaffiliated circuits, the Schine in the Northeast and the Griffith in the Southwest, were ordered to divest themselves of theatres too. For further details on the revival of the Paramount case and its effects in the postwar era, see Conant, *Antitrust in the Motion Picture Industry*, 112–53; Schatz, *Boom and Bust*, 323–28; Simon N. Whitney, 'Antitrust Policies and the Motion Picture Industry', in Kindem (ed.), *The American Movie Industry*, 172–95.

5 Allied Artists was formed as a subsidiary of Monogram Pictures in 1946. Monogram was renamed Allied Artists in 1952. Initially designed to produce low- to medium-budget exploitation films such as *The Riot in Cell Block 11* (1954) and *Invasion of the Body Snatchers* (1956), it went on to produce top line features such as *Friendly Persuasion* (1956) and to distribute expensive epics such as *El Cid* (1961) in the US. Republic Pictures Corporation, which had been formed in 1935, sought to upgrade its productions in the postwar era with little real success. It was acquired by industrialist Victor M. Carter in 1959 and effectively ceased to exist from that point on.

6 As reported in *Variety*, 17 October 1956, 7; 31 October 1956, 18; 14 November 1956, 7; 25 December 1957, 7, these 'specialised providers' included National Screen Services, which was increasingly used to repair and transport film prints (which were by now no longer flammable) in the postwar era, thus displacing some of the work undertaken hitherto by in-house employees at studio-owned or shared exchanges.

7 Universal had been acquired by Decca Records in 1952. General Teleradio, a subsidiary of the General Tire and Rubber Company, acquired RKO in 1955. When RKO ceased production, its remaining features were distributed by Universal and its studio was sold to a television

company, Desilu Productions. For more on corporate takeovers in the 1950s and 1960s and their effects on the nature of the industry, see Gomery, *The Hollywood Studio System*, 198–287; Maltby, *Hollywood Cinema*, 173–76; Paul Monaco, *The Sixties, 1960–1969* (New York: Scribners, 2001), 30–39. Janet Wasko details the acquisition of Decca/Universal in 'Hollywood and Television in the 1950s'.

8 See Balio, *United Artists: The Company that Changed Hollywood* (Madison: University of Wisconsin Press, 1987).

9 Staiger notes that MGM was among the last to engage in 'outside deals' with independent producers. So too were Twentieth Century-Fox and Universal, which resisted the trend toward blockbuster production and studio-financed independents until 1958.

10 The postwar boom in production in New York was augmented by the presence of the advertising industry and a number of television production companies. According to *Variety*, 7 December 1955, 5, 'each of the major film companies shot two or three pictures, in whole or in part, in the N.Y. Metropolitan area' in 1955. *The Tender Trap* and *Marty* (both 1955) and *The Eddy Duchin Story* and *Miracle in the Rain* (both 1956) were among those cited. They were followed, among others, by *The Wrong Man, The Garment Jungle, 12 Angry Men* and *A Face in the Crowd* (all 1957). This trend continued for the remainder of the decade and beyond.

11 Given the view advanced by some that the Hollywood majors were unaware of the importance of the teenage market or were unable to exploit it in the way that independents such as Sam Katzman and AIP (American International Pictures) did, it is worth pointing out that *The Wild One* (1954) and *Rebel without a Cause* (1955), teen-friendly 'creature features' such as *The Beast from 20,000 Fathoms* (1953) and *Them!* (1954) and, indeed, *Rock Around the Clock* (1956) and other Katzman films were all produced, released or financed by major Hollywood companies. It is also worth quoting Sedgwick, who points out in 'Product Differentiation at the Movies', 188, that: 'Whilst it may be true that demographic and other social changes lowered the average age of the audience over the period, leading to the rise in the number of films directed towards teenage audiences, it was rare for one of these niche films to occupy an annual top-ten berth'. As Sedgwick goes on to stress, nearly all the highest grossing films were costly productions aimed at adult and family audiences, and this trend increased in the period between 1957 and 1965.

12 The films involved in these as well as other trends were important sources for musical synergies and hence of much-needed income in the postwar era, especially given the advent of long-playing albums and stereophonic sound, and especially given the rapidly expanding market for singles and the new musical styles aimed at teenage consumers. As a result, in addition to Decca's acquisition of Universal, the acquisition of Dot Records by Paramount, the establishment of two new subsidiaries by United Artists in 1957, and the establishment of similar subsidiaries by Warner Bros., Columbia and Twentieth Century-Fox in 1958 (along with the subsidiary established by MGM in the mid 1940s) were to prove particularly significant, not least when it was reported in 1958 that MGM made more money from its music and broadcasting subsidiaries than from the production and distribution of its films. In addition to the Elvis Presley films and most of the musicals that featured other rock 'n' roll and pop stars, *High Noon* (1952), *Lady and the Tramp* and *Love is a Many-Splendored Thing* (both 1955), *Around the World in Eighty Days* (1956), *Tammy and the Bachelor* (1957), *Ben-Hur* and *A Summer Place* (both 1959), *The Alamo* and *Exodus* (both 1960), *Breakfast at Tiffany's* (1961) and *Lawrence of Arabia* (1962) were among the films whose soundtracks, songs or instrumental themes were popular as albums or as singles in the 1950s and early 1960s. Film versions of Broadway musicals such as *Oklaloma!*, *Guys and Dolls* (1955), *My Fair Lady* (1964) and *The Sound of Music* gave rise to successful musical synergies as well. For most of these points, and for a detailed analysis of the score for *Breakfast at Tiffany's*, see Smith, *The Sounds of Commerce*, 24–27, 32–99.

13 *Variety*, 8 August 1956, 3.

14 For more on acting and the Method, see Sharon Marie Carnicke, 'Lee Strasberg's Paradox of the Actor', in Lovell and Krämer (eds), *Screen Acting*, 75–87; Foster Hirsch, *A Method to their Madness: The History of the Actors Studio* (Norton: New York, 1984); Maltby, *Hollywood Cinema*, 393–401; James Naremore, *Acting in the Cinema* (Berkeley: University of California Press, 1988), 193–212; Lee Strasberg, *A Dream of Passion: The Development of the Method* (Boston: Little, Brown, 1987). For more on jazz- and modernist-influenced scores, see Burt,

The Art of Film Music, 184–203; Cooke, *A History of Film Music*, 183–225; David Cooper, *Bernard Herrmann's Vertigo: A Film Score Guide* (Westport, CT: Greenwood Press, 2001); Darby and Du Bois, *American Film Music*, 398–462; Annette Davidson, *Alex North's A Streetcar Named Desire: A Film Score Handbook* (Metuchen, NJ: Scarecrow Press, 2009); Prendergast, *Film Music*, 104–45; James Wierzbicki, *Louis and Bebe Barron's Forbidden Planet: A Film Score Handbook* (Metuchen, NJ: Scarecrow Press, 2005).

15 Among the writers were Edward Anhalt, John Michael Hayes and Ernest Lehman. Among the producers were Buddy Adler, John Houseman and Albert Zugsmith. Zugsmith wrote and directed films as well, as did Blake Edwards, Samuel Fuller, Stanley Kramer, Jerry Lewis and Ida Lupino. Like Lewis, whose on-screen partnership with Dean Martin came to an end in 1956, Lupino also continued to act. Working as an independent with her own production company, she replaced Dorothy Arzner, who had retired in 1943, as Hollywood's only female director. Other new directors included Robert Aldrich, Stanley Donen, Nicholas Ray, Don Siegel, John Sturges and Arthur Penn. Penn was one of a number of directors who began his career in television. Others included John Frankenheimer, Sidney Lumet and Martin Ritt. In addition to those from overseas, who will be noted elsewhere, the new stars included Marlon Brando, Yul Brynner, Jeff Chandler, Tony Curtis, Doris Day, James Dean, Kirk Douglas, Richard Egan, Ava Gardner, Charlton Heston, William Holden, Rock Hudson, Grace Kelly, Burt Lancaster, Marilyn Monroe, Kim Novak, Gregory Peck, Sidney Poitier, Elizabeth Taylor and Richard Widmark. They were joined in the late 1950s by James Garner, Steve McQueen and others who had already appeared in films but who first achieved stardom on television. New stars were seen by many in the industry as a solution to its problems, and MGM and Universal in particular continued to groom new ones under contract. However, like a number of producers, directors, writers, and like a number of other new or well-established stars, many worked on non-exclusive or short-term contracts, many worked film by film on a fee-and-profit-sharing basis, and many set up their own production companies. According to Maltby, *Hollywood Cinema*, 162, there were 742 actors under contract in 1947 and only 229 in 1956, 'most of them at MGM and Universal'. Among the stars who remained on long-term contract were Robert Taylor (at the former) and Rock Hudson (at the latter). Among those who worked on a fee-and-profit-sharing basis were William Holden, James Stewart and Frank Sinatra. Along with Kirk Douglas, Burt Lancaster, Randolph Scott, John Wayne and a number of others, Sinatra also set up his own production company. For more on postwar stars and on their industrial significance, see R. Barton Palmer (ed.), *Larger than Life: Movie Stars of the 1950s* (New Brunswick, NJ: Rutgers University Press, 2010); Sedgwick, 'Product Differentiation at the Movies', 207–12; Weinstein, 'Profit-Sharing Contracts in Hollywood', 84–95.

16 Although they base their findings on government statistics, the figures cited by scholars tend to vary. According to Finler, *The Hollywood Story*, 15, average weekly attendances fell from 82 million in 1945 to 55 million in 1950, 30 million in 1960 and 20 million in 1965. According to Jowett, *Film*, 475, average weekly attendances fell from 90 million in 1946 and 1947 to 46 million in 1953, rose to 49 million in 1954, fell to 46 million in 1955, then rose again to 47 million in 1956. After that they fell to 45 million in 1957 and 40 million in 1958, rose to 42 million in 1959, fell to 40 million in 1960, then rose again to 44 million in 1965. It is worth pointing out here that there was a recession and a credit squeeze in the US in the latter half of the 1950s. These developments affected consumer spending, cinema attendances and the availability of capital for film productions. See *Variety*, 3 October 1956, 5, 22, and 29 January 1958, 3, 18.

17 According to Finler, *The Hollywood Story*, 280, the number of features released by all the major Hollywood studios fell from 320 in 1951 to 189 in 1958 and to 142 in 1963, while the number of features they themselves produced fell from 277 to 220 to 86. The last remaining Hollywood serial, *Blazing the Overland Trail*, was released by Columbia in 1956. Columbia stopped producing animated shorts in 1949, though it distributed those produced by UPA from 1948 to 1959. Fox stopped producing B films in 1952 and Movietone News in 1963 and distributing cartoon shorts in 1968. MGM stopped distributing News of the Day in 1967 and producing of cartoon shorts in 1957 (though it revived their distribution in the 1960s). Paramount ceased production of Paramount News in 1957 and cartoon shorts in 1967. The last edition of Universal Newsreel was released in 1967, and Warner Bros., which had stopped

producing B films in 1952, stopped distributing Pathé News in 1956 and producing cartoon shorts in 1969. It is important to note here that although in numerical decline since the mid-1950s, cartoon shorts continued to be produced or distributed by a number of the major companies during the late 1950s and 1960s, largely because, although they were originally aimed at adults too, there was a growing market for them on children's television.

18 These employees included senior executives and staff involved in marketing, distribution and sales as well as those involved in production. Following a change of senior management in 1956, Loew's 'launched one of the most drastic economy sweeps in the history of the industry', according to *Variety*, 1 January 1958, 3. Fifty per cent of its staff at company headquarters in New York had already lost or were expected to lose their jobs.

19 *Variety*, 25 February 1959, 1. Most of the remaining majors sold off or reduced the size of their lots in the late 1950s and early 1960s, among them Twentieth Century-Fox, Universal and Warner Bros. (which sold off its ranch and went on to consolidate its production facilities at Burbank). As a further sign of the times, *Psycho* (1960) was distributed by Paramount but filmed on what had until recently been Universal's lot. For a detailed account of *Psycho*'s production, see Stephen Rebello, *Alfred Hitchcock and the Making of Psycho* (London: Marion Boyars, 1990).

20 For an account of a studio-based production that bears this out, see Donald L. Rathgeb, *The Making of Rebel Without a Cause* (Jefferson, NC: McFarland, 2004).

21 Despite the decline in the number of hardtops, the growth in the number of drive-ins resulted in an increase in the number of seats in the 1950s. According to *Variety*, 14 August 1957, 3, seating capacity in 1946 had been 11,961,000. By 1956, it had risen to 26,676,000. When coupled with double billing (especially in drive-ins and neighbourhood hardtops), and despite the fact that the long-term decline in cinema attendances had at this point been briefly reversed, it is hardly surprising that so many exhibitors and so many films were unable to make a profit.

22 Hall and Neale, *Epics, Spectacles and Blockbusters*, 135–86.

23 Excluding intermissions, *The Ten Commandments, South Pacific, Ben-Hur, Exodus, Spartacus, Lawrence of Arabia, Cleopatra, The Sound of Music* and *The Greatest Story Ever Told* ran initially for 220, 151, 212, 213, 184, 243, 179, 174 and 260 minutes, respectively, though *The Greatest Story Ever Told* and a number of others were cut for subsequent release following their premiere screenings or their initial roadshow runs.

24 *Variety*, 28 March 1956, 7, 22; 24 October, 1956, 5, 28; 4 February 1959, 14; 17 June 1959, 3. Indeed *Variety*, 16 January 1957, 17, went on to report that Twentieth Century-Fox, Paramount and Warner Bros. had all signed or were planning to sign deals with independent producers to produce low-budget films for double bills. Meanwhile longer films such as *The Long Gray Line* (138 mins) and *Guys and Dolls* (150 mins) (both 1955) and *Giant* (201 mins) and *The Man in the Gray Flannel Suit* (153 mins) (both 1956) had already been released. They were followed by *Peyton Place* (157 mins) (1957), *The Big Country* (166 mins) and *The Young Lions* (167 mins) (both 1958), and *Anatomy of a Murder* (160 mins) and *The Nun's Story* (149 mins) (both 1959).

25 *Variety*, 7 January 1948, 5.

26 *Variety*, 7 January 1948, 58.

27 In addition to Eastman Color, these processes included Ansco Color, Cinecolor and Trucolor. The Technicolor Corporation signed an anti-monopoly consent decree in 1950. It continued to provide a production service until 1954, at which point it served as a lab for processing colour prints. As Peter Lev points out in *The Fifties: Transforming the Screen, 1950–1959* (New York: Scribners, 2003), 108, Eastman Kodak licensed its processing technologies to other companies, 'so that film companies could use their in-house facilities for color laboratory work. Warnercolor, for example, is Eastmancolor film developed with Eastmancolor technology in Warner Bros.' own laboratory'. For more on these processes, see Salt, *Film Style and Technology*, 241–43.

28 Brad Chisolm, 'Red, Blue, and Lots of Green: The Impact of Color Television on Feature Film Production', in Tino Balio (ed.), *Hollywood in the Age of Television* (Boston: Unwin Hyman, 1990), 213–34; Finler, *The Hollywood Story*, 281; Lev, *The Fifties*, 107–09. Chisolm, 226, argues that the decline in colour film production in the late 1950s 'was not simply a direct response to the absence of color television. Hollywood studio executives ... seem to have

concluded after their experiment with more color that audiences would not stay away from a film just because it was in black and white. The popularity of black and white television and the lack of a public outcry for color video no doubt reinforced this'. The use of black and white for a number of adult-oriented feature films at this time may also have been prompted by its continuing association with foreign art films and, indeed, with adult-oriented television dramas such as *Marty* and *12 Angry Men* (both which were made into black and white feature films).

29 Douglas Gomery, *Shared Pleasures*, 239–40. Four projectors were required because two projectors at a time were required to project two prints of each reel in synchronisation and thus could not be used in alternation. Gomery notes as well that the system worked best when projected onto a metallic screen.

30 David Bordwell, *Poetics of Cinema* (London: Routledge, 2008), 281–325; Harper Cossar, *Letterboxed: The Evolution of Widescreen Cinema* (Lexington: University of Kentucky Press, 2011), 95–184; William Paul, '"Breaking the Fourth Wall": "Belascoism", Modernism, and a 3-D *Kiss Me Kate*', and Sheldon Hall, '*Dial M for Murder*', *Film History*, vol. 16 no. 3 (2004), 229–42 and 243–55, respectively; Salt, *Film Style and Technology*, 246–47. Salt, 244, also notes the introduction and use of new zoom lenses. Zoom lenses had been occasionally used by Paramount in the mid- to late 1920s and early 1930s both to enlarge the image on screen when used on projectors and to substitute for crane shots and dollies when used on cameras. As then, zooms revivified the perception and experience of space in the 1950s and early 1960s too, though in a rather different way.

31 As Lev points out in *The Fifties*, 109–12, 3-D in the 1950s is principally associated with low-budget horror and science-fiction films such as *Bwana Devil, House of Wax* (1953) and *Creature from the Black Lagoon* (1954), but it was also used for bigger-budget Westerns, musicals and thrillers such as *Taza: Son of Cochise* (1953), *Kiss Me Kate* (1953) and *Dial M for Murder* (1954) too.

32 Hall and Neale, *Epics, Spectacles and Blockbusters*, 110–11.

33 For more on these formats in the 1920s and early 1930s, see Belton, *Widescreen Cinema*, 34–68; Cossar, *Letterboxed*, 61–94; Hall and Neale, *Epics, Spectacles and Blockbusters*, 69–75.

34 For more on CinemaScope, Vista Vision, Todd-AO and other systems and trade names, see Belton, *Widescreen Cinema*, 113–79; Robert E. Carr and R.M. Hayes, *Wide Screen Movies* (Jefferson, NC: McFarland, 1988); Hall and Neale, *Epics, Spectacles and Blockbusters*, 140–55; Lev, *The Fifties*, 120–25, Salt, *Film Style and Technology*, 244–48. It should be noted that in addition to ersatz wide-screen formats, the introduction of CinemaScope was accompanied in its early years by the simultaneous production of alternative, non-anamorphic versions in ratios that were closer to previously established Academy norms or to subsequently established ersatz ones. See Sheldon Hall, 'Alternative Versions in the Early Years of CinemaScope', in John Belton, Sheldon Hall and Steve Neale (eds), *Widescreen Worldwide* (New Barnet: John Libbey Publishing, 2000), 113–31.

35 Actors such as June Havoc, Zero Mostel and Lionel Stander did not reappear in Hollywood films until later on in the 1960s; directors such as Jules Dassin, Cy Endfield and Joseph Losey stayed in Europe; and writers such as Ben Barzman, Bernard Gordon and Albert Maltz continued to write under pseudonym for years.

36 *Motion Picture Herald*, 25 April 1959, 21.

Chapter 19

Janet Staiger

INDIVIDUALISM VERSUS COLLECTIVISM: THE SHIFT TO INDEPENDENT PRODUCTION IN THE US FILM INDUSTRY

WHEN HISTORIANS SPEAK OF THE STRUCTURE of the United States film industry after World War II, they describe it as a period of the ending of mass production of films and the diffusion of independent production. Following the detailed and influential study by Michael Conant of the 1948 Paramount case,[1] scholars generally assume that the decree allowed independent film-making to flourish and eventually to dominate the industrial structure. I would propose, however, that the 1948 consent decree was only one factor, and merely a reinforcing one, in a general transition away from a regular output and mass production of films to fewer releases and higher-priced product.

From the late 1920s, studios sought to avoid stereotypical product by changing the method of top management of the mass production of films and by giving some workers greater control over special projects. This was symptomatic of an *ideology* that greater control of a film by specialists and talented individuals would produce more variety (of a certain kind) in the films. In the early 1940s, *economic* incentives enhanced the changeover to independent production. These incentives included the elimination by a 1940 consent decree of blind selling and block booking, certain effects of World War II, and an apparent tax advantage. After the war, the movement intensified because of income losses, the divorcement decree, and new distribution strategies. Furthermore, although we might term the new industrial structure 'independent', as it was eventually organised, it retained much of the earlier (studio) mode of production: work was still highly divided and in a structural hierarchy and financing was still determined by the same factors (e.g., a particular narrative form, stars, a proven director, and a good distribution contract). Although there has been a transformation in the industrial structure and mode of production, it is important to note what has been retained from the earlier systems.

The definition of US commercial independent production has not been based on its organisation of the mode of production: in fact, an independent firm could use any of several systems of production. Firms such as David O. Selznick's, Samuel Goldwyn's, and Charles Chaplin's, as well as lesser known independents, had work structures much like the major studios: a producer or director topped a hierarchy of labourers. These subordinate workers were positioned within a divided labour structure.[2] Instead, an independent production firm has been defined as a company which was not owned by, or did not own, a distribution organisation.

Commercial independent production has always been part of the industrial structure in the United States. In fact, the producers which distributed through United Artists in the 1930s had an excellent record of box office successes. Furthermore, during that period, some major studios such as Paramount, RKO, and Columbia invited independent deals. In the late 1930s, Universal added such arrangements, and in early 1940 Warner Brothers did too, setting up one with Frank Capra and Robert Riskin to make *Meet John Doe*. Even tacitly, all the studios supported some independent work. MGM distributed (on quite profitable terms) Selznick's *Gone with the Wind* (1939). United Artists' producers organised one project (*So Ends the Night*, 1941) in which they bought the story from MGM, borrowed a MGM-contract cinematographer, and rented space from Universal.[3]

In the early 1930s, the management organisation in the major studios changed. Rather than one central producer supervising 30 to 50 films per year, a group of associate producers each specialised in particular types of films and managed only six to eight films. The proponents of the change argued that the new management system would improve the quality of the product. The industry believed that certain workers concentrating on fewer and special films produced greater economic returns than the mass production of the firm's entire product.[4] This attitude – itself an ideological position about artistic creation – continued through the 1930s. In 1939, William de Mille wrote: 'Today two conflicting theories of picture-making are struggling to determine which one shall control the future system of production; they are individualism and collectivism'. Others agreed. In the House of Representatives hearings of 1940, Capra and Leo McCarey claimed that they and others had increased freedom in decision-making. George Stevens, representing the Screen Directors Guild, attributed this trend to the firms' wanting pictures with 'the most box office'.[5]

If 'individualism' was viewed as potentially a more profitable system, what held the industry in check from an outright changeover to it? The affiliated majors (those that were fully integrated, owning theatre chains) were still tied to a release-schedule to supply their own houses. For them, a regular output and mass production were necessary. Columbia and Universal, although without the burden of self-generated demand, had weaker product. These companies had found their place in the market by supplying cheaply produced 'B' films, more efficiently made in a mass production set-up. As a result, none of these industry leaders saw a profit advantage in abandoning mass production.

Early economic incentives for independent production

A major incentive to move away from mass production developed indirectly out of a 1940 consent decree in the federal government's antitrust suits against the majors. The state had been involved in the industry from film's beginning: censorship, taxes, regulation of exhibition sites, labour acts and rulings, suits to investigate and prevent pools and monopolies. By the mid-1930s, public pressure groups for federal censorship of films had allied with independent theatre exhibitors to fight blind selling and block booking. The pressure groups argued that these trade practices forced independent exhibitors to show undesirable films. In a typical instance, the Senate Committee report on bill S 3012 (1936) recommending its passage claimed that one of the bill's purposes was to 'establish community freedom in the selection of motion-picture films'. Among other proposals, these bills included provisions to require the release of a 'complete and true synopsis of each picture' prior to exhibitor leasing and to prohibit films booked in groups. The majors successfully fought off each of these bills, although votes in Congress indicated strong sentiment to move in that direction; the 1940 Senate vote went in favour of the latest bill.[6]

Two years earlier, the antitrust division of the Department of Justice filed charges against the majors. While the entire history of such suits is important to the structure of the industry, here we are interested primarily in the effect of that suit. After a two-day opening session in June 1940, the trial recessed while the participants wrote a consent decree for the affiliated majors. The agreed-upon decree eliminated blind selling by requiring trade shows of the films and prevented block booking in groups larger than five films. The decree was to remain in effect until 1943 or until June 1942 if the 'little three' (Columbia, Universal, and United Artists) did not sign it.[7]

According to the *Motion Picture Herald*, the early effect of the new trade practices was to place more value on talent and spectacle, with each company loading up every film to try to make all blocks-of-five attractive.[8] This made economic sense. If motion pictures were now to be sold in groups of five rather than a large block of 30 or more, filling more films with A or Super-A production values became more necessary for competitive reasons. Differentiation – always a primary element in the economic system – intensified. Stars, directors, and stories became even more important ingredients in the product while selling by brand name decreased in value since the entire output of a firm was no longer a marketing point. This placed top talent in even more demand, allowing them greater bargaining power for contracts that included higher salaries and more desirable working conditions – and, in some instances, more production independence.

In addition, the average cost-per-film already seemed to be the best predictor of the box office success of a firm's product: concentrating on fewer, higher-priced films rather than spreading costs across a larger number of motion pictures augured greater profits. This was particularly pertinent once exhibitors could pick and choose their favourite blocks-of-five. Moreover, with implicit collusion between the affiliated companies, these integrated firms could fill each other's theatre chains with the top product, skimming off the cream (and some of the milk) of the earnings.

Even when the 'little three' did not sign the decree in 1942, ending its validity, the firms did not uniformly return to the prior practices. Perhaps the firms anticipated further attacks from government, pressure groups and exhibitors; perhaps they also realised that limiting output had economic advantages. Paramount, RKO, and Twentieth Century-Fox stayed with blocks of five, MGM went up to ten films per block, Columbia and Universal (with the weakest product) went back to the full-season block, while Warners (like United Artists) used unit sales. Thus, the move to fewer, higher-cost films was encouraged.[9]

Another incentive to move away from mass production developed out of the effects of World War II. During the war, theatres experienced increased attendance, with many people having more disposable income and fewer things on which to spend it. Not only did admissions rise, but 'top budget' films commanded longer runs while Bs played out more quickly.[10] *American Cinematographer* noted in June 1943 that the next production season portended fewer films, higher budgets, and longer shooting times because all films were making money: '... productions given the extra time and money which make them more than just ordinarily adequate ones can command sufficient additional playing time in the theatres so that a comparatively few genuine "A's" can earn a larger total profit than the conventional program of a few "A's" and a larger number of cheaper "B's"'.[11] Monogram announced that month a 'new policy of fewer and higher-budgeted films', including an A starring Jackie Cooper with Sam Levene.[12]

In early 1945 another effect of the war forced the industry into fewer releases. War demands produced a shortage of raw film stock, and distributors were required to limit the number of exhibition prints. Although the major studios had a backlog of negatives, fewer films were released until positive film stock was more available.

That the overall number of films released by the majors declined during and after the war years is evident from US government statistics.[13] The following is a Congressional count of features released in the 1937–55 period by Columbia, MGM, Paramount, RKO, Twentieth Century-Fox, United Artists, Universal, and Warner Bros:

1937	408	1944	262	1951	320
1938	362	1945	270	1952	278
1939	388	1946	252	1953	301
1940	363	1947	249	1954	225
1941	379	1948	248	1955	215
1942	358	1949	234		
1943	289	1950	263		

Into this gap between the current and former product-supply levels stepped independents.

In 1943, as the first wave of wartime consumption patterns became apparent, signs of a resurgence of financing for independent projects developed. The *Motion Picture Herald* explained: 'The nation's box office performance of the past two years, resulting from the public's increased income and greater ability to spend money on entertainment and amusement outlets has made it possible to obtain financial support. As one producer put it, "Not since the golden days of the '20s has film money on Wall Street been so free."'[14] Cagney Productions signed a five-year agreement with United Artists to deliver up to fifteen films with at least five starring James Cagney. Hal Wallis and Joseph Hazen left Warner Brothers, formed Hal Wallis Productions, and contracted with Paramount to deliver two films a year for five years. Paramount bought a partnership in the firm.[15]

By early 1945, other workers were leaving studio-contract employment to participate in partnerships or profit-sharing. Frank Borzage completed a deal with Republic in which he would 'join that studio as a producer-director with his own unit. Borzage will have all authority over selection of stories and casts and purchases of his own vehicles, as well as a financial interest in them'. Sam Jaffe and Lloyd Bacon, who had already independently produced *The Sullivans* (1944) for Twentieth Century-Fox release, formed a firm to make *Glittering Hill* – distribution was undetermined. Walter Wanger and Fritz Lang organised New World Properties and planned their first film *Scarlet Street* (1945); Universal was to release. With two others, Lewis Milestone started Superior Pictures in order to buy *A Walk in the Sun* (1945) which he had produced and directed for Samuel Bronston's Comstock Productions.[16]

At the conclusion of the war, talents who left the major studios for independent projects included Joan Bennett, Benedict Bogeaus, Capra, Gary Cooper, Lester Cowan, Bing Crosby, Joan Fontaine, John Garfield, Paulette Goddard, Bob Hope, Hedy Lamarr, Fred MacMurray, Pat O'Brien, Ginger Rogers, Stevens, Hunt Stromberg, John Wayne, and Sam Wood. Producers fostered 'de-centralisation of studio control' and gave 'semi-independent status' to some workers – at Paramount to Cecil B De Mille, Wallis, Bill Pine and Bill Thomas; at Warner Brothers to Bette Davis and Errol Flynn; at RKO to Dudley Nichols and McCarey; at Universal to Mervyn Le Roy and Mark Hellinger.[17]

Already a number of methods of financing and distributing independent work were in operation. Some firms held long-term contracts; others found distribution upon completion of the film. Some firms were financed by banks or private investors; others were supported either partially or wholly by the major companies. As noted, with every film likely to be a success and with financing readily available, moving into independent production tempted many workers.

An additional incentive to move to independent production was a peculiar tax situation of which some firms started taking advantage. During the war, income taxes increased for higher-salaried workers, but capital gains taxes lagged in comparison: 'Under the present United States system [1947] personal incomes are taxable up to 90 per cent on every dollar. But those who forego stipulated salaries for profit-participation in individual corporation ventures can list their assets as capital gains, which are taxable only up to 25 per cent'.[18]

The next move is particularly significant: the one-film deal. One analyst pointed out why that type of independent company seemed very advantageous for a while. Investors would set up a corporation for a one-film venture, purchase stock, make the film, and then sell the stock to underwriters or distributors. Only capital gains taxes were paid and the earnings were immediately available for further speculation. In August 1946, the Bureau of Internal Revenue clamped down on this, demanding payments based on ordinary tax rates retroactive to July 1943. Samuel Goldwyn was informed that five of his 'single picture corporations' were subject to this ruling, and United Artists announced that all 35 of its films released since July 1943 had been financed this way. The Society of Independent Motion Picture Producers (SIMPP), a trade association formed by independents in January 1942, reported however that 'not many' independents would be affected. While taking advantage of the tax disparity, most independents had incorporated with the intention of taking only some profits out for personal use and retaining most earnings in the firm to continue production.[19]

Thus, by the end of the war, several economic incentives (the 1940 consent decree, the effects of the war, and tax laws) were moving the industry from mass production to fewer, higher-priced, longer-running films, and independent productions released through the majors increased. In addition, the concept of 'individualism' supported this. Although still a minor practice, independent deals spread. At the end of 1946 as the country returned to peacetime, every major firm except MGM had some independent projects as part of its regular production schedule.[20]

Later incentives to continue the trend

The unique situation of World War II, particularly increased attendance, raw film shortages, and tax laws, requires that we consider the post-war context. Once war conditions ceased, why not return to mass production? What happened after the war, however, intensified the movement toward fewer, selectively-produced films.

Significantly income losses provided a powerful incentive to continue. Descriptions and explanations of the losses are plentiful. On the domestic scene, attendance declined. In Spring 1947, box office figures indicated a resumption of more normal consumption patterns as higher cost of living developed. Although 'outstanding films' still drew customers the average film was not doing well. Universal chairman of the board Cheever Cowdin announced his firm had turned entirely to A product and warned that all costs would have to be watched. On the foreign scene, European countries moved to restrict exportation of earnings. On August 8, 1947, the US ceased exporting films to Britain in response to Britain's 75 per cent *ad valorem* tax. Five weeks later, statistics indicated that over the last year as a result of both income problems, studio employment of workers had decreased 30 per cent. The number of films shooting likewise declined.[21]

The response by the companies to these losses was typical. The firms reduced short-term costs (costumes, sets, story purchases) and fixed liabilities – particularly term contracts with labourers. Between June 1945 and August 1948 the number of writers on term contracts fell from 189 to 87; contracts for members of the Screen Actors Guild declined from 742 in February 1947 to 463 in February 1948; the number of employed skilled and unskilled workers also moved from a peak of 22,100 in 1946 to 13,400 in 1949.[22]

Unions and guilds recognised the threat to their members. The Screen Actors Guild, the International Alliance of Theatrical Stage Employees, and others constructed a 'separate category' for independent producers with a picture budget of less than 100,000 dollars. Salary scales and the required minimum number of workers on a crew were reduced.[23]

By early 1948, location shooting was also being touted as a method for solving these various problems, and this location shooting continued to reduce the formerly central role of the Hollywood studios. *Close-Up*, a Marathon Production for Eagle-Lion release, was made entirely in New York City, including all laboratory work, at '25 per cent' of what it would cost in Hollywood. As its producers admitted, however, this was possible only because the film was part of the new tendency towards 'realism'. At the end of 1949, New York City alone could boast of 35 films made partially or totally in the town over the last two years.[24]

Location shooting was also promoted as a means to unfreeze revenues in European countries. The deals arranged continued to reinforce onetime only financial agreements. The majors' trade association, the Motion Picture Association of America, announced that 'American producers may liquidate dollars now blocked in at least six foreign countries by producing in or sending pictures on location to Italy, France, Holland, Norway, Sweden and Australia'.[25] When Britain and the United States came to terms over the *ad valorem* tax in March 1948, the agreement restricted the number of American craftsmen who could work on US productions in Britain. France's 1948 agreement also stipulated that while American firms could invest up to 50 per cent of the cost of a French-produced film, the technical staff had to be French. Despite that set-back, and with the lure of potentially cheaper labour as well as an answer to their 'quest for realism' and spectacle, firms started shooting parts of their schedules abroad. In mid-1949, MGM announced seven of its next season's films would have foreign sites in eight countries. Other firms planned locations ranging from Britain, Germany, Italy, and France to the more exotic locales of Tahiti, India, Africa, and Borneo.[26]

If a post-war return to normal consumption patterns and foreign income losses affected the industry by 1948, by 1950 other domestic factors also led to continued losses: population shifts, changing recreational habits, regional unemployment, and particularly competition with television. After 1950, television's effects became more apparent as the number of sets and broadcast stations rose. A 1950 federal government survey indicated that families with television reduced their movie-theatre attendance by 72 per cent and their children's visits were down 46 per cent. A Congressional Committee reported in 1956 that 'during the first 4 years of television's growth [1948–52] in those areas where television reception was good, 23 per cent of the theatres closed compared to only 9 per cent of the theatres closing in those areas where television was not available'. By 1953, film companies were considering specific technological strategies to outdo television – 3-D, widescreen processes, and stereophonic sound.[27]

As income losses continued, the advantage of working in separate deals, one at a time, was clear. As a year-end summary in 1948 indicated: 'Producers, more and more, are willing to hire talent as needed and are prepared, if necessary, to pay more rather than maintain expensive stock companies with those inevitable periods of idleness, the cost of which must be spread over the remainder of their annual programs'.[28] Not only did the studios eliminate fixed salaries (one of the highest portions of the budget) but the companies' risk was reduced. Either they set up the 'package' themselves or they arranged a deal with an independent firm. Both the elimination of fixed costs and the reduction of their risk were attractive incentives to support a project-by-project system of operation.

Furthermore, another incentive, divorcement of exhibition from production and distribution, eased any pressures to return to mass production. No longer concerned with a regular supply to part of its own firm, a company could concentrate on fewer, specialised projects. Divorcement had been rumoured since 1943. In August 1944, the Department of Justice, unsatisfied with the results of the 1940 consent decree, requested a modification. Asking the district court for divorcement of the exhibition sector from the production-distribution sectors, the Department reopened the antitrust suit. As the case neared its conclusion in 1948, the trade papers reported:

> There are heard, for instance, plausible-sounding whispers that the major studios are not going to throw any really big pictures into production until home office executives find out what the Supreme Court is going to decide with reference to divorcement. Local observers summon support for the report by pointing out that all of the major studios have gone more or less extensively into the practice of setting up deals with independent producers. This is readily interpreted as indicating a decision by the majors to place themselves in a position of preparedness to give up production entirely if that becomes advisable, and convert their studios to the status of rental lots.[29]

Although the majors eventually separated the production studios from the theatre chains rather than abandoning production outright, the particular conjuncture of events again added to the overall impetus of the industrial shift.

Smaller firms such as Monogram and Universal had already moved into A production when As started doing so much better than Bs at the box office. With divorcement looming, other minor companies turned in that direction, often through independent deals. Republic released *Specter of the Rose* (1946, produced by Ben Hecht), *Moonrise* (1948, Borzage), *Macbeth* (1948, Orson Welles), *Sands of Iwo Jima* (1949, Allan Dwan), *The Red Pony* (1949, Milestone), *Secret Beyond the Door* (1949, Diana Productions), and *Rio Grande* (1950, John Ford).[30]

A final incentive to move away from mass production also occurred in these post-war years. New distribution strategies played to the highly differentiated film, one which might seek only part of the audience as its clientele. Forerunners of this were Selznick's *Duel in the Sun* (1946) and RKO's release *The Best Years of Our Lives* (1946, Goldwyn). *Duel in the Sun* used the tactic of mass openings across the United States (saturation booking) while both firms had 'advanced admissions' – roadshowing at higher admission prices. Encouraged by the results, majors announced their intention to show selected films at advanced prices. Not only were special 'packages' encouraged in this way, but small independents supplied the second- and third-run theatres, now being neglected by the major distributors.[31]

In addition to aiming films primarily at first-run audiences, the industry witnessed a growth in the art theatre, which also played to a particular clientele. The trade papers speculated in 1947 that the 'art film' surge was due to lack of access to foreign films during the war and to 'the French-culture-conscious "intelligentisia"' as well as the 'G.I.s' whom the war had brought into contact with European languages and customs'.[32] With the Audience Research Institute and other such companies now supplying marketing information and audience analyses, it was much more possible to tailor a film for a select group than to aim it at a mass audience. As it became clearer that only certain age- and income-levels consistently attended certain theatres, such specialisation increased.

The move away from mass production, to fewer films, with independents providing a substantial proportion, was gradual. A system which had been so successful for four decades was not readily discarded. Various factors helped ease the shift, however. One of these was the latest expansion of support firms to provide experts and technology as needed. People who had been long-term workers for major studios struck out on their own as the companies cut down the workforce. Linwood Dunn of the famed RKO special effects group and Charles Berry of Universal formed an optical printing firm, and another company built the first new studio for independents in eighteen years. Other firms moved later into leasing equipment; arguing that rather than using capital to purchase technology, by leasing it, a firm could have the latest equipment without the long-term investment. Certainly, the support-firm supply of state-of-the-art technology, studio space, and technical experts was a long-time characteristic of the industry; firms now recognised they no longer needed a self-contained studio, particularly since they were no longer mass-producing films. Overhead costs were replaced with expenditures for rentals and short-term hiring.[33]

Another factor easing the change was a new flexibility in the industry: not only was its means of production being pooled through the speciality and support firms, but the unions acted as coordinators for the supply of skilled labourers. This labour pooling eventually helped its workers make an easy transition into television. In early 1949, actors who had appeared in 'such ambitious showcases as the Ford Theater and the Philco Playhouse' included Paul Muni, Peter Lorre, Fay Emerson, Raymond Massey, Eddie Albert, Boris Karloff, and John Carradine. At the same time, *American Cinematographer* started preparing its photography experts to move into television as theatrical-exhibition film work decreased. Hal Roach Studios picked up *The Trouble with Father* series in 1951; in 1952, fifteen ASC [American Society of Cinematographers] members were shooting for television, including Lucien Andriot on *Rebound* and Karl Freund on *I Love Lucy*. By then, the cinematographers' union was 'hard pressed' to supply camera operators and assistants for all the television work, and throughout the 1950s as film employment slowed down, cinematographers found work readily available in television. With the shift to the unions as the stock pool of labour, putting together a package for either medium was comparable, and the separate media unions combined into unified ones. In 1950, the Society of Motion Picture Engineers added 'Television' to its title; in 1954, the writers for screen, radio, and television combined into the Writers Guild of America; in 1960 the screen directors and the radio and television directors merged into the Directors Guild of America.[34]

Financing was never simple. Although during the war money had been readily available, with income losses in the late 1940s, independents felt the pinch. In 1949, banks noticeably tightened loan requirements, restimulating other financing tactics. Independent theatre owners formed a company to finance independent projects in the A category; N. Peter Rathvon and Floyd Odlum, former RKO executives and owners, organised the Motion Picture Capital Corporation to do likewise. Companies continued to use capital gains tax advantages.[35] Finally, one of the solutions which was to become extremely common in later decades was proposed in 1949. Film financier A. Pam Blumenthal suggested that independents be financed by the major distributors. As a trade paper explained his 'formula':

> . . . a major studio can reduce its overhead drastically by releasing substantial numbers of its contract talent to individual producers to whom it first will have supplied financing under terms requiring use of the major studio's facilities and with the product going to the major for release. Talent going to the independent company will be on their own time during pre-production periods, and producers operating on this basis will be on their own responsibility to meet the challenge of the current market.[36]

All of these financing strategies were common from the 1950s on.

The initial and later incentives to shift to an industrial structure of independent firms releasing through the majors fostered the development of a system of production in which film projects were set up on a film-by-film arrangement. By 1954, only MGM was still using the formerly-dominant studio system, but in early 1956, the trade papers reported that 'outside deals are being made even by MGM'. While in early 1950 about 25 per cent of the films in production were clearly independent productions (with other films including profit-sharing for certain staff personnel), in early 1956 the figure was about 53 per cent, and in early 1959, it was 70 per cent. Of the 21 top-grossing films in the United States and Canada in 1956, eleven were independently produced. If there was any doubt about the change, the last of the major studio production heads stepped down in 1956: Dore Schary left MGM in December 1956 and Darryl F. Zanuck relinquished his position in February of that year to go into independent production, releasing through Twentieth Century-Fox.

By this time, as well, film companies had transferred their mass production techniques to television series. By January 1956, Paramount, Twentieth Century-Fox, Columbia, Warners, MGM, United Artists, Disney, Allied Artists, and Republic had interests in television. Harold Hecht and Burt Lancaster could celebrate the tenth anniversary of their independent production company in 1957.[37] Robert Wise, in looking back over his career, could date 1957 as the last time he had a multiple-film contract: 'All my films – every one since then – have been made as individual projects, even when I stayed at a particular studio for several years ... I had chosen to remain completely independent and select my own projects and work my own deals out individually on each one'.[38] By the mid-1950s, independent production had become a firm option in the Hollywood mode of production and was dominant by 1960.

In operation, the mode of production for an independent firm differed little from studio production. Middle- and lower-echelon workers were still strictly specialised (a result of unionisation and technological expertise). Only certain top management positions (producers, directors, writers, and actors) had more flexibility than in the earlier Hollywood mode of production. While industry-wide pooling of labour had occurred, work was still divided and structured in a hierarchy. Financing was based on conceptions of what would sell – now more carefully researched but still an effect of a decision to retain profit-making as the controlling goal for the suppliers of capital.[39]

It is important to understand that the shift from studio mass production to independent, specialised production has not necessarily secured a significant change in the factors which could affect the films produced, distributed and exhibited by the US film industry. In fact, the move to independent production was not 'outside' the dominant sectors of the industry. The major studios supported an ideology of 'individualism' from the late 1920s on. Furthermore, from the early 1940s, the majors judged that concentrating on a smaller number of films per year would be more profitable. In addition, independent production reduced the risks for the majors which began to concentrate their profit-making strategies in the areas of wise investments in 'packages' and in distribution of the finished films. Finally, independent production firms in the Hollywood mainstream retained the essential characteristics of a mode of production which has been considered efficient and economical since the early teens. Thus, overall independent production has reproduced the dominant practices of Hollywood. In the conflict between individualism and collectivism, individualism may seem to have won out, but it is an individualism which retains a great number of the characteristics of its predecessor.

Notes

1 Michael Conant, *Antitrust in the Motion Picture Industry: Economic and Legal Analysis*, Berkeley, University of California Press, 1960. As an example of recent adherence to that position, see Chris Hugo, 'The Economic Background', *Movie* 27/28, Winter 1980/Summer 1981, p. 46.

2 For a detailed analysis of the Hollywood system of labour, see Janet Staiger, *The Hollywood Mode of Production: The Construction of Divided Labour in the Film Industry* (Unpublished PhD Dissertation, University of Wisconsin-Madison, 1981).

3 'Outside Producers for Warners: 11 "Names" in New Studio Jobs', *Motion Picture Herald* (hereafter *MPH*) 138, no. 6 (February 10, 1940), p. 60; Albert Lewin, '"Peccavi!" The True Confession of a Movie Producer', *Theatre Arts*, September 1941, reprinted in Richard Koszarksi (ed.), *Hollywood Directors 1941–1976*, New York, Oxford University Press, 1977, pp. 26 –34.

4 Staiger, *The Hollywood Mode of Production*, pp. 268–72; Janet Staiger, 'Crafting Hollywood Films: The Impact of a Concept of Film Practice on a Mode of Production', paper presented at the Society of Cinema Studies Conference, Los Angeles, California, June 28–July 1, 1982.

5 William C. de Mille, *Hollywood Saga*, New York, E. P. Dutton & Co., 1939, p. 310; US Interstate and Foreign Commerce Committee, House, 'Motion-Picture Films (Compulsory Block-Booking and Blind Selling), hearing on S 280, 76th Cong., 3rd sess., May 13–June 4, 1940, Washington, DC, Government Printing Office, 1940, p. 637.

6 The vote in the Senate on S 280 (1940) was yeas, 46; nays, 28; not voting, 22; *US Congressional Record*, 76th Cong., 1st sess., Washington, DC, Government Printing Office, 1939–40, pp. 9247–48. US Interstate Commerce Committee, Senate, 'To Prohibit and to Prevent Trade Practices known as Compulsory Block-booking and Blind Selling in Leasing of Motion-Picture Films in Interstate and Foreign Commerce', report to accompany S 3012, 74th Cong., 2nd sess., Washington, DC, Government Printing Office, June 15, 1936, p. 1; US Interstate Commerce Committee, Senate, 'To Prohibit and Prevent Trade Practices Known as Compulsory Block Booking and Blind Selling in Leasing of Motion-Picture Films in Interstate and Foreign Commerce', report to accompany S 153, 75th Cong., 3rd sess., Washington, DC, Government Printing Office, February 16, 1938, p. 5.

7 Good summaries of the events surrounding *United States* v *Paramount, et al.*, are in Ernest Borneman, 'United States versus Hollywood: The Case Study of an Antitrust Suit', in *The American Film Industry*, Tino Balio (ed.), Madison, Wisconsin, University of Wisconsin Press, 1976, pp. 332–45; Raymond Moley, *The Hays Office*, Indianapolis, Indiana, Bobbs-Merrill Company, 1945, pp. 206–12; US Select Committee on Small Business, Senate, 'Motion-Picture Distribution Trade Practices, 1956', report, July 27, 1956, Washington, DC, Government Printing Office, 1956, pp. 5–6; Conant, *Antitrust.*

8 'Hollywood Places Greater Value on Stars in Blocks-of-5 Selling', *MPH*, 142, no. 9, March 1, 1941, p. 12.

9 'Mandatory Block-of-Five Sales End for 5 Majors', *MPH*, 148, no. 10, September 5, 1942, p. 17; 'Majors To Sell In Blocks With Decree Big "If"', *MPH*, 151, no. 11 June 12, 1943, pp. 17–18.

10 The *Motion Picture Herald* attributed the A pictures' popularity to heavier advertising and exploitation. This was due to new trade practices, a result of the 1940 decree. 'Hollywood in Uniform', *Fortune*, 25, April 1942, pp. 92–95; 'Playdates On Top Films Increase 30 Per Cent', *MPH*, 151, no. 12, June 19, 1943, p. 14.

11 'Through the Editor's Finder', *American Cinematographer* (hereafter *AC*), 24, no. 6, June 1943, p. 213.

12 'Monogram to Offer 40 for 1943–44', *MPH*, 151, no. 12, June 19, 1943, p. 40.

13 US Select Committee on Small Business, 'Motion-Picture Distribution Trade Practices, 1956', report, p. 38.

14 'War Booming Market for Independent Product', *MPH*, 153, no. 7, November 13, 1943, p. 14.

15 'Cagney, UA in Five-Year Pact', *MPH*, 152, no. 12, September 18, 1943, p. 44; 'Wallis Sets Five Year Paramount Production Deal', *MPH*, 155, no. 9, May 27, 1944, p. 36.

16 'Hal Wallis Starts Picture for Paramount Release', *MPH*, 158, no. 6, February 10, 1945, p. 41; 'Seven Films Are Started; 49 Now in Production', *MPH*, 158, no. 9, March 3, 1945, p. 33; 'Four Pictures Are Started As Strike Hampers Work', *MPH*, 158, no. 13, March 31, 1945, p. 45; '"They Were Expendable" Begins Rolling at MGM', *MPH*, 158, no. 10, March 10, 1945, p. 38.

17 Ernest Borneman, 'Rebellion in Hollywood: A Study in Motion Picture Finance', *Harper*'s, 193, October, 1946, pp. 337–43; Frederic Marlowe, 'The Rise of the Independents in Hollywood', *Penguin Film Review*, no. 3, August 1947, p. 72; 'Review of the Film News', *AC*, 26, no. 10, October 1945, p. 33; Frank Capra, 'Breaking Hollywood's "Pattern of Sameness"' in *Hollywood Directors*, Koszarski (ed.), pp. 83–89; 'Paramount: Oscar for Profits', *Fortune*, 35, no. 6, June 1947, pp. 90–95, 208–21; '42 Pictures Are Shooting: "Junior Miss" Started', *MPH*, 158, no. 7, February 17, 1945, 29; 'Talent Planning to Fight Capital Gains Tax Ruling', *MPH*, 164, no. 5, August 3, 1946, p. 73.

18 Marlowe, 'The Rise of the Independents in Hollywood', p. 72. Also see US Select Committee on Small Business, 'Motion-Picture Distribution Trade Practices, 1956', hearings, 84th Cong., 2nd sess., March 21–May 22, 1956, Washington, DC, Government Printing Office, 1956, pp. 349–50.

19 Although threatening to fight the ruling, the offending companies eventually settled with the Bureau. Borneman, 'Rebellion in Hollywood', pp. 337–43; 'No Personal Films', *MPH*, 164, no. 4, July 27, 1946, p. 9; 'Talent Planning to Fight Capital Gains Tax Ruling', *MPH*, 164, no. 5, August 3, 1946, p. 43; 'Capital Gains Net Terms to Radio and Stage', *MPH*, 164, no. 6, August 10, 1946, p. 21; 'Tax Benefit Unit: on Coast Settle Revenue Claims', *MPH*, 171, no. 3, April 17, 1948, p. 14.

20 See, for instance, the lists of films starting in production in the *Motion Picture Herald* for December 1946 and January 1947.

21 'Industry Must Readjust Price and Cost: Cowdin', *MPH*, 168, no. 3, July 19, 1947, p. 13; 'Pressure Force Issues Down on Stock Market', *MPH*, 169, no. 13, December 27, 1947, p. 13; William R. Weaver, 'Independents Save 20%; Major Studios Stymied', *MPH*, 168, no. 11, September 13, 1947, pp. 13–14; 'Production Index Still Down; Metro, 20th-Fox, Warner Start "A" Films', *MPH*, 169, no. 13, December 27, 1947, p. 31.

22 'Editorial', *The Screen Writer*, 4, no. 3, September 1948, p. 4; Anthony A. P. Dawson, 'Hollywood's Labour Troubles', *Industrial and Labor Relations Review*, 1, no. 4, July 1948, p. 642; William R. Weaver, 'Studio Employment – A 12-Year Study', *MPH*, 174, no. 6, February 5, 1949, p. 15; 'Employment at Studios Hit 13-Year Low in 1949', *MPH*, 178, no. 8, February 25, 1950, p. 35.

23 William R. Weaver, 'Independents Save 20%; Major Studios Stymied', *MPH*, 168, no. 11, September 13, 1947, pp. 13–14.

24 William R. Weaver, 'Studios Hold Gains with Six Pictures Starting', *MPH*, 170, no. 6, February 7, 1948, p. 25; Fred Hift, 'Those Big City Scenes Are Really Shot On the Spot', *MPH*, 177, no. 13, December 24, 1949, p. 25.

25 'Hollywood Is Travelling Abroad for Production', *MPH*, 170, no. 3, January 17, 1948, p. 13.

26 'British Pact Signed and US Ready To Deliver', *MPH*, 170, no. 12, March 20, 1948, p. 13; William R. Weaver, 'Impact of British–U.S. Deal on Production Is Worrying Hollywood', *MPH*, 171, no. 7, May 15, 1948, p. 27; 'Report 20th-Fox Will Make 12 Top Films in Europe', *MPH*, 172, no. 10 September, 1948, p. 12; '20th-Fox To Offer 32 In Month-By-Month Schedule', *MPH*, 172, no. 12, September 18, 1948, p. 17; 'Shoot More Films Abroad', *MPH*, 176, no. 4, July 23, 1949 p. 16.

27 'Video Hurts, Survey Says', *MPH*, 178, no. 6, February 11, 1950, p. 22; US Select Committee on Small Business, 'Motion-Picture Distribution Trade Practices, 1956', report, p. 22. This latter source is excellent on the effects of television (see pp. 29–31); the analysis includes the argument that drive-ins had a relatively low impact on four-wall exhibition (pp. 28–29).

28 Red Kann, 'Hollywood in '49', *MPH*, 174, no. 4, January 22, 1 1949, p. 18.

29 William R. Weaver, 'Production Up But Big Pictures Are Held Off', *MPH*, 170, no. 11, March 13, 1948, p. 35.

30 Lincoln, 'Comeback of the Movies', p. 131; Merrill Lynch, *Radio, Television, Motion Pictures*, pp. 21, 24; Charles Flynn and Todd McCarthy, *Kings of the Bs*, New York, E. P. Dutton, 1975, p. 30.

31 'Six Majors Plan 7 Films Sold At Advanced Prices', *MPH*, 168, no. 9, August 30, 1947, p. 23.

32 'Importers Seek Theatre Outlets', *MPH*, 166, no. 6, February 8, 1947, p. 50.

33 'New Special Effects Company Formed', *AC*, 28, no. 2, February 1947, p. 66; 'Motion Picture Center: Hollywood's Newest Studio', *AC*, 28, no. 9, September 1947, pp. 314–15; John Forbes, 'Want to Expand? Leasing Is the Answer', *AC*, 41, no. 2, February 1960, pp. 100, 116.

34 Fred Hilt, 'Coast Talent Poised Over Television Pond', *MPH*, 174, no. 9, February 26, 1946, p. 19. See *AC* through the 1950s but particularly: Victor Milner, 'ASC Inaugurates Research on Photography for Television', *AC*, 30, no. 3, March 1949, pp. 86, 100–102; John De Mos, 'The Cinematographer's Place in Television', *AC*, 30, no. 3, March 1949, pp. 87, 102, 104–5; Walter Strenge, 'In the Best Professional Manner', *AC*, 32, no. 5, May 1951, pp. 186, 200–201; Leigh Allen, 'Television Film Production', *AC*, 33, no. 2, February 1952, p. 69; Frederick Foster, 'The Big Switch Is To TV!', *AC*, 36, no. 1, January 1955, pp. 27–7, 38–40; Arthur Miller, 'Hollywood's Cameramen at Work', *AC*, 38, no. 9, September 1957, pp. 580–81. Estimates for 1955 were that ten times the work for television companies compared to theatrical exhibition was being done in Hollywood: Morris Gelman, 'The Hollywood Story', *Television Magazine*, 20, September 1963.

35 'UA to Study Financing of Producers', *MPH*, 174, no. 1, January 1, 1949, p. 25; Red Kann, 'Hollywood in '49', *MPH*, 174, no. 4, January 22, 1949; p. 18; William R. Weaver, 'Giannini Says Bank's Loans At New Peak', *MPH*, 175, no. 12, June 18, 1949, p. 23; Red Kann, 'Circuit Owners Form Company to Finance Product', *MPH*, 176, no. 1, July 2, 1949, pp. 13–14; William R. Weaver, 'Partnership Aid Helps Financing Unit Pay', *MPH*, 176, no. 7, August 13, 1949, p. 23; J. A. Otten, 'Tax, Credit and Labor Matters Are On Washington Agenda for 1957', *MPH*, 206, no. 1, January 5, 1957, p. 21.

36 William R. Weaver, 'See Problems Eased by Blumenthal Formula', *MPH*, 174, no. 12, March 19, 1949, p. 33.

37 Lincoln, 'Comeback of the Movies', p. 130; 'Outlook for 1956', *MPH*, 202, no. 1, January 7, 1956, p. 8; films listed in production for January through March 1950, 1955, and 1959, *Motion Picture Herald*; 'Top Grossing Pictures of 1956', *MPH*, 206, no. 1, January 5, 1957, pp. 12–13; 'Name Adler to Zanuck Post', *MPH*, 202, no. 6, February 11, 1956, p. 18; Vincent Canby, 'Hollywood Pushes Into TV Production Arena', *MPH,* 198, no 4, January 22, 1955, p. 13; Jay Remer, 'Hollywood Eyes TV As New Production Source', *MPH*, 202, no. 2, January 14, 1956, p. 12; William R. Weaver, 'Hecht-Hill-Lancaster Plan Nine Features', *MPH*, 206, no. 1, January 5, 1957, p. 20.

38 'Robert Wise Talks About "The New Hollywood"', *AC*, 57, no. 7, July 1976, pp. 770–71.

39 Staiger, *The Hollywood Mode of Production,* pp. 319–27.

Chapter 20

Sheldon Hall

OZONERS, ROADSHOWS AND BLITZ EXHIBITIONISM: POSTWAR DEVELOPMENTS IN DISTRIBUTION AND EXHIBITION

AN UNFORTUNATE SIDE EFFECT OF THE otherwise excellent accounts that film historians have given of the system of 'runs, zones and clearances' that characterised distribution and exhibition in the studio era is the impression of a rigid uniformity in release patterns. Similarly, accounts of more recent developments have tended to perpetuate the common assumption that practices characteristic of contemporary Hollywood – notably, wide or 'saturation' releasing, accompanied by intensive mass-media advertising – have been in existence only since the 1970s and were unknown in the 'classical' period. Neither is in fact the case. This chapter examines some of the earlier history of wide releasing, situated in the context of two other key developments in the postwar American film industry: the advent of the drive-in cinema as a major area of growth; and the resumption of the practice of 'roadshowing'.

Drive-ins and suburbanisation

The late 1940s and 1950s saw profound changes in American social life. Among them was a massive population shift, as large numbers of families moved away from urban and rural areas to occupy new homes in the suburban developments which sprang up across the country following the lifting of wartime building restrictions. According to an article published in the August 1953 issue of *Fortune*, 'In 1929, 60% of Americans lived in big cities or on farms; today nearly 60% live in suburbs or small towns'.[1] The population increased by 14 million people between 1940 and 1957, 83 per cent of whom lived in suburbs.[2] As most theatres, both cinemas and legitimate houses, were located in the central business and shopping districts of cities, this relocation found a large proportion of the potential audience now living a considerable distance from places of entertainment. Although most families were equipped with motor transport, private car ownership having risen by more than 50 per cent in the seven years following the war, the difficulties and the cost of parking in city centres were a deterrent. So too was the lack of rapid transit systems in many cities. Americans increasingly preferred to spend their leisure time at home and their disposable income on domestic goods and activities, from kitchen appliances to D.I.Y. In such a context, television proved a potent alternative to theatrical entertainment. Film distributors were slow to adjust to the increased importance of the suburbs in social life, still pinning their hopes for effective competition against TV on traditional downtown first-run theatres. But an alternative became available in the form of new types of theatre which brought cinema to the suburbs and capitalised upon car ownership.

The first 'roofless' cinema in the United States was opened in 1929 on an experimental basis by Richard M. Hollingshead, Jr., in the driveway of his family home in Camden, New Jersey. From this makeshift beginning, Hollingshead registered a patent on the drive-in concept (though in 1949 a court judgment ruled that it was not capable of being patented) and went into partnership with his cousin Willis W. Smith, the owner of a chain of East Coast parking lots, to form Park-In Theatres. The company opened its first permanent theatre in Camden in June 1933, with space for 400 cars arranged in semicircular rows on seven raised plinths or ramps facing the screen. It also licensed the concept to other exhibitors on a franchise basis. The first drive-in theatre on the West Coast opened in September 1934 on Pico Boulevard in Los Angeles, and the first in New York opened in August 1938. Further construction of all types of theatre was halted during the war, with about sixty in existence at war's end (estimates vary), but by March 1948 there were reported to be between 250 and 370 'ozoners' (to use *Variety*'s preferred term). In that year numbers began to increase exponentially as drive-in business rose by 10 per cent while the rest of the trade saw a 20 per cent decline. Many more open-air than enclosed theatres were built in the postwar decade, due to lower construction costs and cheap, undeveloped land available in areas adjacent to suburbs. There was also, predictably, a direct correlation between the spread of drive-ins and the rise in car ownership in each territory.

The majority of drive-ins were independently owned by companies which generally had no stake in traditional film exhibition. Being reliant on warm weather, these theatres' business was usually seasonal: most stayed open only from six to nine months per year. For obvious reasons, they also operated only evening shows, after darkness had fallen. Despite their reputation as 'passion pits with pix', research conducted by the University of Minnesota in 1949 suggested that around 40 per cent of the theatres' patrons were over 30 years of age and only 10 per cent under 20. Their target audience was the family group rather than teenage lovers, and in order to attract parents with children many theatres offered childcare facilities, playgrounds and games such as miniature golf and bowling, and even laundry services. As the evening wore on, young children could be bedded down in the car's back seat while their parents watched the movie up front. Some theatres also provided dance floors and live entertainment, such as circus acts and singers, while revenue from 'concessions' (snack foods and drinks) accounted for up to half of all takings, a far greater proportion than with conventional theatres. To encourage group attendance, tickets were sometimes sold by the carload rather than to individual patrons; theatres charging individual entry often allowed free entry to children under the age of 12. In an age of rising juvenile crime, ozoners had relatively little to fear from vandalism; as one manager put it, 'There's nothing in a Drive-In theatre for rowdies to deface'.[3] But perhaps their most conspicuous advantage over indoor theatres, aside from their much larger seating capacity (with room for more than 2,000 cars in some instances), was the availability of ample parking space at no extra cost.

Nevertheless, the major distributors initially regarded drive-ins as fit only to take subsequent runs, second-rate releases and reissues, except in those situations where no adequate first-run theatres were available. Instead they gave priority to the regular indoor theatres that they had always dealt with, which operated all year round. But legal challenges were mounted against such discrimination. In 1948, the Grayslake Outdoor Theatre Co. of Chicago (a hotbed of antitrust activity) sued for the right to bid for first-run bookings, placing it in direct competition with traditional 'hardtops'. A precedent was set in November 1950 when the district court of Philadelphia ruled that drive-ins were entitled to book films on the same terms and conditions as other types of venue. Yet, despite the court judgment, most maintained their sub-run status. As more and more exhibitors demanded improved product and reduced clearances, the distributors in turn demanded a proportionate increase in rentals and extended runs for top pictures. Rather than meet their demands many drive-ins preferred to deal with independent companies, offering double- and triple-bills to lure habitual customers. Some major films passed up by traditional circuits for one reason or another were taken by ozoners instead, such as RKO's Jane Russell vehicle *The French Line* (1954), whose brush with the censors alienated many circuit bookers.

By 1953 there were nearly 4,000 drive-ins either already open or under construction, making up one-quarter of all exhibition sites and providing up to one-fifth of domestic rental revenues, with a total national seating capacity of over five million. A trade organisation, the International Drive-In Theatre Owners Association, was formed that year to represent exhibitors' interests. By this time the majors, as well as independents, were studying patterns of ozoner attendance in order to determine what types of product were most successful there, and tailoring a proportion of their output accordingly. Drive-ins were particularly successful in both attracting new audiences and retaining regular patronage; their very difference from traditional cinemas offered an advantage in the fight against competition from television and other leisure pursuits, and also posed a threat to rival exhibitors faced with the same challenges. California-based drive-in operator Robert L. Lippert claimed that, 'TV has reduced first-run grosses by 20%, neighbourhoods as much as 50%, drive-ins only 10%'.[4]

The advent of the 'new era' of widescreen technology, intended partly as a riposte to television, brought problems for drive-ins just as it did for many hardtops. Most exhibitors had already installed the largest screen possible for the space available, so altering the aspect ratio for CinemaScope would in some instances mean reducing height rather than extending width. Lower screen height not only reduced visibility for cars parked further back, but the 'immersive' effect on which publicity for the new formats placed so much emphasis went for little when patrons were encased in a glass-and-metal automobile; nor did stereophonic sound when in-car speakers offered only monaural reproduction. For this reason Twentieth Century-Fox initially refused to licence any ozoners to show films made in CinemaScope, holding to its policy of insisting that exhibitors install four-track magnetic stereo for its patented process. Fox backed down when two- and three-channel portable sound units were developed specifically for use in drive-ins, but other distributors had proved more willing to compromise: the first anamorphic release to be screened at a drive-in was reportedly Warner Bros.' Western *The Command* (1954), with its stereo soundtrack reproduced monophonically. Those cinemas that did install new screens and speakers for CinemaScope, sometimes raising their prices like other exhibitors, found themselves enjoying a similar surge in business. However, they were also required to pay rentals at far higher rates than with other product; drive-in operators in Minneapolis refused to install CinemaScope unless Fox agreed to allow late runs of its films to be booked for flat fees rather than a hefty percentage.[5]

By 1960, some 5,000 drive-ins were in operation, compared to around 13,000 hardtops, contributing 23 per cent of the total domestic box-office take. Around one-third were now staying open all year round, offering in-car heaters in the winter months; a court ruling upheld the right of distributors to claim a proportion of the charges made for these and other services, along with ticket sales, as part of their rental. Rather than being regarded as a fringe business, drive-ins had by this time established themselves as a mainstay of the industry and enjoyed their share of major-studio first runs. Distribution patterns had moved away from the rigid observation of established runs and now favoured flexible booking patterns that addressed the needs of exhibitors outside the metropolitan centres as well as downtown. As a result, drive-ins increasingly drew business that would once have gone to theatres in urban districts as well as in outlying areas. There was one category of product from which they remained excluded; however, at least in the early stages of release, despite what would appear to be an etymological affinity: the roadshow film.

Roadshowing

According to industry tradition since the early teens, 'roadshowing' meant the presentation of motion pictures in the manner of live theatre: an exclusive engagement, sometimes in a 'legitimate' venue that had been taken over for the purpose rather than a regular cinema; two separate performances daily, one matinee and one evening show; raised ticket prices, typically ranging from $1.50 to $2.20 top in major cities; reserved seats and advance booking; lavish front-of-house advertising and lobby displays, an illustrated programme brochure available for purchase, and often an intermission during the movie itself. This was how *Quo Vadis?* (1913), *Cabiria* (1914) and *The Birth of a Nation* (1915), among others, had been successfully presented on Broadway and around the country. They had set the pattern for countless subsequent 'big' pictures, especially during the silent era. Roadshowing signified prestige and was designed to build word of mouth throughout both the trade and the public prior to general release on regular terms.

However, roadshowing went out of fashion in the early years of sound everywhere except in New York and Los Angeles, and was revived only sporadically throughout the 1930s. MGM's release of David O. Selznick's *Gone with the Wind* (1939) changed the usual pattern: limiting the availability of reserved seats, often doing away with separate performances, allowing cinemas to increase the number of daily screenings or omit the intermission, and charging ticket prices which, while still higher than the average, were generally lower than custom would normally have demanded. Rather than insisting on exclusivity, MGM also permitted more than one theatre in each territory to book the film concurrently, and even gave neighbourhood theatres ('nabes') access to it while it was still playing downtown. The result was the most widely seen and profitable film ever released. The lesson was not lost on other distributors. Throughout the 1940s, with rare exceptions, 'roadshowing' meant little more than first-run exhibition at raised prices, without most of the other trappings of live theatre. No film was given the reserved-seat, two-a-day treatment on Broadway for nearly three years after Paramount's *For Whom the Bell Tolls* (1943). However, traditional roadshow presentation underwent a revival in the postwar era: initially with the limited release of 'art' or prestige pictures aimed at specialised audiences, and subsequently with what became known as 'blockbusters', big-budget super-productions often showcasing an array of new presentation technologies.

In 1946, United Artists devised a method of releasing Laurence Olivier's British-made Shakespeare adaptation *Henry V* (1944), that capitalised upon its select appeal to upscale audiences. With few exceptions – notably *The Private Life of Henry VIII* (1933), *Pygmalion* (1938) and *In Which We Serve* (1942) – British productions had never found a large audience in America. Initially reluctant to distribute the film at all, UA devised an innovative distribution strategy which was aimed specifically at attracting 'legit' theatregoers, in smaller communities as well as in the big cities, and at the educational sector of schools, colleges and universities. The company arranged to promote the film through the Theatre Guild, the subscription service whose nationwide mailing list could readily be exploited for direct advertising, and which would assist in theatre bookings and block ticket sales in exchange for a cut of the gross and discounts for its members. As the first film to be sponsored by the Guild, *Henry V* was assured of considerable prestige by association, in addition to the inherent prestige value of the film itself and its identification with the British stage tradition, not least through Olivier's presence as both star and director along with Shakespeare's as original author.[6]

UA adopted a general policy of relying as far as possible on the free publicity afforded by word-of-mouth praise, media news stories and reviews, thereby making considerable savings on paid promotion. Press advertisements, when used, were placed alongside those for live theatres, not cinemas, and carried few indications that *Henry V* was in fact a film. The opening engagement at the Esquire, Boston, proved a surprising success and generated a large amount of favourable coverage, including a *Time* cover story. An eleven-week New York engagement at the City Center, a municipally run theatre which had never before played a motion picture, was extended by a move-over to the Golden for thirty-five weeks, followed in 1947 by 'grind' (continuous-performance) revivals at the Broadway and Park Avenue theatres, for a total New York run of fifty-eight weeks. All the initial engagements in key cities, including Los Angeles, San Francisco, Washington, D.C., Chicago and Philadelphia, were booked on a 'four-wall' basis, with UA hiring the theatre outright, paying publicity, staff and operating expenses, and retaining the entire box-office gross. Admission prices ranged from 75 cents (for special student matinees) to $2.40.

Outside the key cities, *Henry V* was mainly shown in college towns for limited runs of between one and four days, depending on the size of the population. UA's agents were instructed to book the film into first-class, centrally located theatres of small capacity wherever possible. Other than vacation spots, the film played few engagements in the summer months when schools and colleges were out. By deliberately underestimating its potential, the company sought 'to pack the house for a short engagement and play return engagements when the situation requires it, thereby not wearing out its welcome, maintaining prestige, and being constantly in demand throughout the country'. No trailers or other supporting programmes were permitted to be screened with the feature, and sales of popcorn and other concessions were discouraged as potentially 'spoiling the dignity of the performance'.[7] The scant distribution, print and advertising costs involved meant that most of the bookings were pure profit for UA. After a year it began booking the film into regular cinemas on a percentage basis, taking between 60 and 75 per cent of the gross as rental and sharing costs with the exhibitor. Advance agents took charge of local advertising and group ticket sales, and reported the daily grosses direct to UA. After two-and-a-half years in continuous distribution and more than 800 engagements, the strict roadshow policy was dropped in favour of advanced-price grind runs at $1.20 top. By this point *Henry V* had earned over $3 million in the United States, with a profit to UA of $1,620,000.[8]

Variations on the marketing methods employed on *Henry V* were later adopted for other prestige pictures. Michael Powell and Emeric Pressburger's *A Matter of Life and Death* (1946), re-titled *Stairway to Heaven* and released in the United States by Universal-International (U-I), was presented in three separate performances daily in small 'sure-seater' (art-house) theatres in New York, Boston and Los Angeles. Olivier's *Hamlet* (1948), also released by U-I, and Powell and Pressburger's *The Red Shoes* (1948), released by Eagle-Lion, both closely followed the pattern set by *Henry V*. *Hamlet* earned $3,250,000 in rentals from roadshowing followed by a general release at advanced prices. *The Red Shoes* achieved a record-breaking run of 108 weeks at the Bijou Theatre on Broadway and may ultimately have earned as much as $5 million in America. *Variety* reported that this was 'the first time in the history of the industry' that the roadshow market had been dominated and led by British pictures, and indeed English-language imports were notably successful in the expanding art-house sector generally.[9] Later British productions roadshown in the United States included Carol Reed's *The Third Man* (1949), Powell and Pressburger's *The Tales of Hoffmann* (1951), Jean Renoir's *The River* (1951), Frank Launder and Sidney Gilliat's *The Story of Gilbert and Sullivan* (1953, re-titled *The Great Gilbert and Sullivan*) and Olivier's *Richard III* (1955). Among American pictures, RKO adopted the policy for its production of Eugene O'Neill's *Mourning Becomes Electra* (1947), as did Republic for Orson Welles's *Macbeth* (1948), MGM for Joseph L. Mankiewicz's adaptation of *Julius Caesar* (1953), and independent producers Stanley Kramer and Louis de Rochement for *Cyrano de Bergerac* (1950) and *Martin Luther* (1953), respectively.

Pictures such as these tended to be most profitable in populous urban areas: business dropped off sharply in the hinterlands, where there was strong resistance to raised prices and reserved seats. They nevertheless revealed the existence of an audience sector composed substantially of people who rarely patronised the cinema, who had more sophisticated tastes and were generally older than the 15-to-25-year-olds that comprised the majority of moviegoers. Prestige pictures such as the British imports demonstrated that these audiences could be attracted by the right kind of product, marketed without traditional Hollywood 'ballyhoo' and presented amid suitable surroundings. Aside from an expansion in the number of dedicated art houses, one trend (begun in Toronto and spreading to the United States in the early 1950s) was the hosting by regular cinemas of an occasional 'let's go arty' evening (as *Variety* described it). U-I's Special Films Division advised exhibitors on suitable booking possibilities, which naturally included its own acquisitions from the Rank Organisation, and tickets were sold not only for single pictures but also for seasons, the latter at a discount rate. Variously dubbed 'Curtain at 8.30' or 'An Evening at the Theatre', such events involved showing a selected foreign (including British) picture, offering reserved seats and separate performances, ushers dressing in tuxedos, and sales of popcorn and candy being replaced by free coffee.[10]

All these ventures, however, were somewhat marginal to the mainstream of the industry. The vast potential earnings from roadshowing were more dramatically demonstrated by two independent releases, both showcasing new photographic and sound technologies, whose impact had a transformative effect on Hollywood. *This Is Cinerama* (1952) earned a box-office gross of $14,767,535 in its first two years of exhibition in only thirteen theatres. *Oklahoma!* (1955), the first feature film shot in Todd-AO, grossed $8,970,087 from thirty-one U.S. and Canadian theatres over eighteen months. The second feature in the 70mm process, *Around the World in Eighty Days* (1956), was released by a major distributor, United Artists, and proved even more successful, with a rental (approximately half the gross) of $17,700,000 from 1,569 domestic roadshow engagements in its first two years. Top ticket prices for these films rose as high as $3.50.[11]

Another 1956 roadshow established a new all-time box-office record, temporarily besting even *Gone with the Wind*. Cecil B. DeMille's *The Ten Commandments*, distributed by Paramount, was produced at a cost of $13,266,491, nearly double that of the previous most expensive film to date. Premiered at New York's Criterion Theatre on 8 November 1956, the film played nearly 1,000 domestic roadshow engagements in its first year. These were followed by regular advanced-price 'pre-release' engagements with continuous performances, booked in 'waves': limited runs of several weeks in a small group of theatres in a given territory, now including drive-ins. General release at normal prices began in 1959 and continued until the end of 1960, by which time the film had earned $34,200,000 in domestic rentals and nearly $60 million worldwide. According to *Variety*, *Around the World in Eighty Days* and *The Ten Commandments* had 'taken in more rentals from fewer theatre engagements than any pictures in history'.[12] They provided a model for the roadshow release of many subsequent large-scale spectacles. Each successive year thereafter saw an increasing number of would-be blockbusters given reserved-seat handling, until the financial crisis at the end of the 1960s brought about a revised economy and the abandonment of roadshowing in favour of alternative distribution and exhibition methods, which had also been pioneered in earlier years.

Saturation releasing

The release of a new picture in a large number of theatres simultaneously (known within the industry as 'day and date' bookings) has long been used as a distribution strategy designed to capitalise on the presumed desire of audiences to see certain attractions as soon as possible. Whereas exclusive roadshowing was designed to build awareness and anticipation slowly, an opening in multiple theatres at once was predicated on the recognition that such awareness and anticipation already existed. This was particularly the case with films showcasing actors with a loyal following eager for their next screen appearance. It is not surprising, therefore, that many of the films of the silent era's most popular star, Charlie Chaplin, were released this way. 130 prints of the two-reel *The Floorwalker* (1916) and 160 of the three-reel *A Dog's Life* (1918) were used in Greater New York alone. According to the *Moving Picture World*, sixty-five copies of Chaplin's first full-length feature *The Kid* (1921) were 'booked solid for thirty days beginning March 7 in Greater New York', including twenty-five theatres along Broadway.[13] These were examples of localised saturation releases. In the early 1930s several studios began employing wide releases on a national basis, exploiting the opportunities for country-wide promotion of sound films using radio tie-ins. RKO's *Check and Double Check* (1930), starring the blackface radio comedians Amos 'n' Andy, and its epic Western *Cimarron* (1931) both opened in around 300 theatres, while Warner Bros. regularly addressed its trade advertisements to exhibitors with the slogan 'Available to you NOW – Day-and-Date with Broadway'.[14]

However, it was in the 1940s that wide releasing came to offer particular advantages, initially with the increased volume of patronage experienced at downtown theatres and subsequently with the decline in popularity of those same theatres in comparison with suburban venues, including drive-ins. The surge in attendances during the Second World War placed a great deal of strain on the distribution system, as films stayed far longer in first run than at any previous time, resulting in a 'bottleneck' of new pictures in city centres. To speed up their distribution, many formerly sub-run theatres were promoted to early-run status, raising their ticket prices accordingly, and clearances between runs were reduced. This led to an additional improvement in rentals for distributors, as pictures earned by far the greatest proportion of their revenue from first run (typically 86 per cent, according to *Variety* in 1947).[15] A model for day-and-date releases was provided by the 'Los Angeles system'. Owing to the city's wide geographical spread, distributors had long practised opening new films in from three to five theatres, each located in a different district. As a result, Los Angeles was regarded by distributors as 'the most profitable situation in the country for [the size of] its population'.[16]

The film which most emphatically indicated the possibilities of wide releasing was David O. Selznick's epic Western *Duel in the Sun* (1946), distributed through his own company, Selznick Releasing Organization. It opened on a traditional two-a-day roadshow basis at two Los Angeles theatres in late December 1946 in order to qualify for Academy Award nominations, with three more engagements opening in January, including a further roadshow. After this the reserved-seat policy was abandoned in favour of continuous performances and raised prices, consistent with the presentation of most big pictures since *Gone with the Wind*. Around sixty other engagements followed in the next few months, wider release being delayed by various factors including a shortage of Technicolor prints and a censorship battle with the Catholic Legion of Decency. Then in early May 1947, some 300 bookings opened in selected territories around the country, including fifty-three in Greater New York and fifty-five in California. SRO claimed that this was 'the biggest day-and-date setup ever arranged by any picture'.[17] These theatres grossed an estimated $2.5 million from three million admissions in seven days. This was reported to be a record for a single week, though few shows had sold out: with so many situations playing the same attraction in adjacent localities, they tended to cut into one another's business and reduce the average take per venue. Furthermore, second-week holdovers tended to perform disappointingly, with word-of-mouth on the picture apparently unenthusiastic. Nevertheless, after only six weeks of wide release SRO had earned a rental of nearly $3 million from a total of 700 bookings. Its ultimate domestic rental was around $10 million, making *Duel in the Sun* one of the three biggest hits of the decade.[18]

This release plan, designed by Selznick himself, was the result of the producer's concern that exclusive, single-theatre engagements were not economic and did not take maximum advantage of expensive pre-release advertising campaigns, the impact of which would largely be dissipated once a picture went into general release. By saturating a particular area first with intensive local promotion, then with a large number of theatre bookings, public awareness and anticipation could be both created and satisfied almost immediately. Rather than give sole priority to downtown venues, Selznick allowed selected suburban, neighbourhood and other outlying theatres (which would ordinarily have had to wait several weeks or months before getting a major new picture) the opportunity to show the film concurrently with city centres, charging the same premium ticket prices (and paying SRO the same rental). At a time when many independent exhibitors were initiating law suits against the major distributors over the preferential treatment and extended clearances given to affiliated theatres, *Duel in the Sun* offered an alternative system which many bookers must have found appealing. The greater number of admissions possible in a short space of time also increased the speed of revenue to SRO, thereby reducing bank interest payments on a production costing over $5 million, with an additional $2 million spent on advertising. Selznick named his strategy 'blitz exhibitionism'.

Though the strategy proved controversial – with exhibitors who did *not* get to play *Duel in the Sun* complaining that they had been overlooked and others that regular patrons of neighbourhood theatres resented the sudden increase in prices – its success stimulated a rush of imitations. There were, however, considerable variations in the instant-release policies adopted by the major companies. Columbia, United Artists and Allied Artists all preferred wide releasing on a national basis, in up to 400 theatres simultaneously across the country. Both Paramount and Universal-International used multiple-theatre bookings on a territorial basis for routine 'bread-and-butter' pictures, which thereby often earned twice the normal gross from a regular staggered release. Fox also mass-released big-budget productions, including the costume epics *Forever Amber* (1947) with 475 prints in circulation and *Prince of Foxes* (1949) with 500, while the comedy *Father Was a Fullback* (1949) was allocated as many as 542 (the normal inventory was around 325). RKO followed the Broadway first run of Walt Disney's *So Dear to My Heart* (1948) with 113 concurrent bookings in Greater New York and opened the giant-gorilla picture *Mighty Joe Young* (1949) simultaneously in 358 theatres in New England and New York. In certain territories some pictures were also allowed to go into general release in nabes directly after the downtown first run, without any intervening clearance, as for instance with the dubbed, re-edited version of the Italian epic *Fabiola* (1948) released by UA in 1951.

Both national and regional saturation policies helped address the demands of subsequent-run theatres for the opportunity to screen films earlier. But they were also widely regarded as devaluing downtown first runs (which found their earnings declining as those of sub-runs rose), limiting the choice of programmes available to audiences, and restricting the time available for films to benefit from word-of-mouth. This, however, may sometimes have been intentional: in the case of a weak picture, the 'hit-and-run' strategy of a wide, rapid release circumvented the possibility of negative recommendations affecting box-office performance. This may well have been the case when MGM – a company not normally disposed to saturation bookings – opened its epic about the *Mayflower* pilgrims, *Plymouth Adventure* (1952), in 400 to

500 theatres over the Thanksgiving holiday. But throughout the 1950s saturation releasing came to be associated primarily with the kind of downmarket product which was the complete opposite of the films given roadshow treatment: low-budget horror, science-fiction and fantasy films, and later rock 'n' roll musicals aimed at teenagers and young adults.

RKO followed *Mighty Joe Young* in the summer of 1952 with a re-release of *King Kong* (1933). The film opened simultaneously in around 400 theatres across five Midwestern cities (Detroit, Cleveland, Cincinnati, Indianapolis and Pittsburgh). RKO's exploitation manager Terry Turner extended the usual media advertising campaign to include a blanket use of television spots on fourteen local stations. The results defied all expectations, out-grossing the studio's most recent new pictures by a wide margin. Similar tactics were used in other territories, including 100 day-and-date bookings in Greater New York, and the release ultimately grossed $1,600,000, an astounding sum for a nearly twenty-year-old reissue that doubled its original earnings. According to Turner, his *King Kong* campaign marked the first time television advertising had been used so extensively for a feature film. He argued that it demonstrated what could be achieved with the new medium and indeed the film reportedly did poorly in areas not covered by television. Turner went on to publicise a number of other (in his own words) 'freak, horror and spectacle-type' pictures, including *The Beast from 20,000 Fathoms* (1953) and *Them!* (1954) for Warner Bros.[19] These two films used some 600 prints apiece to play around 2,000 engagements over a four-week period. *King Kong* was given a further re-release in 1956 following a television showing the previous year (RKO being the first of the major studios to sell its back catalogue for TV distribution) and again performed well, particularly this time in drive-ins.

Terry Turner also collaborated extensively with an independent company, Embassy Pictures Corporation, whose president was Joseph E. Levine. A former art-house exhibitor based in Boston (where he also still operated a drive-in), Levine made his fortune when he bought the New England reissue rights to *Duel in the Sun* for $40,000 and released it in a 250-theatre saturation pattern in 1954 to considerable success. He subsequently acquired the U.S. rights to the Japanese 'creature feature' *Gojira* (1954) and the Italian epics *Sins of Pompeii* (a re-titling of *The Last Days of Pompeii*, 1950), *Attila* (1954), *Hercules* (1958) and *Hercules Unchained* (1959), having them redubbed, re-edited and in the case of *Gojira* – renamed *Godzilla, King of the Monsters* – partially reshot to suit American tastes. All these were given saturation distribution and elaborate exploitation, including TV advertising, often costing four or five times their acquisition price. As a result of his colourful campaigns for the enormously successful *Hercules* films, in particular, and his own self-promotion within the industry, Levine became by 1960 'the most recognisable showman in show business'.[20]

Despite their success in suburbs, neighbourhoods and the hinterlands, these and other films aimed at the increasingly important 'youth' audience rarely attracted the older patrons who frequented downtown picture palaces. Fox's *Love Me Tender* (1956), the first film starring Elvis Presley, opened in some 500 theatres nationwide in Thanksgiving week and did well in the New York nabes but comparatively poorly at the flagship Paramount Theatre in Times Square. Presley's next two films, Paramount's *Loving You* and MGM's *Jailhouse Rock* (both 1957), were given instant wide releases in what would ordinarily have been sub-run theatres, including drive-ins, often bypassing downtown venues entirely. The 'Presley pattern', delivering downmarket product direct to its target constituency, typified mass-appeal pictures that had little to gain from an expensive premiere run in a city-centre showcase; other such films included Columbia's *The Garment Jungle*, Paramount's *Omar Khayyam* and Fox's *Bernardine* (all 1957). In southern and mid-western states, saturation release seemed particularly suited to Westerns, outdoor adventures and stories with rural settings, all of which enjoyed more popular support in these areas than in the more sophisticated urban centres. This practice had its antecedents in the launching of 'grass roots' attractions such as episodes of MGM's Hardy

Family and Universal's Ma and Pa Kettle series in neighbourhood and small-town cinemas. Warners claimed a record 651 simultaneous bookings with its family western *The Boy from Oklahoma* (1954). United Artists used a mass-booking policy on more upmarket pictures, too. It opened Carol Reed's circus drama *Trapeze* (1956) simultaneously in 405 theatres nationwide – mostly key-city first-run houses – and reported a new record one-week box-office gross of $4,112,500. Like Selznick's with *Duel in the Sun*, UA's strategy was designed to maximise the impact of an intensive exploitation campaign, in this case costing an astronomical $1,500,000. *Trapeze* went on to become the company's highest-earning non-roadshow to date, with domestic rentals over $7 million.[21]

The rise of roadshows and drive-ins as well as of saturation releases marked a period of change for the ways in which the American film industry marketed its wares. In 1960, an unidentified distribution executive interviewed by *Variety* anticipated an intermediary release pattern: neither as exclusive as a roadshow nor as wide as a saturation booking, but one which allowed for metropolitan first runs of major films to be split between downtown and a limited number of neighbourhood or suburban houses simultaneously, thereby allowing audiences a choice of venue according to their preference.[22] This would also address the concerns of sub-run exhibitors facing a product shortage as downtown theatres held on longer than ever to big pictures (there were eight extended-run roadshow releases in 1960 alone). The 'wave' bookings of *The Ten Commandments* were cited as a precedent, and indeed there were others: in 1940 *Gone with the Wind* had opened in some New York nabes while its Broadway roadshow engagement was still running. David Selznick modified this policy for the release in selected territories of his production of Ernest Hemingway's *A Farewell to Arms* (1957), distributed by Fox. In Los Angeles, for instance, the film opened concurrently in eight theatres, each in a different area of the city. Further experiments along these lines were followed in 1962 when UA gave the limited-multiple release pattern a name, 'Premiere Showcase'. It subsequently became the standard way of launching the general release of major-studio movies, catering not only to the demands of nabes and drive-ins but to those of another new type of theatre that began appearing in ever-growing numbers: constructed inside the shopping centres now being built near the suburbs and often incorporating two, three or more auditoria under one roof.

Notes

1 Quoted in *Variety*, 5 August 1953, pp. 3, 20.

2 *Variety*, 16 January 1957, p. 12. These figures were cited by Arno H. Johnson, research director of advertising agency J. Walter Thompson. On the impact of suburbanisation on show business, see also ibid., 9 June 1948, p. 7; 23 January 1952, pp. 1, 61; 7 December 1955, p. 9; 1 August 1956, pp. 1, 18; 11 March 1959, pp. 5, 17; 24 August 1960, pp. 5, 16.

3 *Variety*, 4 December 1946, p. 20. On the growth of drive-ins, see also ibid., 13 July 1933, p. 5; 25 December 1934, p. 2; 10 August 1938, p. 23; 4 December 1946, p. 20; 17 March 1948, pp. 7, 18; 21 April 1948, pp. 7, 17; 20 October 1948, p. 4; 5 January 1949, p. 30; 8 June 1949, pp. 5, 14; 15 June 1949, pp. 7, 18; 12 October 1949, p. 20; 26 October 1949, pp. 3, 20; 4 January 1960, p. 21; 5 April 1950, pp. 5, 20; 29 November 1950, pp. 3, 53; 7 May 1952, p. 5; 4 June 1952, pp. 7, 22; 23 July 1952, pp. 3, 23; 14 January 1953, p. 17; 18 March 1953, pp. 7, 17; 8 July 1953, p. 5; 15 July 1953, pp. 5, 20; 12 August 1953, p. 7; 30 September 1953, pp. 5, 24; 7 July 1954, p. 5; 20 July 1955, p. 18; 3 August 1955, p. 23; 25 January 1956, pp. 7, 19; 7 March 1956, pp. 1, 18; 14 August 1957, pp. 3, 20; 28 May 1958, p. 11; 9 April 1958, p. 22; 26 August 1959, pp. 7, 16; 30 December 1959, p. 5; 27 April 1960, p. 5; Douglas Gomery, *Shared Pleasures: A History of Movie Presentation in the United States* (London: BFI Publishing, 1992), pp. 91–93; Anthony Downs, 'Where the Drive-in Fits into the Movie Industry' (1953), in Ina Rae Hark (ed.), *Exhibition: the Film Reader* (London and New York: Routledge, 2002), pp. 123–27; Kerry Segrave, *Drive-In Theaters: A History from Their Inception in 1933* (Jefferson, N.C.: McFarland, 1993); Frank J. Taylor, 'Big Boom in Outdoor Movies' (1956), in Gregory A. Waller

(ed.), *Moviegoing in America: A Sourcebook in the History of Film Exhibition* (Malden and Oxford: Blackwell, 2002), pp. 247–51.

4 *Variety*, 14 January 1953, p. 17.

5 On the arrival of CinemaScope and stereophonic sound in drive-ins, see *Variety*, 24 June 1953, pp. 7, 24; 3 February 1954, pp. 5, 15, 18; 10 February 1954, pp. 10, 11; 10 March 1954, pp. 4, 20; 17 March 1954, pp. 7, 18; 30 June 1954, p. 22; 4 August 1954, p. 3; 25 August 1954, p. 16.

6 On the U.S. release of *Henry V*, see *Variety*, 28 February 1945, pp. 3, 46; 6 February 1946, p. 3; 10 April 1946, p. 6; 24 April 1946, p. 15; 22 May 1946, p. 7; 26 June 1946, p. 4; 17 July 1946, pp. 3, 22; 24 July 1946, p. 29; 1 January 1947, p. 5; 26 February 1947, p. 20; 14 May 1947, p. 3; 18 June 1947, p. 5; 19 May 1948, pp. 5, 18; 6 October 1948, p. 7; *Motion Picture Herald*, 30 March 1946, p. 56; Richard Griffith, 'Where Are the Dollars?', *Sight and Sound*, December 1949, January 1950 and March 1950; Arthur Knight, 'The Reluctant Audience', *Sight and Sound*, April–June 1953; Sarah Street, *Transatlantic Crossings: British Feature Films in the USA* (New York and London: Continuum, 2002), pp. 91–118; Barbara Wilinsky, *Sure Seaters: The Emergence of Art-House Cinema* (Minneapolis and London: University of Minnesota Press, 2001), pp. 73–79.

7 'Instructions as to Booking and Handling of "Henry V"', unsigned, undated document; in *Henry V* files, United Artists Collection, Wisconsin State Historical Society, University of Wisconsin, Madison.

8 Memo from Harold Auten, 2 September 1948, United Artists Collection.

9 *Variety*, 13 October 1948, p. 4. On these and other art-house roadshows, see also ibid., 22 September 1948, p. 9; 10 November 1948, pp. 3, 18; 8 December 1948, pp. 7, 18; 24 August 1949, p. 29; 19 October 1949, p. 4; 7 December 1949, p. 16; 15 November 1950, p. 4; 12 December 1951, p. 5.

10 *Variety*, 4 April 1951, pp. 1, 55; 5 December 1951, pp. 1, 72; 5 November 1952, p. 19; 9 January 1952, pp. 7, 53.

11 *Variety*, 29 September 1954, p. 20; 6 February 1957, p. 3; 12 November 1958, p. 71. On the Cinerama and Todd-AO films, see also John Belton's chapter in the present volume (Chapter 21) and his *Widescreen Cinema* (Cambridge, Mass.: Harvard University Press, 1992), pp. 85–112, 158–82. On the roadshowing of these and other big pictures, see Sheldon Hall and Steve Neale, *Epics, Spectacles, and Blockbusters: A Hollywood History* (Detroit: Wayne State University Press, 2010).

12 *Variety*, 8 January 1958, p. 6. On the release of *The Ten Commandments*, see also ibid., 10 July 1957, p. 14; 14 May 1958, p. 5; 7 January 1959, p. 48; 28 December 1960, p. 7; 10 November 1965, p. 1; Bosley Crowther, 'Screen Phenomenon', *New York Times*, 10 November 1957; Robert S. Birchard, *Cecil B. DeMille's Hollywood* (Lexington: University Press of Kentucky, 2004), pp. 351–64.

13 *Variety*, 19 May 1916, p. 24; *Moving Picture World*, 11 May 1918, p. 875; 12 March 1921, p. 184; 19 March 1921, p. 266; *Wid's Daily*, 8 March 1921, p. 1.

14 *Variety*, 26 February 1930, pp. 18–19; 1 February 1931, p. 19.

15 *Variety*, 21 July 1943, pp. 5, 16; 18 August 1943, p. 7; 25 August 1943, p. 7; 1 September 1943, p. 7; 6 October 1943, p. 5; 27 October 1943, p. 5; 10 November 1943, pp. 3, 46; 17 November 1943, pp. 5, 25; 2 February 1944, p. 5; 21 June 1944, p. 9; 29 October 1947, pp. 5, 18.

16 *Variety*, 27 November 1946, pp. 3, 33.

17 *Variety*, 7 May 1947, pp. 10, 21.

18 On the release of *Duel in the Sun*, see *Variety*, 11 September 1946, p. 4; 8 January 1947, pp. 4, 33; 26 February 1947, p. 18; 19 March 1947, pp. 4, 18; 7 May 1947, pp. 10, 20, 21; 14 May 1947, pp. 3, 20, 21, 25; 21 May 1947, p. 22; 18 June 1947, pp. 3, 20; 2 July 1947, p. 19; 30 July 1947, p. 6; 3 September 1947, pp. 4, 18; 24 January 1951, p. 14; Rudy Behlmer (ed.), *Memo from David O. Selznick* (Los Angeles: Samuel French, 1989), pp. 345–59; Ronald Haver, *David O. Selznick's Hollywood* (London: Secker and Warburg, 1980), pp. 352–68.

19 *Variety*, 13 January 1960, p. 3. On saturation releasing and advertising, see also ibid., 11 June 1947, pp. 5, 18 October 1947, pp. 5, 49; 29 October 1947, p. 5; 3 December 1947, p. 6; 2 February 1949, p. 5; 6 April 1949, pp. 3, 22; 4 May 1949, p. 15; 15 June 1949, pp. 4, 6; 13 July 1949, pp. 7, 18; 20 July 1949, p. 7; 21 September 1949, pp. 5, 20; 28 September 1949, p. 3; 5 October 1949, p. 9; 19 October 1949, p. 20; 7 December 1949, pp. 7, 22; 15 March 1950, pp. 3, 17; 7 March 1951, pp. 5, 16; 11 June 1951, pp. 5, 29; 27 June 1951, pp. 5, 22; 22 August 1951, p. 20; 7 December 1951, pp. 7, 22; 14 May 1952, p. 17; 25 June 1952, p. 17; 16 July 1952, p. 4; 8 October 1952, p. 18; 7 January 1953, p. 59; 21 January 1953, p. 20; 8 April 1953, pp. 3, 20; Cynthia Erb, *Tracking King Kong: A Hollywood Icon in World Culture* (Detroit: Wayne State University Press, 1998), pp. 121–54; Dade Hayes and Jonathan Bing, *Open Wide: How Hollywood Box Office Became a National Obsession* (Hyperion/Miramax Books: New York, 2004), pp. 145–50; Kevin Heffernan, *Ghouls, Gimmicks, and Gold: Horror Films and the American Movie Business, 1953–1968* (Durham, N.C., and London: Duke University Press, 2004).

20 Anthony T. McKenna, *Joseph E. Levine: Showmanship, Reputation and Industrial Practice 1945–1977* (PhD dissertation, University of Nottingham, 2008), p. 99. On Levine's promotional campaigns for these

and other films, see ibid., pp. 87–116; *Variety*, 14 July 1954, p. 7; 20 October 1954, p. 3; 10 August 1955, p. 3; 22 February 1956, p. 18; 15 January 1958, p. 25; 18 June 1958, p. 18; 6 October 1958, p. 15; 10 December 1958, pp. 7, 19; 4 March 1959, pp. 3, 16; 25 March 1959, pp. 7, 19; 1 April 1959, pp. 5, 18; 3 June 1959, pp. 7, 17; 15 July 1959, p. 32; 5 August 1959, pp. 3, 24; 9 September 1959, p. 7; 14 October 1959, pp. 3, 18; 18 November 1959, p. 7; Robert Alden, 'Advertising: Hard Sell for Motion Pictures', *New York Times*, 29 May 1960; Sheldon Hall, *Zulu: With Some Guts Behind It – The Making of the Epic Movie* (Sheffield: Tomahawk Press, 2005), pp. 129–33.

21 *Variety*, 7 December 1938, p. 4; 26 April 1950, p. 24; 15 June 1953, p. 5; 27 January 1954, p. 21; 24 February 1954, p. 7; 2 June 1954, p. 5; 15 February 1956, p. 10; 2 May 1956, p. 7; 20 June 1956, p. 13; 11 July 1956, pp. 12, 13, 18; 15 August 1956, p. 4; 21 November 1956, pp. 3, 62; 28 November 1956, p. 4; 16 January 1957, pp. 7, 12; 8 May 1957, p. 18; 26 June 1957, pp. 5, 18; 3 July 1957, p. 7; 15 April 1959, p. 14; 31 August 1960, pp. 7, 62; 9 November 1960, p. 4.

22 *Variety*, 30 October 1957, p. 24; 18 December 1957, p. 14; 1 January 1958, p. 13; 15 January 1958, p. 16; 2 March 1960, pp. 7, 20. See also ibid., 29 June 1960, p. 4; 17 August 1960, p. 15; 24 August 1960, p. 13; 31 August 1960, pp. 7, 62; 16 May 1962, pp. 3, 11; 30 May 1962, pp. 3, 16; *Hollywood Reporter*, 5 November 1957, p. 3; 10 January 1958, p. 1; Behlmer, *Memo from David O. Selznick*, pp. 423–43; Bosley Crowther, 'New York's Movie Showcase: 2-Year Exercise in Frustration', *New York Times*, 7 July 1964.

Chapter 21

John Belton

GLORIOUS TECHNICOLOR, BREATHTAKING CINEMASCOPE AND STEREOPHONIC SOUND

THE FIRST CINERAMA FEATURE, *THIS IS CINERAMA* (1952), begins with a roller coaster ride. Filmed with the Cinerama camera mounted on the front of an actual Coney Island roller coaster, the sequence effectively lilts the audience out of their seats in the movie theater and plunges them headlong into the thrill-packed world of the amusement park. The opening credits of *The Robe* (1953), the first CinemaScope film, appear over an enormous red velvet theater curtain, which slowly parts to reveal an ever-expanding wide-screen vista of ancient Rome. Here, the customary space of the movie theater, symbolized by the curtain, opens up, breaking down the narrow confines of the standard motion picture screen. The almost-square image of earlier cinema gives way to the panoramic space of spectacle, into which the audience is slowly drawn.

Both Cinerama and CinemaScope directly engage audiences in a new kind of motion picture experience, an experience characterized, in part, by a rejection (in the case of Cinerama) or a redefinition (in the case of CinemaScope), of the traditional pre-wide-screen motion picture experience. Before Cinerama and CinemaScope, the movies entertained their audiences, holding them entranced in front of a screen whose dimensions averaged 20 feet, by 16 feet. With Cinerama and CinemaScope, the movies *engulfed* their audiences, wrapping images as great as 64 by 24 feet around them. At the same time, the movies began to engage their audiences more actively, creating for them a compelling illusion of participation in the action depicted on the screen, an illusion that proved far greater than that achieved by previous, narrower exhibition formats. With *This Is Cinerama*, going to the movies became similar to going to an amusement park, a theme park like Disneyland (which opened in July 1955 but had been in preparation for over three years), or a World's Fair. With *The Robe*, the audience was invited to cross over the barriers of the footlights and pass beyond the theater curtain, which separate the spectacle from the spectator, and to "enter" into the space of the spectacle.

From entertainment to recreation

The wide-screen revolution that took place during the early 1950s clearly involved more than a change in the size and shape of motion picture theater screens. It represented a dramatic shift in the film industry's notion of the product that it was supplying to the public. During this period, the industry redefined, in part, what a motion picture was or should be, shifting its primary function of providing entertainment to the public to include another function as well—that of recreation. Motion picture audiences were no longer conceived of as passive spectators but as active participants in a film experience—at least, this is how the industry described the new relationship it sought to establish with its patrons.

Though movies continued to be shown in the same theaters that they had played in for decades and, for the most part, the same sorts of stories were being told on the screen, there was something decidedly different about movie-going in the early 1950s. In contrast to the routine distribution treatment of typical motion picture productions of the past, wide-screen films were treated as special theatrical events. Road show exhibition—launched with *Queen Elizabeth* (1912), *Quo Vadis?* (1913), and *The Birth of a Nation* (1915), and periodically revived for certain one-of-a-kind, highly touted releases—became the dominant release pattern for every Cinerama and Todd-AO film (roadshowing, as a regular practice, however, was not used in the case of CinemaScope). In other words, virtually every wide-screen film was treated as a potential blockbuster, booked in exclusive engagements into a small number of large, first-run theaters.

The theaters, especially the large, first-run, downtown movie palaces, underwent an expensive facelift; old prosceniums were torn down and replaced with wall-to-wall curved screens. These old movie palaces finally found, with wide-screen films, an image size commensurate with the overall proportions of their auditoriums. With the installation of multi-track stereo sound systems, theaters not only looked but also sounded different than they had in the past. And, if the same old stories were still being told, there were now fewer of them being told and they were being told more expensively and at greater length. Production budgets increased to accommodate filming in color, wide-screen, and stereo sound. The greater the production values, the industry reasoned, the greater the potential profits. Or, as Cole Porter noted in 1954, "if you want to get the crowds to come around, you've got to have glorious Technicolor, breathtaking CinemaScope, and stereophonic sound."[1]

The reasons for change were obvious; box office statistics, detailing a sharp decline in average weekly admissions from a postwar high of 90 million in 1948 to an all-time low of 46 million in 1953, told much of the story.[2] Behind the statistics lay a dramatic change in the postwar U.S. economy, which the motion picture industry suddenly discovered it had to address. During the 1930s and 1940s (except for the war years), the average workweek had dropped from 60 to 40 hours.[3] At the same time, disposable personal income had doubled between 1940 and 1949, soaring from just under $80 billion to $190 billion.[4] Given greater leisure time and greater disposable income, more and more money was spent for recreation; from just under $4 billion in 1940, recreational expenditure rose to $10 billion in 1948.[5] With more money to spend on recreation and more time to spend it in, Americans could now afford more expensive forms of recreation than they had in the past. They could afford to spend more on entertainment than the relatively inexpensive cost of a movie ticket. They also had not only more time but larger blocks of it. A brief evening at the movies gave way to a day (or even a weekend) of golfing, fishing, or gardening.

In other words, the redefinition of the motion picture as a form of participatory recreation took place against the background of its competition for audiences with other leisure-time activities. The motion picture sought a middle ground between the notion of passive consumption associated with at-home television viewing and that of active participation involved in outdoor recreational activities. The model chosen by Hollywood for purposes of redefining its product, however, was less that of the amusement park, which retained certain vulgar associations as a cheap form of mass entertainment, than that of the legitimate theater.[6] The theater emerged as the one traditional form of entertainment that possessed the participatory effect of recreational activities and, at the same time, retained a strong identification with the narrative tradition within which the cinema remained steadfastly rooted. In aligning itself with the experimental quality of the theater, wide-screen cinema sought to erase the long-standing distinction between spectatorship in the theater and in the cinema, which viewed the former as active and the latter as passive.

Advertisements for CinemaScope films even told the audiences that they would be drawn into the space of the picture much as if they were attending a play in the theater. Publicists compared the panoramic CinemaScope image to the oblong "skene" (or stage) of the ancient Greek theater and insisted that "due to the immensity of the screen, few entire scenes can be taken in at a glance, enabling the spectator to view them . . . as one would watch a play where actors are working from opposite ends of the stage," the wide-screen format enhancing the sensation of the actors' presence and giving spectators the illusion that they could reach out and touch the performers.[7]

The wide-screen revolution proved to be more of a revolution in the nature of the movie-going experience than it was in the narrative content of motion picture films. Shooting in large-screen formats transformed the psychology of both filmmaking and film viewing. Wide-screen filmmakers treated every project as if it were a special, big-budget production. Fewer films were being made (the total dropping from 391 domestic releases in 1951 to 253 in 1954) and greater care was lavished on them in production (with negative costs rising from $1 million to $2 million during this same period). The new formats provided even fairly conventional stories with an appeal, as film spectacle, that the narrow 1.33:1 format could not exploit as fully or as successfully as the panoramic vistas of CinemaScope, VistaVision, and Todd-AO did. Film audiences, as well as filmmakers, fell under the spell of the new era of the wide-screen blockbuster, regarding films made, in any of the new wide-screen formats as special events well worth the increased admission price that first-run exhibitors charged to see these pictures on a big screen and to hear them in stereo sound. As the average weekly attendance declined, audiences became more and more selective in terms of what they chose to see, demanding more bang for their buck—a demand that the over-produced, wide-screen blockbusters sought to satisfy.

While Hollywood insisted that bigger was also better, audiences, though not necessarily equating bigger with better, definitely perceived bigger as, at the very least, different. Though the industry continued to tell the same old familiar stories, the larger canvas on which they were presented transformed old into new. The new format obscured the fact that wide-screen extravaganzas retained a good deal in common, especially in terms of the sorts of subjects deemed suitable for spectacularization, with prewide-screen spectacles. Biblical epics, for example, remain a staple of prewide-screen and wide-screen cinema. *Samson and Delilah* (1949), *Quo Vadis* (1951), *David and Bathsheba* (1951) were answered by *The Robe* (CinemaScope, 1953), *The Ten Commandments* (VistaVision 1956), and *Ben-Hur* (MGM Camera 65, 1959). Wide-screen cinema also drew heavily on other traditional genres that rely upon spectacle, such as the period costumer and the musical. Even the low-budget, nonspectacular genre of the Western was called upon to provide story material, which, through big-screen treatment, helped transform that genre, capitalizing upon its postwar renaissance as adult movie fare and solidifying its elevation from "B" to "A" picture status.

Screenwriters joked that they would have to learn to write differently for the new screen dimensions, adapting their craft to fit the new format. Fox's Nunnally Johnson reportedly quipped that now he would have "to put the paper in the typewriter sideways when writing for CinemaScope."[8] But, in fact, the wide-screen revolution had little or no immediate impact upon the craft of the screenwriter. Indeed, all of the scripts for Twentieth Century-Fox's first batch of CinemaScope spectacles were completed long before the studio acquired the CinemaScope process. What distinguished *The Robe* from Fox's earlier *David and Bathsheba* was not so much story or cast but presentation and perception. The wide-screen format produced a radically different perception of generically similar material: biblical spectacle remained biblical spectacle, but became, for the spectator, markedly more spectacular.

Cinerama: a new era in cinema

The wide-screen revolution began in earnest with the opening of *This Is Cinerama* at the Broadway Theatre in New York City on September 30, 1952. Cinerama had been developed independently, outside traditional industry channels, by a jack-of-all-trades inventor, Fred Waller, who had earlier invented water skis, a still camera that took a 360-degree picture and a PhotoMetric camera that could measure a man for a suit of clothes in a fiftieth of a second.[9] Waller perfected the Cinerama camera and projection system in the late 1940s, basing it, in part, on his prior invention of the multiple-camera/projector Vitarama system, which was developed for (but never actually exhibited at) the New York World's Fair in 1939, and on the five-camera, five-screen Flexible Gunnery Trainer, which was used by the Air Force in World War II to train gunners. Though Waller repeatedly tried to interest the majors in his new process, inviting studio heads to demonstrations staged at a converted tennis court on Long Island, the industry rejected Cinerama because of the expense involved in its installation, which necessarily excluded its widespread deployment in the nation's movie theaters, and because of its technical flaws, which ranged from the visibility of the joints or seams where its multiple images overlapped to the optical distortion produced by its wide-angle lenses. To some extent, subsequent history vindicated the reservations of the majors. At the height of its success, Cinerama, which cost from $75,000 to $140,000 to install, played in only a handful of theaters—there were only 22 Cinerama sites in 1959.[10] By comparison, CinemaScope, which cost from $5,000 to $25,000 to install, was available in over 41,000 theaters worldwide by the end of 1956. However, although *This Is Cinerama* (which cost only $1 million to produce) played, on its initial run, in only 17 theaters, it earned over $20 million, becoming the third largest grossing picture of all time.[11] Subsequent reissues of *This Is Cinerama* increased its total earnings to over $32 million, and though the first five Cinerama features could be seen in only a handful of theaters, they grossed more than $82 million.[12]

Cinerama discovered what the industry had not yet realized—there was a growing market for a new kind of motion picture entertainment, and, given the sort of specialized distribution and exhibition that the industry as a whole, geared for traditional mass market practices, was not capable of providing, this new product could return substantial profits.

Cinerama was a wide-screen process involving the use of three interlocked cameras (rather than one), which recorded an extremely wide field of view on three separate strips of 35mm film, and which were, in turn, projected by three projectors (in three separate booths) onto a 25-foot high by 51-foot wide, deeply curved screen, resulting in an aspect ratio of roughly 2.6 to 2.77:1. Stereo sound was recorded magnetically on seven separate channels (on a sound system pioneered by Hazard Reeves), which were combined on a separate strip of 35mm film and played back in the theater on five speakers behind the screen and two surround speakers. In collaboration with the technical talents of Waller and Reeves were the creative contributions of radio commentator and globe-trotting journalist Lowell Thomas, Hollywood producer Merian C. Cooper, and Broadway showman Mike Todd, who collectively produced *This Is Cinerama*.

Cinerama as event

Cinerama transformed the nature of the movie-going experience, which had gradually become unconsciously automatic, habitual, and routinized over the years. Cinerama defamiliarized the cinema, restoring to it an affective power that, as pure experience, it had lost long ago. In his review of *This Is Cinerama*, *New York Times* critic Bosley Crowther compared Cinerama to the first large-screen projection of motion picture at Koster and Bial's Music Hall in 1896, when shots of sea waves rolling in on a beach supposedly sent naive viewers running for cover. Cinerama recaptured the *experience* of those early films. During Cinerama's aerial sequences, audiences are said to have "leaned sharply in their seats to compensate for the steep banks of the airplane. Others became nauseated by their vicarious ride in a roller-coaster. Still others ducked to avoid the spray when a motor boat cut across the path of the canoe in which they were apparently riding."[13] Even veteran airmen reacted to Cinerama's illusion of reality. Lowell Thomas reports that war ace Gen. James Doolittle clutched his chair when stunt pilot Paul Mantz flew through the Grand Canyon. Drugstores near the Cinerama theater in New York did a land-office business during intermission, selling Dramamine to spectators who either became airsick in the first half or wanted to prepare themselves for the film's finale.[14]

Like early cinema, Cinerama was more than cinema; it was a special phenomenon. People went to it the way they went bowling or golfing. Unlike conventional narrative cinema, Cinerama was not so much "consumed" as it was "experienced." Cinerama was as much "recreation" as "entertainment." As a social phenomenon, it both reflected and generated popular culture as a whole; in fact, it serves as a remarkable index of postwar leisure-time activities. For example, Cinerama's travelogue format not only takes advantage of the increased interest by Americans in domestic sightseeing and travel abroad but functions itself to stimulate tourism.

> Almost immediately after the New York premiere, transcontinental airlines began to report requests for flights that "go over those canyons" seen in Cinerama. The British Travel Bureau was flooded with inquiries about the date of the Rally of Pipers, which is featured in the film.... Closer to home, the roller coaster at Rockaways' Playland, which provides the first sensation in the film, did a record-breaking business all last winter, when amusement parks are generally deserted.... Cypress Gardens in Florida ... reported that the number of visitors jumped to 40 per cent after Cinerama was released.[15]

Like the sights seen within its films, Cinerama itself became a tourist attraction, becoming something that visitors from out of town had to see.

Cinerama advertising capitalized on its own status as one of the seven wonders of the modern world, suggesting to the Midwestern populace that "when you visit Detroit, Cinerama is a must stop." One Texas columnist even commented that "it was particularly worth a trip to New York to see Cinerama."[16] Going to see Cinerama became a special event in itself, like a trip to the World's Fair. Cinerama theaters were situated in cities with the largest populations, and the theaters themselves were frequently the largest theaters in those cities. Unlike the rest of the film industry, which distributed films as widely and in as many theaters as possible, Cinerama was not an everyday experience for its audiences—only eight films were ever made in the three-strip Cinerama process, each playing only in Cinerama theaters for runs that (like those of successful Broadway plays) lasted from one to two years. Since only a few spectators could enjoy these films at any one time, there was tremendous demand for tickets, which were sold on a reserved-seat basis weeks, and even months, in advance.

Like the first films, Cinerama presented "life as it is," rather than narratives. The initial advertising campaign for *This Is Cinerama*, devised by the McCann-Erickson agency, promoted the film as an incomparable motion picture experience. Unlike 3-D and CinemaScope, which stressed the dramatic content of their story material and the radical new means of technology employed in production, Cinerama used a saturation advertising campaign in the newspapers and on radio to promote the "excitement aspects" of the new medium. Cinerama sold itself through appeals to neither content nor form but to audience involvement. Though it failed to duplicate the experience of the theater in terms of telling a story, Cinerama did attempt to mimic the theater in terms of its sense of participation. Ad copy promised that with Cinerama, "you won't be gazing at a movie screen—you'll find yourself swept right *into* the picture, surrounded with sight and sound."[17] Publicity photos literalized this promise, superimposing images of delighted spectators in their theater seats onto scenes from the film. And while 3-D slowly alienated its audiences by throwing things at them, Cinerama drew them into the screen—and the movie theater—in droves. Despite the absence of story or characters, Cinerama enthralled its customers.

The first five Cinerama features were travelogues.[18] Though Cinerama distinguished itself from traditional Hollywood fare by choosing the travelogue as its major format, its basic technology gave it little choice in the matter. Cinerama initially had difficulty telling stories. Traditional close-ups were impossible, given its lens technology. Cinerama relied on three 27mm lenses, mounted at angles to one another. The use of extreme wide-angle lenses resulted in noticeable distortion, which was magnified in close-up cinematography. Unlike the single focal length required by Cinerama, traditional motion pictures employ a variety of different lenses. The optical restriction in the focal length of Cinerama lenses further reduced the medium's options, denying it what Bosley Crowther referred to as the basis of narrative cinema—"flexibility of the images and the facility of varying the shots."[19] Without close-up, which has become the linchpin of film narrative, the medium was reduced to the nonnarrative, presentational techniques of early cinema and virtually forced to choose the travelogue as the format for its first five features. Ten years after the opening of *This Is Cinerama*, the Cinerama system was used by MGM to film two narrative features, *How the West Was Won* (1963) and *The Wonderful World of the Brothers Grimm* (1962), but until that time Cinerama found the travelogue to be the best format for exploiting the Cinerama effect. Unlike traditional cinema, Cinerama established itself as pure experience, unadulterated by the Original Sin of storytelling. In discarding narrative, Cinerama, during the first ten years of its independence from traditional Hollywood production, distribution, and exhibition practices, cut itself off from mainstream narrative cinema and became itself something of a non-traditional, leisure-time activity.

Ersatz wide-screen

The success of Cinerama changed the level of expectations for film audiences; from Cinerama on, motion pictures had to be at least projected in, if not made in, wide-screen. Its success made obsolete traditional film formats (that is, the 1.37:1 aspect ratio, which was now identified with television). The advent of Cinerama, which threatened to "dilute the value of millions of dollars worth of 'flat' films in circulation and in Hollywood vaults," startled the majors into frenetic activity.[20]

The immediate response on the part of a number of studios was to produce an ersatz wide-screen image by cropping films shot in the standard 1.37:1 format to an aspect ratio of from 1.66:1 to 1.85:1 and enlarging the image on the screen by using a wide-angle projection lens. In the spring of 1953, Paramount released *Shane*, suggesting that it be projected 1.66:1.[21] Shortly thereafter, Universal promoted *Thunder Bay* as a wide-screen film, encouraging exhibitors to project it on a slightly curved screen, which was an obvious imitation of the Cinerama screen.[22] *Time* magazine, noting that cropping provided only a bargain-basement solution to the challenge posed by Cinerama, dryly observed that "the wide-screen revolution was looking more and more like an inventory sale."[23]

What distinguished Cinerama from these ersatz wide-screen processes was Cinerama's radically new perception of events based not only on a deeply curved screen that wrapped around the audience, engulfing it, but also on its dramatically increased angle of view, which multiplied the participation effect suggested by the curved screen, creating an illusion not only of engulfment but of depth as well.

According to Waller, Cinerama owes its sense of audience participation to the phenomenon of peripheral vision. Waller's own experiments with depth perception led him to the conclusion that the successful illusion of three-dimensionality derived as much from peripheral as from binocular vision. The three lenses of the Cinerama camera, set at angles of 48 degrees to one another, encompass a composite angle of view of 146 degrees by 55 degrees, nearly approximating the angle of view of human vision, which is 165 degrees by 60 degrees. When projected on a deeply curved screen, this view tends to envelop the spectator sitting in the center of the theater. Stereo sound, broadcast from five speakers behind the screen and from one to three surround horns, reinforces the illusion of three-dimensionality.

In order to duplicate the effect of Cinerama, rival systems had to duplicate its angle of view as well as its peripheral field. Cinerama's first and most successful rival, CinemaScope, played, in the largest theaters, on slightly curved screens as large as 64 feet wide by 24 feet high; in similar sites, VistaVision, developed by Paramount in early 1954, could reach a screen dimension of 55 by 27 feet, while Todd-AO, the last successful, Cinerama-inspired, wide-screen system to reach the theaters (in 1955), filled a deeply curved 60-by-25-foot screen. Unlike the ersatz wide-screen "processes," CinemaScope used an optical system that actually doubled the angle of view of traditional lenses; VistaVision and Todd-AO both achieved an improvement in angle of view by using wide film, necessitating a wider camera aperture, which automatically increased angle of view. As a result of both this and the use of wide-angle lenses, VistaVision achieved a horizontal angle of view roughly one and two-thirds that of standard lenses. Todd-AO, led by optics expert Dr. Brian O'Brien, developed a bug-eye lens, possessing an angle of view of 128 degrees, bringing it remarkably close to Cinerama's 146 degrees, but Todd-AO, inspired by Mike Todd's experiences filming the European footage for *This Is Cinerama*, achieved this angle of view using only one lens rather than three.

CinemaScope: a poor man's Cinerama

The CinemaScope process employed cylindrical lenses that take in an angle of view twice that of normal lenses and that anamorphically distort the image in the horizontal plane, squeezing it (by a compression factor of 2) in order to fit it onto a single strip of 35mm film. In projection, an anamorphic lens unsqueezes the compressed image, expanding it to an aspect ratio of 2.66:1 (which was subsequently reduced to 2.55:1, then 2.35:1 when soundtracks were added to the film strip). Stereo sound was recorded and played back on four magnetic tracks that, shortly after the opening of *The Robe* in September 1953, were fixed to the same strip of 35mm film that bore the anamorphic image. In effect, CinemaScope provided the industry with a simplified version of Cinerama, approximating its 2.77:1 aspect ratio and its seven-track stereo sound system but using one strip of 35mm film (instead of Cinerama's four) to do so.

CinemaScope optics are based on the "Hypergonar" anamorphic lens invented in the late 1920s by French scientist Henri Chretien. Engineers at Twentieth Century-Fox, the studio that innovated CinemaScope, adapted Chretien's invention to suit the demands of the contemporary motion picture production and exhibition marketplace. Chretien's lens permitted Fox to obtain a wide-screen image without the expense of wide-film or multifilm production and exhibition processes, thus maintaining the 35mm standard observed by the industry as a whole. In this way, Fox sought to satisfy the demands of the industry for innovative formats while keeping to a minimum the expenditure for new production and exhibition technology.[24] In other words, unlike Cinerama, which sought to exploit a new, specialized market for wide-screen entertainment, CinemaScope offered itself as a compromise between traditional, mass-market production and exhibition practices and the new redefinition of the motion picture experience introduced by Cinerama.

Unlike the early Cinerama features, which foregrounded the participation effect through the use of attention-grabbing, forward-tracking movements with the camera mounted on the front end of a roller coaster, CinemaScope feature films sought to achieve a balance between narrative and spectacle, using the format to expand the range of the narrative. As Fox executive Darryl Zanuck suggested, implicitly comparing CinemaScope with rival systems (like Cinerama) that indulged in gimmickry, "CinemaScope is not a substitute for a story nor for good actors. But it does give us more octaves in which to tell our story and gives our actors greater range for their talents."[25] Unlike Cinerama, which was left free by the travelogue format to exploit the effect of engulfment and to present itself as pure spectacle, CinemaScope's participation effect remained regulated by the conventions of narrative, which sought to hold the wide-screen format's spectacular qualities in check. CinemaScope enhanced the movie-going experience but carefully avoided foregrounding itself as a process, as did Cinema and 3-D.

Mise-en-scene in CinemaScope

In *The Robe*, Cinemascope expands the scope of the drama, intensifying the emotional mood or atmosphere of individual scenes while also—somewhat paradoxically—effacing itself before the demands of the narrative. It spectacularizes action while, at the same time, it naturalizes the spectacle. For example, when Diana (Jean Simmons) bids Marcellus (Richard Burton) farewell as he departs for the Near East, the space between them gradually expands as his boat pulls away, until each stands at the opposite end of the CinemaScope frame—the spatial dynamics echoing the emotional drama of their separation. Their moment of separation has been drawn out, expanded in both time and space, yet the camera movement and wide-screen framing in no way call attention to themselves as effects. The sequence exists as an experience unique to CinemaScope, yet that experience is not foregrounded as an experience of technology or technique; it has been regulated, integrated into the fabric of the film through its narrativization.

The Robe, though celebrated for introducing a new quasi-theatrical style of filmmaking that permitted filming in long takes and staging action in a way that eliminated the need for shot/reverse shot and/or cause/effect editing, remains something of a conventional film. Cinematographer Leon Shamroy, referring to a scene in which a Roman archer on extreme screen left hits a leader of the Christians (played by Dean Jagger) with an arrow on extreme screen right, proclaimed that "no longer must the cinematographer cut from one bit of action, showing cause, to another bit of action showing effect. In one big scene, the CinemaScope camera shows both."[26] But moments later, in the same scene, when the hero, Marcellus (Richard Burton), challenges a Roman centurion to do battle, the film intercuts their hand-to-hand sword fight with reaction shots of those looking on. Though the CinemaScope *River of No Return* (1954) composes Marilyn Monroe's gold-camp rendition of a song to include Robert Mitchum's intransigent response (within the same frame) as he circles the barroom in search of his son, the film regularly peppers Monroe's songs with reaction shots of groups of beer-guzzling miners lasciviously eyeing her performance. Though CinemaScope's expansive width offered filmmakers new options for staging sequences and audiences new perspectives for viewing them it relied, at times, on fairly conventional techniques for telling its stories.

Unlike Cinerama, which engaged its viewers in the experience provided by the new technology, quite often grabbing them by the scruff of the neck and dragging them into it, CinemaScope explored that new experience in familiar ways, often relying upon conventions of the theater to achieve its effect. CinemaScope, rather than exploiting the Cinerama-effect achieved by camera movement into the screen, relied more often upon character movement than camera movement for its dramatic impact. Aware of the way in which excessive camera movement in CinemaScope called attention to itself, Zanuck cautioned Fox directors to move the actors more and the camera less.[27]

In this respect, CinemaScope mise-en-scene looked back to that found in the theater, but it did so in a way that introduced greater complexity into the traditional approach to motion picture composition. As a result, the theatrical nature of the staging of events in CinemaScope was not a reversion to the filmed theater of the *film d'art*, in which the camera merely recorded a performance designed for the stage. Rather, CinemaScope marked a dramatic step forward in the evolution of *cinematic* mise-en-scene. The introduction of deep-focus cinematography in the late 1930s and early 1940s "converted the screen into a dramatic checkerboard," as André Bazin once put it.[28] But the narrow 1.37:1 frame limited the development of that checkerboard to staging in depth, which reached its limit in the work of Welles and Wyler, or to staging spread successively across a surface (through reframing and lateral camera movement), a practice that culminated in Hitchcock's *Rope.* CinemaScope permitted a staging in width, in which adjacent actions could be seen simultaneously rather than successively. With the wide-screen format, filmmakers could stage action in width as well as in depth, doubling their options for compositional strategies and enabling mise-en-scene to evolve as a means of expression on two different axes. Given the larger canvas of wide-screen, filmmakers found new ways of staging action and of directing the spectator's attention through it. Though they continued to rely on camera movement and upon analytical editing to "read" the action, they also discovered ways to articulate space by working within the expanded borders of the frame. Filmmakers were forced to rethink their craft from the different vantage point of the wide-screen frame, and audiences were forced to adjust to conditions of spectatorship in which traditional ways of reading the image gave way to new, less-familiar practices. CinemaScope, like other wide-screen techniques adopted by narrative cinema, emerged as an amalgam of new and old, combining the new methods of organizing compositions made possible by the wide-screen with the old practices of conventional storytelling. The advent of the wide-screen format disturbed, if only momentarily, the smooth surface of classical cinema's equilibrium profile, causing the whole system to change slightly in order to readjust itself.

VistaVision

In contrast, Paramount's VistaVision, with its 1.66:1 aspect ratio, marked less of a departure from prewide-screen cinema than Cinerama or CinemaScope did. Its chief virtues, according to Paramount president Barney Balaban, lay in its "compatibility" with traditional modes of production and exhibition and in its "flexibility"—that is, in its adaptability to a variety of different theater situations.[29] The VistaVision camera used 35mm film but exposed it horizontally (like 35mm still camera film), producing an image that occupied the space of two traditional 35mm frames (that is, eight perforations in width). This two-frame exposure generated a wide-area negative without the use of wide film. When this large image was rotated 90 degrees and reduced to standard 35mm, it produced an image that possessed tremendous sharpness and depth of field. VistaVision thus solved many of the problems inherent in ersatz wide-screen, providing greater image resolution and a better angle of view for wide-screen projection, without committing itself to the radical redefinition called for by the more extreme wide-screen processes cited above. The conservative nature of VistaVision, in contrast to the radical difference of Cinerama, might be understood in terms of a major studio's—Paramount's—conformity to traditional production and exhibition practices versus the non-Hollywood independent's—Cinerama's—deliberate violation of those same practices.

Paramount publicly criticized Fox for the excessiveness of the CinemaScope aspect ratio, which they argued produced a wider image than any theaters could possibly accommodate. They also refused, as a matter of principle, to pay a rival studio for the use of its technology (especially since that technology was, as Fox freely admitted, in the public domain), thus becoming the only major studio to refuse to make films in CinemaScope. Paramount saw VistaVision less as a wide-screen process than as a big-screen process. The VistaVision image could be blown up to fill an enormous 62-by-35-foot screen without the loss in clarity or sharpness that plagued ersatz wide-screen (and without the blurring or fuzziness that occasionally appeared at the extreme edges of CinemaScope films). As one Paramount executive put it, "height was equally important as width."[30] Knowing that they could not compare with CinemaScope in terms of width, Paramount sought to capitalize on the height of the VistaVision image, and the first few VistaVision films, such as *White Christmas* and *Strategic Air Command*, did not so much stress width in their compositions as height.[31] While actors in CinemaScope films reclined (like Marilyn Monroe in *How to Marry a Millionaire* [1953]) or sprawled across the full width of the frame (like Marilyn Dean in the credit sequence of *Rebel Without a Cause* [1955]), actors in VistaVision films were shown erect, their full figures visible within the frame (like Bing Crosby and Danny Kaye in *White Christmas*).

Though Paramount converted its own production schedule to VistaVision, releasing over 65 films in that format during the next ten years, only a handful of VistaVision films were ever made by other studios (though several major works by major directors, including John Ford's *The Searchers* [Warner Bros., 1956], and Alfred Hitchcock's *North by Northwest* [MGM, 1959], number among them). Fox, on the other hand, quickly succeeded in a matter of months in selling CinemaScope to the rest of the industry, licensing it in 1953 to MGM (March), United Artists (June), Disney (June), Columbia (October), and Warners (November). In February, Fox had announced that all of its future films (which amounted to roughly 15 to 18 pictures a year) would be made in CinemaScope. These films, together with the CinemaScope productions of the other studios, assured theaters a steady supply of CinemaScope product, encouraging them to convert to CinemaScope. Indeed, more than six hundred CinemaScope features were released over the next 12 years—not to mention the more than two thousand anamorphic films made in compatible processes (such as Panavision, Techniscope, Tohoscope, Dyaliscope), which have been and continue to be released since 1953.[32]

Todd-AO

If CinemaScope could be seen as an answer to Cinerama, it was only a partial answer; to its critics, it was only a poor man's Cinerama. Todd-AO, introduced with the premiere of *Oklahoma!* at the Rivoli Theatre in New York on October 10, 1955, answered Cinerama in full, even surpassing it in several respects. Like Cinerama, Todd-AO was developed by independents working outside of the film industry. Mike Todd, dissatisfied with the cumbersome Cinerama process that he had worked with in producing the European sequences of *This Is Cinerama!* secured the assistance of Dr. Brian O'Brien, formerly professor of optics at the University of

Rochester and currently serving as vice-president of research and development at the American Optical Company, in designing a single lens that could duplicate the combined angle of view of Cinerama's three 27mm lenses. With Todd-AO, Todd sought to re-create the experience of Cinerama for spectators in the theater by using a single film and projector rather than the three strips of film and projectors required by Cinerama. As Todd himself put it, he wanted a projection system "where everything comes out of one hole."[33]

Todd-AO emerged—in comparison with the industry's best wide-screen system, CinemaScope—as a process associated with product that was even more exclusive, since it was used, as was Cinerama, to make only one lavishly produced film a year instead of a studio's entire output for the year, as was the case with CinemaScope at Fox. In fact, Todd-AO was used in the production of only 15 films over a period of 16 years.[34] Todd-AO not only duplicated the "Cinerama effect," but it did so at about half the cost. Todd-AO's 128 degree, "bug-eye" lens provided an angle of view that closely approximated that of Cinerama, which was 146 degrees. The opening shots of Passepartout (Cantinflas) bicycling through the narrow streets of London in *Around the World in 80 Days*, for example, use this lens to exploit the "Cinerama effect," engulfing the spectator within the tight confines of the stone walls of buildings on either side and plunging the viewer into the midst of the diverse movements of surrounding traffic.

American Optical also designed three other lenses for the Todd-AO system, whose angles of view were 64, 48, and 37 degrees, thus introducing a greater optical variety into the system than Cinerama possessed and permitting the full range of expression traditionally found in 35mm production, which relied on a variety of lenses each possessing different focal lengths. This facilitated producers' use of the system for narrative films. Indeed, the first seven Todd-AO features—*Oklahoma!*, *Around the World in 80 Days*, *South Pacific*, *Porgy and Bess*, *The Alamo*, *Can-Can*, and *Cleopatra*—fared quite well in competition with other narrative films of the period, winning a total of 11 Academy Awards.

Todd-AO's film speed of 30 frames per second matched and surpassed Cinerama's 26 frames per second, providing a more detailed and more steady image in projection. The higher film speed and the use of a larger filming and projection aperture produced an image that was not only sharper but brighter. Brian O'Brien anecdotally recalls that the woman with whom he attended the premiere of *Oklahoma!* put sunglasses on to watch the picture because the image was so bright.[35]

Similarly, Todd-AO's six-track stereo magnetic sound system closely approximated Cinerama's seven-track stereo system, and its use of 65mm film and of a five-perforation frame resulted in an image quality that approached the three-strip, six-perforation frame of Cinerama. Robert Surtees, director of photography on *Oklahoma!*, argued that Todd-AO's use of a 65mm negative made it, "from the standpoint of optics alone, a superior picture process" to all other 35mm wide-screen processes.[36]

Cinerama introduced the curved screen—which served to amplify the sense of audience participation by surrounding spectators with the image—to motion picture exhibition. However, deeply curved screens also introduced distortion into the image. Anytime that a flat image is projected upon a curved screen the horizontal lines of the image are necessarily distorted. CinemaScope minimized distortion by opting for a slightly curved screen—a screen that curved at the approximate rate of 1 inch per foot, resulting, in the case of a 62-foot-wide screen, in a 5-foot curve. Todd-AO used a screen that curved to a depth of 13 feet over a width of 52 feet, compensating for distortion by means of an "optical correcting printing process."[37]

Like Cinerama, Todd-AO was conceived of as a specialized form of entertainment that would be restricted to a finite number of theaters. Like Cinerama, Todd-AO was based on the notion of the creation of a supercircuit in theatrical exhibition—that is, on the existence of a small number of large first-run urban theaters that could afford the forty thousand dollar conversion cost and that would be rewarded, in turn, with an exclusive product that would be guaranteed, as a result of the 70mm, multitrack stereo technology, not to play in non-Todd-AO houses until the first-run market had been virtually exhausted. In this way, Todd-AO reinstituted, in a legal way, the system of runs and clearances that had been outlawed as a result of the 1948 Consent Decree (also known as the *Paramount* Case). Indeed, *Oklahoma!* ran for over a year in Todd-AO in several cities, rivaling the two-year run of *This Is Cinerama*, before it was released in its 35mm CinemaScope format. Unlike Cinerama, however, Todd-AO succeeded in converting a good number of the best theaters in the country to show their product, securing more than 60 new installations in its first two years of operation.[38] Todd's partnership with the United Artists Theater Circuit served to assure the success of the process by providing him with a built-in market (of over 100 theaters located in 50 major cities) for Todd-AO productions.

Again, like Cinerama, Todd-AO exhibition relied on a "theatrical," roadshow pattern, scheduling only two or three screenings a day and making tickets available on an advance sale, reserved-seat basis. Todd's copartners and creators of *Oklahoma!* for the Broadway stage, Richard Rodgers and Oscar Hammerstein, insisted, as well, that the integrity of the original stage musical be respected and forbade the inclusion of any sensational "audience participation type of screen action," such as the roller coaster ride used in Cinerama.[39] This did not, however, prevent Todd and director Fred Zinnemann from indulging in a modified "Cinerama effect" when the camera dollies forward through an enveloping field of corn "as high as an elephant's eye" during the opening, "Oh, What a Beautiful Morning" song sequence or to introduce a roller coaster-like thrill with a sequence filmed on a runaway carriage. The "theatrical" quality of the presentation of Todd-AO even extended, in the case of *Oklahoma!*, to a ban on the sale of popcorn in the theater.[40]

Todd-AO drew on the theater in other ways as well. Todd, a Broadway producer, launched Todd-AO in collaboration with the Broadway playwright and composer team of Rodgers and Hammerstein, whose theatrical properties provided the basic dramatic material (*Oklahoma!*, *South Pacific*) to be filmed in the system. Even though *Around the World in Eighty Days* superficially resembles a Cinerama travelogue, it, too, has theatrical origins, deriving from Mike Todd's collaboration with Orson Welles on a Broadway version of the Jules Verne novel in the early 1940s. Todd-AO producers continued to ransack Broadway for material, adapting *Porgy and Bess*, *Can-Can*, and *The Sound of Music* for the screen and spearheading a trend in which producers began to draw more and more regularly upon Broadway for material for their big-budget, wide-screen spectacles.

As a result of the creation of a theatrical-style supercircuit, producers of Todd-AO (or other special-format films) gained a control over exhibition that recent government antitrust lawsuits had attempted to deny them. And since more than half of a film's total income during this period came from first-run exhibition in only a handful of theaters (that is, from approximately 700 first-run theaters in 92 cities), Todd-AO producers realized fantastic profits, as did the stockholders of Todd-AO and of Magna Theatres, who received a combined royalty of 5 percent of the admission price for each spectator who saw the film in Todd-AO but who received no royalty for the 35mm CinemaScope release.

The ultimate success of the Todd-AO process can be measured, in part, by its triumph over rival wide-film systems and its emergence as the premier system for filming big-budget blockbusters. Aware of the limitations of 35mm film as a format and of the potential challenge posed by Todd-AO, shortly after early demonstrations of Todd-AO, Fox began development of its own wide-film process, dubbing it CinemaScope 55. Though Todd had signed Rodgers and Hammerstein, two of their Broadway musicals, *Carousel* and *The King and I*, had already been optioned to Fox, giving the studio two extravaganzas to film in CinemaScope 55. Though Fox had been experimenting with wide-film processes for 25 years (since their 70mm Grandeur system in 1929), CinemaScope 55 emerged less as an innovative than as a derivative process—an apparent attempt to answer Todd-AO's success with *Oklahoma!*, which played in theaters a year before.

Fox shortly thereafter abandoned CinemaScope 55 and capitulated to Todd-AO. Fox even privately acknowledged that by 1958, when *Around the World in Eighty Days* was experiencing record profits at the box office, "CinemaScope had lost much of its novelty and that customers were no longer drawn to theatres solely because the picture was in the CinemaScope process."[41] In 1958, Fox made an investment of six hundred thousand dollars in the Todd-AO company, securing the right to make pictures in that process, and eventually, after Mike Todd's death in an airplane crash in 1958, secured control of the Todd-AO Corporation.

Casualties of the wide-screen wars

By the end of the 1950s, CinemaScope had been eclipsed as a special-event process by Todd-AO and other wide-film processes, such as Ultra Panavision (which was used to film single-strip Cinerama features) and Super Panavision 70, both of which provided sharper and brighter big-screen images. CinemaScope had, in effect, been destroyed by its own success. By the mid-1950s, CinemaScope had become a new motion picture standard; it was used in the production of all kinds of films, ranging from color spectacles to black-and-white Elvis Presley movies. Its initial technology had also been seriously compromised. The refusal of small, independent exhibitors to install stereo sound systems had forced Fox, in the spring of 1954, to announce that it would alter its stereo-only policy, making monaural prints available to theaters. Though first-run theaters continued to present CinemaScope films in four-track magnetic stereo sound, the majority of theaters did not, tarnishing somewhat the image of CinemaScope as a high-quality process seen (and heard) in only high-quality theaters. CinemaScope's widespread adoption weakened its initial association with exclusive, lavishly made, big-budget productions—which were now linked in the industry's and the public's minds with Todd-AO, MGM Camera 65, Technirama, Super Technirama 70, Ultra Panavision and Super Panavision 70, which continued to be seen in only a handful of first-run theaters and which continued to provide audiences with stereo sound. Even within the 35mm wide-screen market, CinemaScope had given way, as an anamorphic standard, to Panavision, which producers preferred for its superior close-up lenses.

During the 1960s, Cinerama joined the 70mm bandwagon. In 1963, for the filming of *It's a Mad, Mad, Mad, Mad World*, Cinerama abandoned its three-strip process for Ultra Panavision 70, a 70mm system that employed anamorphic optics to squeeze a wide angle of view onto a single film strip and that, when unsqueezed in projection, produced a Cinerama-like image with an aspect ratio of 2.76:1. Throughout the 1960s Cinerama continued to compete, though with varying degrees of success, with other wide-film systems, faring well with *Battle of the Bulge*, *Grand Prix*, and *2001: A Space Odyssey* and not so well with *The Greatest Story Ever Told* and *Ice Station Zebra*. With the collapse of the Cinerama theater chain (with its deeply curved screens) in the early 1970s, Cinerama became indistinguishable as a presentational process from other wide-film systems and disappeared (though its assets were not finally liquidated until 1978).

During the 1960s, many first-tier theaters around the country screened films exclusively in 70mm, a practice that continues today in the more than nine hundred theaters equipped to show 70mm films.[42] However, the motion picture marketplace has taken its toll on 70mm production as well. The expense of filming in 65/70mm has, since the early 1960s, resulted in the gradual elimination of original production in 70mm. In 1963, Panavision's Robert Gottschalk introduced an optical system that enabled producers to show 35mm film up to 70mm for purposes of theatrical, roadshow exhibition, virtually pulling the rug out from under wide-film production.[43] Now any 35mm film could be blown up to 70mm for roadshow presentation without the expense involved in wide-film production. Since then, less than 30 films have been shot on wide film and more than 230 have been shot in 35mm and blown up to 70mm for first-run exhibition.[44]

Notes

The chapter title is taken from the 1954 lyrics by Cole Porter for the "Stereophonic Sound" number in *Silk Stockings*. The complete text runs as follows: "Today to get the public to attend a picture show/It's not enough to advertise a famous star they know/If you want to get the crowds to come around/You've got to have/Glorious Technicolor/Breathtaking CinemaScope and/Stereophonic sound."

1 See note above.
2 Cobbett Steinberg, *Reel Facts: The Movie Book of Records* (New York: Random House, 1978), p. 371.
3 Paul Lazersfeld and Robert Merton, "Mass Communication, Popular Taste and Organized Social Action," in Mass *Culture: The Popular Arts in America*, ed. Bernard Rosenberg and David Manning White (Glencoe, Ill.; The Free Press, 1957), p. 460.
4 Leo A. Handel, *Hollywood Looks at Its Audience* (Urbana: University of Illinois Press, 1950), p. 97, chart 5.
5 Handel, *Hollywood Looks*, p. 97, chart 5.
6 Comparison of the participation effect produced by wide-screen cinema to the theatrical experience constitutes a major thread of the industry discourse surrounding CinemaScope, as James Spellerberg points out in "CinemaScope and Ideology," *The Velvet Light Trap*, no. 21 (Summer 1985): 30–31.
7 Spellerberg, "CinemaScope and Ideology," p. 30. See also Thomas Pryor, "Fox Films Embark on 3-Dimension Era," *New York Times*, February 1953.
8 *The Letters of Nunnally Johnson*, ed. Dorris Johnson and Ellen Leventhal (New York: Knopf, 1981), p. 103. Director Jean Negulescu also claims authorship for this wisecrack. See "New Medium—New Methods," in *New Screen Techniques*, ed. Martin Quigley (New York: Quigley Publishing, 1953), p. 174.
9 For more information on Waller and Cinerama, see my "Cinerama: A New Era in the Cinema," *The Perfect Vision* 1, no. 4 (Spring/Summer 1988): 78–90.
10 *Film Daily*, January 21, 1963, p. 26.
11 Michael Todd, Jr. and Susan McCarthy Todd, *A Valuable Property: The Life Story of Michael Todd* (New York: Arbor House, 1983), p. 244.
12 *Film Daily*, January 21, 1963.
13 *Fortnight*, October 13, 1952.
14 *Magazine Digest* (September 1953): 40
15 *New York Herald-Tribune*, September 27, 1953.
16 Clifford M. Sage in the *Dallas Times Herald*, November 4, 1952.
17 Lynn Farnol, "Finding Customers for a Product," in *New Screen Techniques*, ed. Martin Quigley, p. 143.
18 They were *This Is Cinerama* (1952), *Cinerama Holiday* (1955), *Seven Wonders of the World* (1956), *Search for Paradise* (1957), and *Cinerama South Seas Adventure* (1958). *Windjammer* (1958), filmed in the CineMiracle process (a three-camera system similar to Cinerama), was also a documentary and properly belongs on this list as well.
19 *New York Times*, October 5, 1952. As late as May 1960, *American Cinematographer* noted that "closeups of people shot with Cinerama cameras show typical wide-angle closeup distortion. Noses appear very long and foreheads slant back. Ears are very small in proportion and chins recede" (p. 304).

Early CinemaScope, which employed an anamorphic lens that was unable to focus clearly at distances under four feet from the camera, faced a somewhat different problem involving not distortion but sharpness of focus. However, though extreme close-ups remained difficult to achieve in CinemaScope, standard close-ups were easily accomplished.

Three-strip Cinerama was ultimately used to film narratives, such as *The Wonderful World of the Brothers Grimm* and *How the West Was Won*, but narrative information was conveyed in these films without the use of close-ups, relying on narrative techniques borrowed from the theater (and which CinemaScope had earlier drawn upon).

20 "Cinerama—The Broad Picture," *Fortune* (January 1953): 146.
21 *Daily Variety*, April 6, 1953. Two months later, they attempted to extend the life of their 1952 hit, *The Greatest Show on Earth*, by rereleasing it "in 1.66:1" (*Daily Variety*, June 11, 1953).
22 *Daily Variety*, May 20, 195, p. 1
23 *Time* (June 8, 1953), p. 70.
24 See my "CinemaScope: The Economics of Technology," *The Velvet Light Trap*, no. 21 (Summer 1985): 35–36.
25 Darryl Zanuck, "CinemaScope in Production," in *New Screen Techniques*, ed. Quigley, p. 156.
26 Leon Shamroy, "Fuming *The Robe*," *New Screen Techniques*, ed. Quigley, p. 178.
27 Zanuck memo, CinemaScope: General Information, August 12, 1954, Sponable Collection, Columbia University Libraries.
28 Andre Bazin, "The Evolution of the Language of Cinema," in *What Is Cinema?* Vol. 1, p. 34.
29 *Daily Variety*, March 3, 1954, p. 14.
30 Thomas Pryor, "Hollywood Expands," *New York Times*, March 7, 1954.
31 In the first public demonstration of VistaVision, Paramount compared it with CinemaScope, projecting CinemaScope at the same width as VistaVision, thus diminishing CinemaScope's wide-screen edge over VistaVision along with its apparent height. Fox, of course, objected to this manipulation. See Leonard Spinrad, *Motion Picture Newsletter*, May 3, 1954. 32.
32 See the list of anamorphic titles in Robert E. Carr and R. M. Hayes, *Wide Screen Movies: A History and Filmography of Wide Gauge Filmmaking* (Jefferson, N.C.: McFarland, 1988), pp. 91–143.
33 Todd, *Valuable Property*, p. 245.
34 Carr and Hayes, *Wide Screen Movies*, pp. 187–88.
35 Dr. Brian O'Brien, in telephone conversation with the author, November 1988.
36 Arthur Rowan, "Todd-AO—Newest Wide-screen System," *American Cinematographer* (October 1954): 495.
37 *Film Daily*, May 2, 1956.
38 Todd-AO/TCFF Corp. folder, Box 120, Sponable Collection, Columbia University Libraries.
39 *Film Daily*, May 2, 1956.
40 Reported by James Limbacher in his *Four Aspects of the Film* (New York: Brussell and Brussell, 1968).
41 Internal Fox memo by Donald A. Henderson on The Todd-AO Matter, September 16, 1963, Box 120, Sponable Collection.
42 Statistics courtesy of Dolby Laboratories, Inc., as of March 24, 1988.
43 Douglas Turnbull, in conversation, October 14, 1988.
44 Carr and Hayes, *Wide Screen Movies*, pp. 187–89, 200–206.

Chapter 22

Janet Wasko

HOLLYWOOD AND TELEVISION IN THE 1950s: THE ROOTS OF DIVERSIFICATION

THE FILM AND BROADCASTING INDUSTRIES have shared a "symbiotic relationship" since the 1920s, with the major Hollywood companies attempting to develop and control television as a new distribution outlet. In the 1950s, the film companies produced programming for much of the prime-time TV schedule, and they also experimented with alternatives to broadcast television. By the end of the 1950s diversification was well under way—the Hollywood film companies were becoming media companies.

Early interactions between the film industry and television

As early as the 1920s the Motion Picture Producers and Distributors Association (MPPDA), the industry's trade organization, was preparing reports on television, followed by investigations in the 1930s of the film industry's role in the forthcoming television business.[1] Mortimer Prall, son of the FCC [Federal Communications Commission] chairman, was hired by the MPPDA to prepare one report, in which he noted that "the motion picture industry has its greatest opportunity for expansion knocking on its door."[2]

In 1938, the Academy of Motion Picture Arts and Sciences requested its research council to study the film industry's preparation for the inevitable introduction of television, while numerous articles appeared that discussed the subject. One of the recommendations from the Academy was for the industry to pursue theater television, which will be discussed below.[3] Furthermore, both the MPPDA and the Academy formed ongoing committees to monitor television developments.

After World War II, there was a good deal of attention to the potential for the film industry to provide programming for the emerging television business, especially in light of the declining box office. Hollywood film producers and labor organizations, in particular, anticipated a new market for filmed products and employment opportunities in television production.[4]

Beyond the general industry interest in television, a few of the major studios had grand plans to control television through the ownership of distribution outlets, both individual stations and networks. Paramount was especially active in the evolution of television through its ownership of television properties, and, as outlined in the following sections, through attempts to innovate alternative television systems. In 1938, Paramount purchased substantial ownership interests in the Alan B. DuMont Laboratories Company. Over the next ten years, DuMont operated two experimental television stations in New York and Washington.[5] Paramount also was involved in another experimental license in Chicago around 1940 through its theater subsidiary, Balaban and Katz, as well as forming a subsidiary called Television Productions, Inc.

By 1948, Paramount owned four out of the first nine TV stations in the United States and had applied for licenses in six additional cities. Paramount owned KTLA, Los Angeles's first television station, and WBKB in Chicago, while DuMont owned and operated WABD in New York, WTTG in Washington, D.C., and WDTV in Pittsburgh. By the end of the 1940s, Paramount was distributing filmed television programs to a few stations through its Paramount Television Network, with plans to develop a full-fledged network. Although additional station ownership was planned, the FCC denied Paramount and DuMont's claim that they were separate companies and they were forced to adhere to the FCC limit of five stations.[6]

But Paramount wasn't the only film studio interested in television outlets.[7] MGM, Warners, Disney, and Twentieth Century-Fox actively vied for early TV outlets. Several of these companies applied for licenses in the Los Angeles area, but lost out to Paramount, probably because of the Paramount–DuMont connection.

Interestingly, some theater exhibitors in addition to Balaban and Katz also were interested in station ownership. Several theater chains applied for licenses in 1945, and the majority of stations owned by film interests in 1956 were held by theater owners.[8] Mitchell Wolfson, for example, owned a chain of South Florida theaters and the first TV station in Miami. He apparently felt that TV was not a threat and told fellow exhibitors that closed theaters should be converted to TV studios.[9]

However, Hollywood's bid to own television outlets mostly failed at this point, impeded chiefly by forces outside the film industry, especially the U.S. government. The government's hostility towards the film industry was apparent as early as 1940, when the FCC held hearings on technical standards for television. Another example of the government's attitude was evident at a meeting held in spring 1945, when the chairman of the FCC warned a group of Hollywood executives not to count on control or extensive ownership in the developing television business.[10] Furthermore, the FCC freeze on licensing new television stations (September 1948 to April 1952) for the announced purpose of establishing allocation policies and technical standards had the additional effect of preventing ownership of television stations by the Hollywood studios.

But most importantly, many of the studios' applications (especially Paramount's) were denied or withdrawn after the government's successful anti-trust suit against the majors. The FCC's policy was established in the Communications Act of 1934, which authorized the agency to refuse licenses to individuals or companies convicted of monopolistic activities. In addition, numerous statements by the agency confirmed a hostile attitude towards applicants with connections to the film industry.

Although the Hollywood studios most often were thwarted in their attempts to own broadcast outlets at this time, some companies still managed to own a few stations or align with networks. Several decades later, corporations associated with the film industry gained full control over television networks, as well as numerous cable channels and systems.

Meanwhile, during the 1940s and 1950s Hollywood also was trying to influence television's development through the innovation of two alternatives to TV broadcasting: theater television and subscription television.

Theater television

Theater television was one of the ways that Hollywood tried to "fight television with television."[11] The process involved screening television programming in motion picture theaters using two different systems: direct projection (television transmission projected onto screens) or "instantaneous projection" using a film intermediary (in other words, television signals transferred to film in less than a minute and then projected).

All of the major Hollywood companies were interested and active in developing theater television during the late 1940s and early 1950s; however, Paramount (again) was the most heavily involved. Hollywood's enthusiasm over theater TV was not surprising, since the Big Five studios owned extensive theater chains at the time. But theater television also provided the opportunity to control distribution, as well as offering a way to differentiate Hollywood's products from broadcast television. Theater TV could feature more costly programming and a larger format than provided by "free" television viewed in homes.[12] In addition, exhibitors were interested in theater television, feeling that it might compete well with home television systems. In 1946, *Film Daily* reported that over half of the 350 theater owners participating in a survey were anticipating using theater television.[13]

Theater television can be traced back to the 1930s when RCA developed a large screen system utilizing a tube similar to those in home receivers, but with greater light output. The company demonstrated the system in 1930 at RKO–Proctor's 58th Street Theater in New York City.[14] By 1941, RCA's system had been installed in a few test theaters at a cost of $15,800 per theater.

Paramount and Fox invested in theater television as early as 1941, using an intermediary film system produced by Scophony, a company owned by Paramount. The Scophony system involved the use of 35 mm. film and an installation cost of $25,000. According to film historian Douglas Gomery, Scophony Ltd. (as well as Baird Television) had experimented with theater television technology in Great Britain, but without too much success. Paramount helped organize the Scophony Corporation of America in 1943 to "protect its position vis-à-vis RCA."[15] However, in 1945, an anti-trust suit claimed that Paramount and Fox, through control of Scophony patents, were conspiring to prevent the development of television technology.[16] From this point on, then, Paramount focused on its Paramount Intermediate Film System, an expensive system (around $35,000) that produced a 35 mm. print that could be edited and reused.[17]

After World War II, Warner and Fox aligned with RCA and arranged several demonstrations in 1948 and 1949. The relationship turned out to be short-lived, but Fox later became involved with several other systems. General Precision Laboratories, one of Fox's main stockholders, developed a method that used 16 mm. film and cost $33,000 to install. Fox also was associated with direct projection systems, including one produced by General Precision for $15,600, as well as a Swiss system called Eidophor that used carbon arc lights to project color images and cost around $25,000 to install. In 1951, Fox arranged with General Electric to produce equipment for use with the Eidophor system. Although the system could not compete with RCA's lower-price equipment, Fox did not give up on theater TV until the late 1950s.[18]

While theater television technologies attracted the interest of many production companies and theater chains before World War II, there was little development during the war. As noted above, activities continued in the late 1940s. The time seemed right: few television sets were in homes, the FCC freeze prevented expansion of television stations, and the studios were withholding films from television.

Paramount introduced its system in New York at the Paramount Theater in 1948 and continued to feature political news coverage, prizefights, and other sporting events through

1951. Paramount introduced theater TV in Chicago and Los Angeles in 1948 and 1949. In Chicago, Paramount's Balaban and Katz experimented with live programming at its theaters in Chicago in 1949 and 1950, but found that it was too expensive. Although Balaban and Katz had given up on theater television by mid-1951, other theaters were still adding the technology.

By the end of 1952, over one hundred theaters nationwide had installed or were installing theater television, with RCA controlling 75 percent of the market. A network called the Theater Television Network had formed, featuring sports events (such as boxing and collegiate games), public affairs, and entertainment events. During 1952, the Walcott–Marciano prizefight was presented in fifty theaters in thirty cities, with revenues totaling more than $400,000.[19]

However, by 1953, the high hopes for theater TV had faded. Paramount finally abandoned the project that year with losses at many of the theaters that had been equipped with its system.[20] Theater television's ultimate failure involves many interrelated factors that were identified by some industry sources at the time and have been discussed more recently by film and broadcast historians.

First, there were ongoing issues involved in securing effective and cost-efficient methods of transmission. Both telephone wires and the broadcast spectrum were used, but there were problems with both approaches. Initially some systems used telephone wires, or more specifically, intercity links from AT&T. However, the wire transmission system proved to be too expensive and insufficient for video transmission. Theater owners requested a hearing from the FCC in 1948 to review the issue of costs, but the petition was denied. Meanwhile, requests had been made to the FCC for radio frequencies to use for theater TV experimentation. In 1944, the Society of Motion Picture Engineers applied for space, and, though some frequencies were provided in 1945, they were deleted by 1947. From 1947 to 1949, Paramount and Fox were temporarily awarded frequencies for experimentation.[21]

Then, in 1949, exhibitors filed another FCC petition requesting 10 to 12 channels in the UHF spectrum for a "movie band." Hearings were held October 1952 through 1953, but the FCC decided to deny special frequencies, suggesting that the petitioners use common-carrier services or reapply. For various reasons, the companies did not reapply.[22] If these frequencies had been allocated, future development of film, cable, and television industries might have been quite different. However, without spectrum space, theater TV systems were forced to use expensive telephone lines, which affected the quality as well as the cost of operation.

Thus, another general problem was the FCC's resistance. Historical evidence indicates that the government did not share the film industry's enthusiasm for theater television. As film and radio historian Michele Hilmes noted, "the FCC, with an unerring eye for the maintenance of the status quo, rejected this vision. . . ."[23] The rejection in many ways was connected with the film industry's monopolistic tendencies, exemplified in the Paramount decrees, as well as the anti-trust suit dealing with the Scophony patents involving Paramount and Fox.

It is important to note that the broadcasting industry had similar inclinations, as evidenced in the case against the radio networks in the late 1930s, as well as a later anti-trust suit against the television networks for monopolizing program supply and distribution (1972).[24] Nevertheless, at this point it was clear that the government favored the "public interest" orientation of the broadcast industry against the "crass commercial" interest of Hollywood companies.

Meanwhile, other developments contributed to the doomed theater television project. With the lifting of the FCC's freeze in April 1952, television exploded on the scene, as millions of Americans turned to "free" television in the convenience of their homes. By 1954, there were 233 commercial stations and 26 million TV homes.[25] In addition, the size of home screens increased and color standards eventually were established. With these changes it became nearly impossible for theater television to compete.

During this time period, the industry was undergoing profound structural changes that ultimately separated production and distribution from exhibition. Thus, the interests of the studios (or the production/distribution companies) at times were different from those of exhibitors. This became even more significant when it came to selling products to the newly developing television industry, but also ultimately affected the support for theater television by the different sectors of the film industry. For instance, several unions were against theater television, especially the International Alliance of Theatrical Stage Employees (IATSE), but also trade organizations representing actors and musicians.[26]

In the end, the major studios and theaters shifted their focus to new formats or gimmicks that would compete with home television rather than continuing to pursue theater television. Various widescreen systems were adopted, as well as experimentation with 3-D. Theaters with financial problems found that these systems were less expensive, yet more profitable, than the equipment needed for theater television. Gomery concluded, "if theater television had proven profitable, it no doubt would have spread quickly to all parts of the United States."[27]

Meanwhile, the Hollywood companies were exploring another alternative television system through their active interests in developing subscription TV. Although there was some discussion of theater and subscription television co-existing, in the end, theater television was abandoned and the subscription TV battle began.

Subscription television

Various experiments with subscription or pay television in the 1940s and 1950s involved companies connected to the film industry in one way or another. But it also must be noted that some film interests were involved in introducing pay systems, while others opposed those efforts. Though theater television was (at least, initially) welcomed by exhibitors, pay television was another thing altogether. Exhibitors not only feared it, they vigorously fought against it.

Again, Paramount took the lead in attempts to develop a viable pay television system. Timothy R. White pointed out that Paramount had a form of subscription TV in mind when it bought Scophony in 1942, and continued these efforts with DuMont. In the mid-1940s, Paramount planned to form a mobile system using DuMont equipment to transmit programming to theaters. However, these ideas ultimately were abandoned.[28]

By the late 1940s, three competing pay-TV sysems had been introduced, which then were tested in the early 1950s. Zenith's Phonevision was introduced in 1947, using telephone lines to unscramble a broadcast signal. Indeed, Zenith was the first company to ask for FCC permission to experiment with pay television in 1949, and tested its system in Chicago in 1950. After 1954, the company shifted to using a coin box or punch card system, but still was having problems obtaining programming. Paramount became involved with a system that used a scrambled broadcast signal through its 50 percent ownership of International Telemeter. Films were viewed by placing coins in a box on a television set, which then descrambled the picture. By 1957, Telemeter's system featured three channels using either wires or broadcast signals.[29] The Skiatron Corporation owned by Matthew Fox, an entrepreneur with some connections to the film industry, developed yet another system. Fox received the help of IBM to develop a system called Subscribervision, which used a punch card and a scrambled broadcast signal. RKO became involved with this system when a Chicago television station, WOR-TV, tested the Subscribervision system in 1950.[30]

Again, there was a good deal of enthusiasm by some Hollywood companies for a pay television system as an alternative to the advertiser-supported system adopted by broadcasters. However, there were only a few actual experiments with pay systems during the 1950s.

The system that attracted the most attention was Telemeter in 1953, when it provided a community antenna system, plus special programs for extra fees, to 274 homes in Palm Springs, California. Charges included a $150–$450 installation fee, plus additional fees for cable and coin box use, plus program charges. In addition to sporting events and other live programming, the service offered the same film that was playing at the local theater for a slightly higher fee.

Apparently, the aim was to attract viewers who never went to theaters, and some theater owners even cooperated with the experiment. However, one of the Palm Springs theater owners charged that Paramount was in violation of the recent anti-trust suit against the majors. Although Telemeter claimed to be a success with over 2,500 subscribers, the system apparently buckled under the threat of governmental restriction. And despite the connection to Paramount, the system seemed to be unable to procure an adequate inventory of Hollywood films.[31]

Meanwhile, Southwestern Bell Telephone and Jerrold Electronics in Bartlesville, Oklahoma, introduced another system, called Telemovies. This featured a first-run movie channel and a rerun movie channel. First-run movies were shown concurrently at the local theater chain, thus avoiding one potential source of opposition. But even though the experiment received a good deal of press attention, the service apparently had financial problems. The use of telephone lines was costly, the flat monthly fee to customers was very high, and the company had some difficulties developing a system for paying Hollywood companies for the use of their films.[32] The Southwestern Bell system ceased operation in 1958.

Meanwhile, Paramount maintained its faith in pay TV, increasing its interest in Telemeter to 88 percent by 1958 and announcing that it would open systems in New York, on the West Coast, and in Canada. The company operated a service near Toronto, Canada, for a few years, but had lost $3 million by 1965, when Paramount's subscription TV activities ended.[33]

Despite the enthusiasm expressed for "pay-see" and "toll-video" (as *Variety* called them), most of these experiments had failed by 1965. One might wonder why a direct pay system of television that became successful two decades later failed at this time. Again, the reasons are multi-faceted, interrelated, and similar to the reasons that theater television failed. First, systems that relied on phone lines found that the costs were prohibitive. However, it seems possible that such technical problems eventually could have been overcome. Even more problematic was the attack on pay-TV that came from the theater sector of the film industry, as well as broadcasters and other groups, especially after the Paramount/Telemeter Palm Springs experiment in 1953. Ironically, Paramount Pictures was attacked by United Paramount Theaters, which had recently been divested from Paramount and merged with ABC television in 1953.

While additional experiments were carried out in the early 1960s, broadcast and theater forces continued to lobby extensively to defeat pay television. A statewide referendum specifically directed at former NBC President Pat Weaver's Subscription Television was passed in California in 1964.[34] Although the referendum was eventually declared unconstitutional, the extensive publicity around the ballot measure and its success seemed to seal the fate of pay television, at least during this period in history.

There were also serious obstacles due to delay and resistance from the federal government. Hilmes argues that pay television failed because of "slow strangulation by federal regulation."[35] It seems clear that the FCC understood its mandate was to protect the existing broadcast system, and it found encouragement as well from those who lobbied against pay-TV, namely broadcast and exhibition interests. Congressional representatives who had broadcast investments joined the anti-pay television movement as well. In fact, at least six bills were introduced to ban pay television, with hearings held on the topic by the FCC, the Senate, and the House.[36] Although the FCC agreed to a temporary subscription TV trial in 1957, during the next year the Senate and House requested a delay and finally forbid the commission to allow such trials.[37] Undoubtedly, the film industry's efforts to establish pay-TV suffered from anti-trust litigation, not only the Paramount decrees, but other cases as well.

Another huge problem was the competition from "free" TV. It might be argued that the film industry mostly moved in the wrong direction with early subscription experiments. Perhaps it was too early for pay systems that did not offer a regular schedule of special programming for which audiences would pay an extra fee. But the idea of paying for television programming also suffered notably from the campaign to save "free" TV. Pay television succeeded in later decades when the systems merged with cable television and offered programming not available on over-the-air broadcasting.

In addition, Hilmes has pointed out that "the FCC's public interest mandate, adopted and reinterpreted by broadcast television and theater interests, became equated in the public mind with the unchallengeable supremacy of the 'free TV' system."[38] For example, an article in *Consumer Reports* from 1949 made a case against the film industry's involvement in television, citing the large size and monopolistic tendencies of the film industry and the likely deterioration of public service with Hollywood's involvement in television programming.[39] It might be noted that both industries exchanged barbs relating to the quality of their products, as can be seen in this quip from *Variety*: "Television is only indirectly show business. Mostly it's advertising business."[40] And during testimony by an NBC executive at a government hearing, the "film-come-latelys" were accused of conspiring to unleash on television, "the lowest common Hollywood denominator ... a continuing flow of stale and stereotyped film product."[41]

In summary, then, it is indisputable that Hollywood actively attempted to become involved in the evolving television system in the United States. Despite its inability to own or control broadcast outlets at an early stage or to develop successful alternative systems, the film industry eventually was able to profit from television in other ways.

Strategies for coexistence with television

While the events surrounding theater television and pay television unfolded, Hollywood was developing specific strategies for selling its products to the emerging television industry. By the 1950s, the film industry had firmly established a key role in the supply of the majority of television programming. Before discussing these developments, it is important to establish the backdrop for Hollywood's eventual triumph.

The evolving economic structure of television programming had shifted rather quickly from commercial sponsorship controlled by advertising agencies, with one main sponsor per program, to a magazine format with different sources of advertising controlled by the networks. As Mullen explained: "The degree of flexibility telefilm and videotape production techniques brought to television programming complemented the flexibility of magazine-format sponsorship. By the 1960s, virtually every component of the television schedule was both interchangeable and recyclable."[42]

The transition from live to recorded programming was another important factor in the evolution of television programming. The networks clung to live television, which was one way of maintaining control. (Live television meant that stations had to receive live feeds from the networks according to a specific schedule. Taped or filmed programs could be aired by stations whenever they chose to run them.) The networks and their critics also insisted that live television programming was creatively superior to filmed fare.[43] Nevertheless, the lower cost and reliability of recorded programs finally prevailed. As Hilmes explained: "by breaking down their own restrictions against the use of recorded programs, the major networks paved the way for the gradual disappearance of unwieldy and unpredictable live programming and the rapid rise in the use of film, as Hollywood wedged a toe in the door by means of syndicated film series."[44]

The ongoing structural changes within the film industry due to the Paramount decrees meant that the interests of producers eventually became different from those of exhibitors. Additional changes included the decline of studio-produced feature films and the growth of independently produced films, as well as fewer, more expensive films with increased promotion costs. Generally, the production-distribution companies benefited from these developments, while exhibition mostly suffered. In a 1985 article on Hollywood and television, William Boddy concluded: "The cumulative effect of these changes within the theatrical film industry in the early 1950s was a shift of power from exhibition to production-distribution. . . ."[45] In an overview of the industry in the late 1950s, a *Variety* writer echoed these sentiments:

> TV has had a terribly divisive effect on the film industry. It has accomplished what even divorcement could not really accomplish, i.e. a split between Hollywood (in the production sense of the word) and exhibition-distribution. It is simply a fact that, today, with the TV mart looming so importantly and the electronic medium advancing ever further, the basic interests of the Coast and the rest of the industry are not necessarily the same.[46]

Hollywood television production

As indicated by the numerous investments in production companies in the late 1930s and 1940s, the major studios clearly intended to become involved in television programming. As new stations opened, the demand for filmed programs increased. However, some of the studios delayed production for broadcasting, as they hoped to develop the alternative television systems discussed above. In addition, it wasn't until the late 1950s that the major integrated companies were fully divorced from their theater chains. Thus, during much of the decade, they were forced to avoid conflict with theater owners and moved ever so cautiously into television production, as well as resisting sales of their new theatrical films to television. (In fact, Boddy noted that in 1950 the FCC actually issued a warning to the Hollywood studios for withholding talent and products from broadcast television.)[47] This is undoubtedly one of the reasons why many observers have concluded that Hollywood was hostile to television.

Nevertheless, a relatively large number of Hollywood independent producers created programming for broadcast television during the late 1940s and early 1950s. Jerry Fairbanks seems to have been the first Hollywood producer to sell a series to television when he marketed *The Public Prosecutor* to NBC in 1948.[48] Other programs that followed were mostly low-budget shows, including numerous Westerns (*Hopalong Cassidy*, *Roy Rogers*, *The Lone Ranger*, and *Cisco Kid*, to name a few), as well as crime, mystery, science fiction, and situation comedies. The business was highly competitive, with over 800 producers claiming to be involved during this early period, which came to be known as the "gold rush" period of television production. However, the business involved high speculation and slow paybacks on programs that averaged $12,000 to $15,000 per half-hour episode. Even at this early stage, a distinct financing system was emerging, with programs usually produced at a deficit and profits emerging in syndication, international distribution, and other products.[49]

By 1952, the production costs were rising (up to $20,000 per program) and the competition declining. The most successful telefilm production companies were those that had resources available and could provide commercially oriented, profitable productions. Among the few surviving independents were Fairbanks's company, Bing Crosby Productions, Hal Roach Studios, Ziv Television, and a few others.[50] Thus, early television production provides another example of an industry that moved through an initial period of robust competition to a concentrated market structure. As Dennis Dombkowski concluded:

> "Efficiency" and industry stability were thus gained at the cost of variety in program sources. It should be pointed out that it was no economic law that made this necessary for the television program supply industry, but rather the inevitable requirements of a commercial system which was being designed for a nationwide service of networks and their correspondingly greater capital requirements, rather than a decentralized and less "efficient" system.[51]

Initially, the networks resisted filmed programming, aiming to control program supply and national advertising distribution. The networks insisted that their live, dramatic programming was higher quality than the cheaply made action-adventure shows and situation comedies produced by film companies. At this point, relatively low prices were offered for TV programming, even though high costs were anticipated for Hollywood-produced programming.[52]

However, even at the networks, filmed programming ultimately prevailed. The market for subsequent release of television programming (reruns, syndication, and foreign sales) was a major factor in the acceptance of filmed TV programming by the networks and the entrance of the Hollywood majors into telefilm production.

Only a few of the major companies produced television series, commercials, and news during the late 1940s and early 1950s. By 1952, television subsidiaries were formed by Columbia (Screen Gems), Universal (United World Films), and Monogram (Interstate Television). It might be noted that none of these companies were major integrated studios that owned theaters.

Most historians agree that the other major studios became much more active in television production after 1953. The reasons often cited are the lifting of the freeze in 1952, the increase in advertising money, the decline in the theatrical box office, and the actual divorcement of production/distribution from exhibition.[53]

At first, the majors offered programs that were designed to promote their film products or that featured their own names: *Disneyland*, *Warner Bros. Presents*, *MGM Parade*, and *Twentieth Century-Fox Hour.* They also tended to draw from their film resources for some program ideas (for instance, *Rin Tin Tin*, a series of movies since the 1920s that became a popular TV series in the mid-1950s). But the studios moved on to produce a wide range of programs, including prestigious dramatic shows (such as *Playhouse 90*) and especially half-hour series that were quite profitable in network and syndicated markets.

Each major company moved into television at a somewhat different pace. In 1956, Columbia produced thirteen half-hour series and one ninety-minute program, with television representing one-quarter of the company's revenues. Around the same time, Twentieth Century-Fox produced four half-hour shows and one one-hour series. Meanwhile, Warner Bros. received nearly one-third of its profits from television by the end of the decade. Paramount was much slower than the other studios, and it wasn't until the 1970s that television production became important. By 1964, six Hollywood companies represented 45 percent of the domestic market for syndicated TV series.[54]

Although Hollywood had to make adjustments for television production, the pattern had been set. Commercially oriented filmed programming became the main staple of American television and the large Hollywood studios came to dominate the market. Erik Barnouw identified more than 100 Hollywood-produced television series on the air or in production by the end of 1957.[55] By the end of 1959, 78 percent of network evening programming originated in Hollywood and over 88 percent of that was filmed.[56] And, although a 1956 congressional study of the film industry found that 170 producers were involved in making television programs of various lengths,[57] a handful of well-endowed Hollywood companies dominated the program supply. At the beginning of 1959, *Variety* estimated that $105 million would be dedicated to telefilm production during the year.[58]

Anderson concluded: "By the end of the 1950s, with the fates of the networks and studios deeply entwined, filmed television series emerged as the dominant product of the Hollywood studios and the dominant form of prime-time programming—a pattern that has remained unchanged for more than thirty years."[59]

Feature films on television

Hollywood's major line of business—feature-length films—would seem to have been an obvious program source for television, with its insatiable need to fill airtime. Paramount actually televised some of its films in the early 1930s, but through an experimental station in Los Angeles that only reached few receivers.[60] A number of British films were sold to television stations as early as 1948, followed by features from small Hollywood companies such as Republic and Monogram. Yet, the major studios held their films back for various reasons, including their relationship with exhibitors and talent unions, their involvement in alternative exhibition schemes such as theater and subscription TV, and their dissatisfaction with the prices offered by the networks for quality films. Based on information from government documentation, Dombkowski argued that low prices offered by the networks was the primary reason. (Statements in the trade press confirm this point. For instance, in 1952, Spyros Skouras stated that TV, "can only pay us buttons for our old films, and certainly can't even begin to compete for the new product.")[61]

By 1955, the divorcement of exhibition freed the distributors from their commitments to theaters and broadcast television had become more established and profitable. Thus, the major studios began releasing their older films to television. An agreement between the producers and trade unions required negotiation for the licensing of films produced after 1948. Thus, pre-1948 films were the first ones sold to television stations. The majors either sold their films to other companies or set up their own television distribution divisions. RKO started the licensing deluge in 1955 when it sold its features and shorts to C. and C. Television. Meanwhile, Columbia sold its own feature films through its Screen Gems subsidiary, which also handled Universal's older films. Fox sold its films to National Telefilm Associates, but then purchased 50 percent of the company in 1956. Eventually, Paramount sold its library to an affiliate of the Music Corporation of America (MCA), which later was to merge with Universal. (White notes that Paramount delayed the sale of pre-1948 films until much later than the other studios because of their hope that pay television would succeed.)[62]

In his 1960 book, Michael Conant cited government documentation claiming that in 1954 there were eighty motion picture exchanges dedicated to the distribution of films to television, with total receipts of $24 million. He further estimated that by February 1958, around 3,700 pre-1949 feature films had been sold or leased to television for an estimated $220 million.[63] As predicted, the appearance of feature films on television ate into the theatrical box office. A specially commissioned study by the theater owners found that towards the end of 1957, one-quarter of television programming represented recycled movies.[64] Independent stations (not affiliated with the networks) found feature films to be a major source of programming, sometimes providing up to 48 percent of their schedules.[65]

But the licensing of feature films was still predominantly to individual television stations, not to the networks. Even though over thirty prime-time network programs consisted of feature films between 1948 and 1957, these offerings were sporadic and inconsistent.[66] The networks resisted the regular use of feature films in their schedules, but also were unable (or unwilling) to pay enough to satisfy the majors. Meanwhile, the studios found the syndication market for pre-1948 films to be quite profitable, but resisted issuing newer films because of the potential loss of revenues from theater reissues.

Of course, the networks eventually came around to featuring Hollywood films in their prime-time schedules. But it wasn't until the early 1960s, after a new labor agreement bad been negotiated with the talent unions, that the majors began seriously releasing their newer films to television. When they did, however, there was a plethora of post-1948 films on the market. In fact, so many newer films were licensed by the majors during 1961 that some feared supplies would be exhausted in only a few years.[67] This was one of the factors that influenced the evolution of Hollywood-produced made-for-TV movies and mini-series that started emerging in the mid-1960s.[68]

By 1960, television had become the hot new medium; 90 percent of American homes were equipped with sets. And a good deal of the televised programming came from Hollywood. By this time, over 40 percent of network programming was produced by the Hollywood majors.[69]

It is nearly impossible to estimate the actual revenues that film companies earned from television during the 1950s, as few companies specifically identified television revenues in their annual reports.[70] An overview of the diversification trends in *Variety* at the end of the decade included estimates of television's contributions to total revenues for a few companies for 1958: Columbia, 25 percent, Disney, 20 percent, Loew's, 11 percent, and Warners, 30 percent.[71]

It is clear that, at least by the mid-1960s, television was heavily dependent on Hollywood-produced programming and Hollywood had become dependent on television as a crucial source of revenues. This relationship continued to grow through the 1970s, as documented by Thomas H. Guback and Dennis J. Dombkowski in a 1976 article, and indeed through the end of the century.[72] Even though the major studios were not able to control television or develop alternative forms, they eventually found television to be a valuable new market for their old products, and they also developed new products for the expanding new medium.

The new diversification

By the late 1950s and early 1960s, the Hollywood majors represented newly diversified corporations, reporting sizable profits. The next two sections will look more closely at two companies that were particularly successful with these new diversification strategies: Walt Disney and Universal. The Walt Disney Company was an independent production company that benefited greatly from diversification in the 1950s. At the same time, Universal was a "second rank" studio though the 1940s, but through its alliance with MCA in the late 1950s it became one of the Hollywood majors by the mid-1960s. Both companies moved into television during the 1950s, thus setting the foundation for their roles as diversified entertainment conglomerates at the end of the century.

Walt Disney productions

The Walt Disney Company was a relatively small independent that mainly produced animated cartoons from the early 1920s, adding animated features in the 1930s. The extensive merchandising campaigns accompanying these films contributed to the company's revenues and made it possible to continue production of cutting-edge animation. However, Disney did not represent the kind of integration and diversification typical of the major studios until the 1950s.

Walt Disney is often acknowledged as the first executive in Hollywood to recognize the potential of television. While this claim ignores other film companies' ongoing attempts to get into the television business in various ways, Disney still deserves credit for recognizing television's potential value in promoting and diversifying the film business. Disney explained, "Through television I can reach my audience. I can talk to my audience. They are the audience that wants to see my pictures."[73] It seems likely that the Disney Company had television in mind for its products from the mid-1930s, when it changed from United Artists to RKO as distributor; the crucial factor was United Artists' refusal to allow Disney to retain television rights to films.[74]

At first, the Disney Company produced a few Christmas specials, beginning with "One Hour in Wonderland," broadcast on NBC in 1950. Apparently, all three of the networks tried to convince Disney to produce a weekly series. Finally, in October 1954 the weekly series *Disneyland* appeared on ABC, moving to NBC seven years later as *Walt Disney's Wonderful World of Color*. The television series allowed the studio to recycle its already released products, similar to the technique of continuously re-releasing its animated features in theaters every few years, thus reaping further profits with little additional cost.

But television also proved helpful in several ways for Disney's most cherished project—an amusement park that would appeal to adults as well as children. The arrangements with ABC for the Disney television series apparently were prompted by Disney's need for capital to build Disneyland, which eventually opened in Anaheim, California, in 1955. ABC invested $500,000 in the park and became a 35 percent owner, plus guaranteeing loans of up to $4.5 million. Walt Disney apparently received little support from the Disney Company itself, but managed to raise the funds from the ABC deal and loans on his insurance policies. In 1952, he formed a separate company called Walt Disney Inc., later to become WED Enterprises, to develop the park without involving company funds. Disneyland ultimately cost $17 million, but was an instant success with one million visitors during its first seven weeks of operation.

In addition to providing financial backing, *Disneyland*, the television series, became a terrific promotional vehicle for the park even before it opened. The show was organized around the same four divisions as the park—Fantasyland, Adventureland, Frontierland, and Tomorrowland—and constantly featured updates on the new park. Of course, new content also was developed for the show. For instance, Davy Crockett started as a three-part episode, inspired a national merchandising sensation, and was then recycled as two feature films. (The company obviously underestimated the success of the series, as Disney explained: "We had no idea what was going to happen to 'Crockett.' Why, by the time the first show finally got on the air, we were already shooting the third one and calmly killing Davy off at the Alamo. It became one of the biggest overnight hits in TV history, and there we were with just three films and a dead hero.")[75] However, a good deal of the show featured the studio's recycled cartoons and feature films. The Davy Crocket phenomenon—with coonskin caps purchased by young boys all over the country—was another reminder of the Disney Company's successful merchandising business, which had been established in the early 1930s.

In 1955, the company introduced a daily afternoon television show designed exclusively for children and proclaimed as a "new concept in television programming." *The Mickey Mouse Club* featured the mouse-eared singing and dancing Mousketeers, plus other segments that involved Disney's products (such as Disney cartoons and news about Disneyland). Despite its enormous popularity in 1955–56 (at one point, reaching 75 percent of the television sets in the United States and attracting lots of advertising), and the endurance of its theme song, *The Mickey Mouse Club* only lasted four seasons. The main reasons for its cancellation were: 1) the high cost of producing the show; 2) a significant drop in ratings from 1955 levels; and 3) a dispute between Disney and ABC over control of *The Mickey Mouse Club* and another series, *Zorro.*[76]

In addition to these developments, the company also finally decided to distribute its own films, creating Buena Vista Distribution. The move was attributed to Walt Disney's deep concern about maintaining control over his own products, as had been seen in an earlier incident when the Disney character, Oswald the Lucky Rabbit, was "stolen" by another company. To a large extent, the move into distribution signaled the Disney Company's transition from a marginal independent film company to one of the Hollywood majors.

Even though the company may have been slow to control its own film distribution, it might be argued that Disney led the way in the diversification that would characterize the industry for the next few decades. As writer Richard Schickel argued, the Disney Company had a head start on the rest of the industry. While the larger, integrated majors were dealing with the rising competitive threat of television, as well as the loss of their theaters due to the Paramount decrees, the Disney Company was diversifying its film products, as well as its over-all business.[77] For instance, in 1960 Disney reported the following income sources: film rentals, $18.4 million, amusement park, $18.1 million, television, $4.9 million and other (publications, comic strips, licensing cartoon characters and music), $4.9 million.[78] Furthermore, by the beginning of the 1960s, the company was integrating these businesses, thus laying the foundation for the Disney synergy that blossomed in the 1980s and 1990s. In addition, the Disney Company was especially successful at selling and promoting its products globally. By 1954, it was estimated that one-third of the world's population had seen at least one Disney film.

MCA/Universal

Another current major entertainment conglomerate to develop important diversified activities (especially in television) during the 1950s was MCA/Universal. The story involves two different companies that came together during this decade.

Universal began diversification activities in 1951 when Decca Records acquired a 38 percent share and soon thereafter controlled the company. Universal had encountered serious losses from a few expensive independent films from 1946 to 1949. Thus, the new management attempted to cut costs and increase revenues through low-budget productions, including situation comedies such as the Ma and Pa Kettle series, and a number of science fiction films, including CREATURE FROM THE BLACK LAGOON (1954) and THE INCREDIBLE SHRINKING MAN (1957).

The company had moved into television production in the late 1940s in hopes of attracting revenues, but also in an effort to keep its facilities open during production lulls. Universal's subsidiary, United World Films, was especially active in producing television commercials and had produced over 5,000 commercials that attracted $3 million in annual revenues by 1958.[79] The company also was releasing theatrical films to television, with *Variety* dubbing United World "a dominant factor in the home movies field" in 1958.[80]

Decca/Universal's relationship with Music Corporation of America (MCA) began in December 1958, when Universal sold its 367-acre studio lot for $11.25 million to MCA, then leased back studio space at $1 million a year.

MCA dates back to 1924, when Dr. Jules Stein and William R. Goodheart Jr. founded it as a talent agency for live music and later radio. In 1935, in addition to Lew Wasserman joining the company, MCA received a "blanket waiver" from the American Federation of Musicians, permitting the agency to act in a dual capacity as talent booker and program producer, and allowing it to package radio shows with its own bands. MCA added live show production in 1943 through a subsidiary called Revue Productions. The company also expanded its agency business with the addition of Hayward–Deverich Agency in Beverly Hills and the Leland Hayward Agency in New York. New clients included top Hollywood stars such as Greta Garbo, Fred Astaire, Jimmy Stewart, Henry Fonda, and Gregory Peck.

As the company grew to become a dominant force in the entertainment industry, it began attracting the attention of the federal government for anti-trust violations. In the first of many suits filed against the company, a judge found MCA guilty of restraint of trade in 1946, calling the agency "the Octopus ... with tentacles reaching out to all phases and grasping everything in show business." However, what followed during the 1950s extended the Octopus's reach much further than the judge might have imagined.

MCA, as the largest talent agency in the business, began its move into television production in 1952. Revue Productions re-emerged as the company's television subsidiary, with a TV version of radio's *Stars over Hollywood*, called *Armour Theatre*. The series was recorded on film, which increased the production expense, but MCA arranged for its clients to appear on the programs for minimum scale.

As Revue continued to produce other television series, the issue of conflict of interest again emerged. And, again, the agency was able to obtain a "blanket waiver" from the talent guilds that enabled it to gain a considerable advantage in television production. The unprecedented and exclusive waiver from the Screen Actors Guild [SAG] allowed Revue to produce television shows and simultaneously represent talent in those shows. The Writers Guild granted MCA a similar waiver, allowing it to represent screenwriters and to produce TV programs through Revue.

By 1954, MCA was earning 57 percent of its revenues from television programming. During the 1950s, Revue Productions became incredibly successful in television, with such high-ranking shows as *Wagon Train*, *Alfred Hitchcock Presents*, *The Jack Benny Show*, *Ozzie and Harriet*, *Dragnet*, *This Is Your Life*, *Leave It to Beaver*, and many, many others. By most accounts, MCA was the dominant force in television by the end of the decade, with 85 percent of its revenues from television sales and producing more programming than any other company, including the television networks themselves.

Although MCA produced programs for every network, the company developed a special relationship with NBC that was similar to Disney's with ABC. One report cited an NBC executive showing an MCA vice president plans for its 1957 season, saying, "here are the empty spots, you fill them." An assortment of MCA programs was featured on NBC that year, including *M Squad*, *Wagon Train*, *Tales of Wells Fargo*, and others.[81] Meanwhile, MCA also produced a number of shows directly for syndication, including *Mike Hammer*, *Biff Baker, U.S.A.*, and *Famous Playhouse*.

MCA expanded in other areas of the entertainment business, as well. In 1958, the company purchased Paramount Studios' pre-1950 sound film library for $50 million, the richest television syndication deal to date. The same year, the company incorporated as MCA Inc., a neatly diversified enterprise encompassing film and television production and distribution, talent management, music production and distribution, and a variety of other entertainment fields. By this time, it also had moved into the tourist/theme park business through a deal with Grey Line Bus Tours called "Dine with the Stars." Buses drove around the lot with visitors, who would then have lunch in the commissary. By 1964, the Universal Studios Tour opened in conjunction with Glamour Trams, including two drivers, two guides, and one ticket seller; the price of admission was set at $6.50 for two adults and a child.

While these diversification activities were proving to be quite lucrative, at the end of the decade the government again began to focus attention on MCA. One of the problems was the blanket waiver that was issued in 1952 (and extended in 1954), while Ronald Reagan (an MCA client) was president of SAG. Although some argued that the waiver was issued to encourage employment, the government alleged that it was a conspiracy that allowed MCA to monopolize talent and television program production. (Other agencies had not been given such waivers.) Reagan's involvement as both SAG negotiator and MCA client suggested a conflict of interest. Reagan remembered few of the details during government hearings, but it seems that a year after MCA's waiver was granted, he had become the host and star (and later, producer) of Revue's *General Electric Theater* on CBS. The program contributed to Reagan becoming a multi-millionaire by the time the show closed nine years later.

Other allegations involved MCA's restraint of trade, extortion, discrimination, blacklisting, and use of predatory business practices. One of MCA's practices was to develop television programs or motion pictures as "packages" of the agency's clients. The agency received a commission on the sale of the package to various outlets. While other agencies used this technique, MCA was particularly successful in its implementation, and had become known for its practice of refusing to let a company employ one of its clients unless a complete MCA "package" was involved.

Around the same time as the government investigations, MCA announced plans to purchase Decca Records, which included Universal Pictures. Interestingly, the growth of Universal Pictures into a major studio was due to sizable profits from increasingly bigger films that were put together as "MCA packages." Examples included three of Universal's biggest moneymakers of 1959: OPERATION PETTICOAT ($18.6 million), IMITATION OF LIFE ($13 million), and PILLOW TALK ($15 million).

Although the government initially moved to block the purchase of Decca and Universal, MCA finally agreed to a consent decree with the Justice Department in 1962, clearing the way for the merger. The company chose to divest its talent management activities, but gained even more clout in the lucrative film and television business, as well as in the music industry. Though the company continued to pursue other forms of diversification during subsequent years, these developments during the 1950s were consequential in the rise of MCA/Universal as a significant player in the entertainment industry at the end of the twentieth century.[82]

Conclusion

Although the film industry failed to dominate the emerging television industry in the 1950s, Hollywood actively participated in television's evolution during the decade and established a strong relationship that eventually led to the integration of these two industries. It is clear that Hollywood companies kept a close watch on the evolving television technology and even became involved in its early growth. However, strategies that involved hardware development and ownership of broadcast outlets were mostly unsuccessful. Station and network ownership, as well as alternative systems, such as theater television and early forms of pay-TV, were thwarted, especially by government protectionism, but also by resistance from exhibitors and other factors. Notably, the Hollywood majors' monopolistic tendencies proved detrimental to some of their efforts to expand into the newly emerging television industry.

However, Hollywood did succeed in becoming involved in the television business through program supply. During a period of box-office decline, the newly emerging television industry provided the opportunity for the major studios to draw upon their expertise and resources to create new products (like prime-time series and made-for-TV films) for the growing television market. In the end, the majors were able to dominate that market. In addition, the film industry was able to recycle and profit yet again from feature films that had already produced sizable profits. Eventually, television also served as another market for newly produced theatrical films.

These activities ultimately led to the diversification of the film industry, as well as the eventual integration of the film and television industries later in the century. As Anderson fittingly concluded:

> Since the movie studios began producing television, the diversification of media corporations into related fields and the consolidation of capital through corporate mergers have produced an environment in which media industries are increasingly interwoven. Although these tendencies existed before the 1950s, the impulse toward integration rose markedly during that tumultuous decade and has become more pronounced in subsequent years.[83]

In other words, television provided the film industry with new opportunities that laid the groundwork for the diversification and concentration that characterized the entertainment industry at the end of the century.

Notes

1 Douglas Gomery, "Failed Opportunities: The Integration of the U.S. Motion Picture and Television Industries," *Quarterly Review of Film Studies,* summer 1984, p. 324; Dennis J. Dombkowski, "Film and Television: An Analytical History of Economic and Creative Integration," Ph.D. dissertation, University of Illinois, 1982, pp. 50–51.

2 "Film Industry Advised to Grab Television," *Broadcasting*, 15 June 1937, p. 7. Also see Eric Smoodin, "Motion Pictures and Television, 1930–45: A Pre-History of Relations between the 'Two *Media*,'" *Journal of the University Film and Video Association*, Vol. 34, No. 3, summer 1982.

3 "Television from the Standpoint of the Motion Picture Producing Industry," *Film Daily Yearbook* (New York: Film Daily Publications, 1939), p. 797; Timothy R. White, "Hollywood's Attempt at Appropriating Television: The Case of Paramount Pictures," in Balio, *Hollywood in the Age of Television*, pp. 150; Michele Hilmes, *Hollywood and Broadcasting: from Radio to Cable* (Urbana: University of Illinois Press, 1990), p. 72.

4 Al Steen, "Television Developments of 1944," *Film Daily Yearbook* (New York: Film Daily Publications, 1945), pp. 66–68; Dombkowski, pp. 52–57.

5 See Gary N. Hess, *An Historical Study of the DuMont Television Network* (New York: Arno Press, 1979).

6 Hilmes, pp. 118–19; White, pp. 146–50; Dombkowski, pp. 31–32.

7 This section draws especially on the following discussions of Hollywood's attempts to own broadcast outlets: Hilmes, pp. 118–19; White, pp. 146–49; Gomery, pp. 219–27; Christopher Anderson, *Hollywood TV: The Studio System in the Fifties* (Austin: University of Texas Press, 1994), pp. 33–45; Dombkowski, pp. 29–37. See also, David Alan Larson, "Integration and Attempted Integration between the Motion Picture and Television Industries Through 1956," Ph.D., dissertation, Ohio University, 1979.

8 Al Steen, "Television Developments," *Film Daily Yearbook* (New York: Film Daily Publications, 1946), p. 75; Dombkowski, pp. 57–59.

9 Mitchell Wolfson, "Report of Theatre Owners of America Television Committee, 1950," pp. 4–5, TOA Collection, Box 41, folder 2, BYU. Wolfson, "Outmoded Theaters as TV Studios," *Film Daily Yearbook of Motion Pictures 1953* (New York: The Film Daily, 1953) pp. 717, 719.

10 Michael Conant, *Anti-trust in the Motion Picture Industry* (Berkeley: University of California Press, 1960), pp. 86–88.

11 John E. McCoy and Harry P. Warner, "Theater Television Today (Part I)," *Hollywood Quarterly*, Vol. 4, winter 1949, p. 160.
12 White, pp. 150–51. See also Orrin E. Dunlap Jr. *The Future of Television* (New York: Harper and Brothers Publishers, 1942) for a representative discussion of the enthusiasm over theater television.
13 Steen (1946), p. 75. The exhibitors' trade organization, Theatre Owners of America, published material promoting theater television, while its executive director toured the country "pushing his pet project." "TV Line Forming 'Rapidly': Sullivan," *Variety*, 31 May 1950, p. 3.
14 McCoy and Warner, p. 161.
15 Gomery, p. 221.
16 United States v. Scophony Corp., 69 F. Supp. 666 (S.D.N.Y., 1946). See also Al Steen, "Television in 1943," in *Film Daily Yearbook* (New York: Film Daily Publications, 1944), p. 65.
17 McCoy and Warner, pp. 161–63; White, p. 152.
18 White, p. 152; Hilmes, pp. 137–38; Larry Goodman, "Television," *Film Daily Yearbook 1953* (New York: Film Daily Publications, 1953), pp. 139–44.
19 Goodman, pp. 140–43; Hilmes, p. 122; Amy Schnapper, "The Distribution of Theatrical Feature Films to Television," Ph.D. dissertation, University of Wisconsin, 1975, p. 70.
20 White, p. 154.
21 White, p. 151. See also McCoy and Warner, pp. 262–66.
22 Donald La Badie, "TV's Threshold Year," *Film Daily Yearbook 1955* (New York: Film Daily Publications, 1955), pp. 23–26.
23 Hilmes, p. 125.
24 See Hilmes, pp. 187–88.
25 Stats from *Television Factbook*, cited in Christopher H. Sterling and Timothy R. Haight, *The Mass Media: Aspen Guide to Communication Industry Trends* (New York: Praeger Publishers, 1978), p. 49.
26 Gomery, p. 223.
27 Ibid.
28 White, p. 155.
29 Goodman, pp. 141–42; Dombkowski, pp. 35–36.
30 Hilmes, p. 126.
31 Fred Hift, "Telemeter to Invade Two Cities," *Variety*, 4 May 1955, p. 1.
32 H. H. Howard and S. L. Carroll, *Subscription Television: History, Current Status, and Economic Projection* (Knoxville: University of Tennessee, 1980); "Suspension of Bartlesville Trial No Surprise to NY Execs," *Variety*, 28 June 1958, p. 23; Hilmes, pp. 127–28.
33 Howard and Carroll, pp. 44–48; White, pp. 159–60.
34 See Mark Freed, "An Analysis of the Failure of Subscription Television in California in 1964," Masters thesis, University of Oregon, December 1969.
35 Hilmes, p. 133.
36 U.S. Congress, House, Committee on Interstate and Foreign Commerce, 1956; U.S. Congress, Senate, Committee on Interstate and Foreign Commerce, 1956. See also White, p. 157; "Five Bills Pend in Congress to Ban Tollvision from American Waves," *Variety*, 5 February 1958, p. 1.
37 White, p. 158.
38 Hilmes, p. 134.
39 Hilmes, p. 136.
40 Fred Hift, "Adjustment to TV Is Biggest Issue," *Variety*, 9 January 1957, pp. 7, 66.
41 U.S. Congress, Senate, Committee on Interstate and Foreign Commerce, *The Television Inquiry*, volume 4: *Network Practices, Hearings before the Committee on Interstate and Foreign Commerce*, Senate, 84th Cong., 2d sess., 1956, p. 2,280, cited in William Boddy, "The Studios Move into Prime Time: Hollywood and the Television Industry in the 1950s," *Cinema Journal*, Vol. 24, No. 4, summer 1985, p. 29.
42 Megan Mullen, *"The Revolution Now in Sight"? Cable Programing in the United States,* 1948–95 (Austin, Texas: University of Texas Press, in press).
43 See William Boddy, *Fifties Television: The Industry and Its Critics* (Urbana: University of Illinois Press, 1990), Chapters 4 and 5.
44 Hilmes, p. 150; Dombkowski, pp. 93–95. See also Robert Vianello, "The Power Politics of 'Live' Television," *Journal of Film and Video*, Vol. 37, summer 1985, pp. 26–40.
45 Boddy, "The Studios," p. 26.
46 Hift, p. 7.
47 For more discussion of these points, see Boddy, *Fifties*, pp. 65–76; Sohnapper, pp. 23–30.
48 Anderson, *Hollywood TV*, pp. 54–56.

49 See Frederick Kugel, "The Economics of Film," *Television*, July 1951, pp. 10–15; "Hollywood Can Grind Out Firm Fare for TV," *'Business Week*, 24 November 1951, pp. 122–26; Bob Chandler, "TV Films: An Updated Version of Freewheeling Picture Pioneers," *Variety*, 4 January 1956, p. 157; Anderson, pp. 56–59.

50 Boddy, *Fifties*, pp. 69–70. For more on these producers, see Anderson, pp. 57–68; Dombkowski, pp. 96–105; Barbara Moore, "The Cisco Kid and Friends: The Syndication of Television Series from 1948–52," *Journal of Popular Film and Television*, Vol. 8, spring 1980, pp. 26–33.

51 Dombkowski, p. 110.

52 Boddy, "The Studios," pp. 26–29; Boddy, *Fifties*, pp. 73–76.

53 Hilmes, p. 153.

54 Data from Dombkowski, pp. 111–23, derived from Irving Bernstein, *The Economics of Television Film Production and Distribution* (Screen Actors Guild, 1960); and Arthur D. Little, Inc., *Television Program Production* (Cambridge, Mass.: Arthur D. Little, Inc., 1969), p. 113.

55 Erik Barnouw, *The Golden Web: A History of Broadcasting in the United States 1933–1933* (New York: Oxford University Press, 1968), pp. 80–84.

56 "Hollywood in a Television Boom," *Broadcasting*, 26 October 1959, p. 90.

57 U.S. Congress, Senate Committee on Small Business, *Motion Picture Trade Practices-1956, Report*, 83rd Cong., 2nd Sess., 1956, p. 37.

58 Dave Kaufman, "Telefilms Production Bill for 59 at $105,000,000 in Banner Year for Giant Biz," *Variety*, 7 January 1959, pp. 100, 174.

59 Anderson, p. 7.

60 Dombkowski, pp. 40–41.

61 Ibid., pp. 164–73; "Skouras Sees Pic Future in Theatres; Big-Screen to Give That Extra Plus," *Variety*, 12 March 1952, p. 1.

62 White, pp. 158–59.

63 Conant, p. 137.

64 *New York Times*, 23 January 1958, p. 1, cited in Conant, p. 137.

65 Dombkowski, p. 180, using data from *Broadcasting Yearbook.*

66 Tim Brooks and Earle Marsh, *Complete Directory to Prime Time Network and Cable TV Shows* (New York: Ballantine Books, 1999), pp. 569–70.

67 "Will First-Run Films Be Extinct?" *Broadcasting*, 27 November 1961, p. 27.

68 See Douglas Gomery, "*Brian's Song:* Television, Hollywood, and the Evolution of the Movie Made for Television," in *American History/American Television*, ed. John E. O'Connor (New York: Frederick Ungar, 1983); Dombkowski, pp. 183, 195–216.

69 Hilmes, p. 166.

70 Even though some of the major Hollywood companies started to include television revenues on their balance sheets, accounting practices varied as to how these revenues were actually reported. For example, while MCA reported ownership of 1,657 television negatives, co-ownership of an additional 525 television negatives, and 700 pre-1948 films from Paramount, television revenues were not broken out of the revenues reported in *Moody's Industrial Manual*, 1961, p. 756. For a discussion of this problem, see Dombkowski, pp. 189–91.

71 Don Carle Gillette, "Self Deception in Diversity," *Variety*, 12 August 1959, pp. 3, 20, used these estimates as evidence that television wasn't proving to be a significant source of revenues for film companies. However, the rest of the article details the range of diversification activities already evident at the major Hollywood companies.

72 See Thomas H. Guback and Dennis J. Dombkowski, "Television and Hollywood: Economic Relations in the 1970s," *Journal of Broadcasting*, Vol. 20, No. 4, fall 1976; William Kunz, "A Political Economic Analysis of Ownership and Regulation in the Television and Motion Picture Industries," Ph.D. dissertation, University of Oregon, 1998.

73 Quoted in Jackson, *Walt Disney: A Bio-Bibliography*, pp. 49–50.

74 Balio, pp. 136–38; Bob Thomas, *Building a Company: Roy O. Disney and the Creation of an Entertainment Empire* (New York: Hyperion, 1998), p. 183.

75 Leonard Maltin, *The Disney Films* (New York: Crown Publishers, 1973), p. 122.

76 Bill Cotter, personal communication with the author, 6 September 2002; Barbara Jenkins, personal communication with the author, 16 September 2002; Bill Cotter, *The Wonderful World of Disney Television* (New York: Hyperion, 1997), pp. 185–87; Wesley Hyatt, *The Encyclopedia of Daytime Television* (New York: Billboard Books, 1997), p. 288.

77 Richard Schickel, *The Disney Version* (New York: Avon Books, 1968), p. 28.

78 *Moody's Industrial Manual*, 1961, p. 390.

79 "The Feature Is the Commercial," *Broadcasting*, 13 January 1958, p. 46.
80 Gillette, p. 20.
81 *Fortune*, July 1960, cited in Barnouw, p. 64.
82 This section is drawn from Universal's website, www.universalstudios.com; *Moody's Industrial Manual*, 1950–61; Clive Hirschhorn, *The Universal Story* (New York: Crown Publishers, 1983); Dan E. Moldea, *Dark Victory: Ronald Reagan, MCA, and the Mob* (New York: Viking, 1986); Bernard F. Dick, *City of Dreams: The Making and Remaking of Universal Pictures* (Lexington: University of Kentucky Press, 1997); and Dennis McDougal, *The Last Mogul: Lew Wasserman, MCA, and the Hidden History of Hollywood* (New York: Crown Publishers, 1998).
83 Anderson, p. 5.

Chapter 23

Brian Neve

HOLLYWOOD AND POLITICS IN THE 1940s AND 1950s

Anti-fascism and the role of the OWI in wartime Hollywood

The 1940s and 1950s represented an intensely political period for Hollywood as the US fought a global war against fascism then embarked on a Cold War against 'international communism'. For much of the 1930s, despite the politicisation of a number of writers, actors and directors and despite pressure from the guilds and a number of anti-fascist organisations, Hollywood was slow to deal with the emerging threat posed by Nazi Germany. With the exception of Warner Bros., which responded early on to Nazi anti-Semitism, the interest of the studios in the march of fascism tended only to increase in step with the closing of markets in fascist countries. Following on from *Confessions of a Nazi Spy* (1939), a personal project of Harry Warner's that dramatised the domestic threat posed by the German–American Bund, came a trickle of anti-Nazi films, among them *Foreign Correspondent*, *The Great Dictator* and *The Mortal Storm*, all of which were produced and released in 1940. At the same time, though, a Congressional Committee on 'Un-American Activities', which had first been created in 1934, began to shift its attention away from American fascists and towards those on the Left. Reflecting growing political opposition to Roosevelt and the New Deal, Martin Dies, a conservative Texas Democrat, proposed investigating Communism in Hollywood as early as 1939. The following year, his House Committee on Un-American Activities (HUAC, though then called the Dies Committee) visited Los Angeles and conducted a series of hearings. A good deal of publicity was created and a number of prominent film stars, among them Fredric March and James Cagney, appeared before Dies to clear themselves from highly speculative and poorly researched charges that they harboured Communist Party sympathies.[1]

Meanwhile, an increasing flow of pro-interventionist pictures led to complaints from isolationist Congressmen that Hollywood was creating war hysteria and producing propaganda in support of US involvement in what they saw as a European war. In 1941, Warners released the enormously popular *Sergeant York*, which dealt with a World War One pacifist turned war hero, and Twentieth Century-Fox released *A Yank in the RAF* and *Man Hunt*, two films that appeared to advocate American involvement overseas. Also that year, at the request of two isolationist Senators, Burton Wheeler and Gerald Nye, a Senate subcommittee on Interstate

Commerce was created to investigate 'Motion Picture Screen and Radio Propaganda'; Charles Lindbergh, a leading member of the America First movement, spoke in hostile terms of the role of Jewish interests, the British government, and the Roosevelt administration in pushing the country towards war; and British films such as *Convoy* (1940) and *That Hamilton Woman* (1941) were cited as propaganda for intervention. However, along with studio executives Harry Warner and Darryl F. Zanuck, Wendell Wilkie, the studios' attorney (and a former Republican Presidential candidate) trenchantly defended the industry at the Nye–Wheeler hearings. Key members of the subcommittee were discredited by their ignorance and their anti-Semitic remarks, and proceedings were abandoned when the Japanese attacked Pearl Harbor on 7 December 1941, an event that led to American involvement in the war and to a period of unprecedented unity between government and the motion picture industry. Although Jack Tenney's Joint Fact-Finding Committee on Un-American Activities continued to investigate the Hollywood Left on behalf of the California Legislature, there was a lull in this particular form of cultural conflict until the end of the war.[2]

In terms of later developments in Hollywood, the other significant event of this period was the Nazi–Soviet Pact. During the 1930s, Communists had played a growing role both in the struggle for studio recognition of the guilds and in anti-Nazi organisations such as the Hollywood Anti-Nazi League. The alliance between Communists and liberals to which these developments gave rise was abruptly halted by the Pact (which was signed in August 1939), and liberals were particularly shocked that their former allies had shifted their line overnight and now regarded the war between Britain and Nazi Germany as one of imperial rivalry, with both parties on an equal moral footing. While liberal–Communist differences were papered over during the war years, the breakdown in Popular Front politics after Pearl Harbor would later become damaging to the American Communist Party (CPUSA) when it most needed liberal support after the war.[3]

While the attack on Pearl Harbor transformed the nature of Hollywood's outlook, growing criticism of the government's information services led President Roosevelt to create the Office of War Information (OWI). Roosevelt went on to appoint Lowell Mellett, a close friend and advisor, as head of the Bureau of Motion Pictures (BMP), which was part of the OWI's domestic branch. In July 1942, the Bureau issued a Government Information Manual for the Motion Picture Industry, a document written by Mellett's appointee Nelson Poynter, head of the BMP's Hollywood liaison office. This manual has been described as 'the clearest possible statement of New Deal, liberal views on how Hollywood should fight the war'. It stressed that the 'people's war' was not just a matter of self-defence, but a fight for democracy and the 'Four Freedoms' against the forces and values of fascism. By late 1942 the OWI began to make an impact on studio production, with many films and genres stressing the importance of inclusiveness, self-sacrifice and cooperation. Summing up its early impact, the BMP felt that there had been a decided change in the industry's outlook. From regarding the war primarily as a 'dramatic vehicle upon which to build successful box office productions', leading producers were beginning to consider the effect of their films on the task of winning the war.[4]

Drawing on its reputation as the studio that was closest to the New Deal and the working classes, Warner Bros. made a particularly prominent contribution to war propaganda. Writing in the *Daily Worker* in 1943, David Platt described Warners as the '100 per cent pro-New Deal studio', while Jack Warner cited his studio's *Yankee Doodle Dandy* (1942), *Casablanca* (1943), *Edge of Darkness* (1943) and *Watch on the Rhine* (1943) as helping the people 'to understand the peace and the victory'. Also released in 1943 was *Mission to Moscow*, a film that Warner saw as showing 'that Russia is doing much more than just pouring out its blood to stop Nazis'. Those who worked on it, among them screenwriter Howard Koch, believed that the venture had the specific endorsement of the President, although its historical distortions, justified in terms of bolstering the wartime alliance with the Soviet Union, led to controversy, and to later references to the film as an example of pro-Soviet propaganda.[5]

However, despite the apparent consensus on US war aims, ideological changes were afoot. Anti-New Deal forces had been strengthened by the November 1942 Congressional elections and in 1943 there was increasing opposition in Congress to the Administration and to what was characterised as New Deal propaganda. In June 1943 drastic cuts were imposed on the budget of the Overseas Branch of the OWI, and Poynter and Mellett both left the organisation. Thereafter Ulric Bell, who had moved to the Hollywood office of the BMP in November 1942, exercised what Koppes and Black call 'an influence over an American mass medium never equalled before or since by a government agency'. Bell advised the Office of Censorship on films that were suitable for export. He thus wielded an unprecedented level of governmental power. Home front dramas risked transgressing the Office of Censorship code if they depicted class or racial conflict or provided ammunition for enemy propagandists, while portrayals of corruption, crime and criminals were generally viewed as at odds with wartime assertions of patriotic unity. From an emphasis on the themes of the Government Information Manual in 1942 and 1943, the later war years were marked by a ruling out or a toning down of topics that might imply criticism of the U.S. and its institutions, though by war's end, Hollywood clearly possessed a stronger sense of its social role and responsibilities than it had done before the war.[6]

Post-war: the origins of the Hollywood blacklist

In 1945, at the prompting of John Rankin, a notoriously anti-Semitic Congressman from Mississippi, the House Committee on Un-American Activities became a standing committee of the House of Representatives. The November 1946 elections returned Republican majorities to both houses of Congress, thus strengthening the anti-New Deal coalition. Grounds for ideological conflict within the film industry had already been augmented by the founding of the Motion Picture Alliance for the Preservation of American Ideals (Motion Picture Alliance) in 1944.[7] The Hollywood blacklist had its origins in HUAC's decision to hold hearings on Communism in the film industry in 1947. In a period of emerging Cold War, HUAC became a vehicle for politicians who opposed New Deal policies and who saw the film industry as vulnerable to charges of Communist subversion.

The Motion Picture Alliance, which included such figures as Sam Wood, Walt Disney and Lela Rogers, sought to publicise what they saw as left-wing propaganda in wartime films and the left-leaning influence of Hollywood screenwriters. Many members of the Alliance had links with William Randolph Hearst, who had used his media outlets to pursue an anti-Communist agenda before the war. In early 1947, with much of Europe in economic crisis, President Truman publicly committed the United States (in what became known as the 'Truman Doctrine') to a policy of containing Soviet influence world-wide. Attorney General Clark's talk of domestic Communist plots and Truman's executive order of March 1947 establishing a loyalty programme for the executive branch helped foster the idea of internal subversion, while bitter industrial conflicts in 1945–46 helped polarise Hollywood politics. Roy Brewer, head of the dominant International Alliance of Theatrical Stage Employees (IATSE) – in conflict at the end of the war with a Red-baited rival, the Conference of Studio Unions – was to become another leading member of the Motion Picture Alliance. Alliance members encouraged the House Committee to investigate Hollywood Communism, while the new chairman of HUAC, J. Parnell Thomas, began to develop closer relations with the FBI, which, with the aid of informers, had been gathering information on Hollywood Communists since 1943. Among the Hollywood moguls, it was Jack Warner (whose studio had been riven by strikes, pickets and violence in 1945) who stoked the fires of the anti-Communist crusade by providing the Committee with horror stories of Communists working at his studio.[8]

The first formal post-war HUAC hearings took place in Washington DC in October 1947. In the first week, there was testimony from 'friendly' witnesses, mainly studio bosses and figures from the Motion Picture Alliance, while the second week was reserved for the testimony of nineteen witnesses (a majority of them Jewish) considered as having a case to answer. These nineteen agreed to stand on the principle of the First Amendment on free speech. Liberals Philip Dunne, William Wyler and John Huston organised the Committee for the First Amendment, and arranged for a planeload of sympathetic Hollywood stars to visit Washington. However, the First Amendment strategy was implicit rather than explicit in the testimony of those of the ten 'unfriendly' witnesses – eight writers, a producer and a director – who actually testified, and most of them became involved in a shouting match with Chairman J. Parnell Thomas that served only to alienate a number of their supporters. These unfriendly witnesses – Alvah Bessie, Herbert Biberman, Lester Cole, Edward Dmytryk, Ring Lardner Jr, John Howard Lawson, Albert Maltz, Samuel Ornitz, Adrian Scott and Dalton Trumbo – became known as the 'Hollywood Ten'. They all followed the example of the initial witness, screenwriter and key Hollywood Communist Party leader John Howard Lawson, in refusing to answer questions regarding their membership of the Party while attempting, with little success, to read their prepared statements. As a result, the hearings were suspended and the Ten were cited for contempt of Congress. With the Committee publicising details of the witnesses' Communist affiliations, liberal support for their First Amendment rights quickly declined. Pressure on stars and their agents may have also been applied. Either way, many Hollywood liberals abandoned their protests or went to work abroad.[9]

Eric Johnston, President of the Motion Picture Producers Association (MPAA), and key spokesman for the producers and their boards of directors, had originally resisted any blacklist. But in light of the evidence of Communist membership supplied by the Committee he sought to protect the studios from possible boycotts and threats of censorship. The Ten were cited for contempt, and on November 24 the House and Senate voted overwhelmingly to confirm this judgement. On the same day, Johnston convened a meeting of studio producers and executives at the Waldorf-Astoria Hotel in New York. Their discussions led to the issuing of the so-called Waldorf Statement, declaring that the studios would dismiss those of the Ten under contract and would not re-employ any of them 'until such time as he is acquitted or has purged himself of contempt and declared under oath that he is not a Communist'. The Statement continued:

> We will not knowingly employ a Communist or a member of any party or group which advocates the overthrow of the Government of the United States by force or by any illegal or unconstitutional methods.[10]

This statement essentially instituted a blacklist, although its full implementation by the studios would have to wait three years for the legal appeals of the Ten against their convictions to be considered.

In the wake of this decision, it soon became apparent that the studios were becoming more wary of tackling social topics. Eric Johnston stated that 'We'll have no more films that show the seamy side of American life', while William Wyler suggested that in the 'current climate' he would not have been allowed to make *The Best Years of Our Lives* (1946).[11] However, the picture in Hollywood in the late 1940s was mixed. Anti-Semitism was addressed in films such as *Crossfire* (1947) and *Gentleman's Agreement* (1948), and cycles of what would later be categorised as 'semi-documentaries', 'films noirs' and 'films gris' facilitated Left-liberal critiques of American society in crime-film form. In addition, the unexpected re-election of Harry Truman in November 1948 was taken as evidence that 'liberal days were here again', encouraging studio heads to give the go-ahead to a group of films that raised and articulated (however belatedly) a set of progressive positions on racial equality. Writer and director Abraham Polonsky referred to this period as 'interesting times' for a new generation of socially minded filmmakers. Having experienced the Depression and the New Deal, these filmmakers, among them John Berry, Joseph Losey, Jules Dassin, Cy (Cyril) Endfield (and, of course, Polonsky himself) were gaining a foothold in the industry, and their distinctive social and aesthetic aims were evident.[12]

What is more difficult to assess is the impact of the darkening national and international political situation, with the emergence of the Cold War and the rise of the spectre of 'Red Fascism' (the parallel that was drawn at this time between the political system and foreign policy of the defeated Nazi state and that of the Soviet Union). In the wake of a series of legal and financial blows to the studio system, established industry leaders saw themselves as needing to protect their industry. The increasing importance of foreign markets in an era of declining domestic attendances, and the industry's need for State Department help in exploiting them, tended to encourage an 'optimistic portrayal of the American way of life', while the growing identification of the CPUSA with the Soviet Union meant that the alliance between radicals and liberals more or less broke down. When Albert Maltz wrote to John Huston calling for an alliance against the forces of reaction based on 'common Jeffersonian principles', there was no response. Instead, Arthur Schlesinger Jr, standard bearer for a new anti-Communist liberalism, a 'vital center' against the totalitarianism of left or right, emerged as a more representative figure on the reconstituted left of American politics. When in 1950 the Supreme Court declined to review the cases of Dalton Trumbo and John Howard Lawson, the blacklist already implicit in Eric Johnson's statement was ready to be implemented. Whatever the niceties of the Constitution, members of the domestic Communist Party were not seen as deserving First Amendment guarantees.[13]

The early 1950s: renewed HUAC hearings

The Hollywood Ten were sent to prison in 1950. In 1951, the Supreme Court upheld the conviction of the leaders of the American Communist Party under the 1940 Alien Registration Act (the 'Smith Act') for advocating the overthrow of the United States government.[14] The country was in the grip of a national panic. In 1949 the Soviet Union had tested an atomic bomb and Communists had come to power in China. When Senator Joe McCarthy began his four-year period in the limelight the following year, he was in part reacting to a series of spying cases, and in particular to the conviction of Alger Hiss, the former State Department official and Roosevelt aide who had been accused by HUAC in 1948 of passing secrets to a Communist spy ring. To Richard Pells, the conviction of Hiss lent credence to 'the theory that all communists should be regarded as potential foreign agents'. Later on that year, the Korea war began and Ethel and Julius Rosenberg were arrested on a charge of conspiracy to commit espionage, further fuelling an atmosphere in which ex-Communists (and in some cases liberals) were called upon to demonstrate their patriotism.[15]

In 1950, amid signs of an impending second wave of Congressional interest in Communism in Hollywood, the Motion Picture Alliance and the American Legion called for a 'complete delousing' of the motion picture industry. In September, Harry Warner addressed 2,000 Warner Bros. employees and executives, warning of the Communist threat and making it clear that he wanted no Communists at the studio. Earlier that summer, there had been a protracted battle between liberal and conservative factions of the Screen Directors Guild over the plan, urged on the Guild membership by arch anti-communist Cecil B. DeMille, to introduce a mandatory oath for new and existing members affirming their lack of allegiance to the Communist Party. To DeMille, who was finally forced to back down in an epic membership meeting, the 'question we are asking is, are you on the American side or on the other side'.[16]

As anticipated, in March 1951 the House Committee resumed its hearings on Communism in the film industry. From that point on until 1953, the blacklist that had been foreshadowed in 1947 was fully enforced. The core group of those who now found themselves unemployable at the major studios were those known by the authorities to be present or former Communist Party members, and those who declined to cooperate fully with the Committee. Given the fate of those who pleaded the First Amendment in 1947, those who were now unwilling to cooperate and 'name names' were advised to invoke the Fifth Amendment against self-incrimination, though doing so usually led to blacklisting too. The great bulk of Communists were well known to the FBI and Committee, so that cooperation became a kind of clearance ritual, what Victor Navasky later called a 'degradation ceremony'. From 1951 to 1953 over 200 Communists were named. Larry Parks, Richard Collins, Budd Schulberg, Edward Dmytryk (reversing his position as one of the original Hollywood Ten), Elia Kazan, Clifford Odets and Martin Berkeley were among those who became 'friendly' witnesses, while Howard Da Silva, Abraham Polonsky, Paul Jarrico, Carl Foreman and Lillian Hellman were among those who took the Fifth or avoided appearing (sometimes by leaving the country). Robert Rossen testified in 1951 about his own party membership but declined to name others, However, he returned to the Committee as a friendly witness two years later. Rossen was one of a number of film-makers in this period who could not leave the country because they were unable to obtain or renew a passport. Some blacklisted writers continued to work anonymously, usually at a fraction of the salary and often on unpromising assignments. Others moved abroad. They included John Berry, Joseph Losey and the other young socially minded filmmakers mentioned above.[17]

A number of private organisations exploited and expanded the blacklist process in the early 1950s for a mixture of commercial and ideological motives. American Business Consultants, formed in 1947 by ex-FBI men, marketed books and newsletters, among them *Red Channels* and *Counterattack*, to employers and advertisers. Ayn Rand's 'Screen Guide for Americans', a pamphlet which in particular urged favourable screen treatment of business, was published by the Motion Picture Alliance in 1950. And in addition to expanding the blacklist, the American Legion helped create a 'greylist'. The Legion had 3,000 branches, some of which picketed theatres that were playing pictures involving 'suspect' artists. Through publications such as *Firing Line* (Facts for Fighting Communism'), it had a direct effect on the studios and their hiring policies. Around 300 film industry employees were blacklisted or grey listed in the early to mid 1950s.[18]

In December 1952, the Legion published an article by professional anti-communist J.B. Matthews entitled 'Did the Movies Really Clean House?' thus adding to the woes of an industry that was already laying off workers in response to declining audiences. The Legion reported a 'fruitful' meeting with Eric Johnston and representatives of the key producing companies in March 1952, and it was as a result of this meeting that procedures were established for those under pressure to clear themselves. Meanwhile an article in *The Nation* suggested that the investigations were seen by film executives as a blessing in disguise. Workers and unions had now been 'pacified', enabling the industry to cut its wage bill. This argument seems to support the view that the blacklist was, at least in part, 'good business', enabling the studios to regain control of the entertainment marketplace.[19]

The Eisenhower years: Hollywood and the blacklist legacy

When Joe McCarthy was censured by the Senate in December 1954, and with a Republican, Dwight D. Eisenhower, now in the White House, fears of an internal Communist threat began to abate. HUAC continued to hold occasional sessions on Communism in the film industry and the blacklist remained in place, but a number of writers, most notably Dalton Trumbo, used fronts in order to work in the industry, albeit on less prestigious pictures. While inferences can be made about the effect of the blacklist on the content of motion pictures, it is difficult to distinguish between the effects of the blacklist, the effects of the Cold War, and effects of other socio-cultural factors. In a survey of the content of American films made between 1947 to 1954, Dorothy Jones observed a marked decline in the number of social problem films made between 1949 and 1952, a greater emphasis on 'pure entertainment', and a greater number of anti-communist films and war films of the 'sure-fire patriotic variety'.

Other interpretations are generally consistent with this analysis, stressing a shift toward an increasing number of Westerns, war films and biblical epics, and a perspective that was more often psychological than social (though some 1950s Westerns, among them *High Noon* [1952], *Johnny Guitar* [1954] and *Silver Lode* [1954], appear to make metaphoric reference to the blacklist, while others dealt with racial prejudice or featured corrupt or dictatorial ranch owners, cattle barons or businessmen). Larry May charts a decline in 'Noir' (i.e. unhappy) endings in the 1950s, a rise in the focus on youth as an alternative to the adult world, and a fall in the incidence of depictions of the rich as a moral threat and of big business as villainous.[20]

Certainly the mid 1950s saw a tailing off in the number of crime melodramas that are now either classified as 'noir' or 'gris', partly because many of the writers and directors associated with them were affected by the second wave of Congressional investigations. Nicholas Ray avoided the blacklist, but his late 1940s films, which included *Knock on Any Door* and *They Live By Night* (both 1949), were more overtly sceptical about the American Dream and more overtly concerned with the 'psychological injuries of class' than were some of his later films. Two of the more well known victims of the blacklist were John Garfield and Abraham Polonsky. Garfield had helped to set up Enterprise Pictures. He appeared in two key Enterprise productions in the late 1940s, *Body and Soul* (1947) and *Force of Evil* (1949), both scripted and directed by Polonsky, and it was Garfield who had insisted on reinstating Hemingway's key black character in *The Breaking Point* (1951), providing what Thomas Cripps calls a 'bold stride towards a humane portrayal of interracial comradeship'. Although the FBI knew that he was not a Communist, he was subpoenaed in March 1951 and harassed further when he talked of his own political affiliations but refused to discuss others. Having been forced to recant his political views, he died of a heart attack in May 1952. Tracked and wiretapped by the FBI, Polonsky took the Fifth Amendment in April 1951. He subsequently worked anonymously on a number of films and on the television series 'You Are There'.[21]

For a time in the early 1950s the CIA, with its emerging concern for the mobilisation of art and culture in the service of America's foreign policy goals, became interested in Hollywood and began to encourage more appearances by black Americans and other ethnic minorities as a means of improving foreign perceptions of American democracy. Other 1950s films reflect the changing politics in other ways. Elia Kazan, a Communist for two years in the mid 1930s, had twice been called before HUAC in 1952. On the second occasion he gave the names of fellow members of his Communist Party unit in the Group Theatre in New York some sixteen years earlier. *Viva Zapata!*, which he directed in 1952, provided a mixed political message, combining an unusually sympathetic Hollywood account of Emiliano Zapata's role in the Mexican revolution with some pointed contemporary references to Communist intellectuals and their exploitation of popular revolt. Post-testimony, Kazan teamed up with another friendly witness, novelist Budd Schulberg, to make *On the Waterfront* (1954). Some have seen the film's treatment of the central character's testimony against mob rule among the New York longshoreman unions as making metaphoric reference to the Congressional investigations.[22]

One film that clearly represented an attempt at resistance by blacklisted filmmakers was *Salt of the Earth* (1954). Producer Paul Jarrico, writer Michael Wilson and director Herbert Biberman sought to present the class, ethnic and gender issues at stake in a strike in New Mexico, though various bodies, from the FBI to the Projectionists Union, ensured that the film (which was made with union finance) gained very limited distribution. By the mid 1950s the political climate was beginning to moderate, and the cycles of anti-Communist and Korean war films were in decline; over fifty films with an anti-Communist theme were produced, but most were box-office failures. The blacklist as an institution began to collapse at the end of the 1950s. It took thirty years or more for Michael Wilson and Carl Foreman to receive a credit for their contributions to the script for *The Bridge on the River Kwai* (1957), but in 1960 Dalton Trumbo's name appeared on the credits of *Spartacus* and *Exodus*. This was not the end of blacklist. But it was the beginning of the end.[23]

Looking back at the period, Victor Navasky has written of the blacklist in terms of a moral distinction between those who named names and those who resisted. While most current and dedicated Party members had little doubt as to what to do, others, especially those who had been out of the Party for some time, clearly balanced the unpleasantness of informing with a reluctance to be a martyr for a cause that they no longer believed in. Navasky's perspective has rather over-simplified the moral dynamics of the period, or at least under-examined the political dimension of the moral choices artists, performers and others faced at the time. The issues always appeared more morally clear-cut to those – the great majority of resisters – who were still committed to the Party. Paul Jarrico, a screenwriter who was prominent in the Hollywood Communist Party and who was himself blacklisted, argued that, 'for those who were generally pissed off at the Party but reluctant to name names, the choice must have been difficult. For a person like me, a true blue red, the choice was easy'. He also speculated on the reasons why America in this period had been susceptible to such an obvious transgression of the Constitution:

> What we underestimated was the direct connection between the Cold War abroad and repression at home. Looking back on it now, it just seems very obvious. If you are going to call on people to give their lives in a fight against Communism internationally, you can certainly raise logically the question of why we should allow Communists or Communist sympathizers to express themselves domestically. There was a logic to the reactionary position that we underestimated.[24]

Liberal anti-Communists, corporate visionaries, and bigoted right wingers were for a time thrown together in a project that none of them totally controlled. The blacklist itself was part of an intense period of history in which American national identity was remade for a new era of US Cold War leadership, an era marked by a prosperity that was unimaginable in the Depression years and that changed attitudes and identities in a number of ways. A new liberalism compatible with America's new national and international status emerged gradually, and was to find particular expression (on and off the screen) in the Kennedy era.

Notes

1. Michael E. Birdwell, *Celluloid Soldiers: Warner Bros.'s Campaign Against Nazism* (New York: New York University Press, 1999); Larry Ceplair, 'The Film Industry's Battle against Left-wing Influences, from the Russian Revolution to the Blacklist', *Film History*, vol. 20 no. 4 (2008), 401–4; Larry Ceplair and Steven Englund, *The Inquisition in Hollywood: Politics in the Film Community, 1930–1960* (Berkeley: University of California Press, 1983), 155; Eric Sandeem, 'Anti-Nazi Sentiment in Film: *Confessions of a Nazi Spy* and the German-American Bund', *American Studies*, vol. 20 no. 2, 1979, 69–73, David Welky, *The Moguls and the Dictators: Hollywood and the Coming of World War II* (Baltimore: Johns Hopkins Press, 2008), 11–292.
2. Steven Carr, *Hollywood and Anti-Semitism: A Cultural History up to World War II* (Cambridge: Cambridge University Press, 2001), 242–47, 250ff.; Thomas Doherty, *Projections of War: Hollywood, American Culture and World War II* (New York: Columbia University Press, 1993), 40–42; Welky, *The Moguls and the Dictators*, 293–324.
3. Larry Ceplair and Steven Englund, *The Inquisition in Hollywood*, pp. 175–77.
4. K.R.M. Short, 'Documents (B): Washington's Information Manual for Hollywood, 1942', *Historical Journal of Film, Radio and Television*, vol. 3 no. 2 (1983), 171, 179.
5. Platt, *Daily Worker*, 2 June 1943, 7; Warner Bros. press release (undated), OWI files, Washington National Records Center, Suitland, Maryland; David Culbert (ed.), *Mission to Moscow* (Madison: University of Wisconsin Press, 1980), 31–41.
6. Clayton R. Koppes and Gregory D. Black, 'What to Show the World: The Office of War Information and Hollywood, 1942–45', *Journal of American History*, vol. 64 no. 1 (1977), 103; see also Clayton R. Koppes and Gregory D. Black, *Hollywood Goes to War: How Politics, Profits and Propaganda Shaped World War II Movies* (New York: The Free Press, 1987).
7. Thomas Schatz, *Boom and Bust: The American Cinema in the 1940s* (New York: Scribners, 1997), 307; Walter Goodman, *The Committee* (London: Secker and Warburg, 1968), 173–74.
8. Schatz, *Boom and Bust*, pp. 307–13; Louis Pizzitola, *Hearst Over Hollywood: Power, Passion and Propaganda in the Movies* (New York: Columbia University Press, 2002), 405, 409–11; on the key role of the FBI and its Director, J. Edgar Hoover, see John Sbardellati, 'Brassbound G-Men and Celluloid Reds: The FBI's Search for Communist Propaganda in Wartime Hollywood', *Film History*, vol. 20 no. 4 (2008), 412–36; Ellen Schrecker, *Many are the Crimes: McCarthyism in America* (Boston: Little, Brown and Company, 1998), 157.
9. Howard Suber, 'The Anti-Communist Blacklist in the Hollywood Motion Picture Industry', Ph.D. Dissertation (UCLA: 1968), 25–26.
10. John Cogley, *Report on Blacklisting, 1 – Movies* (New York: The Fund for the Republic, 1956), 11; Ceplair and Englund, *The Inquisition in Hollywood*, 209–25, 261ff; *Variety*, November 26, 1947, 3.
11. Johnston, in Murray Schumach, *The Face on the Cutting Room Floor: The Story of Movie and Television Censorship* (New York: Da Capo Press, 1975), 139.
12. Abraham Polonsky, interview with the author, Los Angeles, August 20, 1988.
13. Carol Traynor Williams, *The Dream Beside Me: The Movies and the Children of the Forties* (Rutherford: Fairleigh Dickinson University Press, 1980), 203; Les K. Adler and Thomas G. Paterson, 'Red Fascism: The Merger of Nazi Germany and Soviet Russia in the American Image of Totalitarianism, 1930s–1950s', *American Historical Review*, vol. 75 no. 4 (1970), 1046–64; Richard Maltby, 'Film Noir: The Politics of the Maladjusted Text', *Journal of American Studies*, vol. 18 no. 1 (1984), 64; Albert Maltz to John Huston, undated, Huston Collection (1949 HUAC folder), Academy of Motion Picture Arts and Sciences; Arthur Schlesinger, Jr, *The Vital Center: The Politics of Freedom* (Boston: Houghton Mifflin, 1949); Abraham Polonsky, interviewed in Eric Sherman and Martin Rubin (eds), *The Director's Event: Interviews with Five American Film-Makers* (New York: New York American Library, 1970), 10.
14. Ellen Schrecker, *Many are the Crimes*, 190.
15. Schrecker, *Many are the Crimes*, p. 154; Richard Pells, *The Liberal Mind in a Conservative Age: American Intellectuals in the 1940s and 1950s* (New York: Harper and Row, 1985), 272.
16. Suber, 'The Anti-Communist Blacklist in the Hollywood Motion Picture Industry', 37; Harry Warner, *New York Times*, 10 September, 1950; Kenneth L. Geist, *Pictures Will Talk: The Life and Films of Joseph L. Mankiewicz* (New York: Scribners, 1978), 173–206; DeMille, October 18, 1950, in George Stevens Collection, Folder on Screen Directors Guild, 1950, Academy of Motion Picture Arts and Sciences, Los Angeles.
17. Victor S. Navasky, *Naming Names* (New York: The Viking Press, 1980), 319; Suber, 'The Anti-Communist Blacklist in the Motion Picture Industry', 38.

18 Ayn Rand, *Guide for Americans*, The Motion Picture Alliance for the Preservation of American Ideals, Beverly Hills, California, undated.

19 J. B. Matthews, 'Did the Movies Really Clean House?', *The American Legion Magazine*, December 1951; Suber, 'The Anti-Communist Blacklist in the Hollywood Motion Picture Industry', 44–48; 'X: Hollywood Meets Frankenstein', *The Nation*, 28 June 1952, 628–31; Jon Lewis, '"We Do Not ask You to Condone This": How the Blacklist Saved Hollywood', *Cinema Journal*, vol. 39 no. 2 (2000), 4–12, 17–18.

20 Dorothy Jones, 'Communism and the Movies: A Study of Film Content', in Cogley, *Report on Blacklisting*, 220, 282; Richard Slotkin, *Gunfighter Nation: The Myth of the Frontier in Twentieth-Century America* (New York: Athenaeum, 1992), 347–48; Walter Bernstein, *Inside Out: A Memoir of the Blacklist* (New York: Knopf, 1996), 177; Larry May, *The Big Tomorrow: Hollywood and the Politics of the American Way* (Chicago: University of Chicago Press, 2000), 204, 273–93.

21 On the politics of 'noir' and 'gris', see Thom Andersen, 'Red Hollywood', in Frank Krutnik, Steve Neale, Brian Neve and Peter Stanfield, eds, *"Un-American" Hollywood: Politics and Film in the Blacklist Era* (New Brunswick, NJ: Rutgers University Press, 2007), 257–61; on Garfield, see Robert Sklar, *City Boys: Cagney, Bogart, Garfield* (Princeton, NJ: Princeton University Press, 1992), 203ff.; on Polonsky, see Abraham Polonsky, 'Introduction', in Howard Gelman, *The Films of John Garfield* (Secaucus, NJ: The Citadel Press, 1975), 8, and Paul Buhle and Dave Wagner, *A Very Dangerous Citizen: Abraham Lincoln Polonsky and the Hollywood Left* (Berkeley: University of California Press, 2001), 10–11. A number of other blacklistees wrote scripts for television series. Among them were Ian McLellan Hunter, Ring Lardner Jr, Maurice Rapf, Waldo Salt and Adrian Scott. See Steve Neale, 'Swashbucklers and Sitcoms, Cowboys and Crime, Nurses, Just Men and Defenders: Blacklisted Writers and TV in the 1950s and 1960s', *Film Studies*, no. 7 (2009), 83–103.

22 David N. Eldridge, '"Dear Owen": The CIA, Luigi Luraschi, and Hollywood, 1953', *Historical Journal of Film, Radio and Television*, vol. 20 no. 2 (2000), 155; Brian Neve, *Elia Kazan: The Cinema of an American Outsider* (London: I. B. Tauris, 2009).

23 James J. Lorence, *The Suppression of Salt of the Earth: How Hollywood, Big Labor, and Politicians Blacklisted a Movie in Cold War America* (Albuquerque: University of New Mexico Press, 1999), 54–55. On the collapse of the blacklist, see Suber, 'The Anti-Communist Blacklist in the Hollywood Motion Picture Industry', 124ff.

24 Paul Jarrico quotations, in Patrick McGilligan and Paul Buhle, *Tender Comrades: A Backstory of the Hollywood Blacklist* (New York: St Martin's Press, 1997), 345–46. See also Dan Georgakas, 'The Hollywood Reds: Fifty Years Later', *American Communist History*, vol. 2 no. 1 (June 2003), 63–76.

Chapter 24

Steve Neale

ARTIES AND IMPORTS, EXPORTS AND RUNAWAYS, ADULT FILMS AND EXPLOITATION[1]

THE LATE 1940s WITNESSED THE ADVENT of a number of key developments in the US film industry. In 1945, World War Two came to an end, the MPPDA became the MPAA (Motion Picture Association of America), Eric Johnstone replaced Will Hays as its president, and another industry organisation, the Motion Picture Export Association (MPEA), in essence a merger between the MPAA and the overseas arm of the Office of War Information, was 'created to facilitate trade' in a 'postwar global marketplace'.[2] Thus as domestic attendances and profits surged, so too did profits from the unrestricted release of US films in previously occupied or enemy countries in Europe and elsewhere abroad. In October 1946,

> the *Motion Picture Herald* reported that the 'lid' on the foreign market had been 'pried open' and that the MPEA seemed to be fending off protectionism overseas. At year's end, the studios reported that their overseas income of $125 million was virtually identical to their overall net profits – a situation that many in the industry considered ideal, with the domestic market on a break-even basis and overseas income amounting essentially to pure profit.[3]

The traffic at this point was nearly all one way. However in 1946 Britain's J. Arthur Rank, who had established links with Universal and United Artists in the 1930s and the early 1940s respectively, and who was determined to build what *Variety* called a 'worldwide film empire', went on to orchestrate a merger between Universal and International Pictures and the setting up of a production and distribution company called Eagle-Lion to handle the distribution of his company's film in the US.[4] The first of these films was *Brief Encounter*, which was released through a new Universal subsidiary called Prestige Pictures and which opened at the Little Carnegie Playhouse on 25 August 1946. Six months earlier, on 25 February, an Italian film entitled *Open City* in the US opened at the World Theatre. Directed by Roberto Rossellini and released in Italy as *Roma, città aperta* in 1945, *Open City* was distributed in the US by Arthur Mayer and Joseph Burstyn.[5] The Little Carnegie and the World Theatre were Manhattan art houses. Rank was reportedly unhappy that *Brief Encounter* was released in this market and in this way, but it performed reasonably well and went on to profit from subsequent showings in ordinary commercial theatres. A number of Rank's subsequent medium-range productions, among them *I Know Where I'm Going!* and *Great Expectations* (both 1947), followed suit.[6] Meanwhile, the success of *Open City* was unprecedented. It ran for nearly two years at the World; it was subsequently released in a number of mainstream cinemas as well as in other art houses; and it reportedly grossed $5 million (a record for a foreign film).[7] Voted Best Foreign Language Film in 1946 by the New York Film Critics Circle, it demonstrated not only that there was a market for art-house films but that at least some of them possessed wider commercial appeal.

Art films and imports in the 1920s and 1930s

Art-house cinemas and successful foreign imports were not new. As Tino Balio, Douglas Gomery and Barbara Wilinsky all point out, and as has been already been noted elsewhere in this Reader, art-house and ethnic cinemas, film societies and museums, and upscale city-centre venues all provided sites for the exhibition of domestic and foreign art films, amateur, avant-garde and independent films, and retrospectives and mainstream foreign-language films in the mid- to late 1920s and 1930s.[8] While most of these films were imported and distributed by independent organisations of one sort or another, others were imported and distributed by the majors, occasionally showcased in their big-city theatres, and sometimes released more widely.

As Wilinsky points out, it was the financial success of *Passion* (*Madame Dubarry*) and the critical success of *The Cabinet of Dr. Caligari* in the US in 1921 that prompted the majors and others to import more films from Europe in the 1920s. By the end of April 1921, 'Adolph Zukor alone had imported 129 German features for Famous Players-Lasky to distribute in the United States'.[9] Others followed suit, importing films from Italy, Sweden, France and Britain too. At this point there were no art-house cinemas in the US. Imported films were either shown in ordinary commercial cinemas or left on the shelf. However, there did exist a number of emergent discourses on the art of film. One was centred on the Better Films Movement, an offshoot of the National Committee for Better Films, which had been established by the National Board of Review in 1914. By 1923, members received subscriptions to *Photoplay Guide*, *Film Progress* and *Exceptional Photoplays*, which offered appraisals of 'films adjudged to have unusual merit or significance in the development of motion picture art', and which was thus 'the first journal of serious film criticism in the United States'.[10]

The Board went on in 1923 to organise screenings of 'exceptional screenplays' at New York's Town Hall. Two years later, the Shadowbox, the first full-time art house in the US, opened on West 12th Street in New York, 'an area populated by bohemians, artists and other cultural sophisticates'.[11] Clearly modelled on the 'little theatre' movement, the Shadowbox was programmed by an organization called the Screen Guild and operated on a subscription-only basis. Its first film was *Kriemhild's Revenge* (1925). The following year, alongside the foundation of the Amateur Cinema League (ACL), the newly formed International Film Arts Guild leased the Cameo Theatre in New York, dubbed it 'The Salon of the Cinema', interspersed screenings of Hollywood classics with those of European films such as *The Last Laugh* (1925) and *Crainquebille* (1922), and mounted a series of special evenings that mixed both kinds of films with experimental shorts such as *Ballet Mécanique* (1926). With the Hollywood majors importing more foreign films in order to comply with quota agreements, the 'little cinema' movement, spurred on by interest in *Cruiser Potemkin* (*Battleship Potemkin*) (1926) and other films from the USSR, began to grow, and an 'art-of-film' discourse, much of it focused on the specificities of film as a medium, began to flourish.[12] This discourse was heterogeneous. Some of it encompassed Hollywood films and Hollywood directors. But most of it contrasted Hollywood films, which were increasingly perceived as commercially driven, factory-produced and aesthetically formulaic, with those made elsewhere. However, despite the success of *Variety* (1926), it was the imports that adhered most closely to Hollywood norms and values, imports such as *Passion*, *Gypsy Blood* (*Carmen*) and *Theodora*, all of which were released in 1921, that proved the most successful.

The coming of sound led to an increase in the number of imports, but most of them were shown in ethnic theatres rather than ordinary commercial or art-house cinemas.[13] Alongside the ACL, the newly formed Workers' Film and Photo League began to flourish, and German, French and Soviet films continued to be shown in art-house cinemas. However, the number of German imports fell with the coming to power of Hitler in 1933 and the number of Soviet imports fell with the advent of the Nazi–Soviet pact in 1939, at which point the number of art houses in New York fell from six to three. The most commercially successful imports were Alexander Korda's prestige British films (which despite constant debates about British accents did not require dubbing), and some of the films produced by Gaumont-British. British imports (and US films with British settings) were also popular during the war.

The late 1940s

The number of art-house cinemas in the US in the late 1940s is unknown. Industry estimates varied from 365 in 1946 and 400 in 1947, though these figures probably include ordinary cinemas that occasionally showed foreign films.[14] One distributor estimated that 250 theatres regularly played foreign films and *Variety* 'reported a similar total in late 1949 (226 theaters) and put the total of "strictly artfilm theatres" in the United States at 57'.[15] However, what probably matters most is the fact that along with the profile of Italian films such as *Shoe Shine* (1947) and *Paisan* (*Paisà*) (1948), British films such as *Passport to Pimlico* and *Tight Little Island* (*Whiskey Galore*) (both 1949), and French films such as *Beauty and the Beast* (*La Belle et la Bête*) (1947) and *Devil in the Flesh* (*Le Diable au Corps*) (1949), the art-film trend was on the rise and the number of imports was steadily increasing (from 118 in 1947 to 123 in 1949), while the export market for US films was shrinking and becoming much more difficult to navigate.[16]

As Thomas Schatz points out, there were three main reasons for this: 'first, cold war tensions, which rendered Americans' access to many countries behind the Iron Curtain difficult if not impossible; second, the trend toward tariffs, frozen revenues, and other protectionist policies in nations like Britain, France, and Italy that were determined to build up their own film industries and to prevent Hollywood from completely dominating their markets; and third, Britain's deepening postwar crisis'.[17] Thus although foreign revenues amounted to an estimated $120 million in 1947 and an estimated $100 million in 1948 and 1949 (figures sustained, at least in part, by an increase in the acquisition of foreign films for release in foreign markets),[18] they had fallen from an estimated $125 million in 1946. Britain was still Hollywood's largest overseas market. But its economy was rapidly deteriorating and its government imposed a tax of 75 per cent on all film imports. As a result, 'remittances fell from $70 million in 1946 to $56 million in 1947, $35 million in 1948 and $17 million in 1949'.[19]

However, although Spain, parts of South America, and much of the emerging Communist bloc were more or less closed to US films, the number of overseas cinemas began to increase, and although most Western European countries adopted protectionist measures in order to support their economies, they were subsequently boosted by the availability of Marshall Aid and their cinema industries by various forms of government subsidy.[20] In response to the freezing of revenues and the introduction of tariffs and quotas, there was an increase in the number of Hollywood companies that began to establish (or re-establish) overseas studios and subsidiaries, to engage in overseas production, and to set up deals with foreign producers. There was also an increase in the number of US producers who based themselves abroad. Among the companies were MGM, Paramount, Twentieth Century-Fox and Warner Bros., among the producers Irving Allen, Sam Bischoff, David Selznick and Columbia's Gregor Rabinovich, who embarked on the production of a series of opera and costume films which were all filmed in Italy and subsequently shown in US art houses. Among the more mainstream productions were *Foreign Affair*, *So Evil My Love* and *To the Victors* (all 1948) and *I Was a Male War Bride*, *The Man on the Eiffel Tower* and *Prince of Foxes* (all 1949).[21] US companies were not alone. Quotas and currency restrictions applied to European countries too, and a number of French companies began making films in Italy on an experimental coproduction basis as early as 1946. Among them was *Fabiola* (1948),[22] which anticipated the trend toward ancient-world spectacles that is usually said to have begun with *Samson and Delilah* (1949) and *Quo Vadis* (1951) and which was released in cut-and-dubbed form in the US by United Artists a few months prior to the release of *Quo Vadis* itself. (Complete foreign-language versions were subsequently shown in Italian- and Spanish-speaking ethnic theatres.)[23]

In the meantime, although they were a product of a specific set of circumstances, the production of US films abroad augmented a growing trend toward shooting films in authentic locales. This particular trend was initially confined to 'semi-documentary' crime, cop and spy films set and filmed in the US itself.[24] However, along with broader trends toward more vivid depictions of violence and pain, stories of criminal transgression that were more heavily influenced by pulp conventions than ever before, social problem films that sought to address racial prejudice and the social and psychological problems faced by returning veterans; and along with costume films of various kinds, among them *The Outlaw* (1943), *Duel in the Sun* (1946) and *Forever Amber* (1947), which were each heavily marked by the 'provocative' ways in which their leading female characters dressed and behaved, they all contributed to what were increasingly labelled as 'adult' films in the postwar era.[25] Contextualised by wartime and postwar liberalism, the publication of the first of the Kinsey reports on human sexuality, and a growing market for lurid paperbacks and softcore pornography;[26] and contextualised as well by wartime experiences, Cold War concerns, the passage of the G.I. Bill of Rights and the growth in adult and higher education, the increase in audiences for adult and art films is hardly surprising, particularly given the extent to which the latter as well as the former were accompanied by publicity campaigns that stressed (or simply invented) their sexual content.[27] It is also unsurprising that they often courted the ire of local censorship boards, the Production Code Administration (PCA), and the Catholic Legion of Decency.[28]

The principal importers and distributors of art-house films were neither obliged nor always inclined to obtain a Seal of approval from the PCA because they were not members of the MPAA, because their films were seen as catering for a 'sophisticated' middle-class minority, and because they knew that many of their films would be denied a Seal at a time when the PCA was tightening up rather than relaxing its regulations.[29] However, they were keen to widen the commercial potential of their films if they could, and they were obliged to have them examined by US Customs and reviewed by the Motion Picture Division of the New York State Board of Regents (otherwise known as the New York Board of Film Censors). They were also obliged to have them viewed by other local censorship boards and reviewed by the Legion of Decency, which rated films on a scale that ranged from A-I ('Morally Unobjectionable for General Patronage') to A-II ('Morally Unobjectionable for Adults') to B ('Morally Objectionable in Part for All') to C ('Condemned'). Along with *Forever Amber* and *Duel in the Sun*, *Open City* was rated B; among those rated C were *The Outlaw*, *La Ronde* (1951) and *The Miracle* (1950). *The Outlaw* and *La Ronde* were condemned for sexual immorality, *The Miracle* for blasphemy.

The early 1950s

The Miracle was a 47-minute Italian film directed by Rossellini. It was acquired by Burstyn in 1948 and passed twice by the New York Board of Censors, initially as an individual film then as part of a trilogy of shorts assembled together under the title *Ways of Love* (1950).[30] However, the Legion condemned it (along with *Ways of Love* as a whole) and the New York Board of Censors reversed its position and banned the showing of the film and the programme. Burstyn took the case to the Supreme Court, which in 1952 'handed down a landmark decision that extended to the movies the constitutional protections long enjoyed by newspapers, books, magazines, and other organs of communication'.[31] By then the movie landscape had begun to change in significant ways. The number of art houses had begun to double, foreign films were increasingly shown on television, attendances at mainstream cinemas were still declining and the number of adult films was on the increase. As more and more films were imported or produced abroad, and as the number of independents continued to grow, the members of the Society of Independent Motion Picture Producers established an export association and United Artists and others released some of their films without a Seal.[32]

The adult films included *The Glass Menagerie* and *No Way Out* (both 1950), *Detective Story* and *A Streetcar Named Desire* (both 1951), *Clash By Night* and *Viva Zapata!* (both 1952), *From Here to Eternity* and *Julius Caesar* (both 1953), *The Barefoot Contessa* and *On the Waterfront* (both 1954), and *Bad Day at Black Rock* and *Marty* (both 1955). The films released without a Seal included *The Moon is Blue* (1953), *The French Line* (1954) and *The Man with the Golden Arm* (1955). Taking their cue from *The Moon is Blue* (a film that used the word 'virgin'), adult romantic comedies such as *The Tender Trap* and *The Seven Year Itch* (both 1955) were emerging too, as were adult musicals such as *It's Always Fair Weather* and *Love Me or Leave Me* (both 1955). Along with adult science-fiction films such as *The Day the Earth Stood Still* (1951), they were joined by the trend toward adult Westerns such as *Broken Arrow* (1950) and *Apache* (1954), which increasingly dealt with racial prejudice, and *High Noon* (1952) and *The Naked Spur* (1953), which increasingly dealt with social alienation and the psychology and physiology of frustration, pain and violence, a trend also evident in crime films such as *The Big Heat* (1953) and *Kiss Me Deadly* (1955) and war films such as *Halls of Montezuma* (1950).[33]

Meanwhile the art films and imports began to include Japanese productions such as *Rashomon* (1951) and *Gate of Hell* (1954), which were distributed in the US by RKO and Edward Harrison, respectively. They also continued to include British, Italian and French ones such as *The Third Man* (1950), *Bitter Rice* (1950) and *The Wages of Fear* (1955), which were distributed in the US by Selznick International, Trans-Lux Distribution and the Distributors Corporation of America (DCA), respectively. *The Wages of Fear* and *Bitter Rice* were initially exhibited in art houses in their original languages then dubbed into English and released more widely.[34] Along with *Marty*, they all won awards at the Cannes or Berlin Film Festivals, which by then had become key sites for the emergence of art and adult trends.[35]

By 1956 the number of US-produced feature films released in the North American market had declined by 28 per cent since 1946 and the number of imports had risen by 233 per cent and now accounted for 43 per cent of the total, at which point 'all the major circuits at least occasionally played subtitled pictures'.[36] At the same time, with quota and remittance regulations still in place in most key countries, the number of US films produced or wholly or partly filmed abroad, a phenomenon increasingly known as 'runaway production', had continued to grow.[37] According to Irving Bernstein, 314 runaways were produced by US companies between 1949 and 1956.[38] In addition to the Cinerama travelogues, they included costume adventure films such as *Captain Horatio Hornblower* (1951) and *Ivanhoe* (1952), which were both filmed in Britain;[39] colonial dramas and adventure films such as *King Solomon's Mines* (1950), *The Snows of Kilimanjaro* (1952) and *Elephant Walk* (1954), which were all filmed in

British colonies; international romances such as *Roman Holiday* (1953), *Three Coins in the Fountain* and *The Last Time I Saw Paris* (both 1954), which were all filmed in Italy or France; and Biblical films and epics such as *Quo Vadis* (1951), which was filmed in Italy, and *Valley of the Kings* (1954), *Land of the Pharoahs* (1955) and *The Ten Commandments* (1956), which were all partly filmed in Egypt.[40] They also included Westerns and costume pictures such as *Sitting Bull* and *The Black Pirate* (both 1954), which were filmed in Latin America.[41]

As these trends continued, international coproductions and the production of foreign films for release by the majors as well as by art-house importers was becoming more common and a new generation of European performers, among them Rossano Brazzi, Richard Burton, Vittiorio Gassman, Stewart Grainger, Audrey Hepburn, Gina Lollobrigida, Sophia Loren, Silvana Mangano and Rosanna Podestà, were beginning to feature in Hollywood films. Alongside them, a new generation of Hollywood performers, among them Richard Basehart, Kirk Douglas and Anthony Quinn, were beginning to feature in European ones. At various points in the early mid-1950s, Basehart, Douglas and Quinn were all in Rome. So too were US actress Shelley Winters and US directors Robert Rossen and Robert Wise. Along with Gassman and Mangano, Winters and Rossen were making *Mambo*, Quinn and Loren were acting in *Attila*, Basehart and Quinn in *La Strada*, and Douglas, Mangano, Podestà and Quinn in *Ulysses*. Podestà and Wise (and an extensive British cast) were shortly to be involved in *Helen of Troy* and Rossen (and an equally extensive British cast) in *Alexander the Great*.[42]

Attila, *La Strada* and *Ulysses* were Italian productions. *Mambo* was a coproduction. *Alexander the Great* and *Helen of Troy* were runaways. *Mambo* was released in Italy 1954 and in the US in 1955 and *Attila* was released in Italy in 1954 and in the US (as *Attila the Hun*) in 1958. *Ulysses* was released in 1955 and *Alexander the Great* and *Helen of Troy* in 1956, the year in which *La Strada* (which had been released in Italy in 1954) was released in the US. These productions contributed in various ways to recent trends: coproductions and runaways, romances set in the present and spectacles set in the past, avowedly commercial productions, art films aimed at international markets, and the increasing involvement of European producers such as Carlo Ponti and Dino De Laurentiis (who were at this point planning *War and Peace* with Paramount for release in 1956) in these and other developments. The involvement of Hollywood companies in the domestic as well as the international distribution of these and other Italian films was a trend as well. In 1954, Columbia was involved in the distribution of *Hanno rubato un tram* in Italy, Twentieth Century-Fox and Universal in the distribution of *Una Donna Libera* and *L'Oro di Napoli* (*Gold of Naples*), respectively. Paramount, which coproduced and helped distribute *Mambo*, helped distribute *La Strada* and *Totó a Colori* (1953) in Italy too.[43] However, *Gold of Naples* and *La Strada* were distributed in the US not by Universal and Paramount but by DCA and Trans-Lux, respectively. Aside from British productions, which were variously distributed by nearly all the major companies in the US in the early mid-1950s, the only foreign films distributed in the US by a Hollywood major at this time were distributed by United Artists.[44]

The late 1950s and early 1960s

In 1957, following its successful art-house run, Trans-Lux hired Richard Basehart and Anthony Quinn (who had been dubbed into Italian in Italy) to dub their parts in *La Strada* into English in preparation for a wider commercial release. Later that year, on October 21, *And God Created Woman* (*Et Dieu créa la femme*), which had been produced in France in 1956 and imported by Kingsley International Pictures, premiered in French with English subtitles at the Paris Theatre art house in New York. *La Strada* was reasonably successful. According to *Daily Variety*, it grossed a million dollars during the course of its New York run alone.[45] But thanks largely to the appearance and performance of Brigitte Bardot, *And God Created Woman* caused a sensation. One theatre manager was quoted as saying that the film was nothing but a sex picture; Arthur Knight and Hollis Alpert went on to argue that Bardot 'personified the youth of the Fifties'.[46] Whatever the case may be, the commercial success of *And God Created Woman* led to the widespread recycling of previous Bardot films and to the widespread distribution of later ones.[47] In an already expanding import market, it helped spark interest in new French and British imports such as *The Lovers* (*Les Amants*) and *The Four Hundred Blows* (*Les Quatre Cents Coups*) (both 1959) and *Room at the Top* (1959) and *Saturday Night and Sunday Morning* (1961) too.[48] It also helped pave the way for the ironic depiction of relationships between older men and much younger women in films such as *Bonjour Tristesse* (1958), a Columbia runaway, and *Lolita* (1962), a British-based production released by MGM. *And God Created Woman* and *Lolita* encountered censorship problems and were condemned by the Legion of Decency. In this as in other respects, they had both been preceded by *Baby Doll* (1956).

Baby Doll had been produced by Warner Bros. and scripted by Tennessee Williams, a major source of adult films in the 1950s and early 1960s. Although condemned by the Legion of Decency, it had been released with a PCA Seal. Depicting as it did the sexual frustration experienced by an older man whose young teenage bride wished to delay the consummation of their marriage, the awarding of a Seal had been controversial. But the provisions of the Code had just been relaxed in response to the adult trends and the commercial pressures that lay behind them, and to that extent the awarding of a Seal was a sign of the times.[49] Either way the adult trends continued. In addition to adult Westerns such as *The Last Hunt* (1956) and *Day of the Outlaw* (1959), adult crime films such as *Touch of Evil* and *Screaming Mimi* (both 1958), adult science-fiction films such as *Forbidden Planet* (1956) and *On the Beach* (1959) and adult war films such as *Attack* (1956) and *The Naked and the Dead* (1958); and alongside adult musicals such as *Invitation to the Dance* (1956) and *Gigi* (1958) and adult romantic comedies such as *Pillow Talk* and *Some Like It Hot* (both 1959), examples include *Lust for Life* and *The Rose Tattoo* (both 1956), *Bachelor Party* and *Written on the Wind* (1957), *Desire Under the Elms* and *The Long Hot Summer* (both 1958), and *Anatomy of a Murder* (1959) and *Butterfield 8* (1960).[50] Following the example set by *Marty* and *Lili* (1953), *Bachelor Party, Desire Under the Elms, Invitation to the Dance, Lust for Life* and *Gigi* were all premiered in art-house venues as well as or instead of conventional showcase theatres in New York and subsequently shown in art-house as well as ordinary cinemas elsewhere.[51]

In the meantime another adult trend involved the revival of films that focused on black characters as well as on white ones, which sought to deal with racial prejudice, and which, thanks partly to the revisions of the Code, were able to focus in a positive way on interracial romance. Among them were *Edge of the City* and *Island in the Sun* (both 1957) and *The Defiant Ones* (1958).[52] These films were released in the midst of the struggle for Civil Rights and while a number of Southern theatres showed them, others refused. In the past the production of films like these might have been curtailed for commercial reasons, especially given the downturn in ticket sales in the US. However, African-American audiences in the US were on the increase, African-American actors and multiracial films were popular abroad, and overseas markets, which accounted for a higher proportion of the industry's earnings than ever before, were now seen as crucial.[53] The British market was declining as TV sales there increased. Italy and France were stalling too. But frozen earnings were at a postwar low, subsidies were still available in Britain and elsewhere, markets in Japan and West Germany were expanding, and Italy soon recovered to become not only a major site for international production but the possessor of the second-largest film industry in the world and the second-largest presence in the global market.[54] For all these reasons, US coproductions and runaways were on the increase too. Those produced or coproduced and wholly or partly filmed in Britain and its colonies included *The Bridge on the River Kwai* (1957), *Gideon of Scotland Yard* (1959) and *Sink the Bismark!* (1960); those in Italy, *Ben-Hur* (1959), *It Started in Naples* (1960) and *Barabbas* (1961); those in France, *Gigi* and *Can-Can* (1960); those in Germany, *A Time to Love and a Time to Die* (1958) and *Town Without Pity* (1961); and those in Japan, *House of Bamboo* (1955), *Teahouse of the August Moon* (1956) and *Sayonara* (1957). While Westerns such as *They Came to Cordura* (1959) and *The Magnificent Seven* (1960) continued to be filmed in Latin America, films like these were augmented in the late 1950s and early 1960s by *Never on Sunday* (1960), which was produced and filmed in Greece, and *The Pride and the Passion* (1957), *King of Kings* and *El Cid* (both 1961) and *55 Days at Peking* (1963) and *The Fall of the Roman Empire* (1964), which were all produced and filmed in Spain.[55]

The Magnificent Seven starred Yul Brynner, who was born on an island off the coasts of Japan and the USSR. In addition to a number of up-and-coming US performers, it also featured Horst Buchholz, a German-born actor who had already appeared in a British film as well as a number of German ones, the Russian-born Alexander Sokoloff, and a supporting cast of Mexican actors who included Jorge Martínez de Hoyos and Rosenda Monteros. As is well known, *The Magnificent Seven* was based on *Schichinin no samurai*, a Japanese film directed by Akira Kurosawa which was first shown in the US in 1955 as *The Magnificent Seven* and which was later known as *The Seven Samurai*. The 1960 version of *The Magnificent Seven*, which was more successful abroad than it was in the US, drew on an emerging template for international productions or for productions designed for international markets, that of a heterogeneous, often multi-national group engaged in a collective endeavour of one sort or another (usually a military campaign, a mission or a heist). Along with the trend toward international romances, this template was to mark a number of late 1950s and early 1960s films, among them *Spartacus*, *The Longest Day* (1962) and *Topkapi* (1964). But there is another aspect of *The Seven Samurai* itself that is worthy of note, and that is the fact that it was imported and distributed in the US by Columbia Pictures, a Hollywood major.

It was Columbia that set up Kingsley International Pictures in order to secure the distribution of *And God Created Woman* without having to submit it to the PCA for a Seal (which as a member of the MPPA it would have been obliged to do, and which would almost certainly have been refused). And it was Columbia that had earlier set up a distribution deal with Warwick Films in Britain (which was managed by two expatriate US independents, Irving Allen and Cubby Broccoli, and which contributed successfully to a number of ongoing trends). However, it was Joseph E. Levine of Embassy Pictures who imported and distributed *Attila the Hun*. It was Levine who later imported *Hercules* (1959) and *Hercules Unchained* (1960) from Italy and devised their mass saturation campaigns for Warners. And it was Levine who had imported, cut and dubbed *Godzilla, King of the Monsters*, a Japanese 'creature feature', for a mass saturation release to a burgeoning teenage market in 1956. *Godzilla* was the first of many. Coinciding with the production of similar films by American International Pictures and others, prompted further by the distribution of UK horror films such as *The Curse of Frankenstein* (1957) and *Horror of Dracula* (1958) by Warner Bros. and Universal respectively, and spurred on by the screening of old horror films on syndicated TV shows such as 'Shock' and 'Son of Shock', a horror craze with specific appeal to youthful audiences was well under way.[56]

In the midst of all this, and in the midst of a thaw in relations between the US and the USSR that involved an agreement to widen the distribution of each other's films in each other's countries,[57] Paramount released *Psycho* (1960), Universal released *The Birds* (1963) and Allied Artists released *Shock Corridor* (1963) and *Naked Kiss* (1964), all of which challenged or ignored the benign and providential tenets of the Production Code. At this point along with *La Dolce Vita* (1961), a major international hit that played both in art-house and mainstream theatres, a series of avowedly modernist European imports such as *The Seventh Seal* (1958), *Hiroshima Mon Amour* (1960), *Breathless* and *L'Avventura* (both 1961) and *Last Year at Marienbad* (1962) were or had all recently been in circulation in art houses.[58] Horror-oriented exploitation films were still on the increase in drive-ins and neighbourhood theatres. Along with roadshows and runaways, coproductions and mainstream foreign imports, adult films, teen films, children's films and family films, albeit in declining overall numbers, were all on offer too. With the number of film clubs and art-house theatres still on the increase, and with a new set of avant-garde trends joining a number of well-established ones, a broader mix of films was available for audiences in at least parts of the US in the early 1960s than had ever been available before.[59]

Notes

1 The term 'artie', like the term 'sure seater', was coined by *Variety*. The former was used to refer to art films, the latter to art-house theatres. The term 'sure seater' implied that art-house theatres were so small that they were guaranteed to fill their seats. The term 'exploitation' referred in general to the promotion of films via publicity campaigns, unusual modes of distribution and exhibition and other forms of ballyhoo. It became associated with low-budget features, promotional gimmicks and saturation bookings in particular in the 1950s and 1960s and thus with horror, sensation and sex.

2 Schatz, *Boom and Bust: American Cinema in the 1940s* (New York: Scribners, 1997), 289.

3 Schatz, *Boom and Bust*, 297. According to Thomas H. Guback in *The International Film Industry: Western Europe and America Since 1945* (Bloomington: Indiana University Press, 1969), 39–63, the number of US films imported or registered for distribution or release in 1946 in Austria, Finland, Italy, the Netherlands, Norway and Spain were 52, 157, 600, 199, 60 and 201 respectively.

4 Tino Balio, *United Artists: The Company that Changed the Film Industry* (Madison: University of Wisconsin Press, 1987), 32–34, and *The Foreign Renaissance on American Screens, 1946–1973* (Madison: University of Wisconsin Press, 2010), 62–64; Robert Murphy, 'Rank's Attempt on the American Market' in James Curran and Vincent Porter (eds), *British Cinema History* (London: Weidenfeld and Nicholson, 1983), 164–78; Sarah Street, *Transatlantic Crossings: British Feature in the USA* (New York: Continuum, 2002), 43–69, 93–110. In *Boom and Bust*, 156, Schatz notes that in addition to his deals with United Artists and Universal and his ownership of Eagle-Lion, Rank signed 'a five-year coproduction and global distribution deal with 20th Century-Fox; a two-picture production and distribution deal with RKO . . . and a coproduction deal with David Selznick'.

5 For a detailed account of the production of *Open City*, its acquisition by Mayer and Burstyn, and its reception in the US, see Tag Gallagher, *The Adventures of Roberto Rossellini: His Life and Films* (New York: Da Capo Press, 1998), 115–79.

6 Street, *Transatlantic Crossings*, 94; Barbara Wilinsky, *Sure Seaters: The Emergence of Art House Cinema* (Minneapolis: University of Minnesota Press, 2001), 74–75. According to Street, other British films that performed well in the US in the mid- to late 1940s included the prestige roadshows discussed by Hall elsewhere in this Reader (Chapter 20) and *Stairway to Heaven* (*A Matter of Life and Death*), *Caesar and Cleopatra* and *The Seventh Veil*, all of them released in 1946. *Caesar and Cleopatra* was distributed by United Artists, *Stairway to Heaven* and *The Seventh Veil* by Universal-International.

7 Balio, *The Foreign Renaissance on American Screens*, 3. As Balio himself goes on to point out, this may well be an exaggerated figure. *Open City* does not appear in the list of 'All-Time Foreign-Language Films in North America' in *Variety*, 21–27 February 2000, 16, which ranks all such films to have grossed $4 million or more in the US and Canada. Of the films referred to later on in this article, *La Dolce Vita*, which is cited as grossing $19.5 million, is the only one to appear on the list, which is republished with further details in Balio, *The Foreign Renaissance on American Screens*, 309–12.

8 Balio, *The Foreign Renaissance on American Screens*, 25–39; Douglas Gomery, *Shared Pleasures: A History of Movie Presentation in the United States* (London: BFI, 1992), 155–93; Wilinsky, *Sure Seaters*, 46–62. As Gomery points out in *Shared Pleasures*, 137–38, 141–54, other specialised venues included newsreel theatres and theatres that catered for devotees of action films and Westerns.

9 Wilinsky, *Sure Seaters*, 47–48.

10 Richard Koszarski, *An Evening's Entertainment: The Age of the Silent Feature Picture, 1915–1928* (New York: Scribners, 1990), 208.

11 Anthony Henry Guzman, 'The Exhibition and Reception of European Films in the United States during the 1920s', PhD Dissertation, UCLA (1993), 208.

12 Tony Guzman, 'The Little Theatre Movement: The Institutionalization of the European Art Film in America', *Film History*, vol. 17 no. 2/3 (2005), 261–84; Wilinsky, *Sure Seaters*, 47.

13 Gomery, *Shared Pleasures*, 176–79.

14 Wilinsky, *Sure Seaters*, 79.
15 Schatz, *Boom and Bust*, 295. It is worth noting in this context that according to *The Hollywood Reporter*, 12 January 1949, 1, Fox-West Coast theatres, a mainstream commercial chain, had already booked *Paisan* into 260 of its theatres in California and planned to play it in a further five or six hundred.
16 Balio, *The Foreign Renaissance on American Screens*, 44–75; Schatz, *Boom and Bust*, 295; Wilinsky, *Sure Seaters*, 60, 63–65, which notes that there were approximately 5,000 film societies in the US in 1949.
17 Schatz, *Boom and Bust*, 297.
18 *Variety*, 28 January 1948, 7, 22.
19 Schatz, *Boom and Bust*, 298.
20 Guback, *The International Film Industry*, 16–34; Kerry Segrave, *American Films Abroad: Hollywood's Domination of the World's Movie Screens* (Jefferson, NC: McFarland, 1997), 140–85; Schatz, *Boom and Bust*, 299–303.
21 The location scenes in *I Was a Male War Bride* were filmed in those parts of Germany under US and British control and the interiors were filmed at Fox's studio in Britain. Both were paid for with unremitted earnings. For further details on Germany, see Guback, *The International Film Industry*, 124–41.
22 Anne Jäckel, 'Dual Nationality Film Productions in Europe after 1945', *Historical Journal of Film, Radio and Television*, vol. 23 no. 3 (2003), 231.
23 *Variety*, 9 May 1951, 6; 9 April 1952, 5.
24 Matthew Bernstein, 'A Reappraisal of the Semi-Documentary Film, 1945–48', *The Velvet Light Trap*, no. 20 (Summer 1983), 22–26; Schatz, *Boom and Bust*, 379–80, 386–92.
25 For an early instance, see *Variety*, 3 November 1948, 4. These trends are discussed in greater detail in Thomas Cripps, *Making Movies Black: The Hollywood Message Movie from World War II to the Civil Rights Era* (New York: Oxford University Press, 174–249); Frank Krutnick, *In a Lonely Street: Film Noir, Genre, Masculinity* (London: Routledge, 1991); Stephen Prince, *Classical Film Violence: Designing and Regulating Brutality in Hollywood Cinema, 1930–1968* (New Brunswick, NJ: Rutgers University Press, 2003), 146–80; Schatz, *Boom and Bust*, 381–86.
26 Lawrence Alloway, *Violent America: The Movies, 1946–1964* (New York: The Museum of Modern Art, 1971); Kenneth C. Davis, *Two-Bit Culture: The Paperbacking of America* (Boston: Houghton Mifflin, 1982); Geoffrey O'Brien, *Hardboiled America: The Lurid Years of Paperbacks* (New York: Van Nostrand-Reinhold, 1981); Lee Server, *Over My Dead Body: The Sensational Age of the American Paperback: 1945–1955* (Chronicle Books: San Francisco, 1994); Eric Schaeffer, *'Bold! Daring! Shocking! True!': A History of Exploitation Films, 1919–1959* (Durham, NC: Duke University Press, 1999), 197–252, 299–324; Peter Stanfield *Maximum Movies – Pulp Fictions: Film Culture and the Worlds of Samuel Fuller, Mickey Spillane, and Jim Thompson* (New Brunswick, NJ: Rutgers University Press, 2011). The first of the Kinsey Reports, on *The Sexual Behaviour of the Human Male*, was published in 1948. *The Sexual Behaviour of the Human Female* was published in 1953.
27 Wilinsky, *Sure Seaters*, 122–27. The advertising of art and adult films in this way was a practice that became particularly widespread in the late 1950s, when it involved the majors' films and thus courted the concern of the MPAA. See *Variety*, 1 October 1958, 5.
28 Gregory D. Black, *The Catholic Crusade Against the Movies, 1940–1975* (Cambridge: Cambridge University Press, 1997), 29–65; Thomas Doherty, *Hollywood's Censor: Joseph I. Breen and the Production Code Administration* (New York: Columbia University Press, 2007), 213–82; Schatz, *Boom and Bust*, 320–23; Wilinsky, *Sure Seaters*, 99–104.
29 Doherty, *Hollywood's Censor*, 299–301. Most of these revisions were introduced in 1951.
30 The other films in the trilogy were *A Day in the Country* (*Un Partie de Campagne*) and *Joyfroi*.
31 Balio, *The Foreign Renaissance on American Screens*, 54. The most detailed account of the *Miracle* case can be found in Laura Wittern-Keller and Raymond J. Haberski Jr, *The Miracle Case: Film Censorship and the Supreme Court* (Lawrence: University of Kansas Press, 2008).
32 Balio, *The Foreign Renaissance on American Screens*, 54–61, 95–96, 100–110, 118–23; Peter Lev, *The Fifties: Transforming the Screen, 1950–1959* (New York: Scribners, 2003), 89; 97; Street, *Transatlantic Crossings*, 140–57; Wilinsky, *Sure Seaters*, 65, 87–89, 131–32. *The Hollywood Reporter*, 14 May 1953, 1, 4, noted that approximately '300 foreign-made features – an all-time record – are in distribution at the present time by companies other than the 11 major and minor distribution firms, which have about 50 on their present schedules . . . Many of these films normally play only the art circuit, but the regular houses have begun to use them not only as supporting features but as top-of-the-bill material'.
33 Barbara Klinger, *Melodrama and Meaning: History, Culture, and the Films of Douglas Sirk* (Bloomington: Indiana University Press, 1994), 38–40, 71–74. As Klinger points out, these trends drew on a mix of highbrow, lowbrow and middlebrow sources. It is worth noting that this mix was also evident in the field of publishing, where source novels by Nelson Algren, James Jones and John Steinbeck (and later

on by John O'Hara, Norman Mailer and Françoise Sagan) rubbed shoulders with Grace Metalious and Mickey Spillane in the field of the lurid paperback. (See the article entitled 'Adult Books into Films Can Bring Back That "Lost Audience"' in *Variety*, 8 January 1958, 46.) As Klinger also points out, aside from and in addition to *The French Line, The Moon is Blue* and *The Man with the Golden Arm*, a number of contributors to these trends experienced opposition from the Legion of Decency, problems with the PCA, or both. Among them were *Detective Story*, *A Streetcar Named Desire*, *The Big Heat* and *Kiss Me Deadly*. For details see Black, *The Catholic Crusade Against the Movies*, 109–42; Leonard J. Leff and Jerrold R. Simmons, *The Dame in the Kimono: Hollywood, Censorship and the Production Code from the 1920s to the 1960s* (London: Weidenfeld and Nicholson, 1990), 162–213; Prince, *Classical Film Violence*, 180–90. For discussion of the adult Western, see John H. Lenihan, *Showdown: Confronting Modern America in the Western Film* (Urbana: University of Illinois Press, 1985).

34 Balio, *The Foreign Renaissance on American Screens*, 59–62. Japan was occupied by the US in the immediate postwar era and the production and circulation of films was heavily policed. It was not until 1953 that restrictions were eased and when they were a number of Japanese producers set their sights on foreign markets. In addition to Balio, *The Foreign Renaissance on American Screens*, 118–24, see Hiroshi Kitamura, *Screening Enlightenment: Hollywood and the Cultural Reconstruction of Defeated Japan* (Ithaca: Cornell University Press, 2010); Kyoko Hirano, *Mr. Smith Goes to Tokyo: Japanese Cinema under the American Occupation, 1945–1952* (Washington: Smithsonian Institution Press, 1992).

35 See Vanessa R. Schwartz, *It's So French! Hollywood, Paris, and the Making of a Cosmopolitan Film Culture* (Chicago: Chicago University Press, 2007), 57–99, for a discussion of Cannes.

36 Irving Bernstein, *Hollywood at the Crossroads: An Economic Study of the Motion Picture Industry* (Hollywood: American Federation of Labor Film Council, 1957), 8; *Motion Picture Herald*, 26 January 1957, 30.

37 For details on quotas, remittances, subsidies and the interventions of the MPAA, the MPEA and the US government, see Ian Jarvie, 'Free Trade as Cultural Threat: American Films and TV Exports in the Post-War Period', Jean-Pierre Jeancolas, 'From the Blum-Byrnes Agreement to the GATT Affair', and Christopher Wagstaff, 'Italian Genre Films in the World Market' in Geoffrey Nowell-Smith and Steven Ricci (eds), *Hollywood and Europe: Economics, Culture and National Identity* (London: BFI, 1998), 36–39, 47–54 and 74–81, respectively. While the industries in Italy and France were on their way to recovery, Hollywood's foreign policies were bearing fruit. According to *The Hollywood Reporter*, 15 December 1953, 1, US films earned approximately $200 million abroad in 1953, 'a 10 to 15 percent increase over 1952 and a record for the industry'.

38 Bernstein, *Hollywood at the Crossroads*, 48–49.

39 For a detailed account of the ways in which blocked funds and overseas production facilities were deployed in the making of *Captain Horatio Hornblower* and other Hollywood films produced in Britain in the late 1940s and early mid-1950s, see Jonathan Stubbs, '"Blocked" Currency, Runaway Production in Britain and *Captain Horatio Hornblower* (1951)', *Historical Journal of Film, Radio and Television*, vol. 28 no. 3 (2008), 335–51. In 'The Eady Levy: A Runaway Bribe? Hollywood Production and British Subsidy in the Early 1960s', *Journal of British Cinema and Television*, vol. 6 no. 1 (2009), 1–20, Stubbs provides an equally detailed account of the making of *Lawrence of Arabia* (1962) and the extent to which the Eady levy (a levy on ticket sales designed to produce funding for British films) was drawn on by Hollywood companies and producers in the late 1950s and 1960s.

40 Many of these films and trends are well known but have rarely been discussed in detail. Exceptions include Monica Silveira Cyrano, *Big Screen Rome* (Oxford: Blackwell, 2005), 7–33; Sheldon Hall and Steve Neale Neale, *Epics, Spectacles and Blockbusters: A Hollywood History* (Detroit: Wayne State University Press, 2010), 135–40; Peter Krämer, '"Faith in the Relations Between People": Audrey Hepburn, *Roman Holiday* and European Integration' in Diana Holmes and Alison Smith (eds), *100 Years of European Cinema: Entertainment or Ideology?* (Manchester: Manchester University Press, 2000), 195–206; Robert R. Shandey, *Runaway Romances: Hollywood's Postwar Tour of Europe* (Philadelphia: Temple University Press, 2009). According to *The Hollywood Reporter*, 25 August 1949, 1, 11, MGM was the first of the majors to engage systematically in 'world-wide' production.

41 On 15 September 1953, 3, *The Hollywood Reporter* noted that out of a total of '118 pictures now being written and in pre-production stages at six major studios, only 31, or 26 percent, will depict the modern American scene. The remainder divides into 13 period westerns, 17 other American period pieces, 28 foreign period or Biblical pictures, and 29 with modern settings abroad'.

42 Hollywood films and Hollywood personnel were not the only ones involved in production in Italy. According to *The Hollywood Reporter*, 11 January 1955, 3, '152 pictures went into production in Italy or on location abroad in Italian coproductions during the first 11 months of 1954'. Of the 43 coproductions, 36 were with France, 3 with Spain, 2 with Germany, and 1 each with Egypt and Japan. According to an earlier edition, 5 March 1954, 4, Italy was already exporting its films to over 80 countries, thus increasing its earnings from abroad by over 70 per cent. Italy was also an early

adopter of CinemaScope and other widescreen technologies. See Federico Vitella, 'Before Techniscope: The Penetration of Foreign Widescreen Technology in Italy (1953–59)', in John Belton, Sheldon Hall and Steve Neale (eds), *Widescreen Worldwide* (New Barnet: John Libbey Press, 2010), 163–73.

43 *Variety*, 17 August 1955, 14.

44 For a listing of British films and their US distributors in the early mid-1950s, see Street, *Transatlantic Crossings*, 235–40. For a listing of the foreign films distributed by United Artists, see Balio, *United Artists: The Company That Changed the Film Industry*, 349–56. Frustrated by what it considered to be the inadequate handling of its films, Rank established its own distribution company in the US in 1957.

45 *Daily Variety*, 5 July 1957, 1.

46 Both sources are quoted by Balio, *The Foreign Renaissance on American Screens*, 116, during the course of a detailed account of the film's distribution, reception and impact.

47 According to Guback, *The International Film Industry*, 86, *And God Created Woman* grossed $4 million in the US.

48 Balio, *The Foreign Renaissance on American Screens*, 168–77, 145–57, respectively. For more on Bardot, see Schwartz, *It's So French!*, 103–54. According to *Variety*, 9 April 1958, 1, 119, a total of 373 foreign feature films grossed $10,132,392 in the US in 1956; the following year, a total of 832 foreign features grossed $59,907,769. At this point the volume and the earnings of French films were on the increase while those of Italian films had declined.

49 For detailed accounts of *Baby Doll*, revisions to the Code, and the response of the Legion of Decency and local censorship boards to this and other films, see Black, *The Catholic Crusade Against the Movies*, 154–75; Lev, *The Fifties*, 90–97; Jon Lewis, *Hollywood v. Hard Core: How the Struggle over Censorship Saved the Modern Film Industry* (New York: New York University Press, 2000), 115–34. For discussion of *Lolita*, see Leff and Simmons, *The Dame in the Kimono*, 214–40. As these accounts all make clear, the 1956 revisions were triggered by *Tea and Sympathy* (1956), which involved adultery and male homosexuality and which was subject to extensive negotiations with the PCA. In its revised version, the Code's previous prohibitions on miscegenation were dropped and topics such as abortion, childbirth and drug addiction could be dealt with more openly. 'Excessive and inhumane acts of cruelty and brutality' and 'sex perversion' were still forbidden (though in practice still alluded to in well-worn if now more knowing and self-conscious ways).

50 The contributions made by independents to these and other adult films are discussed in Denise Mann, *Hollywood Independents: The Postwar Talent Takeover* (Minneapolis; University of Minnesota Press, 2008), 87–241, which draws attention to the number of adult films that focused critically on US corporate culture in the late 1950s, among them *The Man in the Gray Flannel Suit* (1956), *The Sweet Smell of Success* (1957) and *The Apartment* (1960). On 5 December 1956, 19, *Variety* argued that 'unusual, off-beat films with adult themes' would be a means of competing with television. However, teleplays written by Robert Alan Arthur, Paddy Chayevsky and Reginald Rose were all sources for adult adaptations in the mid- to late 1950s, among them *Edge of the City* (1957), *Marty* and *12 Angry Men* (1957).

51 *Variety*, 21 September 1955, 13; 1 August 1956, 7; 24 October 1956, 13; 12 March 1958, 13, 15; 19 March 1958, 3, 13; 14 May 1958, 10. See *Variety*, 16 March 1960, 7, 79, for a number of later examples.

52 Cripps, *Making Movies Black*, 250–94.

53 *Variety*, 9 May 1956, 5; 1 May 1957, 3, 20; 8 May 1957, 3, 14; 7 August 1957, 7. For similar reasons, the presence of Asian actors was increasing and the representation of Asian characters and interracial romance were under revision too according to *Variety*, 8 February 1956, 1, 53. However, Native American characters with extensive speaking parts were still rarely played by Native American actors and interracial romances were still more likely to end in tragedy than happiness.

54 Tim Bergfelder, *International Adventures: German Popular Cinema and European Co-Productions in the 1960s* (New York: Berghan Books, 2005), 39–52; David Desser, 'Japan', and Marc Silberman, 'Germany', in Gorham Kindem (ed.), *The International Movie Industry* (Carbondale: Southern University of Illinois Press, 2000), 17–18 and 217–18, respectively. See also *Variety*, 2 July 1958, 14; 9 July 1958, 11; 16 July 1958, 12. West German and Japanese exports to the US were on the increase too. According to *Variety*, 15 January 1958, 5, and 30 July 1958, 5, the former were usually screened in ethnic theatres catering to German-speaking audiences, though the award-winning *Captain of Koepenick* (1958) was a success both with ethnic and art-house audiences. According to Balio, *The Foreign Renaissance on American Screens*, 120–29, some of the Japanese films were successful too. But although a Japanese film festival was held at the Little Carnegie theatre in 1959, and although Toho opened showcase theatres for its films in the early 1960s, the only consistent commercial successes were its horror films. Italy's status and recovery is noted in *Variety*, 6 January, 1960, 166; 10 February 1960, 27; 20 April 1960, 1, 65, 67, 78. Having joined the European Union, Italy, France and West Germany were at this point all

engaged extensively in coproduction. This was a policy that the British film industry had been unwilling or unable to adopt, but according to *Variety*, 20 April 1960, 47, 50, 168, one it now sought to pursue.

55 For more on those filmed in France and on 'cosmopolitan films' in general, see Schwartz, *It's So French!*, 18–53 and 159–98, respectively. For more on Spain, see Bernard Eisenschitz, *Nicholas Ray: An American Journey* (London: Faber and Faber, 1993), 360–89; Hall and Neale, *Epics, Spectacles and Blockbusters*, 177–81; Neal Moses Rosendorf, '"Hollywood in Madrid": American Film Producers and the Franco Regime, 1950–70', *Historical Journal of Film, Radio and Television*, vol. 27 no.1 (2007), 77–109. With domestic production in decline, the majors not only played a part in distributing most of these films abroad (sometimes in cooperation with each other), but in distributing those made for international markets in other countries too. For a number of early examples, see *Variety*, 25 June 1958, 13; 16 July 1958, 5; 23 July 1958, 4, 7; 30 July 1958, 4.

56 Hall and Neale, *Epics, Spectacles and Blockbusters*, 170–72; Kevin Heffernan, *Ghouls, Gimmicks, and Gold: Horror Films and the American Movie Business, 1953–1968* (Durham NC: Duke University Press, 2004), 43–179; Street, *Transatlantic Crossings*, 157–63.

57 Following extensive negotiations and the eventual signing of a cultural exchange agreement with the Soviet Union in 1958, the Hollywood majors agreed to distribute seven Soviet films in the US in return for the distribution of ten of the majors' films. It should be noted here that a number of US films had been exported to Hungary, Yugoslavia, and one or two other Communist countries in the postwar era. It should also be noted that Artkino had been importing and distributing Soviet films in the US since the mid-1920s. See James H. Krukones, 'The Unspooling of Artkino: Soviet Film Distribution in America, 1940–75', *Historical Journal of Film, Radio and Television*, vol. 29 no. 1 (2009), 91–112.

58 The best account of these particular films can be found in András Bálint Kovács, *Screening Modernism: European Art Cinema, 1950–1980* (Chicago: Chicago University Press, 2007), 290–309. It is important to underline the fact that the films shown in art houses in the US were aesthetically heterogeneous and cannot all be called modernist. Modernism was just one of many art-house trends.

59 According to *Variety*, 11 May 1960, 5, there were approximately 550 art houses and 450 film societies in the US in 1960. For an overview of the avant-gardes in the US in the 1950s and 1960s, see Greg S. Faller, '"Unquiet Years": Experimental Cinema in the 1950s', in Lev, *The Fifties*, 280–302, and Walter Metz, '"What Went Wrong?": The American Avant-Garde Cinema of the 1960s', in Paul Monaco, *The Sixties, 1960–1969* (New York: Scribners, 2001), 231–60. Films of all kinds were also increasingly shown and discussed in colleges and universities. According to *Variety*, 29 January 1958, 12, 71, approximately 18,000 film courses were on offer in universities and colleges in the US in the late 1950s.

Steve Neale

EPILOGUE

A NUMBER OF CRITERIA CAN BE USED to pinpoint the demise of classical Hollywood, some of them stylistic or representational, others industrial. In *The Classical Hollywood Cinema*, David Bordwell cites the decline in in-house production by in-house staff on long-term contracts, the trend toward independent and unit production, the acquisition of the Hollywood studios by large-scale conglomerates, the growth in drive-ins and multiplex cinemas, and the place of cinema in the new media landscape during the course of the late 1950s and early 1960s.[1] He also notes the increasing adoption of some of the traits of modernist art-house cinema: 'the jump-cut used for violence and comedy, the sound bridge for continuity or shock effect, the elimination of the dissolve as a transition, and the freeze-frame used to signify finality.'[2] And he acknowledges the extent to which a number of new Hollywood directors later drew on art-house styles in more fundamental ways:

> the new directors sometimes flaunt the act of narration, as in the parallel-time structure of *Godfather II* (1974), the opening credits of *Nashville* (1975), or the gratuitous tracking shot in *Taxi Driver* that leaves the protagonist behind. In *McCabe and Mrs. Miller* (1971), the Leonard Cohen songs serve as a symbolic commentary on the action; at the end of *Blue Collar* (1978), an earlier line from the film is repeated nondietically over a freeze-frame. Altman's *Three Women* (1977) is a veritable orgy of art-film narration, creating dream/reality confusion and symbolic transmutations derived from *Persona*.[3]

Nevertheless for Bordwell, as for Kristin Thompson, Hollywood cinema and its genres usually 'swallow-up art-house borrowings, taming the (already limited) disruptiveness of the art cinema'.[4] Continuity editing may now be 'intensified', new technologies may now be involved, but the dominant paradigms of Hollywood storytelling and its genres remain in place.[5] All that has really changed is the mode of production and even that is not so very different. However, there is another factor and that is the demise of the Production Code.

As we have seen in the previous section, adult trends placed the Code and its administration under severe strain in the late 1950s and early 1960s. This was not just a matter of language, sex and violence. It was also a matter of the ways in which certain sexual or violent acts, certain types of character motivation and behaviour, and certain moral or dramatic dilemmas could be conveyed to spectators in inexplicit ways. Issues of this sort had been faced before. What was new was the extent to which Hollywood's representational conventions and the strictures of the Code were used to undermine the certainties that had nearly always marked Hollywood narratives in previous years. Thus we never know whether or not Laura (Lee Remick) in *Anatomy of Murder* was raped and beaten by a bartender prior to his death at the hands of her husband (Ben Gazzara), who is put on trial for murder. Given the central role of judges, juries and trials in determining the facts and administering justice in classical Hollywood films, the 'absence of absolute truth', as Chris Fujiwara calls it, is all the more unusual.[6] What was also unusual was the extent to which the Code's benign and providential tenets were transgressed. Thus although in keeping with the fantasy scenarios that govern many of Alfred Hitchcock's earlier films, the excessive, unanticipated and distinctly non-providential fate suffered by Marion Crane (Janet Leigh) in *Psycho*, and the almost equally excessive fate suffered by Melanie Daniels (Tippi Hedren) in *The Birds* (1963), who is not killed but who is rendered catatonic, each undermine the security hitherto provided by the Code's conventions, especially insofar as they end in uncertainty or unresolved tension.[7] As Paul Monaco points out, the violent set pieces in these films helped herald a new 'cinema of sensation'.[8] They were accompanied in this and in other respects by *Shock Corridor* (1963), whose protagonist is a reporter who enters an asylum in order to track down a criminal and who ends up insane, and *The Naked Kiss* (1964), whose moral centre is a former prostitute. Both were written, produced and directed by Samuel Fuller. They were released with a Seal but widely viewed as reprehensibly exploitative.[9]

Thanks to these and other films and factors, the Code was now in crisis. It was suspended in 1966 and replaced by a system of ratings two years later. Hollywood's films were no longer all officially shaped for universal consumption. The immediate result was a group of very violent Westerns, gangster films and horror films, a group of films that were much more sexually explicit (both visually and verbally) than any other previous Hollywood films, a group of pornographic imports, and a freer use of hitherto forbidden verbal epithets.[10] For some, the demise of the Production Code heralded the arrival of a truly adult cinema, a Hollywood Renaissance, especially now that the restrictions hitherto placed on the overt representation of sexual minorities were now removed.[11] For others, though, it meant the replacement of a densely metaphorical cinema by a cinema of literalisms, one in which telling meant showing, one in which the limitations of explicit narration, however much they were accompanied by modernist ticks or however much they may have been marked by indeterminate endings, were all the more exposed. Either way, the classical Hollywood cinema was dead and something new had taken its place.

Notes

1 The first multiplex was opened in 1963. The first cinema to be located in a suburban shopping mall opened in Massachusetts in 1951. Together these developments transformed the exhibition sector during the course of the 1960s and early 1970s. See Gary R. Edgerton, *American Film Exhibition and an Analysis of the Motion Picture Industry's Market Structure, 1963–1980* (New York: Garland Publishing, 1983).

2 Bordwell, Staiger and Thompson, *The Classical Hollywood Cinema*, 374.

3 Bordwell, Staiger and Thompson, *The Classical Hollywood Cinema*, 375. In *Hollywood Incoherent: Narration in Seventies Cinema* (Austin: University of Texas Press, 2010), Todd Berliner argues that a number of 1970s Hollywood films are marked by other non-classical modes and features as well. He suggests, 53, that three 'key modes of perverse narration permeate Hollywood cinema of the 1970s'. These modes he labels 'narrative frustration' (avoiding straightforward narration and the fulfilment of narrative promises 'in conventionally satisfying ways'); 'genre deviation' (exploiting and modifying

'conventional genre devices in order to create unsolved narrative incongruities'); and 'conceptual incongruity' (the generation of 'ideas and emotions counterproductive to the films' essential concepts and overt narrative purposes').

4 Bordwell, Staiger and Thompson, *The Classical Hollywood Cinema*, 375.

5 David Bordwell, *The Way Hollywood Tells It: Story and Style in Modern Movies* (Berkeley: University of California Press, 2006); Kristin Thompson, *Storytelling in the New Hollywood: Understanding the Classical Narrative Technique* (Cambridge, MA: Harvard University Press, 1999).

6 Chris Fujiwara, *The World and Its Double: The Life and Work of Otto Preminger* (New York: Faber and Faber, 2008), 247.

7 *The Birds* ends with images that suggest a fragile and vulnerable equilibrium, one that could be shattered by the birds at any moment. The final scene in *Psycho* ends with the sight of Norman Bates (Anthony Perkins) smiling malevolently and the sound of Mother's voice disavowing any responsibility for Norman's killings, thus contradicting the theories of Norman's psychiatrist – unless of course Mother's words are actually voiced in Norman's head by Norman himself rather than by Mother. It is impossible to be certain either way. For more on the 'unsafe' nature of *Psycho*, see Maltby, *Hollywood Cinema*, 353–55.

8 Paul Monaco, *The Sixties*, 188–92.

9 Lisa Dombrowski, *The Films of Samuel Fuller: If You Die I'll Kill You!* (Middletown CT: Wesleyan University Press, 2008), 148–68.

10 Drew Caspar, *Hollywood Film: 1963–1976* (Oxford: Wiley-Blackwell, 2011), 107–29; Heffernan, *Ghouls, Gimmicks and Gold*, 181–228; Jon Lewis, *Hollywood v. Hardcore*, 135–266; Monaco, *The Sixties*, 56–66; Stephen Prince, *Classical Film Violence*, 252–89, and *Savage Cinema: Sam Peckinpah and the Rise of Ultraviolent Movies* (Austin: University of Texas Press, 1998); Steven Jay Schneider (ed.), *New Hollywood Violence* (Manchester: Manchester University Press, 2004). The classifications introduced in 1968 were 'G' ('Suggested for General Audiences'), 'M' ('Suggested for Mature Audiences, Adults and Mature Young People') and 'X' ('Persons Under Sixteen Not Admitted). The 'X' rating was insisted upon by exhibitors and the minimum age and was raised was to 17 or 18, depending upon local state and city statutes, in 1970. The ratings were administered by the Classification and Rating Administration which was established in 1968 and which replaced the PCA.

11 David A. Cook, *Lost Illusions: American Cinema in the Shadow of Watergate and Vietnam, 1970–1979* (New York: Scribners, 2000); Diane Jacobs, *Hollywood Renaissance* (Cranbury, NJ: Barnes, 1977); Jon Lewis, *American Film: A History* (New York: W. W. Norton, 2008), 281–349.

BIBLIOGRAPHY

Film catalogues

The American Film Institute Catalog of Motion Pictures Produced in the United States: Feature Films, 1911–1920 (Berkeley: University of California Press, 1988).

The American Film Institute Catalog of Motion Pictures Produced in the United States: Feature Films, 1921–1930 (New York: Bowker, 1971).

The American Film Institute Catalog of Motion Pictures Produced in the United States: Feature Films, 1931–1940 (Berkeley: University of California Press, 1993).

The American Film Institute Catalog of Motion Pictures Produced in the United States: Feature Films, 1941–1950 (Berkeley: University of California Press, 1999).

The American Film Institute Catalog of Motion Pictures Produced in United States: Feature Films, 1961–1970 (New York: Bowker, 1976).

The American Film Institute Catalog of Motion Pictures Produced in the United States: Within Our Gates, Ethnicity in American Feature Films, 1911–1960 (1997).

While most of these volumes use contemporary period terms to identify the generic affiliations of the films they catalogue, it should be noted that *Feature Films, 1941–1950* uses the term 'film noir'. This term was not in use in the US in the 1940s and has been subject to dispute and variable application. See Steve Neale, *Genre and Hollywood* (London: Routledge, 2000), 151–77.

Film archives

The Federation of Film Archives (FIAF) provides information on the whereabouts of prints of Hollywood films.

Film industry publications

American Cinematographer, *American Projectionist*, *Daily Variety*, *Exhibitors Herald and Motography*, *Exhibitors Trade Review*, *The Film Daily*, *The Film Daily Yearbooks*, *The Film Spectator*, *Harrison's Reports*, *The Hollywood Reporter*, *The International Motion Picture Almanac*, *The Motion Picture Herald*, *Motion Picture News*, *Motography*, *Motion Picture Almanac, Moving Picture Weekly*, *The New York Dramatic Mirror*, *Photoplay*, *The Silent Picture*, *Technicolor News and Views*, *Transactions* (later *Journal*) *of the Society of Motion Picture Engineers*, *Universal Weekly* (sometimes also called *Motion Picture Weekly*), *Variety*, *Wid's Daily*, *Wid's Yearbooks*.

Indexes of periodical articles

The Film Index, 3 vols (New York: Kraus International, 1985); Richard Dyer McCann and Ted Perry (eds), *The New Film Index* (New York: Dutton, 1975); John Gerlach and Lana Gerlach (eds), *The Critical Index* (New York: Teachers College Press, 1974); Linda Batty (ed.), *Retrospective Index to Film Periodicals, 1930–1971* (New York: Bowker, 1971); Mel Schuster (ed.), *Motion Picture Directors: A Bibliography of Magazine and Periodical Articles, 1900–1972* (Metuchen, NJ: Scarecrow Press, 1973); Patricia King Hansen and Stephen L. Hansen (eds), *Film Review Index*, vol. 1, *1882–1949* (Phoenix: Oryx, 1986); annual editions of *International Index to Film Periodicals* (since 1972) and *Film Literature Index* (since 1973).

Surveys of research collections

Linda Mehr (ed.), *Motion Pictures, Television, and Radio; A Union Catalogue of Manuscript and Special Collections in the Western United States* (Boston: Hall, 1977); Kim N. Fisher (ed.), *On the Screen: A Film, Television and Video Research Guide* (Littleton, CO: Libraries Unlimited, 1986); *International Directory of Film and Television Documentation Centers* (Chicago: St James Press, 1988).

Other publications

Christopher R. Sterling and Timothy R. Haight, *The Mass Media: Aspen Institute Guide to Communication Industry Trends* (New York: Praeger, 1978).

The following books contain outline histories of the major Hollywood companies and complete listings of the films they distributed and produced in the classical era:.

Bergan, Ronald, *The United Artists Story* (London: Octopus Books, 1986).
Eames, John Douglas, *The Paramount Story* (London: Octopus Books, 1985).
— *The MGM Story* (London: Octopus Books, 1975).
Hirschorn, Clive, *The Columbia Story* (London: Octopus Books, 1989).
— *The Universal Story* (London: Octopus Books, 1983).
— *The Warner Bros. Story* (London: Octopus Books, 1979).
Jewell, Richard and Vernon Harbin, *The RKO Story* (London: Octopus Books, 1982).
Thomas, Tony and Aubrey Solomon, *The Films of Twentieth Century-Fox* (Secaucus, NJ: Citadel Press, 1985).

Books

In addition to those cited in this Reader, this bibliography lists a number of other books.

Abel, Richard, *Americanizing the Movies and 'Movie Mad' Audiences, 1910–1914* (Berkeley: University of California Press, 2006).

— (ed.), *Encyclopedia of Early Cinema* (London: Routledge, 2005).

— *The Red Rooster Scare: Making Cinema American, 1900–1910* (Berkeley: University of California Press, 1999).

Abels, Richard and Rick Altman (eds), *The Sounds of Early Cinema* (Bloomington: Indiana University Press, 2002).

Adorno, Theodore W. and Hanns Eisler, *Composing for the Films* (London: Continuum, 2007; originally published in 1947).

Albrecht, Donald, *Designing Dreams: Modern Architecture in the Movies* (New York: Harper & Row/ Museum of Modern Art, 1986).

Aleiss, Angela, *Making the White Man's Indian: Native Americans and Hollywood Movies* (New York: Greenwood Press, 2005).

Allen, Robert C. and Douglas Gomery, *Film History: Theory and Practice* (New York: Knopf, 1985).

Alloway, Lawrence, *Violent America: The Movies, 1946–1964* (New York: The Museum of Modern Art, 1971).

Altman, Diana, *Hollywood East: Louis B. Mayer and the Origins of the Studio System* (New York: Birch Lane Press, 1992).

Altman, Rick, *Silent Film Sound* (New York: Columbia University Press, 2004).

Altomara, Rita, *Hollywood on the Palisades* (New York: Garland, 1983).

Anderson, Christopher, *Hollywood TV: The Studio System in the Fifties* (Austin: University of Texas Press, 1994).

Anderson, Gillian, *Music for the Silent Films, 1894–1929* (Washington, DC: Library of Congress, 1988).

Arthur D. Little, Inc., *Television Program Production* (Cambridge, MA: Arthur D. Little, Inc., 1969).

Bacher, Lutz, *Max Ophuls in the Hollywood Studios* (New Brunswick, NJ: Rutgers University Press, 1996).

Bachman, Gregg and Thomas J. Slater (eds), *American Silent Film: Discovering Marginalized Voices* (Carbondale: Southern Illinois University Press, 2002).

Bakker, Gerben, *Entertainment Industrialised: The Emergence of the International Film Industry, 1890–1940* (Cambridge: Cambridge University Press, 2008).

Balio, Tino, *The Foreign Renaissance on American Screens, 1946–1973* (Madison: University of Wisconsin Press, 2010).

— *Grand Design: Hollywood as a Modern Business Enterprise, 1930–1939* (New York: Scribners, 1993).

— (ed.) *Hollywood in the Age of Television* (Boston: Unwin Hyman, 1990).

— *United Artists: The Company that Changed the Motion Picture Industry* (Madison: University of Wisconsin Press, 1987).

— (ed.), *The American Film Industry* (Madison: Wisconsin University Press, 1985 edition).

— *United Artists: The Company Built by the Stars* (Madison: University of Wisconsin Press, 1976).

Balshofer, Fred and Arthur Miller, *One Reel a Week* (Los Angeles: University of California Press, 1967).

Bansak, Edmund G., *Fearing the Dark: The Val Lewton Career* (Jefferson, NC: McFarland, 1995).

Baran, Paul A. and Paul M. Sweeney, *Monopoly Capital: An Essay on the American Economic and Social Order* (New York: Modern Reader Paperbacks, 1968).

Barnouw, Erik, *The Golden Web: A History of Broadcasting in the United States 1933–1953* (New York: Oxford University Press, 1968).

Barrier, Michael, *Hollywood Cartoons: American Animation in its Golden Age* (New York: Oxford University Press, 1999).

Barrios, Richard, *Screened Out: Playing Gay in Hollywood Cinema from Edison to Stonewall* (New York: Routledge, 2003).

Barsacq, Léon and Elliott Stein, *Caligari's Cabinet and Other Grand Illusions* (Boston: New York Graphic Society, 1976).

Basinger, *The Star Machine* (New York: Vintage Books, 2007).

— Jeanine, *Silent Stars* (Hanover: Wesleyan University Press, 1999).

— *The World War II Combat Film: Anatomy of a Genre* (New York: Colombia University Press, 1986).

Basten, Fred, *Glorious Technicolor: The Movies' Magic Rainbow* (Cranbury, NJ: A.S. Barnes, 1980).

Batchelder, C.W., *Multiplane Manual* (Burbank: Walt Disney Studio, *c.*1939).

Baxandall, Michael, *Patterns of Intention: On the Historical Explanation of Pictures* (New Haven: Yale University Press, 1985).

Baxter, John, *The Hollywood Exiles* (New York: Taplinger, 1976).

Baxter, Peter, *Just Watch! Sternberg, Paramount and America* (London: BFI, 1993).

Bazin, André, *What is Cinema?*, vol. 1 (Berkeley: University of California Press, 1971).

Bean, Jennifer and Diane Negra (eds), *A Feminist Reader in Early Cinema* (Durham, NC: Duke University Press, 2002).

Beauchamp, Cari, *Without Lying Down: Francis Marion and the Powerful Women of Early Hollywood* (New York: Scribners, 1997).

Beauchamp, Cari and Mary Anita Loos, *Anita Loos Rediscovered: Film Treatments and Fiction* (Berkeley: University of California Press, 2003).

Behlmer, Rudy, *Memo from Darryl Zanuck: The Golden Years at Twentieth Century-Fox* (New York: Grove Press, 1993).

— *America's Favorite Movies* (New York: Ungar, 1982).

— *Memo from David O. Selznick* (Los Angeles: Samuel French, 1972).

Belton, John, *Widescreen Cinema* (Cambridge, MA: Harvard University Press, 1992).

Belton, John, Sheldon Hall and Steve Neale (eds), *Widescreen Worldwide* (New Barnet: John Libbey Publishing, 2010).

Berg, A. Scott, *Goldwyn: A Biography* (New York: Knopf, 1989).

Bergfelder, Tim, *International Adventures: German Popular Cinema and European Co-Productions in the 1960s* (New York: Berghan Books, 2005).

Berliner, Todd, *Hollywood Incoherent: Narration in Seventies Cinema* (Austin: University of Texas Press, 2010).

Bernstein, Irving, *The Economics of Television Film Production and Distribution* (Hollywood: Screen Actors Guild, 1960).

— *Hollywood at the Crossroads: An Economic Study of the Motion Picture Industry* (Hollywood: American Federation of Labor Film Council, 1957).

Bernstein, Matthew, *Walter Wanger: Hollywood Independent* (Berkeley: University of California Press, 1994).

Berry, Sarah, *Screen Style* (Minneapolis: University of Minnesota Press, 2000).

Birchard, Richard S., *Cecil B. DeMille's Hollywood* (Lexington: University Press of Kentucky, 2004).

Birdwell, Michael E., *Celluloid Soldiers: Warner Bros.' Campaign Against Nazism* (New York: New York University Press, 1999).

Black, Gregory D., *The Catholic Crusade Against the Movies, 1940–1975* (Cambridge: Cambridge University Press, 1997).

Boddy, William, *Fifties Television: The Industry and Its Critics* (Urbana: University of Illinois Press, 1990).

Bogle, Donald, *Bright Boulevards, Bold Dreams: The Story of Black Hollywood* (New York: Random House, 2005).

— *Toms, Coons, Mulattoes, Mammies and Blacks* (New York: Continuum, 2001 edition).

Bondanella, Peter, *Hollywood Italians: Dagos, Palookas, Romeos, Wiseguys and Sopranos* (New York: Continuum, 2004).

Bordwell, David,—*Poetics of Cinema* (London: Routledge, 2008).

— *The Way Hollywood Tells It: Story and Style in Modern Movies* (Berkeley: University of California Press, 2006).

— *Planet Hong Kong* (Cambridge, MA: Harvard University Press, 2000).

— *On the History of Film Style* (Cambridge, MA: Harvard University Press, 1997).

— *Narration in the Fiction Film* (Madison: University of Wisconsin Press, 1985).

Bordwell, David, Janet Staiger and Kristin Thompson, *The Classical Hollywood Cinema: Film Style and Mode of Production to 1960* (London: Routledge and Kegan Paul, 1985).

Bowser, Eileen, *"Intolerance": The Film by David Wark Griffith. Shot by Shot Analysis* (New York: Museum of Modern Art, 1966).

— *The Transformation of Cinema, 1907–1915* (New York: Scribners, 1990).

Bowser, Pearl, Jane Gaines and Charles Musser (eds), *Oscar Micheaux and His Circle: African-American Filmmaking and Race Cinema of the Silent Era* (Bloomington: Indiana University Press, 2001).

Brooks, Tim and Earle Marsh, *Complete Directory to Prime Time Network and Cable TV Shows* (New York: Ballantine Books, 1999).

Browne, Nick (ed.), *Reconfiguring American Film Genres: History and Method* (Berkeley: University of California Press, 1998).

Brownlow, Kevin, *The Parade's Gone By* (Berkeley: University of California Press, 1968).

Brownlow, Kevin and John Kobal, *Hollywood: The Pioneers* (New York: Knopf, 1979).

Buhle, Paul and Dave Wagner, *A Very Dangerous Citizen: Abraham Lincoln Polonsky and the Hollywood Left* (Berkeley: University of California Press, 2001).

Buhler, James, David Neumeyer and Rob Deemer, *Hearing the Movies: Music and Sound in Film History* (New York: Oxford University Press, 2010).

Burt, George, *The Art of Film Music* (Boston: Northeastern University Press, 1994).

Buscombe, Edward, *'Injuns!': Native Americans and the Movies* (Chicago: Reaktion Books, 2006).

Cabarga, Leslie, *The Fleischer Story* (New York: Nostalgia Press, 1976).

Callahan, Vicki (ed.), *Reclaiming the Archive: Feminism and Film History* (Detroit: Wayne State University Press, 2010).

Cameron, Evan William (ed.), *Sound and the Cinema: The Coming of Sound to American Film* (New York: Redgrave Publishing Company, 1980).

Carr, Robert E. and R.M. Hayes, *Wide Screen Movies* (Jefferson, NC: McFarland, 1988).

Carr, Steven, *Hollywood and Anti-Semitism: A Cultural History up to World War II* (Cambridge: Cambridge University Press, 2001).

Carringer, Robert, *The Making of Citizen Kane* (Berkeley: University of California Press, 1996 edition).

Carroll, Noël, *Interpreting the Moving Image* (New York: Cambridge University Press, 1998).

— *Theorizing the Moving Image* (New York: Cambridge University Press, 1996).

Caspar, Drew, *Hollywood Film: 1963–1976* (Oxford: Wiley-Blackwell, 2011).

Ceplair, Larry and Steven Englund, *The Inquisition in Hollywood: Politics in the Film Community, 1930–1960* (Berkeley: University of California Press, 1983).

Clark, Danae, *Negotiating Hollywood: The Cultural Politics of Actors' Labor* (Minneapolis: University of Minnesota Press, 1995).

Clark, Randall (ed.), *American Screenwriters*, Second Series (Detroit: Gale Research Co., 1986).

Cline, William C., *In the Nick of Time: Motion Picture Sound Serials* (Jefferson, NC: McFarland, 1984).

Cogley, John, *Report on Blacklisting, 1 – Movies* (New York: The Fund for the Republic, 1956).

Conant, Michael, *Antitrust in the Motion Picture Industry: Economic and Legal Analysis* (New York: Arno Press, 1978).

Cook, David A., *Lost Illusions: American Cinema in the Shadow of Watergate and Vietnam, 1970–1979* (New York: Scribners, 2000).

— *A History of Narrative Film* (New York: W.W. Norton, 1981).

Cooke, Mervyn (ed.), *The Hollywood Music Reader* (Oxford: Oxford University Press, 2010).

— *A History of Film Music* (Cambridge: Cambridge University Press, 2008).

Cooper, David, *Bernard Herrmann's The Ghost and Mrs Muir: A Film Score Handbook* (Metuchen, NJ: Scarecrow Press, 2005).

— *Bernard Herrmann's Vertigo: A Film Score Guide* (Westport, CT: Greenwood Press, 2001).

Cooper, Mark Garrett, *Universal Women: Filmmaking and Institutional Change in Early Hollywood* (Urbana: University of Illinois Press, 2010).

Corliss, Richard, *Talking Pictures: Screenwriters in the American Cinema, 1927–1973* (Woodstock, NY: The Overlook Press, 1974).

Cossar, Harper, *Letterboxed: The Evolution of Widescreen Cinema* (Lexington: University of Kentucky Press, 2011).

Cotter, Bill, *The Wonderful World of Disney Television* (New York: Hyperion, 1997).

Courtney, Susan, *Hollywood Fantasies of Miscegenation: Spectacular Narratives of Gender and Race, 1903–1967* (Princeton, NJ: Princeton University Press, 2005).

Cowan, Lester (ed.), *Recording Sound for Motion Pictures: AMPAS* (New York: McGraw-Hill, 1931).

Crafton, Donald, *The Talkies: American Cinema's Transition to Sound, 1926–1931* (New York: Scribners, 1997).

— *Before Mickey: The Animated Film, 1898–1928* (London: MIT Press, 1982).

Cripps, Thomas, *Making Movies Black: The Hollywood Message Movie from World War II to the Civil Rights Era* (New York: Oxford University Press, 1993).

— *Slow Fade to Black: The Negro in American Film, 1900–1942* (New York: Oxford University Press, 1977).

Croy, Homer, *How Motion Pictures Are Made* (New York: Harper and Brothers, 1918).

Cubert, David (ed.), *Mission to Moscow* (Madison: University of Wisconsin Press, 1980), 31–41.

Cyrano, Monica Silveira, *Big Screen Rome* (Oxford: Blackwell, 2005).

Darby, William and Jack Du Bois, *American Film Music: Major Composers, Techniques, Trends, 1915–1990* (Jefferson, NC: McFarland, 1990).

Daubney, Kate, *Max Steiner's Now, Voyager: A Film Score Guide* (Westport, CT: Greenwood Press, 2000).

Davidson, Annette, *Alex North's A Streetcar Named Desire: A Film Score Handbook* (Metuchen, NJ: Scarecrow Press, 2009).

Davis, Kenneth C., *Two-Bit Culture: The Paperbacking of America* (Boston: Houghton Mifflin, 1982).

Davis, Ronald L., *The Glamour Factory: Inside Hollywood's Studio System* (Dallas: Southern Methodist University Press, 1993).

deCordova, Richard, *Picture Personalities: The Emergence of the Star System in America* (Urbana: University of Illinois Press, 1990).

De Mille, William C., *Hollywood Saga* (New York: Dutton, 1939).

De Rochefort, Charles, *Le Film de mes souvenirs* (Paris: Société Parisienne, 1943).

Decherny, Peter, *Hollywood and the Culture Elite: How the Movies Became American* (New York: Columbia University Press, 2005).

Dick, Bernard, *City of Dreams: The Making and Remaking of Universal Pictures* (Lexington: University of Kentucky Press, 1997).

— *The Star-Spangled Screen: The American World War II Film* (Lexington: University of Kentucky Press, 1985).

Dmytryk, Edward, *On Film Editing* (London: Focal Press, 1984).

Dodds, John W., *The Several Lives of Paul Fejos: A Hungarian-American Odyssey* (New York: Wenner-Gren Foundation, 1973).

Doherty, Thomas, *Hollywood's Censor: Joseph I. Breen and the Production Code Administration* (New York: Columbia University Press, 2007).

— *Pre-Code Hollywood: Sex, Immorality and Insurrection in American Cinema, 1930–1934* (New York: Columbia University Press, 1999).

— *Projections of War: Hollywood, American Culture and World War II* (New York: Columbia University Press, 1993).

Dombrowski, Lisa, *The Films of Samuel Fuller: If You Die I'll Kill You!* (Middletown, CT: Wesleyan University Press, 2008).

Dunlap Jr, Orrin, *The Future of Television* (New York: Harper and Brothers Publishers, 1942).

Edgerton, Gary R., *American Film Exhibition and an Analysis of the Motion Picture Industry's Market Structure, 1963–1980* (New York: Garland Publishing, 1983).

Ehrenberg, Ilya, *Usine de Rêves* (Paris: Gallimard, 1936).

Eisenschitz, Bernard, *Nicholas Ray: An American Journey* (London: Faber and Faber, 1993).

Emerson, John and Anita Loos, *How to Write Photoplays* (Philadelphia: George W. Jacobs, 1920).

Enticknap, Leo, *Moving Image Technology: From Zoetrope to Digital* (London: Wallflower Press, 2005).

Erb, Cynthia, *Tracking King Kong: A Hollywood Icon in World Culture* (Detroit: Wayne State University Press, 1998).

Erens, Patricia, *The Jew in American Cinema* (Bloomington: Indiana University Press, 1984).

Evans, Mark, *Soundtrack: The Music of the Movies* (New York: Hopkinson and Blake, 1975).

Eyman, Scott, *The Speed of Sound: Hollywood and the Talkie Revolution* (New York: Simon and Schuster, 1997).

Fabe, Marilyn, *Closely Watched Films: An Introduction to the Art of Narrative Film Technique* (Berkeley: University of California Press, 2004).

Fairservice, Donald, *Film Editing: History, Theory and Practice* (Manchester: Manchester University Press, 2001).

Fielding, Raymond, *The March of Time, 1935–1951* (New York: Oxford University Press, 1978).

— *The American Newsreel, 1911–1967* (Norman: University of Oklahoma Press, 1972).

— (ed.), *A Technological History of Motion Pictures and Television* (Berkeley: University of California Press, 1967).

Finler, Joel W., *The Hollywood Story* (London: Octopus, 1988).

Flynn, Charles and Todd McCarthy, *Kings of the Bs* (New York: Dutton, 1975).

Fordin, Hugh, *The Movies' Greatest Musicals Produced in Hollywood USA by the Freed Unit* (New York: Ungar, 1975).

Francke, Lizzie, *Script Girls: Women Screenwriters in Hollywood* (London: BFI, 1994).

Frank, Nino, *Petit Cinéma Sentimental* (Paris: La Nouvelle Edition, 1950).

Friedman, Lester (ed.), *Unspeakable Images: Ethncity and American Cinema* (Carbondale: University of Illinois Press, 1991).

Fujiwara, Chris, *The World and Its Double: The Life and Work of Otto Preminger* (New York: Faber and Faber, 2008).

Fuller, Kathryn H., *At the Picture Show: Small-Town Audiences and the Creation of a Movie Fan Culture* (Charlottesville: University Press of Virginia, 1996).

Fuller-Selley, Kathryn H. (ed.), *Hollywood in the Neighborhood: Historical Case Studies of Local Moviegoing* (Berkeley: University of California Press, 2008).

Gabler, Neil, *An Empire of Their Own: How the Jews Invented Hollywood* (New York: Crown, 1988).

Gaines, Jane (ed.), *Classical Hollywood Narrative: The Paradigm Wars* (Durham, NC: Duke University Press, 1992).

Gallagher, Tag, *The Adventures of Roberto Rossellini: His Life and Films* (New York: Da Capo Press, 1998).

Geist, Kenneth L., *Pictures Will Talk: The Life and Films of Joseph L. Mankiewicz* (New York: Scribners, 1978).

Gish, Lilian, with Ann Pinchot, *The Movies, Mr. Griffith, and Me* (Englewood-Cliffs, NJ: Prentice-Hall, 1969).

Glancy, Mark, *The 39 Steps: A British Film Guide* (London: I.B. Tauris, 2002).

— *When Hollywood Loved Britain: The Hollywood 'British' Film, 1939–45* (Manchester: Manchester University Press, 1999).

Goldmark, Daniel, *Tunes for 'Toons: Music and the Hollywood Cartoon* (Berkeley: University of California Press, 2005).

Goldner, George E. and Orville Turner, *The Making of King Kong: The Story Behind a Classic Film* (London: Ballantine Books, 1975).

Gomery, Douglas, *The Coming of Sound* (New York: Routledge, 2005).

— *The Hollywood Studio System: A History* (London: BFI, 2005).

— *Shared Pleasures: A History of Movie Presentation in the United States* (London: BFI, 1992).

— *The Hollywood Studio System* (London: BFI/Macmillan, 1986).

Goodman, Walter, *The Committee* (London: Secker & Warburg, 1968).

Grieveson, Lee and Peter Krämer (eds) *The Silent Cinema Reader* (London: Routledge, 2003).

Griffin, Sean (ed.), *What Dreams Were Made Of: Movie Stars of the 1940s* (New Brunswick, NJ: Rutgers University Press, 2011).

Guback, Thomas H., *The International Film Industry: Western Europe and America Since 1945* (Bloomington: Indiana University Press, 1969).

Haines, Richard, *Technicolor Movies: The History of Dye Transfer Printing* (Jefferson, NC: McFarland, 1993).

Hall, Ben M., *The Best Remaining Seats: The Golden Age of the Movie Palace* (New York: Da Capo, 1988, originally published in 1975).

Hall, Sheldon, *Zulu: With Some Guts Behind It – The Making of an Epic Movie* (Sheffield: Tomahawk Press, 2005).

Hall, Sheldon and Steve Neale, *Epics, Spectacles and Blockbusters: A Hollywood History* (Detroit: Wayne State University Press, 2010).

Hambley, John and Patrick Downing, *The Art of Hollywood* (London: Thames Television, 1979).

Hampton, Benjamin, *History of the American Film Industry from Its Beginnings to 1931* (New York: Dover Publications, 1970, originally published in 1931).

Handel, Leo A., *Hollywood Looks at Its Audience* (Urbana: University of Illinois Press, 1950).

Hardy, Ursula, *From Caligari to California: Eric Pommer's Life in the International Film Wars* (Providence, RI: Berghahn Books, 1996).

Harmetz, Aljean, *Roundup the Usual Suspects, The Making of Casablanca: Bogart, Bergman, and World War II* (New York: Hyperion, 1992).

— *The Wizard of Oz* (New York: Limelight, 1977).

Haver, Ronald, *David O. Selznick's Hollywood* (London: Secker & Warburg, 1980).

Hayes, Dade and Jonathan Bing, *Open Wide: How Hollywood Box Office Became a National Obsession* (New York: Hyperion/Miramax Books, 2004).

Heffernan, Kevin, *Ghouls, Gimmicks, and Gold: Horror Films and the American Movie Business, 1953–1968* (Durham, NC: Duke University Press, 2004).

Heisner, Beverly, *Hollywood Art: Art Directors in the Days of the Great Studios* (London: St James Press, 1990).

Henderson, Robert M., *D.W. Griffith: His Life and Work* (New York: Oxford University Press, 1972).

Hess, Gary N., *An Historical Study of the DuMont Television Network* (New York: Arno Press, 1979).

Higgins, Scott, *Harnessing the Technicolor Rainbow: Color Design in the 1930s* (Austin: University of Texas Press, 2007).

Higham, Charles, *Hollywood Cameramen* (London: Thames and Hudson, 1970).

Higson, Andrew and Richard Maltby, *'Film Europe' and 'Film America': Cinema, Commerce and Cultural Exchange, 1920–1939* (Exeter: University of Exeter Press, 1999).

Hilmes, Michele, *Hollywood and Broadcasting: From Radio to Cable* (Urbana: University of Illinois Press, 1990).

Hirano, Kyoko, *Mr. Smith Goes to Tokyo: Japanese Cinema under the American Occupation, 1945–1952* (Washington, DC: Smithsonian Institution Press, 1992).

Hirsch, Foster, *A Method to their Madness: The History of the Actors Studio* (New York: Norton, 1984).

Hoberman, J. *Bridge of Light: Yiddish Film Between Two Worlds* (Philadelphia: Temple University Press, 1991).

Hoberman, J. and Jeffrey Shandler, *Entertaining America: Jews, Movies, and Broadcasting* (New York: The Jewish Museum and Princeton University Press, 2003).

Hollis, Richard and Brian Sibley, *The Disney Studio Story* (New York: Crown, 1981).

Horne, Gerald, *Class Struggle in Hollywood: Moguls, Mobsters, Stars, Reds, and Trade Unions* (Austin: University of Texas Press, 2001).

Howard, David and Edward Mabley, *The Tools of Screenwriting: A Writer's Guide to the Craft and Elements of a Screenplay* (New York: St Martin's Press, 1995).

Howard, H.H. and S.L. Carroll, *Subscription Television: History, Current Status, and Economic Projection* (Knoxville: University of Tennessee Press, 1980).

Huettig, Mae D., *Economic Control of the Motion Picture Industry: A Study of Industrial Organization* (Philadelphia: University of Pennsylvania Press, 1944).

Hughes, Laurence (ed.), *The Truth about the Movies* (Hollywood: Hollywood Publishers, 1924).

Hyatt, Wesley, *The Encyclopedia of Daytime Television* (New York: Billboard Books, 1997).

Jackson, Kathy Merlock, *Walt Disney: A Bio-Bibliography* (Westport, CT: Greenwood Press, 1993).

Jacobs, Diane, *Hollywood Renaissance* (Cranbury, NJ: Barnes, 1977).

Jacobs, Lea, *The Decline of Sentiment: American Film in the 1920s* (Berkeley: University of California Press, 2008).

— *The Wages of Sin: Censorship and the Fallen Woman Film, 1928–1942* (Madison: University of Wisconsin Press, 1991).

Jacobs, Lewis, *The Rise of the American Film: A Critical History* (New York: Columbia University Press, 1968).

Jarvie, Ian, *Hollywood's Overseas Campaign: The North American Movie Trade, 1920–1950* (Cambridge: Cambridge University Press, 1992).

Johnson, Doris and Ellen Leventhal (eds), *The Letters of Nunnally Johnson* (New York: Knopf, 1981).

Jowett, Garth, *Film: The Democratic Art* (Boston: Little, Brown, 1976).

Jura, Jean-Jacques and Rodney Norman Bardin II, *Barboa Films: A History and Filmography of the Silent Film Studio* (Jefferson, NC: McFarland, 1999).

Kalinak, Kathryn, *How the West Was Sung: Music in the Westerns of John Ford* (Berkeley: University of California Press, 2007).

— *Settling the Score: Music and the Classical Hollywood Film* (Madison: University of Wisconsin Press, 1992).

Kalmus, Herbert, with Eleonore King Kalmus, *Mr. Technicolor* (Abescon, NJ: MagicImage Filmbooks, 1993).

Karlin, Fred and Rayburn Wright, *On the Track: A Guide to Contemporary Film Scoring* (New York: Schirmer Book, 1990).

Keating, Patrick, *Hollywood Lighting from the Silent Era to Film Noir* (New York: Columbia University Press, 2010).

Kemper, Tom, *Hidden Talent: The Emergence of Hollywood Agents* (Berkeley: University of California Press, 2010).

Kennedy, Joseph P. (ed.), *The Story of the Film* (Chicago: A.W. Shaw, 1927).

Kessler, Frank and Nanna Verhoeff (eds), *Networks of Entertainment: Early Film Distribution, 1895–1915* (Eastleigh: John Libbey Press, 2007).

Kiesling, Barratt, *Talking Pictures: How They are Made and How to Appreciate Them* (Richmond: Johnson Publishing Company, 1937).

Kilpatrick, Jacquelyn, *Cellulioid Indians: Native Americans and Film* (Lincoln: University of Nebraska Press, 1999).

Kindem, Gorham (ed.), *The International Movie Industry* (Carbondale: Southern Illinois University Press, 2000).

— *The American Movie Industry: The Business of Motion Pictures* (Carbondale: Southern Illinois University Press, 1982).

Kitamura, Hiroshi, *Screening Enlightenment: Hollywood and the Cultural Reconstruction of Defeated Japan* (Ithaca: Cornell University Press, 2010).

Klinger, Barbara, *Melodrama and Meaning: History, Culture, and the Films of Douglas Sirk* (Bloomington: Indiana University Press, 1994).

Klumph, Helen and Klumph, Inez, *Screen Acting: Its Requirements and Rewards* (New York: Falk, 1922).

Knietel, Seymour and Izzy Sparber, *Standard Production Reference* (Miami: Fleischer Studios, Inc., 1940).

Knopf, Robert, *The Theater and Cinema of Buster Keaton* (Princeton: Princeton University Press, 1999).

Kobal, John (ed.), *Hollywood Glamour Portraits* (New York: Dover, 1976).

Koppes, Clayton R. and Gregory D. Black, *Hollywood Goes to War: How Politics, Profits and Propaganda Shaped World War II Movies* (New York: The Free Press, 1987).

Koszarski, Richard, *An Evening's Entertainment: The Age of the Silent Feature Picture, 1915–1928* (New York: Scribners, 1990).

— *The Astoria Studio and Its Fabulous Films* (New York: Dover, 1983).

— *The Man You Loved to Hate: Erich von Stroheim and Hollywood* (New York: Oxford University Press, 1983).

— (ed.), *Hollywood Directors, 1941–1976* (New York: Oxford University Press, 1977).

Kovács, András Bálint, *Screening Modernism: European Art Cinema, 1950–1980* (Chicago: Chicago University Press, 2007).

Kreimeier, Klaus, *Die Ufa-Story: Geshichte eines Filmkonzerns* (Munich: Carl Hanser Verlag, 1992).

Krutnick, Frank, *In a Lonely Street: Film Noir, Genre, Masculinity* (London: Routledge, 1991).

Kwolek-Folland, Angel, *Engendering Business: Men and Women in the Corporate Office, 1870–1930* (Baltimore: Johns Hopkins University Press, 1994).

Lahue, Kalton C., *Dreams for Sale: The Rise and Fall of the Triangle Film Corporation* (New York: A.S. Barnes, 1971).

— *Continued Next Week: A History of the Motion Picture Serial* (Norman: University of Oklahoma Press, 1964).

Lane, Tamar, *The New Technique of Screen Writing: A Practical Guide to the Writing and Marketing of Photoplays* (New York: McGraw-Hill, 1937).

Larsen, Peter, *Film Music* (London: Reaktion Books, 2005).

Larson, Randall, *Musique Fantastique: A Survey of Film Music in the Fantastic Cinema* (Metuchen, NJ: Scarecrow Press, 1985).

Lasky, Jessie L., with Don Weldon, *I Blow My Own Horn* (London: Victor Gollancz, 1957).

Lastra, James, *Sound Technology and the American Cinema: Perception, Representation and Modernity* (New York: Columbia University Press, 2000).

Leff, Leonard J., *Hitchcock and Selznick* (New York: Weidenfeld, 1987).

Leff, Leonard J. and Jerrold R. Simmons, *The Dame in the Kimono: Hollywood, Censorship and the Production Code from the 1920s to the 1960s* (London: Weidenfeld and Nicholson, 1990).

Lenihan, John H., *Showdown: Confronting America in the Western Film* (Urbana: University of Illinois Press, 1985).

Lev, Peter, *The Fifties: Transforming the Screen, 1950–1959* (New York: Scribners, 2003).
Lewis, Howard T., *The Motion Picture Industry* (New York: D. Van Nostrand, 1933).
— (ed.), *Cases on the Motion Pictures. With Commentaries* (New York: McGraw-Hill, 1930).
Lewis, Jon, *American Film: A History* (New York: W.W. Norton, 2008).
— *Hollywood v. Hard Core: How the Struggle over Censorship Saved the Modern Film Industry* (New York: New York University Press, 2000).
Limbacher, James L., *Four Aspects of the Film* (New York: Brussel and Brussel, 1969).
Lorence, James J., *The Suppression of Salt of the Earth: How Hollywood, Big Labor, and Politicians Blacklisted a Movie in Cold War America* (Albuquerque: University of New Mexico Press, 1999).
Lovell, Alan and Peter Krämer (eds), *Screen Acting* (London: Routledge, 1999).
Lyons, Timothy James, *The Silent Partner: The History of the American Film Manufacturing Company, 1910–1921* (New York: Arno Press, 1974).
McCarty, Clifford (ed.), *Film Music 1* (New York: Garland, 1989).
McDougal, Dennis, *The Last Mogul: Lew Wasserman, MCA, and the Hidden History of Hollywood* (New York: Crown, 1998).
McLaughlin, Robert, *Broadway and Hollywood: A History of Economic Interaction* (New York: Arno Press, 1974).
McLean, Adrienne J. (ed.), *Glamour in a Golden Age: Movie Stars of the 1930s* (New Brunswick, NJ: Rutgers University Press, 2011).
McMahon, Alison, *Alice Guy Blaché: Lost Visionary of the Cinema* (New York: Continuum, 2002).
Maeder, Edward, *Hollywood and History: Costume Design in Film* (Los Angeles: Los Angeles County Museum, 1987).
Mahar, Karen Ward, *Women Filmmakers in Early Hollywood* (Baltimore: Johns Hopkins Press, 2006).
Maltby, Richard, *Hollywood Cinema* (Oxford: Blackwell, 2003 edition).
Maltby, Richard, Daniel Biltereyst and Philippe Meers (eds), *Explorations in New Cinema History: Approaches and Case Studies* (Oxford: Wiley-Blackwell, 2011).
Maltby, Richard, Melvyn Stokes and Robert C. Allen (eds), *Going to the Movies: Hollywood and Social Experience of the Cinema* (Exeter: University of Exeter Press, 2007).
Maltin, Leonard, *Of Mice and Magic: A History of American Animated Cartoons* (New York: Plume, 1987 edition).
— *The Disney Films* (New York: Crown, 1973).
— *Selected Short Subjects: From Spanky to Our Gang* (New York: Da Capo Press, 1972).
Mandelbaum, Howard and Eric Myers, *Screen Deco* (New York: St Martin's Press, 1985).
Mann, Denise, *Hollywood Independents: The Postwar Talent Takeover* (Minneapolis: University of Minnesota Press, 2008).
Mann, William J., *Behind the Screen: How Gays and Lesbians Shaped Hollywood, 1910–1969* (New York: Viking, 2001).
Marchetti, Gina, *Romance and the 'Yellow Peril': Race, Sex, and Discursive Strategies in Hollywood Fiction* (Berkeley: University of California Press, 1993).
Marion, Francis, *Off with Their Heads!* (New York: Macmillan, 1972).
Marks, Martin Miller, *Music and the Silent Film: Contexts and Case Studies, 1895–1924* (New York: Oxford University Press, 1997).
May, Larry, *The Big Tomorrow: Hollywood and the Politics of the American Way* (Chicago: University of Chicago Press, 2000).
Mayne, Judith, *Directed by Dorothy Arzner* (Bloomington: Indiana University Press, 1984).
Moldea, Dan E., *Dark Victory: Ronald Reagan, MCA, and the Mob* (New York: Viking, 1986).
Moley, Raymond, *The Hays Office* (Indianapolis: Bobbs-Merrill, 1945).
Monaco, Paul, *The Sixties, 1960–1969* (New York: Scribners, 2001).

Morris, Oswald, with Geoffrey Bull, *Huston, We Have a Problem: A Kaleidoscope of Filmmaking Memories* (Landham, MD: Scarecrow Press, 2006).

Mueller, John, *Astaire Dancing: The Musical Films* (London: Hamish Hamilton, 1986).

Mullen, Megan, *The Rise of Cable Programming in the United States: Revolution or Evolution?* (Austin: University of Texas Press, 2003).

Musser, Charles, *The Emergence of Cinema: The American Screen to 1907* (New York: Scribners, 1990).

Musser, Charles, with Carol Nelson, *High-Class Moving Pictures: Lyman H. Howe and the Forgotten Era of Traveling Exhibition, 1880–1920* (Princeton, NJ: Princeton University Press, 1991).

Naremore, James, *Acting in the Cinema* (Berkeley: University of California Press, 1988).

Naumberg, Nancy (ed.), *We Make the Movies* (New York: W.W. Norton, 1937).

Navasky, Victor S., *Naming Names* (New York: The Viking Press, 1980).

Neale, Steve, *Genre and Hollywood* (London: Routledge, 2000).

Neale, Steve and Murray Smith (eds), *Contemporary Hollywood Cinema* (London: Routledge, 1998).

Negra, Diane, *Off-White Hollywood: American Culture and Ethnic Female Stardom* (New York: Routledge, 2001).

Neve, Brian, *Elia Kazan: The Cinema of an American Outsider* (London: I.B. Tauris, 2009).

Nielsen, Mike and Gene Miles, *Hollywood's Other Blacklist: Union Struggles in the Studio System* (London: BFI, 1995).

Nowell-Smith, Geoffrey and Stephen Ricci (eds), *Hollywood and Europe: Economics, Culture and National Identity* (London: BFI, 1998).

O'Brien, Geoffrey, *Hardboiled America: The Lurid Years of Paperbacks* (New York: Van Nostrand-Reinhold, 1981).

O'Connor, John, *I am a Fugitive from a Chain Gang* (Madison: University of Wisconsin Press, 1981).

Palmer, R. Barton (ed.), *Larger than Life: Movie Stars of the 1950s* (New Brunswick, NJ: Rutgers University Press, 2010).

Pathé, Charles, *De Pathé Frères à Pathé Cinéma* (Paris: Premier Plan, 1940).

Peary, Danny and Gerald Peary (eds), *The American Animated Cartoon* (New York: Dutton, 1980).

Pells, Richard, *The Liberal Mind in a Conservative Age: American Intellectuals in the 1940s and 1950s* (New York: Harper & Row, 1985).

Petrie, Graham, *Hollywood Destinies: European Directors in America, 1922–1931* (London: Routledge and Kegan Paul, 1985).

Petro, Patrice (ed.), *Idols of Modernity: Movie Stars of the 1920s* (New Brunswick, NJ: Rutgers University Press, 2010).

Pitts, Michael R., *Poverty Row Studios, 1929–1940* (Jefferson, NC: McFarland, 1997).

Pizzitola, Louis, *Hearst Over Hollywood: Power, Passion and Propaganda in the Movies* (New York: Columbia University Press, 2002).

Powdermaker, Hortense, *Hollywood the Dream Factory: An Anthropologist Looks at the Movie-Makers* (Boston: Little, Brown, 1950).

Prendergast, Roy M., *Film Music: A Neglected Art* (New York: W.W. Norton, 1992 edition).

Prince, Stephen, *Classical Film Violence: Designing and Regulating Brutality in Hollywood Cinema, 1930–1968* (New Brunswick, NJ: Rutgers University Press, 2003).

— *Savage Cinema: Sam Peckinpah and the Rise of Ultraviolent Movies* (Austin: University of Texas Press, 1998).

Prindle, David F., *The Politics of Glamour: Ideology and Democracy in the Screen Actors Guild* (Madison: University of Wisconsin Press, 1988).

Puttnam, David, with Neil Watson, *Movies and Money* (New York: Knopf, 1998).

Quigley, Martin (ed.), *New Screen Techniques* (New York: Quigley Publishing, 1953).

Rainey, Buck, *Serials and Series: A World Bibliography, 1912–1956* (Jefferson, NC: McFarland, 1999).

Rainsberger, Todd, *James Wong Howe: Cinematographer* (La Jolla, CA: A.S. Barnes, 1981).

Rathgeb, Donald L., *The Making of Rebel Without a Cause* (Jefferson, NC: McFarland, 2004).

Rebello, Stephen, *Alfred Hitchcock and the Making of Psycho* (London: Marion Boyars, 1990).

Regester, Charlene, *African American Actresses: The Struggle for Visibility, 1900–1960* (Bloomington: Indiana University Press, 2010).

Roddick, Nick, *A New Deal in Entertainment: Warner Brothers in the 1930s* (London: BFI, 1983).

Roffman, Peter and James Purdy, *The Hollywood Social Problem Film: Madness, Despair and Politics from the Depression to the 1950s* (Bloomington: Indiana University Press, 1981).

Ross, Murray, *Stars and Strikes: Unionization of Hollywood* (New York: Columbia University Press, 1941).

Ross, Steven J., *Working-Class Hollywood: Silent Film and the Shaping of Class in America* (Princeton, NJ: Princeton University Press, 1998).

Rosten, Leo, *Hollywood: The Movie Colony, the Movie Makers* (New York: Harcourt, Brace, 1941).

Rotundo, Anthony, *American Manhood: Transformations in Masculinity from the Revolution to the Modern Era* (New York: Basic Books, 1993).

Rubin, Martin, *Showstoppers: Busby Berkeley and the Tradition of Spectacle* (New York: Columbia University Press, 1993).

Ryan, Roderick, *A History of Motion Picture Color Technology* (New York: Focal Press, 1977).

Sabaneev, Leonid, *Music for the Films* (London: Pitman and Sons, 1935).

Salt, Barry, *Film Style and Technology: History and Analysis* (London: Starword, 1992 edition).

Sarris, Andrew, *The American Cinema: Directors and Directions, 1929–1968* (New York: Dutton, 1968).

Saunders, Thomas J. *Hollywood in Berlin: American Cinema and Weimar Germany* (Berkeley: University of California Press, 1994).

Schaeffer, Eric, *'Bold! Daring! Shocking! True!': A History of Exploitation Films, 1919–1959* (Durham, NC: Duke University Press, 1999).

Schafer, Murray, *The Soundscape: Our Sonic Environment and the Turning of the World* (Rochester: Destiny Books, 1977).

Schartz, Vanessa R., *It's so French! Hollywood, Paris, and the Making of a Cosmopolitan Film Culture* (Chicago: Chicago University Press, 2007).

Schatz, Thomas, *Boom and Bust: American Cinema in the 1940s* (New York: Scribners, 1997).

— *The Genius of the System: Hollywood Filmmaking in the Studio Era* (New York: Pantheon Books, 1988).

Schickel, Richard, *The Disney Version* (New York: Avon Books, 1968).

Schlesinger, Jr, Arthur, *The Vital Center: The Politics of Freedom* (Boston: Houghton Mifflin, 1949).

Schneider, Steve, *That's All Folks! The Art of Warner Bros. Animation* (New York: Henry Holt, 1990).

Schneider, Steven Jay (ed.), *New Hollywood Violence* (Manchester: Manchester University Press, 2004).

Schrecker, Helen, *Many Are the Crimes: McCarthyism in America* (Boston: Little, Brown, 1998).

Schumack, Murray, *The Face on the Cutting Room Floor: The Story of Movie and Television Censorship* (New York: Da Capo, 1975).

Sedgwick, John and Michael Pokorny (eds), *An Economic History of Film* (London: Routledge, 2005).

Segrave, Kerry, *American Films Abroad: Hollywood's Domination of the World's Movie Screens* (Jefferson, NC: McFarland, 1997).

— *Drive-in Theaters: A History from Their Inception in 1933* (Jefferson, NC: McFarland, 1993).

Server, Lee, *Over My Dead Body: The Sensational Age of the American Paperback: 1945–1955* (San Francisco: Chronicle Books, 1994).

— *Screenwriter: Words Become Pictures* (Princeton, NJ: Main Street Press, 1987).

Shandey, Robert R., *Runaway Romances: Hollywood's Postwar Tour of Europe* (Philadelphia: Temple University Press, 2009).

Sheehan, Perley Poore, *Hollywood as a World Center* (Hollywood: Hollywood Citizen's Press, 1924).

Shindler, Colin, *Hollywood in Crisis: American Cinema and Society, 1929–39* (London: Routledge, 1996).

Shipman, David, *The Great Movie Stars: The Golden Years* (London: Angus and Robertson, 1979 edition).

Shipman, Nell, *The Silent Screen and My Talking Heart: An Autobiography* (Boise, ID: Boise State University Press, 1988 edition).

Sintzenich, Hal, 'The Sintzenich Diaries', *Quarterly Journal of the Library of Congress*, Summer/Fall 1980.

Skiles, Martin, *Music Scoring for TV and Motion Pictures* (Blue Ridge Summit, PA: Tab Books, 1976).

Skinner, Frank, *Underscore* (New York: Criterion Music, 1960).

Sklar, Robert, *City Boys: Cagney, Bogart, Garfield* (Princeton, NJ: Princeton University Press, 1992).

— *Movie-Made America: A Cultural History of American Movies* (New York: Random House, 1975).

Slide, Anthony, *Early Women Directors* (New York: A.S. Barnes, 1977).

Slotkin, Richard, *Gunfighter Nation: The Myth of the Frontier in Twentieth-Century America* (New York: Athenaeum, 1992).

Smith, Jeff, *The Sounds of Commerce: Marketing Popular Film Music* (New York: Columbia University Press, 1998).

Solomon, Aubrey, *Twentieth Century-Fox: A Corporate and Financial History* (Lanham, MD: Scarecrow Press, 2002).

Stanfield, Peter, *Maximum Movies – Pulp Fictions: Film Culture and the Worlds of Samuel Fuller, Mickey Spillane and Jim Thompson* (New Brunswick, NJ: Rutgers University Press, 2011).

Steinberg, Cobbett, *Reel Facts: The Movie Book of Records* (New York: Random House, 1978).

Stempel, Tom, *Framework: A History of Screenwriting in the American Film Industry* (New York: Continuum, 1988).

Stephenson, Ralph, *The Animated Film* (New York: A.S. Barnes, 1973).

Sterling, Christopher H. and Timothy R. Haight, *The Mass Media: Aspen Guide to Communication Industry Trends* (New York: Praeger, 1978).

Stewart, Jackie, *Migrating to the Movies: Cinema and Black Urban Modernity* (Berkeley: University of California Press, 2005).

Stones, Barbara, *America Goes to the Movies: 100 Years of Motion Picture Exhibition* (North Hollywood: National Association of Movie Theatre Owners, 1993).

Strasberg, Lee, *A Dream of Passion: The Development of the Method* (Boston: Little, Brown, 1987).

Street, Sarah, *Transatlantic Crossings: British Feature Films in the USA* (New York: Continuum, 2002).

Streible, Dan, *Fight Pictures: A History of Boxing and Early Cinema* (Washington, DC: Smithsonian Institution Press, 2002).

Tan, Ed S., *Emotion and the Structure of Narrative Film: Film as an Emotion Machine* (Mahwah, NJ: Lawrence Erlbaum Associates, 1996).

Taylor, John Russell, *Strangers in Paradise: The Hollywood Émigrés, 1933–1950* (New York: Holt, Rinehart and Winston, 1983).

Thomas, Bob, *Building a Company: Roy O. Disney and the Creation of an Entertainment Empire* (New York: Hyperion, 1998).

Thomas, Frank and Ollie Johnston, *Disney Animation: The Illusion of Life* (New York: Abbeville, 1981).

Thomas, Tony, *Music for the Movies* (New York: A.S. Barnes, 1973).

Thompson, Kristin, *Herr Lubitsch Goes to Hollywood: German and American Film after World War I* (Amsterdam: Amsterdam University Press, 2005).

— *Storytelling in the New Hollywood: Understanding Classical Narrative Technique* (Cambridge, MA: Harvard University Press, 1999).

— *Exporting Entertainment: America in the World Film Market, 1907–1934* (London: BFI, 1985).

Thompson, Kristin and David Bordwell, *Film History: An Introduction* (New York: McGraw-Hill, 2003 edition).

Thomson, David, *Showman: The Life of David O. Selznick* (London: Abacus, 1993).

Thorp, Margaret Ferrand, *America at the Movies* (New Haven, CT: Yale University Press, 1939).

Tibbetts, John, *The American Theatrical Film: Stages in Development* (Bowling Green, OH: Bowling Green State University Popular Press, 1985).

Todd, Michael Jr and Susan McCarthy Todd, *A Valuable Property: The Life Story of Michael Todd* (New York: Arbor House, 1983).

Van Zile, Edward, *That Marvel – The Movie* (New York: Putnam, 1923).

Vasey, Ruth, *The World According to Hollywood, 1918–1939* (Exeter: University of Exeter Press, 1997).

Vertrees, Alan David, *Selznick's Vision: Gone with the Wind and Hollywood Filmmaking* (Austin: University of Texas Press, 1997).

Vidor, King, *A Tree Is a Tree* (New York: Harcourt, Brace, 1953).

Viera, Mark A., *Irving Thalberg: Boy Wonder to Producer Prince* (Berkeley: University of California Press, 2010).

Vizzard, Jack, *See No Evil: Life Inside a Hollywood Censor* (New York: Simon and Schuster, 1970).

Waldman, Harry, *Paramount in Paris: 300 Hundred Films Produced at the Joinville Studios, 1930–1933, with Credits and Biographies* (Lanham, MD: Scarecrow Press, 1988).

Waller, Gregory A. (ed.), *Moviegoing in America: A Sourcebook in the History of Film Exhibition* (Oxford: Blackwell, 2002).

— *Main Street Amusements: Movies and Commercial Entertainment in a Southern City, 1896–1930* (Washington, DC: Smithsonian Institution Press, 1995).

Wasko, Janet, *Understanding Disney: The Manufacture of Fantasy* (Oxford: Polity Press, 2001).

— *Movies and Money: Financing the American Film Industry* (Norwood: Ablex, 1982).

Watts, Stephen (ed), *Behind the Screen: How Films are Made* (London: A. Barkes Ltd., 1938).

Weis, Elisabeth and John Belton (eds), *Film Sound: Theory and Practice* (New York: Columbia University Press, 1985).

Weiss, Andrea, *Vampires and Violets: Lesbians and Films* (New York: Penguin Books, 1993).

Welky, David, *The Moguls and the Dictators: Hollywood and the Coming of War* (Baltimore: Johns Hopkins University Press, 2008).

Whitfield, Eileen, *Pickford: The Woman Who Made Hollywood* (Toronto: Macfarlane Walter & Ross, 1997).

Wierzbicki, James, *Louis and Bebe Barron's Forbidden Planet: A Film Score Handbook* (Metuchen, NJ: Scarecrow Press, 2005).

Wilinsky, Barbara, *Sure Seaters: The Emergence of Art-House Cinema* (Minneapolis: University of Minnesota Press, 2001).

Williams, Carol Traynor, *The Dream Beside Me: The Movies and the Children of the Forties* (Rutherford: Fairleigh Dickinson University Press, 1980).

Williams, Raymond, *Television: Technology and Cultural Form* (New York: Schocken, 1975).

Winokur, Mark, *American Laughter: Immigrants, Ethnicity, and 1930s American Film Comedy* (New York: St Martin's Press, 1996).

Winters, Ben, *Erich Wolfgang Korngold's The Adventures of Robin Hood: A Film Score Handbook* (Metuchen, NJ: Scarecrow Press, 2007).

Wittern-Keller, Laura and Raymond J. Haberski Jr, *The Miracle Case: Film Censorship and the Supreme Court* (Lawrence: University of Kansas Press, 2008).

Wolfstein, Martha and Nathan Leites, *Movies: A Psychological Study* (Glencoe, IL: The Free Press, 1950).

Woll, Allen L., *The Latin American Image in American Film* (Los Angeles: Latin American Center, University of California Press, 1996).

Articles

In addition to those cited in this Reader, this bibliography lists a number of other articles.

Abel, Richard, 'The "Backbone" of the Business: Scanning Signs of US Film Distribution in the Newspapers, 1911–14', in Kessler and Verhoeff (eds), *Networks of Entertainment*.

— 'Nickelodeons', in Abel (ed.), *Encyclopedia of Early Cinema*.

Adamakos, Peter, 'Ub Iwerks', *Mindrot*, no. 7 (1977).

Adler, Les K. and Thomas G. Paterson, 'Red Fascism: The Merger of Nazi Germany and Soviet Russia in the American Image of Totalitarianism, 1930s–1950s', *American Historical Review*, vol. 75 no. 4 (1970).

Allen, Robert C., 'Manhattan Myopia; Or, Oh, Iowa! Robert C. Allen on Ben Singer's "Manhattan Nickelodeons: New Data on Audiences and Exhibitors"', *Cinema Journal*, vol. 35 no. 3 (1996).

Altman, Rick, 'Dickens, Griffith, and Film Theory Today', in Gaines (ed.), *Classical Hollywood Narrative*.

— 'The Evolution of Sound Technologies', in Weis and Belton, *Film Sound*.

Altman, Rick, McGraw Jones and Sonia Tratoe, 'Inventing the Cinema Soundtrack: Hollywood's Multiplane Sound System', in James Buhler, Caryl Flynn and David Neumeyer (eds), *Music and Cinema* (Hanover, NH: Wesleyan University Press, 2000).

Andersen, Thom, 'Red Hollywood', in Frank Krutnik, Brian Neve, Steve Neale and Peter Stanfield (eds), *'Un-American' Hollywood: Politics and Film in the Blacklist Era* (New Brunswick: Rutgers University Press, 2007).

Anderson, Michael, 'The Motion Picture Patents Company: A Revaluation', in Balio (ed.), *The American Film Industry*.

Andrew, Dudley, 'Sound in France: the Origins of a Native School', *Yale French Studies*, no. 60 (1980).

Armatage, Kaye, 'Sex and Snow: Landscape and Identity in the God's Country Films of Nell Shipman', in Bachman and Slater (eds), *American Silent Film*.

Arnold, John, 'Shooting the Movies', in Naumberg (ed.), *We Make the Movies*.

Arvey, Verney, 'Composing for the Pictures by Erich Korngold: An Interview', *Etude*, vol. 55 no. 1 (1937).

Asper, Helmut G. and Jan-Christopher Horak, 'Three Smart Guys: How a Few Penniless German Émigrés Saved Universal Studios', *Film History*, vol. 11 no. 2 (1999).

Bakker, Gerben, 'At the Origins of Increased Productivity Growth in Services: Productivity, Social Savings and the Consumer Surplus of the Film Industry, 1900–1938', *Working Papers in Economic History* (London School of Economics), no. 81 (2004).

— 'Stars and Stories: How Films Became Branded Products', *Enterprise and Society*, vol. 2 (2001), reprinted in Sedgwick and Pokorny, *An Economic History of Film*.

Bauchens, Annie, 'Cutting the Film', in Naumberg (ed.), *We Make the Movies*.

Bazin, André, 'Theater and Cinema – Part One' and 'Theater and Cinema – Part Two', in *What Is Cinema?*

Beauchamp, Clem, 'The Production Takes Shape', in Naumberg (ed.), *We Make the Movies*.

Behlmer, Rudy, 'Tumult, Battle, and Blaze: Looking Back at the 1920s – and Since – with Gaylord Carter, Dean of Theater Organists', in McCarty (ed.), *Film Music 1*.

Belton, John, 'CinemaScope: The Economics of Technology', *The Velvet Light Trap*, no. 21 (1985).

— 'Cinerama: A New Era in the Cinema', *The Perfect Vision*, vol. 1 no. 4 (1988).

Bernstein, Matthew, 'Hollywood's Semi-Independent Production', *Cinema Journal*, vol. 32 no. 3 (1993).

— 'The Producer as Auteur', in Barry Keith Grant (ed.), *Auteurs and Authorship: A Film Reader* (Malden, MA: Blackwell, 2008).

— 'A Reappraisal of the Semi-Documentary Film, 1945–48', *The Velvet Light Trap*, no. 20 (1983).

Bernstein, Matthew and Dana F. White, '"Scratching around a Fit of Insanity": The Norman Film Company and the Race Film Business in the 1920s', *Griffithiana*, vol. 21 no. 62/3 (1998).

Black, Gregory D., 'The Production Code and the Hollywood Film Industry, 1930–40', *Film History*, vol. 3 no. 2 (1989).

Boddy, William, 'The Studios Move into Prime Time: Hollywood and the Television Industry in the 1950s', *Cinema Journal*, vol. 24 no. 4 (1985).

Bodeen, Dewitt, 'Frances Marion', *Films in Review*, vol. 20 no. 2 (1969).

Bordwell, David, 'William Cameron Menzies: One Forceful, Expressive Idea', *David Bordwell's Website on Cinema*, March 2010, http://www.davidbordwell.net/essays/menzies.php.

Borneman, Ernest, 'United States versus Hollywood: The Case Study of an Antitrust Suit', in Balio (ed.), *The American Film Industry*.

Bradley, Scott, 'Conversation with Carl W. Stalling', in Cooke (ed.), *The Hollywood Music Reader*.

— 'Personality on the Soundtrack', in Cooke (ed.), *The Hollywood Music Reader*.

Brewster, Ben, 'Multiple-Reel/Feature Films', in Abel (ed.), *Encyclopedia of Early Cinema*.

Burrows, Jonathan, 'Penny Pleasures: Film Exhibition in London During the Nickelodeon Era, 1904–14', *Film History*, vol. 16 no. 1 (2004).

Carbine, Mary, '"The Finest Outside the Loop": Motion Picture Exhibition in Chicago's Black Metropolis, 1905–28', *Camera Obscura*, no. 23 (1990).

Capra, Frank, 'Breaking Hollywood's "Pattern of Sameness"', in Koszarski (ed.), *Hollywood Directors, 1941–1976*.

Carnicke, Sharon Marie, 'Crafting Film Performances: Acting in the Hollywood Studio Era', in Lovell and Krämer (eds), *Screen Acting*.

— 'Lee Strasberg's Paradox of the Actor', in Lovell and Krämer (eds), *Screen Acting*.

Ceplair, Larry, 'The Film Industry's Battle against Left-wing Influences, from Russian Revolution to Blacklist', *Film History*, vol. 20 no. 4 (2008).

Chell, Samuel, 'Music and Emotion in the Classic Hollywood Film: The Case of *The Best Years of Our Lives*', *Film Criticism*, vol. 8 no. 2 (1984).

Cherchi Usai, Paulo, 'Un Américain à la Conquête de l'Italie: George Kleine à Grugliasco, 1913–14', *Archives*, 22/3 and 26/7 (1989).

Chisolm, Brad, 'Red, Blue, and Lots of Green', in Balio (ed.), *Hollywood in the Age of Television*.

Clair, Réné, 'Interview by John Gillett', *Focus on Film* no. 12 (1972).

Clark, Danae, 'Acting in Hollywood's Best Interests: Representations of Actors' Labour During the National Recovery Administration', *Journal of Film and Video*, vol. 42 no. 4 (1990).

Corliss, Mary and Carlos Clarens, 'Designed for Film: The Hollywood Art Director', *Film Comment*, vol. 14 no. 3 (1978).

Cowie, Elizabeth, 'Storytelling: Classical Hollywood Cinema and Classical Narrative', in Neale and Smith (eds), *Contemporary Hollywood Cinema*.

Cripps, Thomas, 'The Myth of the Southern Box-Office: A Factor in Racial Stereotyping in American Movies, 1920–40', in J.C. Curtis and L.L. Gould (eds), *The Black Experience in America: Selected Essays* (Austin: University of Texas Press, 1970).

Cromwell, John, 'The Voice Behind the Megaphone', in Naumberg (ed.), *We Make the Movies*.

Davis, Paul H., 'Investing in Movies', *Photoplay Magazine*, February 1916, 71–3, 164.

Dawson, Anthony A.P., 'Hollywood's Labour Troubles', *Industrial and Labor Relations Review*, vol. 1 no. 4 (1948).

Desser, David, 'Japan', in Kindem (ed.), *The International Movie Industry*.

Diffrient, David Scott, 'Cabinets of Cinematic Curiosities: A Critical History of the Animated "Package Feature", From *Fantasia* (1940) to *Memories* (1995)', *Historical Journal of Film, Radio and Television*, vol. 26 no. 4 (2006).

Dobbs, Michael, 'Koko Komments', *Animato*, no. 18 (1989).

Doherty, Thomas, 'Documenting the 1940s', in Schatz, *Boom and Bust*.

— 'This Is Where We Came in: The Audible Screen and the Voluble Audience of Early Sound Cinema', in Melvyn Stokes and Richard Maltby (eds), *American Movie Audiences: From the Turn of the Century to the Early Sound Era* (London: BFI, 1999).

Downs, Anthony, 'Where the Drive-in Fits into the Movie Industry' (1953), in Ina Rae Hark (ed.), *Exhibition: The Film Reader* (London: Routledge, 2002).

Dreier, Hans, 'Designing the Sets', in Naumberg (ed.), *We Make the Movies*.

Dreyer, Carl, 'Sound Personnel and Organisation', in Cowan (ed.), *Recording Sound for Motion Pictures*.

Driscoll, John, 'Laurence Stallings', Source Database: *Dictionary of Literary Biography*, vol. 44.

Eckert, Charles, 'The Carol Lombard in Macy's Window', *Quarterly Review of Film Studies*, vol. 3 no. 1 (1978).

Eldridge, David N., '"Dear Owen": The CIA, Luigi Luraschi, and Hollywood, 1953', *Historical Journal of Film, Radio and Television*, vol. 20 no. 2 (2000).

Faller, Greg S., '"Unquiet Years": Experimental Cinema in the 1950s', in Lev, *The Fifties*.

Farnol, Lynn, 'Finding Customers for a Product', in Quigley (ed.), *New Screen Techniques*.

Fletcher, Lucille, 'Squeaks, Slams, Echoes, and Shots', in John D. Kern and Irwin Griggs (eds), *This America* (New York: Macmillan, 1942).

Ford, Greg, 'Warner Brothers [*sic*]', *Film Comment*, vol. 11 no. 1 (1975).

Friedman, Phil, 'The Players Are Cast', in Naumberg, *We Make the Movies*.

Frith, Simon, 'Mood Music: An Enquiry into Narrative Film Music', *Screen*, vol. 25 no. 3 (1984).

Georgakas, Dan, 'The Hollywood Reds: Fifty Years Later', *American Communist History*, vol. 2 no. 1 (2003).

George, Russell, 'Some Spatial Characteristics of the Hollywood Cartoon', *Screen*, vol. 13 no. 3 (1990).

Glancy, Mark, 'MGM Film Grosses, 1924–48: The Eddie Mannix Ledger', *Historical Journal of Film, Radio and Television*, vol. 12 no. 2 (1992).

— 'Warner Bros. Film Grosses, 1921–51: The William Schaefer Ledger', *Historical Journal of Film, Radio and Television*, vol. 15 no. 1(1995).

Gomery, Douglas, '*Brian's Song*: Television, Hollywood, and the Evolution of the Movie Made for Television', in John E. Connor (ed.), *American History/American Television* (New York: Frederick Ungar, 1983).

— 'Cinema Circuits or Chains', in Abel (ed.), *Encyclopedia of Early Cinema*.

— 'Economic Struggle and Hollywood Imperialism: Europe Converts to Sound', *Yale French Studies*, no. 60 (1980).

— 'The Economics of U.S. Exhibition Policy and Practice', *Cine-Tracts*, vol. 3 no. 4 (1981).

— 'Failed Opportunities: The Integration of the U.S. Motion Picture and Television Industries', *Quarterly Review of Film Studies*, vol. 10 no. 2 (1984).

— 'Failure and Success: Vocafilm and RCA Innovate Sound', *Film Reader*, no. 2 (1977).

— 'Hollywood, the National Recovery Administration, and the Question of Monopoly Power', *Journal of the University Film Association*, vol. 31 no. 2 (1979), reprinted in Kindem (ed.), *The American Movie Industry*.

— 'Hollywood Converts to Sound: Chaos or Order?', in Evan W. Cameron (ed.), *Sound and the Cinema* (Pleasantville, NY: Redgrave, 1980).

— 'Margaret Booth', in Christopher Lyon and Susan Doll (eds), *International Dictionary of Films and Filmmakers* (New York: Putnam, 1985), vol. 4.

— 'Problems in Film History: How Fox Innovated Sound', *Quarterly Review of Film Studies*, vol. 1 no. 3 (1976).

— 'The Warner–Vitaphone Peril: The American Film Industry Reacts to the Innovation of Sound', *Journal of the University Film Association*, vol. 28 no. 1 (1976).

— 'Writing the History of the American Film Industry', *Screen*, vol. 17 no. 1 (1976).

Griffith, Richard, 'Where Are the Dollars?', *Sight and Sound*, December 1949, January 1950, March 1950.

Guback, Thomas H. and Dennis J. Dombkowski, 'Television and Hollywood: Economic Relations in the 1970s', *Journal of Broadcasting*, vol. 20 no. 4 (Fall 1976).

Gunckel, Colin, 'The War of the Accents: Spanish Language Hollywood Films in Mexican Los Angeles', *Film History*, vol. 20 no. 3 (2008).

Gunning, Tom, '"Now You See It, Now You Don't": The Temporality of the Cinema of Attractions', *The Velvet Light Trap*, no. 32 (1993), reprinted in Lee Grieveson and Peter Krämer, *The Silent Cinema Reader* (London: Routledge, 2004).

— 'Response to "Pie and Chase"', in Kristine Brunovska Karnick and Henry Jenkins (eds), *Classical Hollywood Comedy* (New York: Routledge, 1994).

Guzman, Tony, 'The Little Theatre Movement: The Institutionalization of the European Art Film in America', *Film History*, vol. 17 no. 2/3 (2005).

Halberda, Marek, 'Polskie Filmy Made of Paramount', *Kino* (Poland) (May 1983).

Hall, Sheldon, 'Alternative Versions in the Early Years of CinemaScope', in Belton, Hall and Neale (eds), *Widescreen Worldwide*.

— '*Dial M for Murder*', *Film History*, vol. 16 no. 3 (2004).

Hansen, Miriam, 'Early Silent Cinema – Whose Public Sphere?', *New German Critique*, no. 29 (Spring–Summer 1983).

Hanssen, F. Andrew, 'The Block Booking of Films Re-Examined', *Journal of Law and Economics*, vol. 43 (2000), reprinted in Sedgwick and Pokorny (eds), *An Economic History of Film*.

— 'Revenue Sharing in Movie Exhibition and the Arrival of Sound', *Economic Inquiry*, vol. 40 (2002), reprinted as 'Revenue Sharing and the Coming of Sound' in Sedgwick and Pokorny (eds), *An Economic History of Film*.

Higgins, Scott, 'Blue and Orange, Desire and Loss: The Color Score for *Far from Heaven*', in James Morrison (ed.), *The Cinema of Todd Haynes* (London: Wallflower, 2006).

— 'Color Accents and Spatial Itineraries', Dossier on Cinematic Space, *The Velvet Light Trap*, no. 62 (2008).

— 'Color at the Center: Minnelli's Technicolor Style in *Meet Me in St. Louis*', *Style*, vol. 32 no. 3 (1998).

— 'Deft Trajectories for the Eye: Bringing Arnheim to Vincente Minnelli's Color Design', in Scott Higgins (ed.), *Arnheim for Media Studies* (New York: Routledge, 2010).

Higson, Andrew and Richard Maltby, '"Film Europe" and "Film America": An Introduction', in Higson and Maltby, *'Film Europe' and 'Film America'*.

Holmes, Sean P. 'And the Villain Still Pursued Her: The Actors' Equity Association in Hollywood, 1919–29', *Historical Journal of Film, Radio and Television*, vol. 25 no. 1 (2005).

— 'The Hollywood Star System and the Regulation of Actors' Labour, 1916–34', *Film History*, vol. 12 no. 1 (2000).

Horak, Jan-Christopher, 'German Exile Cinema', *Film History*, vol. 8 no. 4 (1996).

— 'Maurice Tourneur and the Rise of the Studio System', in Paulo Cherchi Usai and Lorenzo Codelli (eds), *Sulla via di Hollywood: 1911–1920* (Pordenoni: Edizioni Biblioteca dell'Immagine, 1988).

Hugo, Chris, 'The Economic Background', *Movie*, no. 27/8 (1980/81).

Jäckel, Anne, 'Dual Nationality Film Productions in Europe after 1945', *Historical Journal of Film, Radio and Television*, vol. 23 no. 3 (2003).

Jacobs, Lea, 'The B Film and the Problem of Cultural Distinction', *Screen*, vol. 33 no. 1 (1992).

— 'Keeping Up with Hawks', *Iris*, no. 32 (1998).

Jacobs, Lewis, 'Avant-Garde Production in America', in Roger Manvell (ed.), *Experiment in the Film* (London: Grey Walls Press, 1949).

Janson, Henri, 'Cinq Semaines à la Paramount, choses vécues', *Le Crapouillot*, numéro special (November 1932).

Jarvie, Ian, 'Free Trade as Cultural Threat: American Films and TV Exports in the Post-War Period', in Nowell-Smith and Ricci (eds), *Hollywood and Europe*.

Jeancolas, Jean-Pierre, 'From the Blum-Byrnes Agreement to the GATT Affair', in Nowell-Smith and Ricci (eds), *Hollywood and Europe*.

Jewell, Richard B., 'Hollywood and Radio: Competition and Partnership in the 1930s', *Historical Journal of Film, Radio, and Television*, vol. 4 no. 2 (1984).

— 'RKO Film Grosses, 1929–51: The J.C. Tevlin Ledger', *Historical Journal of Film, Radio and Television*, vol. 14 no. 1 (1994).

Jewell, Rick, 'How Howard Hawks Brought Baby Up: An Apologia for the Studio System', in Jim Hillier and Peter Wollen (eds), *Howard Hawks: American Artist* (London: BFI, 1996).

Jones, Dorothy, 'Communism in the Movies: A Study of Film Content', in Cogley, *Report on Blacklisting*.

Karr, Kathleen, 'Hooray for Wilkes-Barre, Saranac Lake – and Hollywood,' in *The American Film Heritage* (Washington, DC: Acropolis, 1972).

Kellogg, Edward W. 'History of Sound Motion Pictures', in Fielding (ed.), *A Technological History of Motion Pictures and Television*.

Kent, Sidney, 'Distributing the Product', in Kennedy (ed.), *The Story of the Films*.

Klapholtz, Ernst, 'Interview with Arthur Lange and Ernst Klapholtz', *Cue Sheet*, vol. 7 no. 4 (1990).

Klaprat, Cathy, 'The Star as Market Strategy: Bette Davis in Another Light', in Balio (ed.), *The American Film Industry*.

Knight, Arthur, 'The Reluctant Audience', *Sight and Sound*, April–June 1953.

— 'Searching for the Apollo: Black Moviegoing and its Contexts in the Small-Town U.S. South', in Maltby, Biltereyst and Meers (eds), *Explorations in New Cinema History*.

Koppes, Clayton R., 'Regulating the Screen', in Schatz, *Boom and Bust*.

Koppes, Clayton R. and Gregory D. Black, 'What to Show the World: The Office of War Information and Hollywood, 1942–45', *Journal of American History*, vol. 64 no. 1 (1977).

Korngold, Erich Wolfgang, 'Some Experiences in Film Music', in Jose Rodriguez (ed.), *Music and Dance in California* (Hollywood: Bureau of Musical Research, 1940).

Krukones, James H., 'The Unspooling of Artkino: Soviet Film Distribution in America, 1940–75', *Historical Journal of Film, Radio and Television*, vol. 29 no. 1 (2009).

Langer, Mark, 'Institutional Power and the Fleischer Studios: The *Standard Practice Reference*', *Cinema Journal*, vol. 30 no. 2 (1991).

— 'Regionalism in Disney Animation: Pink Elephants and Dumbo', *Film History*, vol. 4 no. 4 (1990).

Lasky, Jesse L., 'The Producer Makes a Plan', in Naumberg (ed.), *We Make the Movies*.

Lazersfeld, Paul and Robert Merton, 'Mass Communication, Popular Taste and Organized Social Action', in Bernard Rosenberg and David Manning White (eds), *Mass Culture: The Popular Arts in America* (Glencoe, IL: The Free Press, 1957).

Lee, Robert Edward, 'On the Spot', in Naumberg (ed.), *We Make the Movies*.

Lewin, Albert, '"Paccavi!" The True Confession of a Movie Producer', *Theatre Arts* (September 1941), reprinted in Richard Koszarski (ed.), *Hollywood Directors, 1941–1976*.

Lewis, Jon, '"We Do Not ask You to Condone This": How the Blacklist Saved Hollywood', *Cinema Journal*, vol. 39 no. 2 (2000).

Leyda, Jay, 'Black-Audience Westerns and the Politics of Cultural Identification in the 1930s', *Cinema Journal*, vol. 42 no. 1 (2002).

Lowry, Edward, 'Edwin J. Hadley: Traveling Film Exhibitor', in John L. Fell (ed.), *Film before Griffith* (Berkeley: University of California Press, 1983).

McCoy, John E. and Harry P. Warner, 'Theater Television Today (Part 1)', *Hollywood Quarterly*, vol. 4 (1949).

McKernan, Luke, 'Propaganda, Patriotism and Profit: Charles Urban and British Official War Films in America During the First World War', *Film History*, vol. 14 no. 3/4 (2002).

Maltby, Richard, 'Film Noir: The Politics of the Maladjusted Text', *Journal of American Studies*, vol. 18 no. 1 (1984).

— 'The Production Code and the Hays Office', in Balio, *Grand Design*.

— 'Sticks, Hicks and Flaps: Classical Hollywood's Generic Conception of Its Audiences', in Melvyn Stokes and Richard Maltby (eds), *Identifying Hollywood's Audiences: Cultural Identity and the Movies* (London: BFI, 1999).

— '"To Prevent the Prevalent Type of Book": Censorship and Adaptation in Hollywood, 1924–34', *American Quarterly*, vol. 44 no. 4 (1992).

Maltby, Richard and Ruth Vasey, 'The International Language Problem: European Reactions to Hollywood's Conversion to Sound', in David W. Ellwood and Rob Kroes (eds), *Hollywood in Europe: Experiences of a Cultural Hegemony* (Amsterdam: VU University Press, 1994).

Mantle, Burns, 'The Shadow Stage', in George C. Pratt, *Spellbound in Darkness: A History of the Silent Film* (Greenwich, CT: New York Graphic Society, 1966).

Marlowe, Frederic, 'The Rise of the Independents in Hollywood', *Penguin Film Review*, no. 3 (1947).

Metz, Walter, '"What Went Wrong?": The American Avant-Garde Cinema of the 1960s', in Monaco, *The Sixties*.

Moore, Barbara, 'The Cisco Kid and Friends: The Syndication of Television Series from 1948–52', *Journal of Popular Film and Television*, vol. 8 (1980).

Morgan, Kenneth, 'Dubbing', in Cowan (ed.), *Recording Sound for Motion Pictures*.

Moritz, William, 'Resistance and Subversion in Animated Films of the Nazi Era: The Case of Hans Fischerstein', *Animation Journal*, vol. 1 no. 1 (1992).

Morris, Peter, 'The Taming of the Few: Nell Shipman in the Context of Her Times', in Shipman, *Silent Screen*.

Morsberger, Robert E., Stephen O. Lesser and Randall Clark, 'Frances Marion', in *American Screenwriters* (Detroit: Gale University Press, 1984).

Mottram, Ron, 'American Sound Films, 1926–30', in Weis and Belton (eds), *Film Sound*.

— 'The Great Northern Film Company: Nordisk Film in the American Motion Picture Market', *Film History*, vol. 2 no. 1 (1988).

Murphy, Robert, 'Rank's Attempt on the American Film Market', in James Curran and Vincent Porter (eds), *British Cinema History* (London: Weidenfeld and Nicholson, 1983).

Muscio, Giuliana, 'Clara, Ouida, Beulah, et al.: Women Screenwriters in American Silent Cinema', in Callahan (ed.), *Reclaiming the Archive*.

Musser, Charles, 'Itinerant Exhibitors', in Abel (ed.), *Encyclopedia of Early Cinema*.

Neale, Steve, 'Swashbucklers and Sitcoms, Cowboys and Crime, Nurses, Just Men and Defenders: Blacklisted Writers and TV in the 1950s and 1960s', *Film Studies*, no. 7 (2005).

Negelescu, Jean, 'New Medium – New Methods', in Quigley (ed.), *New Screen Techniques*.

Neupert, Richard, 'Technicolor and Hollywood: Exercising Color Restraint', *Post Script*, vol. 10 no. 1 (Fall 1990).

Ogihara, Junko, 'The Exhibition of Films for Japanese Americans in Los Angeles during the Silent Era', *Film History*, vol. 4 no. 2 (1990).

Pafort-Overduin, Clara, 'Distribution and Exhibition in the Netherlands, 1934–36', in Maltby, Biltereyst and Meers (eds), *Explorations in New Cinema History*.

Pagnol, Marcel, 'Interview', *Cahiers du cinéma*, no. 173 (1965).

Pathé, Théophile, 'Doublage du film étranger', *Le Cinéma* (Paris: Corréa Editeur, 1942).

Pattison, Barrie, 'Thirty Leading Film Editors', in Peter Cowie (ed.), *International Film Guide* (New York: A.S. Barnes, 1973).

Paul, William, '"Breaking the Fourth Wall": "Belascoism", Modernism, and a 3-D *Kiss Me Kate*', *Film History*, vol. 16 no. 3 (2004).

Pearson, Roberta E., 'Biblical Films', in Abel (ed.), *The Encyclopedia of Early Film*.

Pierce, David, 'Costs and Grosses for the Early Films of Cecil B. DeMille', in *L'Eredita DeMille* (Pordenone: Edizione Biblioteca dell'Immagine, 1991).

Pithon, Rémy, 'Presences françaises dans le cinéma italien pendant les dernières années du regimes mussolinien, 1935–43', *Risorgimento*, no. 2/3 (1981).

Pokorny, Michael and John Sedgwick, 'Warner Bros. in the Inter-War Years: Strategic Reponses to the Risk Environment of Filmmaking', in Sedgwick and Pokorny, *An Economic History of Film*.

Polonsky, Abraham, 'Introduction' to Howard Gelman, *The Films of John Garfield* (Secausus, NJ: Citadel Press, 1975).

Pryluck, Calvyn, 'The Itinerant Movie Show and the Development of the Film Industry', *Journal of the University Film and Video Association*, vol. 25 no. 4 (1983).

Quinn, Michael, 'Distribution, the Transient Audience, and the Transition to the Feature Film', *Cinema Journal*, vol. 40 no. 2 (2001).

— 'Paramount and Early Feature Distribution: 1914–21', *Film History*, vol. 11 no. 1 (1999).

— 'USA: Distribution', in Abel (ed.), *Encyclopedia of Early Cinema*.

Raksin, David, 'Holding a Nineteenth Century Pedal at Twentieth Century-Fox', in McCarty (ed.), *Film Music 1*.

Regester, Charlene, 'African American Extras in Hollywood during the 1920s and 1930s', *Film History*, vol. 9 no. 1 (1997).

— 'Stepin Fetchit: The Man, The Image and the African American Press', *Film History*, vol. 6 no. 4 (1994).

Reiniger, Lotte, 'The Adventures of Prince Ahmed or What May Happen to Someone Trying to Make a Full Length Cartoon in 1926', *The Silent Picture*, no. 6 (1970).

Rhodes, Gary D., '"The Double Feature Evil": Efforts to Eliminate the American Dual Bill', *Film History*, vol. 23 no. 1 (2011).

Rosendorf, Neal Moses, '"Hollywood in Madrid": American Film Producers and the Franco Regime, 1950–70', *Historical Journal of Film, Radio and Television*, vol. 27 no. 1 (2007).

Sandeem, Eric, 'Anti-Nazi Sentiment in Film: *Confessions of a Nazi Spy* and the German-American Bund', *American Studies*, vol. 20 no. 2 (1979).

Schmidt, Jackson, 'On the Road to MGM: A History of Metro Pictures Corporation, 1915–20', *The Velvet Light Trap*, no. 19 (1982).

Seale, Paul, '"A Host of Others": Toward a Nonlinear History of Poverty Row and the Coming of Sound', *Wide Angle*, vol. 13 no. 1 (1991).

Sedgwick, John, 'Patterns of First-Run and Suburban Filmgoing in Sydney in the Mid-1930s', in Maltby, Biltereyst and Meers (eds), *Explorations in New Cinema History*.

— 'Product Differentiation at the Movies: Hollywood 1946 to 1965', in Sedgwick and Pokorny, *An Economic History of Film*.

Sedgwick, John and Mike Pokorny, 'Hollywood's Foreign Earnings During the 1930s', *Transnational Cinemas*, vol. 1 no. 1 (2010).

Shamroy, Leon, 'Filming *The Robe*', in Quigley (ed.), *New Screen Techniques*.

Short, K.R.M., 'Documents (B): Washington's Information Manual for Hollywood, 1942', *Historical Journal of Film, Radio and Television*, vol. 3 no. 2 (1983).

Silberman, Mark, 'Germany', in Kindem (ed.), *The International Movie Industry*.

Singer, Ben, 'Manhattan Nickelodeons: New Data on Audiences and Exhibitors', *Cinema Journal*, vol. 34 no. 3 (1995).

Slater, Thomas J., 'June Mathis: The Woman Who Spoke through Silents', in Bachman and Slater (eds), *Discovering Marginalized Voices*.

Slide, Anthony, '*Remodeling Her Husband*', in Frank N. McGill (ed.), *Magill's Survey of Cinema* (Englewood Cliffs, NJ: Salem Press, 1982).

Smith, David R., 'Beginnings of the Disney Multiplane Camera', in Charles Solomon (ed.), *The Art of the Animated Image: An Anthology* (Los Angeles: The American Film Institute, 1987).

— 'Up to Date in Kansas City', *Funnyworld*, no. 19 (1978).

Smith, Murray, 'Theses on the Philosophy of Hollywood History', in Neale and Smith (eds), *Contemporary Hollywood Cinema*.

Smoodin, Eric, 'Motion Pictures and Television, 1930–45: A Pre-History of Relations between the Two Media', *Journal of the University Film and Video Association*, vol. 34 no. 3 (1982).

Spellerberg, James, 'CinemaScope and Ideology', *The Velvet Light Trap*, no. 21 (1985).

Spring, Katherine, 'Pop Go the Warner Bros., et al.: Marketing Film Songs during the Coming of Sound', *Cinema Journal*, vol. 48 no. 1 (2008).

Staiger, Janet, 'Authorship Approaches', in David A. Gerstner and Janet Staiger (eds), *Authorship and Film* (New York: Routledge, 2003).

— 'Blueprints for Feature Films: Hollywood's Continuity Scripts', in Balio (ed.), *The American Film Industry*.

— 'Combination and Litigation: Structures of U.S. Film Distribution, 1896–1917', *Cinema Journal*, vol. 23 no. 2 (1984).

Stamp, Shelley, 'Lois Weber, Star Maker', in Callahan (ed.), *Reclaiming the Archive*.

Starr, Helen, 'Putting It Together', *Photoplay*, July 1918, 52.

Steiner, Max, 'The Music Director', in Bernard Rosenberg and Harry Silverstein (eds), *The Real Tinsel* (London: Macmillan, 1970).

— 'Scoring the Film', in Naumberg (ed.), *We Make the Movies*.

Sternfeld, Frederick W, 'Frederick W. Sternfeld on Hugo Friedhofer's *The Best Years of Our Lives*', in Cooke (ed.), *The Hollywood Music Reader*.

Stills, Milton, 'The Actor's Part', in Kennedy (ed.), *The Story of the Films*.

Stothart, Herbert, 'Film Music', in Stephen Watts (ed.), *Behind the Screen* (London: Barker, 1939).

Streibel, Dan, 'Boxing Films', in Abel (ed.), *The Encyclopedia of Early Film*.

Stubbs, Jonathan, '"Blocked" Currency, Runaway Production in Britain and *Captain Horatio Hornblower* (1951)', *Historical Journal of Film, Radio and Television*, vol. 28 no. 3 (2008).

— 'The Eady Levy: A Runaway Bribe? Hollywood Production in the Early 1960s', *Journal of British Cinema and Television*, vol. 6 no. 1 (2009).

Swartz, Mark E., 'Motion Pictures on the Move', *Journal of American Culture*, vol. 9 no. 3 (1986).

Taves, Brian, 'The B Film: Hollywood's Other Half', in Balio, *Grand Design*.

— 'Robert Florey and the Hollywood Avant-Garde', in Jan-Christopher Horak (ed.), *Lovers of Cinema: The First American Film Avant-Garde, 1919–1945* (Madison: University of Wisconsin Press, 1995).

Taylor, Frank J., 'Big Boom in Outdoor Movies' (1956), in Waller (ed.), *Moviegoing in America*.

Thomas, Tony, 'A Conversation with John Williams', *Cue Sheet: The Journal for the Preservation of Film Music*, vol. 8 no. 1 (1991).

Thompson, Kristin, 'The Formulation of the Classical Style, 1909–1928', in Bordwell, Staiger and Thompson, *The Classical Hollywood Cinema*.

Thompson, Kristin, 'Implications of the Cel Animation Technique', in Teresa De Lauretiis and Stephen Heath (eds), *The Cinematic Apparatus* (New York: St Martin's Press, 1980).

Thomson, Virgil, 'A Little about Movie Music', *Modern Music*, vol. 10 no. 4 (1933).

Varma, Neil, 'Honeymoon Shocker: Lucille Fletcher's "Psychological" Sound Effects and Wartime Radio Drama', *Journal of American Studies*, vol. 44 no. 1 (2010).

Vianello, Robert, 'The Power Politics of "Live" Television', *Journal of Film and Video*, vol 17 no. 3 (1985).

Vitella, Federico, 'Before Techniscope: The Penetration of Foreign Widescreen Technology in Italy (1953–59)', in Belton, Hall and Neale (eds), *Widescreen Worldwide*.

Wagstaff, Christopher, 'Italian Genre Films in the World Market', in Nowell-Smith and Ricci (eds), *Hollywood and Europe*.

Waller, Gregory, 'Hillbilly Music and Will Rogers: Small Town Picture Shows in the 1930s', in Melvyn Stokes and Richard Maltby (eds), *American Movie Audiences: From the Turn of the Century to the Early Sound Era* (London: BFI, 1999).

— 'Robert Southard and the History of Traveling Film Exhibition', *Film Quarterly*, vol. 57 no. 2 (2003/4).

Walsh, Mike, 'From Hollywood to the Garden Suburb (and Back to Hollywood): Exhibition and Distribution in Australia', in Maltby, Biltereyst and Meers (eds), *Explorations in New Cinema History*.

— 'Options for American Foreign Distribution: United Artists in Europe, 1919–30', in Higson and Maltby (eds), *'Film Europe' and 'Film America'*.

Weinstein, Mark, 'Profit-Sharing Contracts in Hollywood: Evolution and Analysis', *Journal of Legal Studies*, vol. 27 no. 1 (1998).

Wescott, Steven D., 'Miklós Rósza's *Ben-Hur*: The Musical-Dramatic Function of *Leitmotiv*', in McCarty, *Film Music 1*.

White, Timothy R., 'Hollywood's Attempt at Appropriating Television: The Case of Paramount Pictures', in Balio (ed.), *Hollywood in the Age of Television*.

Whitney, Simon N., 'Antitrust Policies and the Motion Picture Industry', in Kindem (ed.), *The American Movie Industry*.

Wilkins, Mira, 'Charles Pathé's American Business', *Enterprises et Histoire*, vol. 34 (1994).

Zanuck, Darryl, 'CinemaScope in Production', in Quigley (ed.), *New Screen Techniques*.

Zukor, Adolph, 'Origin and Growth of the Industry', in Lewis (ed.), *Cases on the Motion Picture Industry*.

Ph. D. and masters dissertations

Deneroff, Harvey Raphael, 'Popeye the Union Man: A Historical Study of the Fleischer Strike' (University of Southern California, 1985).

Dombkowski, Dennis J., 'Film and Television: An Analytical History of Economic and Creative Integration' (University of Illinois, 1982).

Freed, Mark, 'An Analysis of Subscription Television in California in 1964' (University of Oregon, 1969).

Guzman, Anthony Henry, 'The Exhibition and Reception of European Films in the United States during the 1920s' (UCLA, 1993).

Herzog, Charlotte Kopac, 'Motion Picture Theater and Film Exhibition' (Northwestern University, 1980).

Holliday, Wendy, 'Hollywood's Modern Women: Screenwriting, Work Culture, and Feminism, 1910–40' (New York University, 1995).

Humphrey, Francis, 'The Creative Women in Motion Picture Production' (University of Southern California, 1970).

Kallmann, Alfred, 'Die Konzenierung in der Filmindustrie: Erläutert an den Filmindstrien Deuschlands und Amerikas' (Würzburg University, 1932).

Kunz, William, 'A Political Economic Analysis of Ownership and Regulation in the Television and Motion Picture Industries' (University of Oregon, 1998).

Larsen, David Alan, 'Integration and Attempted Integration between the Motion Picture and Television Industries through 1956' (Ohio University, 1979).

McKenna, Anthony T., 'Joseph E. Levine: Showmanship, Reputation and Industrial Practice, 1945–77' (University of Nottingham, 2008).

Nelson, Richard Alan, 'Florida and the American Motion Picture Industry' (Florida State University, 1980).

Potamianos, George, 'Hollywood in the Hinterlands: Mass Culture in Two California Communities, 1896–1936' (University of Southern California, 1998).

Schnapper, Amy, 'The Distribution of Theatrical Feature Films to Television' (University of Wisconsin-Madison, 1975).

Staiger, Janet, 'The Hollywood Mode of Production: The Construction of Divided Labour in the Film Industry' (University of Wisconsin-Madison, 1981).

Suber, Howard, 'The Anti-Communist Blacklist in the Hollywood Motion Picture Industry' (UCLA, 1968).

Index

Page numbers in italics refer to frame stills. Most foreign films are listed as titled in the US, followed by the year of their US release in parenthesis. Where appropriate, these are followed in additional parenthesis by their original language titles, their country or countries of origin, and the year in which they were first released abroad if different from that of their release in the US. 'MLV' in parenthesis indicates a multiple-language version. The titles of radio and television programmes in this Reader have been cited either in quotation marks or in italics. These titles have been cited in italics in the index, with quotation marks reserved for individual episodes or one-off programmes.